Foundations of Stress Waves

Foundations of Stress Waves

应力波基础

ISBN: 7–118–04015–0

Author: WANG Li-li (Li-Lih WANG)

Reviewer and Corrector: Zhu Zhao-xiang (Z.-X. ZHU)

(First Edition in Chinese, 1985)

(Second Edition in Chinese, 2005)

Author: WANG Li-li (Li-Lih WANG)

English Translators: YANG Li-ming, ZHOU Feng-hua, WANG Li-li

For the First Edition in English (2007)

ELSEVIER

Amsterdam • Boston • Heidelberg • London • New York • Oxford
Paris • San Diego • San Francisco • Singapore • Sydney • Tokyo

Elsevier
The Boulevard, Langford Lane, Kidlington, Oxford OX5 1GB, UK
Radarweg 29, PO Box 211, 1000 AE Amsterdam, The Netherlands

First edition 2007

British Library Cataloguing in Publication Data
A catalogue record for this book is available from the British Library

Library of Congress Cataloging-in-Publication Data
A catalog record for this book is available from the Library of Congress

ISBN–13: 978-0-08-044494-9
ISBN–10: 0-08-044494-6

For information on all Elsevier publications
visit our web site at books.elsevier.com

Printed and bound in Great Britain

07 08 09 10 10 9 8 7 6 5 4 3 2 1

CONTENTS

Preface xi

1 Introduction 1

2 Elementary Theory of One-Dimensional Stress Waves in Bars 7

2.1 Material Coordinate System and Spatial Coordinate System 7
2.2 Governing Equations of Longitudinal Waves in Bars in Material Coordinates 10
2.3 Characteristic Lines and the Compatibility Relationships along the Characteristic Lines 13
2.4 Elastic–Plastic Longitudinal Loading Waves in Semi-Infinite Bars 17
2.4.1 Linear elastic waves 17
2.4.2 Elastic–plastic loading waves 22
2.5 Governing Equations of Longitudinal Waves in Bars in Spatial Coordinates 26
2.6 Strong Discontinuity and Weak Discontinuity; Shock Waves and Continuous Waves 29
2.7 Conservation Conditions across Wave Front, Rankine–Hugoniot Relations 35
2.8 Dispersion Effects Induced by the Transverse Inertia 42
2.9 Torsion Waves in Cylindrical Bars 51

3 Interaction of Elastic Longitudinal Waves 57

3.1 Coaxial Collision of Two Elastic Bars 57
3.2 Interaction of Two Elastic Longitudinal Waves 58
3.3 Reflection of Elastic Longitudinal Waves at Fixed End and Free End 60
3.4 Coaxial Collision of Two Elastic Bars with Finite Length 62
3.5 Reflection and Transmission of Elastic Longitudinal Waves at the Interface of Two Different Bars 66
3.6 Reflection and Transmission of Elastic Waves in Bars with Varying Cross Sections 69
3.7 Hopkinson Pressure Bar and Flying Piece 74
3.8 Split Hopkinson Pressure Bar 76
3.9 Dynamic Fracture Induced by Reflective Unloading Waves 87

4 Interaction of Elastic–Plastic Longitudinal Waves in Bars 95

4.1 Interaction of Two Longitudinal Elastic–Plastic Loading Waves in Bars 95
4.1.1 Head-on loading interaction of two strong-discontinuous stress waves 96
4.1.2 Head-on loading interaction of two weak-discontinuous stress waves 99
4.2 Reflection of Longitudinal Elastic–Plastic Loading Waves at a Fixed End 102
4.3 Governing Equations and Characteristic Lines of Unloading Waves 105
4.4 Pursuing Unloading by Strong-Discontinuous Unloading Disturbances 110
4.4.1 Sudden unloading of strong-discontinuous loading wave in a linear hardening bar 110
4.4.2 Sudden unloading of continuous loading waves in a linear hardening bar 116
4.4.3 Sudden unloading of centered plastic loading waves 120
4.5 Pursuing Unloading by Weak-Discontinuous Unloading Disturbances 126
4.5.1 Continuously unloading of centered plastic waves 128
4.5.2 Attenuation of shock waves in linearly hardening materials 131
4.6 Shock Wave Attenuation Due to Pursuing Unloading 135
4.7 Propagating Properties of Elastic–Plastic Boundaries in Semi-Infinite Bars 139
4.7.1 An analysis by the characteristic line method 139
4.7.2 Local linearization method 146

4.8 Head-On Unloading 148
4.9 High-Speed Impact of a Finite Bar onto a Rigid Target 153
4.9.1 Linear hardening bar 154
4.9.2 Increasingly hardening bar 161
4.9.3 Decreasingly hardening bar 164
4.10 General Properties of the Loading–Unloading Boundary Propagation 167
4.10.1 Loading boundary and unloading boundary 167
4.10.2 Elastic–plastic boundary as a singular interface 169
4.10.3 Strong-discontinuous elastic–plastic boundary 170
4.10.4 First order weak-discontinuous boundary 171
4.10.5 Second order weak-discontinuous boundary 176
4.10.6 Discussions on higher-than-second order weak-discontinuous boundaries 180
4.10.7 Higher order isolated points on elastic–plastic boundary 183
4.10.8 Supplementary conditions on loading boundary 190

5 Rigid Unloading Approximation 197

5.1 Rigid Unloading in a Semi-Infinite Bar 197
5.1.1 Rigid unloading in linear hardening plastic bars 197
5.1.2 Rigid unloading in linear elastic–linear hardening plastic bars 200
5.1.3 Rigid unloading in linear elastic–decreasing hardening plastic bars 202
5.2 Rigid Unloading in Finite Bars 207
5.3 Rigid Unloading Analyses for Shock Wave Propagation 212
5.3.1 Rigid unloading of shock wave propagating in semi-infinite bars 212
5.3.2 Rigid unloading of shock wave propagating in finite bars 215

6 One-Dimensional Visco-Elastic Waves and Elastic-Visco-Plastic Waves 219

6.1 Linear Visco-Elastic Constitutive Relationship 220
6.1.1 Maxwell model 222
6.1.2 Kelvin–Voigt model 224
6.1.3 Standard linear solid model 226
6.2 Stress Waves Propagating in Linear Visco-Elastic Bars 227
6.2.1 Longitudinal visco-elastic waves propagating in Kelvin–Voigt bars 227

6.2.2 Longitudinal visco-elastic waves propagating in Maxwell bars 229
6.2.3 Longitudinal visco-elastic waves propagating in standard linear solid bars 230
6.2.4 Solutions of linear visco-elastic longitudinal waves propagating in bars by the characteristics method 233
6.2.5 Split Hopkinson visco-elastic bar 237
6.3 Nonlinear Visco-Elastic Constitutive Relationship 246
6.4 Stress Waves Propagating in Nonlinear Visco-Elastic Bars 251
6.5 Elastic-Visco-Plastic Constitutive Relationship 255
6.6 Stress Waves Propagating in Elastic-Visco-Plastic Bars 260

7 One-Dimensional Strain Plane Waves 265

7.1 Governing Equations 265
7.2 One-Dimensional Strain Elastic Waves 267
7.3 Elastic–Plastic Constitutive Relationship in 1D Strain Condition 270
7.4 One Dimensional Strain Elastic–Plastic Waves 276
7.5 Influence of Reverse Yield on the Propagation of 1D Strain Elastic–Plastic Wave 279
7.6 State Equation of Solids under High Pressures 284
7.7 Shock Waves in Solids under High Pressures 292
7.7.1 Shock jump conditions 292
7.7.2 Shock adiabat 297
7.8 Interaction of Shock Waves in Solids under High Pressure 308
7.9 Plane Waves in Hydro-Elasto-Plastic Media 317
7.10 Attenuation of Shock Waves in Hydro-Elasto-Plastic Media 325
7.11 One-Dimensional Strain Elasto-Visco-Plastic Waves 333

8 Spherical Waves and Cylindrical Waves 337

8.1 Continuity Equation and Motion Equation 337
8.2 Elastic Spherical Waves and Cylindrical Waves 339
8.3 Elasto-Plastic Spherical Waves 347
8.4 Approximate Analysis for the Fragmentation of Spherical Shells 354
8.5 Elasto-Visco-Plastic Spherical Waves and Cylindrical Waves 357

9 Elastic–Plastic Waves Propagating in Flexible Strings 363

9.1 Governing Equations 364
9.2 Semi-Infinite Straight String under Abrupt Constant-Velocity Oblique Impact 372
9.3 Infinite Straight String under Abrupt Constant-Velocity Oblique Point-Impact 377
9.4 Prestretched Strings Subjected to Transverse Impact 388
9.4.1 Experimental investigation of wave velocity in a prestretched string 391
9.4.2 Experimental study on constitutive relationship of string materials 394
9.4.3 Determination of transverse wave velocity from longitudinal wave velocity measurements 396

10 Elastic–Plastic Waves Propagating in Beams under Transverse Impact (Bending Wave Theory) 399

10.1 Basic Assumptions and Governing Equations 399
10.2 Elastic Bending Waves 405
10.3 Plastic Bending Waves (Elastic–Plastic Beams) 409
10.4 Rigid-Plastic Analysis 432
10.5 Shear Failure of Beams under Transverse Impact 442

11 General Theory for Linear Elastic Waves 453

11.1 Linear Elastic Waves in Infinite Media 453
11.2 Oblique-Incidence, Reflection, and Transmission of Elastic Plane Waves 457
11.3 Elastic Surface Waves 467

12 Numerical Methods for Stress Wave Propagation 475

12.1 Characteristics Numerical Method 477
12.1.1 Characteristics numerical method for 1D waves 477
12.1.2 Characteristic surfaces numerical method for 2D waves 482
12.2 Finite Difference Method 490
12.2.1 Establishment of finite difference format 491
12.2.2 Convergence of difference formats 494

12.2.3 Stability of difference formats 495
12.2.4 Artificial viscosity 497
12.3 Finite Element Method (FEM) 500
12.3.1 Basic procedures of finite element method 500
12.3.2 Examples 503

Appendices **511**

References **519**

Index **529**

Preface

Due to its practical and scientific importance, the theory of stress waves has been the subject of numerous investigations conducted over the past several decades. It is found that the propagation of stress waves is an important factor in earthquake, high velocity impact, explosion, and weapon effects and describing the responses of materials and structures under impact also needs the theory of stress waves.

Applications of the theory of stress waves in natural exploration and exploitation and in technology development are relatively common, as knowledge that was once confined to a few specialists is becoming more widespread. For instance, the inner structure of the earth is imaged or natural resources (i.e. oil, coal bed) are prospected by using the propagation, reflection, and refraction of wave, which is produced by a blast, underground. Stress waves can also be used to examine the damage in materials or structures without effects on their properties, to be a loading resource to investigate the dynamic properties of materials, to rivet metal plates, to produce the spalling fracture. It can even be used in medicine to destroy calculi in human body. These applications have furthered the in-depth study on the theory of stress waves, and have shown that the investigation on the stress waves is currently an important subject in solid mechanics and the theory of stress waves is the basic of many science and technology branches.

The theory of stress waves was developed based on the study on the elastic waves. Unexpectedly, the investigation on the elastic waves is driven by the study on what nature light was. In 1821, physicist Fresnal first proposed a new concept that transverse wave could propagate in a medium, when he was studying the interference of polarized lights. Two mathematicians Cauchy and Poisson were interested in the new concept, which resulted in Cauchy building up the general governing equations for elastic body and Poisson finding out that there were two kinds of waves, longitudinal wave and transverse wave, propagating in elastic medium. Thus, the theory of stress waves was originated by both the mathematicians.

A collision between two elastic solids was investigated, in earlier days, by using the theory of vibration. However, it was not successful. Boussinesq (1883) and Saint-Venant (1883) first began to apply the concept of stress waves to study the collision between elastic straight bars (and between rigid body and straight bar). The investigation on the earthquake had further driven the development of stress wave theory. In the Second World War, the theory of stress waves had been developed quickly by Taylor (1942), Karman (1942) and Рахматулин (Rakhmatulin) (1945), who had independently built up the theories of plastic waves to satisfy the needs of military technology. This was followed by post-war period, when the theory of stress waves underwent holistic development, which can be referred to in this book.

In the early 1960s, Professor Wang had lectured the course "Dynamics of Plasticity" in University of Science and Technology of China. Based on his lecture script, "FOUNDATION OF STRESS WAVES" (应力波基础), the first edition in Chinese, was published by National Defense Industry Press (Beijing, China) in 1985. Considering the overwhelming response from readers and satisfying their needs, the second edition in Chinese was published in 2005. This book is a translation from the second edition of "FOUNDATION OF STRESS WAVES" in Chinese. For convenience of studying, this book introduces the theory of stress waves with a gradual build-up of complexity, which is its track of development, from simple to complex and from specific to general. Thus, in this book, the theory of the linear elastic wave is first studied, then the nonlinear elastic wave and the plastic waves are discussed; the simple waves are studied earlier than the complex waves; the waves in rate independent materials are analyzed before the rate dependent materials; one dimensional problems are dealt with first, followed by the multi-dimensional wave; and based on the results of study on continuous waves, the propagation of discontinuous wave surfaces with different orders, such as shock wave (first order discontinuous wave) and acceleration wave (second order discontinuous wave), are analyzed.

This book includes five topics. In the first one, discussions are as follows: one-dimensional elastic waves, plastic waves, shock waves, unloading waves, then going with a gradual build-up of complexity to linear visco-elastic waves, nonlinear visco-elastic waves and visco-elastic-plastic waves. In the second, multi-dimensional waves, such as one-dimensional strain plane waves, spherical waves, and cylindrical waves, where shock waves under high pressure, elastic–plastic waves, and visco-elastic-plastic waves are mentioned. In the third, the theory of elasto-plastic waves in flexible strings and beams under transverse impact is studied. The analyses include the propagation in strings, where longitudinal waves and transverse waves couple each other, and the propagation in beams where bending moment disturbances and shear disturbances couple each other also. After that, the general three-dimensional theory of stress wave propagation is introduced. Finally, the numerical methods used to solve the problems of stress wave propagation, including the characteristics method, the finite difference method, and the finite element method, are briefly presented.

This book comprehensively describes the fundamental knowledge of the stress waves propagating in solids, which should appeal to a fairly wide audience. It is important for studies on explosions, earthquake, and the dynamic responses of materials and structures under impact loading. The book is intended to be understandable without being superficial. The fundamental concepts are first described clearly, with a minimum of higher level mathematics. This enables a reader to gain physical insight into the subject. Theoretical analyses and practical applications are covered in detail. However, the studies presented in the book draw on several scientific and engineering disciplines, and require some familiarity with the basic principles of continuum mechanics and constitutive relationship of materials. This book is not only suitable as an undergraduate and graduate text, but also as a reference for engineers and researchers, particularly because it reflects the research results originally published in Chinese by the author and other Chinese colleagues.

If this book is used as a college text, covering all the content in a single semester would be unlikely. The lecturer should select parts of the book that suit the needs and background of the students. The first three chapters should be the foundation of courses. For an undergraduate, part of Chapter 4 could be included. As a graduate text, part of Chapters 4–6 can be selected in addition. Other chapters in this book provide a useful reference for the more advanced readers.

At last, we would like to express our appreciation to Mr. K. Qin, Miss J. Zhu, Mr. Z.J. Sun and Mr. Y.G. Wang in Ningbo University, who assisted us in the translation and the illustrations in this book.

YANG Li-Ming (Li-Ming YANG)

CHAPTER 1

Introduction

In a wide range of engineering sciences, military technology science, research, and even in daily life, people are more and more faced with various problems in relation to explosion or impact loading. It has been observed that the mechanical response of materials and structures under explosive or impact loading differs markedly from that under static loading. For example, when a stone hits onto a glass plate, the fracture usually occurs at the back surface of the plate, called spalling. High-explosive plastic projectile destroying tank armor is similar to this. For another example, when a metal bar is subjected to axial load at one end, in the static case the deformation is almost uniformly distributed along the bar length; while in the impact case (like hammering a drill rod or a pile), the deformation is extremely non-uniformly distributed and the residual deformation is concentrated near the hammered end of the bar. The mushroom shape of the projectile head formed when the projectile impacts against a target is similar to this. The rapid development of dynamic theory of solid mechanics is closely connected with the urgent requirement to solve such sorts of mechanics problems.

Why those unusual phenomena mentioned above only occur under explosive or impact loading? Why those phenomena are unable to be satisfactorily explained and solved on the basis of the static theory of solid mechanics? What are the main differences between the dynamic theory and the static theory of solid mechanics?

First of all, it is known that the subjects studied in the static theory of solid mechanics are in states of static balance, with the prerequisite that the inertia of infinitesimal element of media could be neglected. This is valid only when the applied load does not change markedly with time. However, an explosive or impact loading is characterized by **short duration**. The load itself and consequently other mechanical variables all markedly change with time in the scale of milliseconds (ms), microseconds (μs), and even nanoseconds (ns). Some typical examples are as follows. The pressure at the core zone of a nuclear explosion dramatically increases to the order of $10^3 \sim 10^4$ GPa in few microseconds. The pressure induced by a contact explosion between the explosive and the solid surface suddenly increases to 10 GPa in a few microseconds. When a bullet is shot on a target at a speed of $10^2 \sim 10^3$ m/s, the total duration of loading is about several tens of microseconds and the pressure at the contact interface is as high as about $1 \sim 10$ GPa. Under such intensive dynamic loading, the infinitesimal elements of media are in quickly changed dynamic processes. Such problems are classified as dynamic problems where the inertia of infinitesimal elements should be taken into account, and thus the problems are finally attributed to studies of the propagation of stress waves.

In fact, when a part of a surface of a deformable solid is subjected to an external load, at the beginning only those surface particles directly loaded move away from their initial equilibrium positions. As there is relative motion between surface particles and their neighboring particles (deformation), these two particles are loaded with each other, and the acting force and the reacting force take place between them (stress). This makes the neighboring particles begin to move away from their initial equilibrium positions. However, due to inertia of particles of media, the motion of neighboring particles starts later than surface particles. On this analogy, the disturbances caused by external load on surface gradually propagate from the near to the distant in the media, and the stress waves are formed. The boundary between a disturbed area and an undisturbed area is named as **wave front**; its propagating speed is named as **wave velocity**. Wave velocity of common materials is about $10^2 \sim 10^3$ m/s. It should be noticed that the wave velocity must be distinguished from the particle velocity. The former is the speed that a signal of disturbance propagates in media, while the latter is the speed of particle motion. If their directions are identical, the wave is named as **longitudinal wave**. If their directions are perpendicular, the wave is named as **transverse wave**. According to different geometric shapes of wave front, there are **plane wave**, **cylindrical wave**, and **spherical wave**, etc. Earthquake waves, sound waves, and ultrasonic waves in solids, shock waves in solids as well are all typical examples of stress waves.

All solid materials have **inertia** and are **deformable**. Under an external load varied with time, the deformation process of solid is substantially a process containing the propagation, reflection, and interaction of stress waves. In the static problems of deformable solids, the inertia of particles is neglected, or in other words, the propagation and interaction process of stress waves before achieving the static equilibrium is disregarded or neglected, since only the motion after achieving the static equilibrium is focused on. In rigid-body mechanics, the ability to deform the material is generally neglected. This is, if the wave velocity approaches infinity, then wave propagation is not considered. However, in the dynamic problems of deformable solids under explosive or impact loadings, the wave propagating process must be considered, since the time scale measuring the external load change is in the same or lower order of the time scale, measuring the wave passing through the characteristic length of the subject. In such cases, our main concern is the motion process of deformable solid, and then the propagating process of stress waves must be considered.

Second, for an intense impulsive loading, since the load changes dramatically in a very short duration, it implies a **high loading rate** or **high strain rate**. The strain rates in ordinary static tests are about $10^{-5} \sim 10^{-1}\ \mathrm{s}^{-1}$; while the strain rates in impact test where the propagation of stress waves must be considered, are about $10^2 \sim 10^4\ \mathrm{s}^{-1}$, even to $10^7\ \mathrm{s}^{-1}$, namely they are much higher than that in static tests. A number of experimental results have shown that material behavior is generally different under various strain rates. From the view-point of deformation mechanism of materials, except the ideal elastic deformation which can be regarded as an instantaneous response, various inelastic deformation and fracture are non-instantaneous response all of which develop at finite rate (like the process of dislocation motion, stress-induced diffusion, damage evolution, crack growth, and propagation, etc.). Thus, mechanical behaviors of materials are essentially rate-dependent, which generally are displayed as follows: with increasing strain rate, the increase in yield strength and ultimate strength, the decrease in elongation, and the delay of yielding and fracture, etc. Therefore, in addition to the particle inertia effect mentioned earlier, the rate-dependent character of material constitutive relation is another important reason why the mechanical response of structures under explosive or impact loading much differs from that under static loading. In the view of thermodynamics, the stress–strain process under static loading can be regarded as an isothermal process, and the corresponding stress–strain curves can be regarded as isothermal curves. While the dynamic stress–strain process under high strain rates can be regarded as an adiabatic process accompanied with temperature change, consequently a thermo-mechanical coupled process, and the corresponding stress–strain curves can be regarded as adiabatic curves.

Therefore, to distinguish the dynamic response of structure under explosive or impact loading and that under static loading, it should consider both the inertia effect of media particles and the strain rate effect of material constitutive relation. When we treat dynamic problems of solids under explosive or impact loading, we actually face two kinds of problems. One is to study the motion of media (structures) under given external loadings when the dynamic mechanical behavior is known; this is attributed to the study on propagation of stress waves (direct problems). Another is to study the dynamic behavior of materials under high strain rates by measuring and analyzing the propagation of stress waves (inverse problems). The complexity of the problem lies in the coupling between the inertia effect and the strain rate effect. In fact, on the one hand, no wave propagation can be analyzed without knowing the corresponding dynamic constitutive relation of material, and consequently the basic characteristics of wave propagation inevitably depend on the strain-rate dependence of mechanical behavior of materials. On the other hand, the study on dynamic mechanical behavior of materials generally depends on the analysis based on the theory of stress wave propagation, namely wave propagation effects should not be neglected in the study of dynamic constitutive relations and failure criteria of materials under high strain rates. Therefore, studies on stress waves propagating in structures and studies of dynamic mechanical behavior of materials are interrelated closely.

Essentially, all material constitutive relations are more or less rate-sensitive, but the rate-sensitivity for different material is much different and depends on the stress range and the strain rate range. Under certain conditions, material constitutive relation can be approximately regarded as rate-independent in a certain range of strain rate. Those theories of stress waves based on such approximate assumption are named as **rate-independent theories**. According to whether the stress–strain relation is linear elastic, nonlinear elastic, or plastic, etc., the related theories are named as the theory of linear elastic waves, the theory of nonlinear elastic waves, and the theory of plastic waves, respectively. On the contrary, theories considering rate-dependence of material constitutive relation are named as **rate-dependent theories** of stress waves. According to the different rate-dependent stress–strain relations such as visco-elastic, visco-elasto-plastic, or elasto-visco-plastic, etc., the related theories are named as the theory of visco-elastic waves, visco-elasto-plastic waves and elasto-visco-plastic waves, etc., respectively.

Research of stress waves was originated from the rate-independent theory. The theory of linear elastic waves was developed in the twenties of the nineteenth century by Poisson, Остроградский, Stokes, *et al.* and later Rayleigh *et al.* in connection with the studies on elastic vibration (refer to Kolsky, 1953).

Foundation of the theory of plastic waves is 100 years later than that of elastic wave theory. There are two key difficulties: one is the nonlinear stress–strain relationship induced by plastic deformation, which needs to develop nonlinear loading wave theory. The other is that the stress–strain relation in unloading is different in loading due to the irreversibility of plastic deformation, which needs to develop theory of unloading waves and to take account of the interaction between the loading waves and the unloading waves. Looking from this angle, the earliest research on plastic loading waves may trace to Donnell's work (1930). But the full theories of plastic waves were founded and developed by von Karman (1942, in United States), Taylor (1942, in Great Britain) and Рахматулин, (1945, in Soviet Union), respectively, due to the requirement of military technology during World War II, although their researches were published only after World War II. It is worthwhile to note that the plastic wave theory mentioned above was developed following such an approach, namely, to generalize the linear theory of elastic waves under low stresses to the nonlinear theory of plastic waves under high stresses, including the unloading waves.

On the other side, during World War II, the theory of plastic waves was developed following another approach, namely, to generalize the hydrodynamics theory of solids under very high pressure to the hydro-elastic-plastic theory of solids under medium pressure. In fact, if the pressure applied to the solids is higher enough so that the shear strength of solids can be ignored, then the solid under such high pressure can be regarded as compressible fluids without shear strength. Thereby, the theory of shock waves in compressible fluids was extended as the theory of shock waves in solids under high pressure (hydrodynamic approximation). Typical works in this field include a number of works completed in the forties by Rice *et al.* (1958). Subsequently, this theory was generalized to sub-high pressure or medium pressure in which shear strength of solids is considered (Zheng and Xie, 1965; Lee, 1971; Chou and Hopkins, 1972), and consequently the theory of one-dimensional strain plastic waves (theory of hydro-elasto-plastic waves) was founded and developed.

As for the rate-dependent theory of visco-plastic waves, it was originated from the one-dimensional theory of elasto-visco-plastic waves developed by Соколовский (Sokolovskii) (1948) and Malvern (1951), and then generalized to three-dimensional theory by Perzyna (1963).

In the recent 70 years, studies and applications of stress wave theories develop rapidly. These theories have been extensively applied in many areas, such as seismic research, engineering blasting (mining, road construction, dam construction, etc.), explosive working (forming, bonding, combining, welding, hardening, etc.), explosive synthesizing (synthetic diamond, synthetic BN, synthetic nano-materials, etc.), ultrasonic and sound emission technology, dynamic response and impact strength of structures (buildings, machines, constructions, etc.), ordnance effects (fragmentation of shells, penetration of armor plate, high-explosive plastic projectile destroying tank armor, blast effects of explosives or nuclear weapon and the corresponding protection, etc.), aero-vehicle impacted by dispersed foreign objects, studies of meteorite craters on the earth and the moon, studies of dynamic response of materials under high strain rates and high pressure (including state equations of solids, phase transformation, and changes of other physical properties), studies of dynamic fracture, and the interaction between the materials and the particle beams with high energy intensity like electron beams, X-rays, laser, etc.

In this book, the primary one-dimensional theory of stress waves in bars is first discussed in Chapter 2. Then, from the shallower to the deeper, elastic waves (Chapter 3), elasto-plastic loading and unloading waves (Chapter 4), rigid unloading (Chapter 5), visco-elastic waves and visco-elasto-plastic waves (Chapter 6) are discussed in sequence. Although each chapter is not all-inclusive, its content is elaborately arranged into a system, and described from a consistent logic and unified treatment. Lagrange system is mainly used in this book, while Euler system is also discussed. We sincerely hope that this book will help readers to obtain a primary but systematic knowledge on the theory of stress waves.

For those reading the theory of stress waves for the first time, the first five chapters are elementary. Although contents in those chapters are limited to one-dimensional motion, the mathematical analyses are relatively simple, the basic concepts are readily comprehensible and the problem-solving methods are easy to be mastered. The theory of stress waves mainly focuses on the inhomogeneous and nonsteady motion of media varying with coordinates and time, emphasizes the analyses of local effects and early effects in media induced by dynamic loading. In analyses of stress waves, it stresses to pay more attention to the coupling effect between loads and media and the interdependence between stress waves and dynamic constitutive behavior of materials. These are the main differences between the dynamic theory and the static theory of solids, and can be understood from the easy to the difficult and complicated in the first five chapters.

In Chapter 7, the theory of one-dimensional strain plane waves is discussed. Here, although the strain state is one dimensional, while the stress state is three dimensional; the constitutive relation of material is much complicated. The problems in this chapter are discussed from two sides. On one side, along with increase of stress, the analyses on one-dimensional strain elastic waves under low stresses are extended to the one-dimensional strain elasto-plastic waves under high stresses. On the other side, first the shock wave theory of solids under high pressures, in which the shear strength of materials is neglected, is discussed, and then along with decrease of pressure, the one-dimensional strain elasto-plastic waves (hydro-elasto-plastic model) are discussed. Because of this, the state equations of solids under high pressures are discussed briefly in this chapter.

In Chapter 8, spherical waves and cylindrical waves are discussed together, since they are similar not only in their mathematical description but also in their character of wave propagation. Here, the discussion also begins from elastic waves, then elasto-plastic waves, and elasto-visco-plastic waves in the last. Besides, an example of rigid-plastic analysis is presented.

As an example of interaction of longitudinal waves and transverse waves, the theory of elasto-plastic waves in flexible strings under transverse impact is discussed in Chapter 9. A flexible string is such a simple structure component that only bears tension (stretch) while its bending resistance is disregarded. Under transverse impact, there are two kinds of wave propagating in strings: one is the longitudinal wave which induces the change of stress and strain but no change of the string shape, and another is the transverse wave which induces the change of string shape but no change of the stress and strain, however, they influence each other.

Based on the theory of elasto-plastic waves propagating in flexible strings and considering bending resistance of structural element, the theory of elasto-plastic waves propagating in beams under transverse impact (bending waves) are discussed in Chapter 10. Herein, the propagation of bending moment disturbances and shear disturbances are coupled each other. Although these discussions are limited to the primary theory based on the assumptions of "plane cross-section" of beams and rate-independence of constitutive relations, it is expected that they are helpful to beginners.

To obtain elementary knowledge of the general three-dimensional theory of stress wave propagation, the linear elastic waves in infinite homogeneous media are discussed in Chapter 11, including the wave reflection and transmission under oblique incidence and the Rayleigh surface waves.

Finally, to adapt to the rapid development and extensive application of numerical simulation by computer, the numerical methods used to solve the problems of stress wave propagation, including the characteristics method, the finite difference method and the finite element method, are briefly presented in Chapter 12.

Surely, the contents of stress wave theory should be much more than those mentioned above. The present book only provides a basis for further deeper studies. More specific contents can be found in references.

The trends of current studies of stress waves include the following aspects: further extending the one-dimensional theory to the two-dimensional and three-dimensional theory; developing from the stress wave theory for small deformation to the theory for large deformation; from the rate-independent theory to the rate-dependent theory; from the pure mechanical theory to the thermo-mechanical coupled theory; from the theory for isotropic materials to the theory for anisotropic materials; from the theory for homogeneous materials to the theory for heterogeneous materials (such as composites, porous, and multi-phase materials); wider application of computers in numerical simulation of stress wave propagation; and developing new experimental technologies.

CHAPTER 2

Elementary Theory of One-Dimensional Stress Waves in Bars

2.1 Material Coordinate System and Spatial Coordinate System

A coordinate system should be chosen first for investigating the motion of a bar.

In continuum mechanics there exist two coordinate systems (methods) to study the movement of the medium, namely the **material coordinate system** (**Lagrange method**) and the **spatial coordinate system** (**Euler method**).

One of the basic concepts of continuum mechanics is that the material's realistic (microscopic) structure is not taken into account in the analysis. Instead a medium is mathematically idealized as a system constituted of **continuously distributed mass points**. In other words, a medium is a continuous collection of mass points. A mass point is described by its location in the space. Different mass points have different locations at a time. At a specified time, the locations of all mass points within a medium constitute a **configuration**. To distinguish the mass points, each point must be given a name. Simultaneously each point has a spatial location at a given time. A reference spatial coordinate system is used to specify the location of mass points.

Take the one-dimensional movement of a bar as an example. Consider a mass point specified by X (X is the name of the point), the spatial location of the mass point is expressed by x. The motion of the media is generally represented by a different spatial location x occupied by an arbitrary mass point X at different time t, namely, x is the function of X and t:

$$x = x(X, t) \tag{2.1}$$

For a fixed mass point X, Eq. (2.1) defines the spatial location varying with time, namely, the motion of this mass point with time. At a fixed time t, on the other hand, Eq. (2.1) defines the spatial locations of all mass points. In general, at a fixed time each mass point has a unique location; conversely at each location there exists only one mass point. Therefore at a specified time we can distinguish the mass points by their spatial locations.

In other words, as long as the media motion is continuous and single-valued, we can reverse Eq. (2.1) to give:

$$X = X(x, t) \tag{2.2}$$

A simple method to name a mass point is to use its spatial location x_0 at a reference time t_0, and denote x_0 as X. In this case Eqs. (2.1) and (2.2) give the conversion relationships of mass point locations at time t_0 and time t. We may chose $t_0 = 0$, i.e. the initial time as the reference time. Other proper reference times are also possible. For the convenience of investigation, the reference spatial coordinate system at time t_0 used to define the mass point can be identical to, or different from the coordinate system used to describe the mass point motion.

Therefore, there are two methods to study the motion of media. In the first method, we follow the fixed mass point to observe the motion of media, and what we investigate are how the physical quantities for this fixed mass point vary with time and how the physical quantities vary with different mass point. Mathematically, any physical quantity ψ is viewed as the function of mass point X and time t, $\psi = F(X, t)$. This method is called **Lagrange method**. The independent variable $\boldsymbol{X}$ is called **Lagrange coordinates** or the **material coordinates**.

In the second method, we stay on a spatial point to observe the motions of material. What we investigate is how the physical quantities of different mass points arriving at this observation location at different time vary with time and how the physical quantities vary with different spatial location. Mathematically, any physical quantity ψ is viewed as the function of the spatial location x and time t: $\psi = f(x, t)$. This method is called **Euler method**. The independent variable x is called **Euler coordinate** or the **spatial coordinate**.

Making use of Eqs. (2.1) and (2.2), namely the conversions between the material coordinate and the spatial coordinate at time t, we can transform the physical quantity ψ expressed as a material coordinate function $F(X, t)$ into a spatial coordinate function $f(x, t)$, as:

$$f(x, t) = F[X(x, t), t]$$

or conversely:

$$F(X, t) = f[x(X, t), t]$$

Accordingly there exist two types of time derivatives. One is the time derivative of quantity ψ at a fixed spatial location x, denoted as:

$$\frac{\partial \psi}{\partial t} = \left(\frac{\partial f(x, t)}{\partial t} \right)_x \tag{2.3}$$

which is called the **spatial derivative** (or **Euler derivative**). The other one is the time derivative of quantity ψ observed by following a fixed particle X, denoted as:

$$\frac{d\psi}{dt} \equiv \left(\frac{\partial F(X, t)}{\partial t} \right)_X \tag{2.4}$$

which is called the **material derivative** (or **Lagrange derivative**). We can consider the function $F(X, t)$ in Eq. (2.4) as a compound function of variables (x, t), namely $f[x(X, t), t]$. Using the chain rule to deduce the derivative of compound function, we obtain:

$$\frac{d\psi}{dt} = \left(\frac{\partial f[x(X,t),t]}{\partial t}\right)_x + \left(\frac{\partial f[x(X,t),t]}{\partial x}\right)_t \left(\frac{\partial x}{\partial t}\right)_X$$
$$= \left(\frac{\partial f(x,t)}{\partial t}\right)_x + \left(\frac{\partial f(x,t)}{\partial x}\right)_t \left(\frac{\partial x}{\partial t}\right)_X$$

here $(\partial x/\partial t)_X$ is the material derivative of the spatial location x of particle X, namely the velocity of particle X, denoted as v:

$$v = \left(\frac{\partial x}{\partial t}\right)_X \equiv \frac{dx}{dt} \tag{2.5}$$

Neglecting the subscripts, we obtain the following relationship:

$$\frac{d\psi}{dt} = \frac{\partial \psi}{\partial t} + v\frac{\partial \psi}{\partial x} \tag{2.6}$$

If ψ represents particle velocity v, its material derivative is the acceleration of the particle, a, expressed as:

$$a = \left(\frac{\partial v}{\partial t}\right)_X \equiv \frac{dv}{dt} \tag{2.7}$$

From Eq. (2.6) we have:

$$a = \frac{\partial v}{\partial t} + v\frac{\partial v}{\partial x} \tag{2.8}$$

The first term in the right-hand side of (2.8) is the time derivative of particle velocity at location x with regard to time t. This term is called the **local acceleration**, which is equal to zero in the case of steady field. The second term in the RHS of (2.8) is the change rate of particle velocity with regard to time due to the location change of the mass point, called the **migration acceleration**, which is equal to zero in the case of uniform field.

It is noteworthy to notice that the wave velocity definition is closely related to the coordinate system being used. If we observe the stress wave propagation in a material coordinate system, at time t the wave front arrives at particle X, and $X = \Phi(t)$ represents the locus of wave front propagation in material coordinate system, then

$$C = \left(\frac{dX}{dt}\right)_W = \dot{\phi}(t) \tag{2.9a}$$

is called the **material wave velocity** (**Lagrange wave velocity**), or the **intrinsic wave velocity**. If we observe stress wave propagation in a spatial coordinate system, at time t

the wave front arrives at spatial point x, and $x = \varphi(t)$ represents the locus of wave front propagation in spatial coordinate system, then

$$c = \left(\frac{dx}{dt}\right)_w = \dot{\varphi}(t) \tag{2.9b}$$

is called the **spatial wave velocity** (**Euler wave velocity**). These two wave velocities describe the propagation behavior of one identical wave front. Because they are measured in two different coordinate systems, their values are generally different. An exception is the case when the wave front propagates into a media that is stationary and undeformed before the wave front arrives, then the material wave velocity equals to the spatial wave velocity.

After the definitions of wave velocity, we can discuss a third type of time derivative that is frequently used in stress wave study. This is called the **wave derivative**, defined as the total derivative of an arbitrary physical quantity ψ with regard to time following the wave, namely $(d\psi/dt)_W$. Similar to the expression for the material derivative in a spatial coordinate system (2.6), the wave derivative in the spatial coordinate system is:

$$\left(\frac{d\psi}{dt}\right)_W = \left(\frac{\partial\psi}{\partial t}\right)_x + c\left(\frac{\partial\psi}{\partial x}\right)_t \tag{2.10a}$$

and in the material coordinate system is:

$$\left(\frac{d\psi}{dt}\right)_W = \left(\frac{\partial\psi}{\partial t}\right)_X + C\left(\frac{\partial\psi}{\partial X}\right)_t \tag{2.10b}$$

Eqs. (2.10a) and (2.10b) again are descriptions of a same physical quantity from different coordinate systems. If the quantity ψ specifies the spatial location of a mass point $x(X, t)$, and noticing that in case of one-dimensional material motion,

$$\left(\frac{\partial x}{\partial X}\right)_t = (1 + \varepsilon)$$

where ε is the engineering strain of the material, then we can deduce the relationship between the spatial wave velocity c and the material wave velocity C of a plane wave front:

$$c = v + (1 + \varepsilon)C \tag{2.11}$$

Clearly for the plane wave propagating in a medium with zero initial speed and zero initial strain, its material wave velocity and spatial wave velocity are the same.

2.2 Governing Equations of Longitudinal Waves in Bars in Material Coordinates

In this section we use the material coordinate system (Lagrange system) to study the longitudinal motion of a bar with uniform cross section. The spatial coordinate of material

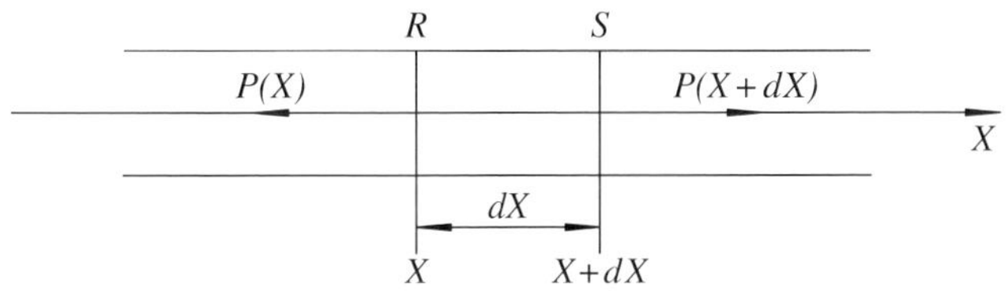

Fig. 2.1. An infinitesimal element of bar with uniform cross section in a Lagrange coordinate system.

before deformation ($t = 0$) is taken as the material coordinate (Lagrange coordinate). The axis of the bar is taken as the X axis (Fig. 2.1). Before deformation, the initial bar cross-sectional area A_0, density ρ_0 and other initial material parameters are independent of the coordinate. In the current problem no specific limitation on the bar's cross-sectional shape is necessary.

A basic assumption is: during deformation the bar's cross section is kept as planar, and the only axial stress is uniformly distributed in the cross section. With this assumption all physical quantities are functions of X and t, and the motion of the bar is reduced to a one-dimensional problem.

In the following discussions, the displacement u, strain $\varepsilon = (\partial u/\partial X)$, particle velocity $v = (\partial u/\partial t)$, and stress σ denote the corresponding components in X direction (except for special cases we do not use the subscript X). Note that the stress and strain used here are engineering stress and engineering strain, and for the one-dimensional case, the Lagrangian formulation strain $\partial u/\partial X$ is no longer limited to small strain.

The governing equations in the current problem include the kinematic condition (called the continuous equation or the mass conservative equation), the dynamic condition (called the motion equation or the momentum conservative equation), and the material's constitutive relationship (material equation). The mathematical forms of these equations are deduced below.

Having noted that the strain ε and the particle velocity v are the first-order derivatives of displacement u with respect to X and t, from the single-value, continuous property of u we obtain a compatible relationship between ε and v. This is the continuous equation, as:

$$\frac{\partial v}{\partial X} = \frac{\partial \varepsilon}{\partial t} \tag{2.12}$$

Consider a small segment dX of the bar (Fig. 2.1): on the left cross section R the segment is subjected to force $P(X, t)$, on the right cross section S it is subjected to force

$$P(X + dX, t) = P(X, t) + \frac{\partial P(X, t)}{\partial X} dX$$

From Newton's law we have:

$$\rho_0 A_0 dX \frac{\partial v}{\partial t} = P(X + dX, t) - P(X, t) = \frac{\partial P}{\partial X} dX$$

Substitute in the engineering stress $\sigma = P/A_0$, the motion equation of this segment is:

$$\rho_0 \frac{\partial v}{\partial t} = \frac{\partial \sigma}{\partial X} \tag{2.13}$$

Note that under the current Lagrange coordinate system, the derivative $\partial/\partial t$ in Eqs. (2.12) and (2.13) by default means the material time derivative on a fixed X value. We do not use d/dt expression further.

Here we discuss the material's constitutive relationship within the framework of the rate-independent theory. A second basic assumption made for the current analysis is: stress is the function of strain. Mathematically the material's **constitutive relationship** can be written as:

$$\sigma = \sigma(\varepsilon) \tag{2.14}$$

Because the stress wave velocity is very high, in the short time, stress wave propagates through a segment, heat exchange of the segment with the neighboring regions is very small. As a result we can consider the material's deformation process to be adiabatic. The constitutive equation listed here, is therefore an adiabatic stress–strain relationship. Based on this concept we do not need to introduce other physical equations such as the energy conservative equation, and the governing Eqs. (2.12)–(2.14) are closed for variables σ, ε, and v. Given adequate initial conditions and boundary conditions we can solve this group of equations for the three unknown functions $\sigma(X, t)$, $\varepsilon(X, t)$, and $v(X, t)$.

In the discussions here and after, the stress and strain are positive for tensile case, and negative for compressive case. The particle velocity is positive when the particle moves in X axis, and *vice versa*.

In general the function $\sigma(\varepsilon)$ is continuously differentiable. Introducing

$$C^2 = \frac{1}{\rho_0} \frac{d\sigma}{d\varepsilon} \tag{2.15}$$

we can eliminate the variable σ from Eqs. (2.13) and (2.14), obtaining:

$$\frac{\partial v}{\partial t} = C^2 \frac{\partial \varepsilon}{\partial X} \tag{2.16}$$

or, eliminate the variable ε from Eqs. (2.12) and (2.14), obtaining:

$$\frac{\partial \sigma}{\partial t} = \rho_0 C^2 \frac{\partial v}{\partial X} \tag{2.17}$$

Therefore the problem is reduced to solving the first-order partial differential Eqs. (2.12) and (2.16) for unknown functions ε and v, or solving the first-order partial differential Eqs. (2.13) and (2.17) for unknown functions σ and v.

If the expressions of ε and v in terms of displacement u are substituted into (2.16), we get a second-order partial differential equation for the unknown function u, which is called the wave equation:

$$\frac{\partial^2 u}{\partial t^2} - C^2 \frac{\partial^2 u}{\partial X^2} = 0 \tag{2.18}$$

In the process of deducing the governing equation, we made two basic assumptions. In the first assumption, we approximately neglected the inertial effect of the particles due to their transverse motion. In other words, the kinetic energy corresponding to the transverse expansion or contraction of the bar is omitted. In fact, the transverse motion of particles will result in the non-uniform axial stress distribution on the cross section, and hence the original planar cross section will be warped. Problem involving this deformation is no longer a one-dimensional one, and is much more difficult to solve. However, according to the known exact solutions for elastic waves, as long as the ratio of the wavelength to the transverse size of the bar is large enough, the approximation neglecting the **transverse inertial effect** is acceptable (refer to the Section **2.8**). In this chapter all discussions on stress wave propagations are based on this assumption. Theories developed are called the **primary theory** or the **engineering theory**.

The second basic assumption is common in all rate-independent stress wave theories. At first glance, this theory seems to be applicable only to the elastic deformation regime, since elastic parameters are generally considered to be strain-rate independent, or to the elastic–plastic materials that are insensitive to the strain-rate. However, as the strain-rate under impact loading, is in general, several orders of magnitude higher than the quasistatic case, and material properties under impact loading, is generally different from that under quasistatic loading, the rate-independent assumption is better understood in the following way: the material behavior under impact loading in a certain strain-rate region is described by a unique stress–strain relationship, which is the average constitutive relationship of the material within this strain-rate region, and may be different from the quasistatic one. In this meaning, the strain-rate effect has been included in the constitutive relationship, but not explicitly. Therefore the second assumption of rate-independence is applicable practically to broader cases, and the rate-independent stress wave theories are useful in many engineering applications. In Chapter 6 we will discuss the rate-dependent stress wave theory.

2.3 Characteristic Lines and the Compatibility Relationships along the Characteristic Lines

Below we discuss further the governing Eq. (2.18).

First, according to the assumption that "stress is a single-valued function of strain", $C^2 = (1/\rho_0)\,(d\sigma/d\varepsilon)$ is also a function of strain $\varepsilon(= \partial u/\partial X)$, so Eq. (2.18) is linear with respect to the second-order derivatives of u, and belongs to the type of second-order quasi-linear partial differential equation of two independent variables. Under the special case that stress is a linear function of strain, C^2 is constant, and (2.18) is a linear partial differential equation.

Second, as long as we exclude the special case of unstable plastic deformation, stress must be a monotonic increasing function of strain, namely $d\sigma/d\varepsilon > 0$. Also the material density ρ_0 is positive. Therefore the quantity C^2 is a positive value ($C^2 > 0$). According to the categorization of second-order partial differential equations (one may refer to the textbooks of mathematical physical equations), Eq. (2.18) is a hyperbolic partial differential equation (PDE) (the so-called wave equation). This equation has two groups of real characteristic lines, i.e. on the plane of the independent variables (X, t), there exist two different real characteristic lines passing through any point.

Not only is the concept of the characteristic lines useful for categorizing PDEs, it is also the fundamentals of one basic PDE solution technique, namely the method of characteristics. Therefore, the concept of the characteristic line is very important to study the propagations of waves, specifically the one-dimensional wave motion. Using the method of characteristics, the problem to solve a PDE with two independent variables is reduced to solving the ordinary differential equations (ODEs) along the characteristic lines.

The characteristic lines can be defined with different but equivalent approaches. There are two major methods for doing this. The first is called the **directional derivative method**. If a second-order PDE (or, equivalently, the linear combinations of a group of first-order PDEs) can be reduced to the directional derivative along a certain curve $\mathscr{C}$ on the independent variable plane (X, t), this curve $\mathscr{C}$ is called a characteristic line. The second method is called the **indeterminate line method**. According to this definition, along a certain curve $\mathscr{C}$ on the independent variable plane (X, t) the initial values are prescribed; if with the prescribed initial values we still cannot determine all partial derivatives, the curve $\mathscr{C}$ is a characteristic line. The two definitions from different angles of view reflect the essential property of the characteristic line, respectively. Either method applied, we will get the same results. In the following, we will mainly use the directional derivative method to discuss the wave Eq. (2.18).

Consider a curve $\mathscr{C}(X, t)$ on the (X, t) plane. Along this curve the directional derivative of the first-order derivatives of displacement u, namely v and ε, are:

$$dv = \frac{\partial v}{\partial X}dX + \frac{\partial v}{\partial t}dt = \frac{\partial^2 u}{\partial X\,\partial t}dX + \frac{\partial^2 u}{\partial t^2}dt \tag{2.19}$$

$$d\varepsilon = \frac{\partial \varepsilon}{\partial X}dX + \frac{\partial \varepsilon}{\partial t}dt = \frac{\partial^2 u}{\partial X^2}dX + \frac{\partial^2 u}{\partial t\,\partial X}dt \tag{2.20}$$

In the above equations dX and dt are the components of a curve segment dS, in X and t axes, so that dX/dt is the slope of curve $\mathscr{C}$ at point (X, t). If curve $\mathscr{C}$ is a characteristic line of Eq. (2.18), the left side of (2.18) shall be reduced to containing only the directional derivatives along the curve, namely a linear combination of Eqs. (2.19) and (2.20):

$$dv + \lambda d\varepsilon = \frac{\partial^2 u}{\partial t^2}dt + (\lambda dt + dX)\frac{\partial^2 u}{\partial X \partial t} + \lambda \frac{\partial^2 u}{\partial X^2}dX = 0 \tag{2.21}$$

where λ is a coefficient to be determined. Comparing (2.21) with (2.18), the two equations are equivalent under following conditions:

$$\frac{1}{dt} = \frac{0}{\lambda dt + dX} = -\frac{C^2}{\lambda dX} \tag{2.22}$$

From the first equation, we have $\lambda = -(dX/dt)$, then by substituting it into the second equation, we obtain the **characteristic direction** dX/dt, as $(dX/dt) = \pm C$. Rewrite the relationship as:

$$dX = \pm C dt \tag{2.23}$$

which is the **characteristic line equation** in **differential form**. An integration of (2.23) leads to the characteristic line. Substitute (2.23) back into (2.22) we obtain $\lambda = \pm C$. So Eq. (2.18), or equivalently (2.21), is reduced to the ordinary differential equations that only contain the directional derivatives along the characteristic lines:

$$dv = \pm C d\varepsilon \tag{2.24}$$

Equation (2.24) regulates a compatible relationship between v and ε on the characteristic line, and is therefore called the **compatibility equations on characteristic lines**. Therefore, solving the quasi-linear PDE (2.18) is equivalent to solving the characteristic line Eq. (2.23) and the corresponding compatibility Eq. (2.24), both are ODEs.

The Eq. (2.23) defines characteristic lines in (X, t) plane. Similarly Eq. (2.24) can be viewed as a characteristic line equation on (v, ε) plane, and the integration of which leads to the **characteristic lines on (v, ε) plane**. In some occasions the (X, t) plane is called the **physical plane**, and the (v, ε) plane is called the **velocity plane**. Therefore, Eq. (2.23) geometrically represents two groups of characteristic lines in (X, t) plane, and Eq. (2.24) geometrically represents two groups of characteristic lines in (v, ε) plane. The characteristic lines in different planes are correlating (mapping) to each other. As shown in Fig. 2.2, the region G in (X, t) plane and the region G' in (v, ε) plane maps to each other, and the same for the characteristic lines $\mathscr{C}$ and $\mathscr{C}'$, as well as the intersections of characteristic lines, Q and Q'. Such corresponding relationships provide the fundamentals of the method of characteristics.

In the above discussions we used the directional derivative method to deduce the equations of characteristics. We can also use the indeterminate line method to conduct the same deductions. Note that the first-order PDEs (2.12) and (2.16) are equivalent to the second-order PDE (2.18). We can rewrite all the first-order governing Eqs. (2.12), (2.16), (2.19),

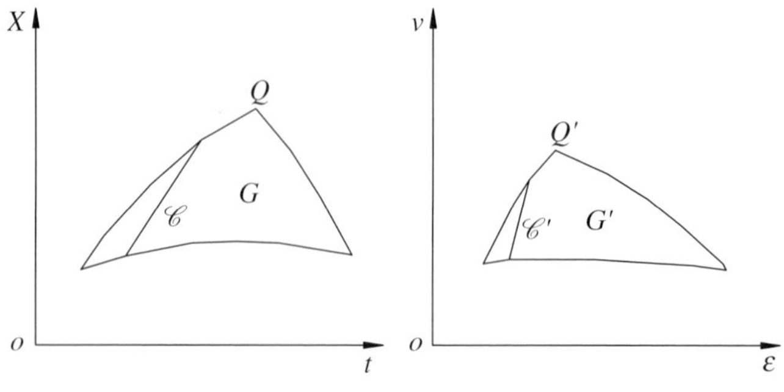

Fig. 2.2. A region G in (X, t) plane is mapped to a region G' in (v, ε) plane.

and (2.20) as below:

$$\begin{cases} \dfrac{\partial v}{\partial X} - \dfrac{\partial \varepsilon}{\partial t} = 0 \\ \dfrac{\partial v}{\partial t} - C^2 \dfrac{\partial \varepsilon}{\partial X} = 0 \\ \dfrac{\partial v}{\partial X} dX + \dfrac{\partial v}{\partial t} dt = dv \\ \dfrac{\partial \varepsilon}{\partial X} dX + \dfrac{\partial \varepsilon}{\partial t} dt = d\varepsilon \end{cases}$$

This group of equations can be viewed as an algebraic equation group of unknown partial derivatives $\partial v/\partial X, \partial v/\partial t, \ \partial \varepsilon/\partial X$, and $\partial \varepsilon/\partial t$. In the matrix form the equations are written as:

$$\begin{bmatrix} 1 & 0 & 0 & -1 \\ 0 & 1 & -C^2 & 0 \\ dX & dt & 0 & 0 \\ 0 & 0 & dX & dt \end{bmatrix} \begin{bmatrix} \dfrac{\partial v}{\partial X} \\ \dfrac{\partial v}{\partial t} \\ \dfrac{\partial \varepsilon}{\partial X} \\ \dfrac{\partial \varepsilon}{\partial t} \end{bmatrix} = \begin{bmatrix} 0 \\ 0 \\ dv \\ d\varepsilon \end{bmatrix} \tag{2.25}$$

If curve $\mathscr{C}$ is a characteristic line, the above equations have indeterminate solutions, namely,

$$\Delta_0 = \Delta_1 = \Delta_2 = \Delta_3 = \Delta_4 = 0$$

where

$$\Delta_0 = \begin{vmatrix} 1 & 0 & 0 & -1 \\ 0 & 1 & -C^2 & 0 \\ dX & dt & 0 & 0 \\ 0 & 0 & dX & dt \end{vmatrix}, \Delta_1 = \begin{vmatrix} 0 & 0 & 0 & -1 \\ 0 & 1 & -C^2 & 0 \\ dv & dt & 0 & 0 \\ d\varepsilon & 0 & dX & dt \end{vmatrix}, \Delta_2 = \cdots, \quad \cdots$$

After the expansions of these determinations, we can obtain again the characteristic line Eq. (2.23) and the characteristic compatibility Eq. (2.24).

If we use the stress σ and velocity v as unknown functions, and start from the first-order PDEs (2.13) and (2.17), we can arrive at the characteristic line Eq. (2.23), and the corresponding compatibility equation as:

$$d\sigma = \pm\rho_0 C dv \tag{2.26}$$

Equation (2.26) is equivalent to (2.24). Actually, substituting (2.15) into (2.24), we obtain Eq. (2.26) immediately.

In the following discussion, we will show that the characteristic line Eq. (2.23) physically represents the propagation of disturbances. In other words, the characteristic lines on (X, t) plane are the propagation loci of the disturbances (wave fronts). The quantity $C = \sqrt{(1/\rho_0)(d\sigma/d\varepsilon)}$ is the material velocity of disturbance propagation. The plus symbol (+) in Eq. (2.23) represents the positive waves that are propagating to the right, and the minus symbol (−) represents the negative waves propagating to the left. Equations (2.24) and (2.26) determine the compatible relationships between the particle velocity v and strain ε or stress σ in the process of disturbance propagation. The quantity $\rho_0 C$ is called the **wave impedance**. Note that in Eqs. (2.24) and (2.26) the (+) symbol corresponds to the rightward wave and the (−) symbol corresponds to the leftward wave. This is converse to the symbols in the compatibility relationships across a wave front [which will be discussed in subsequent sections, see Eq. (2.63) in 2.7].

2.4 Elastic–Plastic Longitudinal Loading Waves in Semi-Infinite Bars

In this section, we discuss the longitudinal stress waves propagating in a semi-infinite long bar. The bar extends from $X = 0$ to $X = \infty$. All stress waves propagate along the positive X direction and no wave reflection is involved. An example of this case is wave propagating unidirectionally in a finite length bar before reflections occur. Also in this section we will consider only the loading case, namely the stress level in the bar is monotonically increasing. At the boundary of the bar, the external loading can be prescribed either in terms of boundary stress $(\partial|\sigma|)/(\partial t) \geq 0$, or boundary speed $(\partial|v|)/(\partial t) \geq 0$.

2.4.1 Linear elastic waves

First, we consider the case when the impact loading is not high so that the bar deforms elastically. In this case the stress–strain relationship is described by Hooke's law, and the constitutive relationship (2.14) is simplified as:

$$\sigma = E\varepsilon \tag{2.27}$$

where E is Young's modulus. In this case the wave Eq. (2.18) is simplified as a linear wave equation:

$$\frac{\partial^2 u}{\partial t^2} - C_0^2 \frac{\partial^2 u}{\partial X^2} = 0 \tag{2.28}$$

where C_0 is a constant that is determined from material parameters ρ_0 and E:

$$C_0 = \sqrt{\frac{E}{\rho_0}} \tag{2.29}$$

From Eqs. (2.23) and (2.24) we see that the characteristic lines and the compatibility equations are two groups of straight lines with slope $\pm C_0$, on (X, t) plane and (v, ε) plane, respectively:

$$\begin{aligned} dX &= \pm C_0 dt \\ dv &= \pm C_0 d\varepsilon \end{aligned} \tag{2.30a}$$

By introducing the integration constants ξ_1, ξ_2, R_1, and R_2, equations in (2.30a) can be represented by:

$$\left.\begin{aligned} X - C_0 t &= \xi_1 \\ v - C_0 \varepsilon &= R_1 \end{aligned}\right\} \text{for rightward waves}$$

$$\left.\begin{aligned} X + C_0 t &= \xi_2 \\ v + C_0 \varepsilon &= R_2 \end{aligned}\right\} \text{for leftward waves} \tag{2.30b}$$

The constants R_1 and R_2 are sometimes called **Riemann's invariables**.

Assume that the bar is at a stationary, stress-free state before loading. At time $t = 0$, an impact is applied at the bar end $X = 0$. If, for example, the boundary condition is prescribed as a particle velocity $v_0(\tau)$, the problem is to solve the partial differential Eq. (2.28) under the initial condition:

$$v(X, 0) = \varepsilon(X, 0) = 0, \quad 0 < X \leq \infty \tag{2.31a}$$

and the boundary condition:

$$v(0, t) = v_0(\tau), \quad t \geq 0 \tag{2.31b}$$

or, to solve the characteristic Eqs. (2.30) for v and ε that satisfy the same boundary and initial conditions. Actually, in solving this problem we deal with two kinds of initial-boundary-value-problems (IBVP), namely the Cauchy problem and the Picard problem.

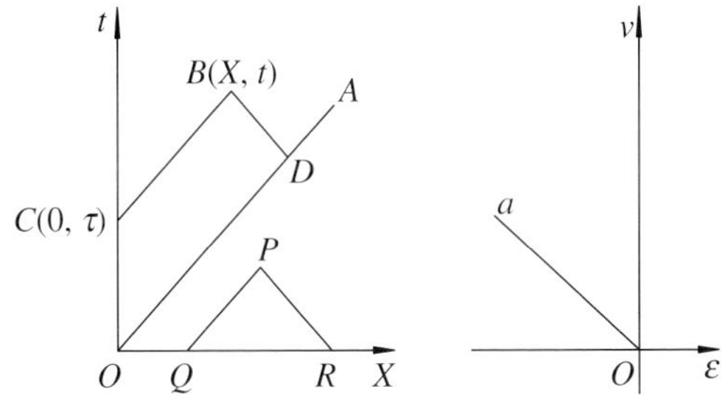

Fig. 2.3. Left: On (X, t) plane, a positive characteristic line and a negative one pass through an arbitrary point; Right: On (v, ε) plane, the line Oa represents the simple wave relationship Eq. (2.33).

As shown in Fig. 2.3, passing through an arbitrary point on (X, t) plane, there exist one positive and one negative characteristic line. The line *OA* in Fig. 2.3 is the positive characteristic line passing through the origin (0, 0). We consider the region *AOX* below line *OA*. Along the *OX* axis the values of v and ε are known as the initial conditions. Since the two characteristic lines (*QP* and *RP*) passing through an arbitrary point (P) in the *AOX* region eventually intersect with the *OX* axis, the Riemann invariables along these two characteristic lines R_1 and R_2 are determined by the initial conditions:

$$\text{Along } QP \qquad v - C_0\varepsilon = v(Q) - C_0\varepsilon(Q)$$

$$\text{Along } RP \qquad v + C_0\varepsilon = v(R) + C_0\varepsilon(R)$$

so that at the intersection point P of lines *QP* and *RP*, the values $v(P)$ and $\varepsilon(P)$ can be solved from the above equations, as:

$$v(P) = \frac{1}{2}\{[v(R) + v(Q)] + C_0[\varepsilon(R) - \varepsilon(Q)]\}$$

$$\varepsilon(P) = \frac{1}{2C_0}\{[v(R) - v(Q)] + C_0[\varepsilon(R) + \varepsilon(Q)]\} \tag{2.32}$$

For the current case with zero initial conditions (2.31), $v(Q) = \varepsilon(Q) = v(R) = \varepsilon(R) = 0$, we therefore get $v(P) = \varepsilon(P) = 0$. Since P is an arbitrary point in the *AOX* region, this means that the whole *AOX* region is a constant-value region with $v = \varepsilon = 0$. Indeed as long as the initial values are constant, namely $v(Q) = v(R) = const.$, $\varepsilon(Q) = \varepsilon(R) = const.$, the region *AOX* is a constant-value region with values: $v(P) = v(Q) = v(R)$ and $\varepsilon(P) = \varepsilon(Q) = \varepsilon(R)$.

In the above discussions, the initial value curve *OX* is not a characteristic line. From any point on this line, two characteristic lines grow with time into the region being considered. All curves have similar properties (not necessarily parallel to the X axis) and are called the **space-like curve**. If on a segment of space-like curve *QR* the velocity v and the strain ε are prescribed, then the curvilinear triangle bounded by *QR* and two characteristic lines *QP* and *RP*, namely the region *QRP* can be exclusively solved. This kind of IBVP is called the **initial value problem** or the **Cauchy problem**.

Now we consider the region above *OA* line, namely the *AOt* region. Observe an arbitrary point *B* in this region, the negative characteristic line passing through point *B*, *BD*, eventually interacts with the *OA* line. Since we have known that $v = \varepsilon = 0$ along *OA*, we have $R_2 = 0$ in the *AOt* region. Namely,

$$v = -C_0\varepsilon = -\frac{\sigma}{\rho_0 C_0} \tag{2.33}$$

On the other hand, the positive characteristic line passing through *B* always intersects with the time axis *Ot*. Notice that on *Ot* the velocity *v* is prescribed as the boundary condition (2.31b), so that along *CB* we have:

$$R_1 = v - C_0\varepsilon = 2v = -2C_0\varepsilon = 2v_0(\tau)$$

Note that the mathematical expression of the positive characteristic line *CB* is:

$$X = C_0(t - \tau)$$

where $C_0\tau$ is the integration constant ξ_1, and τ is the intercept of the characteristic line on *t* axis. Therefore, the *v* and ε values at any point $B(X, t)$ in the *AOt* region are determined as:

$$v = -C_0\varepsilon = v_0\left(t - \frac{X}{C_0}\right) \tag{2.34}$$

This means that the boundary velocity disturbance $v_0(\tau)$ propagates into the bar with velocity C_0, and at time *t* arrives at the cross section *X*. Therefore the characteristic lines are physically the propagation loci of the disturbances (wave fronts). C_0 is the longitudinal elastic wave speed in the bar determined by the material parameters ρ_0 and *E* (2.29).

Similar to the concept of space-like curve, the non-characteristic curves such as *Ot* axis are called the **time-like curve**. From any point on such curve only one characteristic line grows with time into the region being considered. Consider a region that is bounded by a characteristic line and a time-like line. On the characteristic line boundary, the quantities *v* and ε are given, while on the other boundary, that is a time-like curve, one quantity *v* or ε is given. We can solve the region using the approach similar to that used to solve the *AOt* region. This kind of problem is called the **mixed problem** or the **Picard problem**.

As a result, solution of a semi-infinite bar impacted at the end is reduced to a Cauchy problem in the *AOX* region, and a Picard problem in *AOt* region. The solution of the Cauchy problem is completely determined by the initial condition. This means that the region is not influenced by the boundary disturbances. The solution of the Picard problem is determined jointly by the initial condition and the boundary condition. This means that any point *B* in the *AOt* region is influenced by the initial disturbance propagating leftwards, and the boundary disturbance propagating rightwards. Under the current initial-boundary conditions (2.31), the initial disturbances are zero. Therefore the bar cross section *X* is at rest until the first boundary disturbance (propagating along the characteristic line *OA* with velocity C_0) arrives, namely before $t = X/C_0$. The state of region *AOX* is mapped as the origin point in (v, ε) plane. After the time $t = X/C_0$, the boundary disturbances $v_0(\tau)$

arrive at the bar cross section X successively. Because the initial disturbances propagating leftwards are all zero, the boundary disturbances propagating rightwards are not disturbed and reserve the original boundary state (2.34). Such kind of undisturbed waves are called the **simple waves**. For the elastic simple waves propagating into an initially stationary, undeformed region, the mass velocity v, strain ε, and stress σ satisfy the Eq. (2.33), which is called the **simple wave relationship**. Note that the physical states of all the points on a leftward characteristic line DB in the AOt region are mapped onto the (v, ε) plane as line Oa, and the simple wave equation is the mathematical equation of line Oa. Obviously, if the bar has a uniform initial velocity v_0, initial strain ε_0, and initial stress σ_0, then Eq. (2.33) should be rewritten as:

$$v - v_0 = \mp C_0(\varepsilon - \varepsilon_0) = \mp \frac{\sigma - \sigma_0}{\rho_0 C_0} \tag{2.35}$$

where $(-)$ corresponds to the rightward wave and $(+)$ corresponds to the leftward wave. This equation determines the basic relationship between particle velocity and strain or stress, and is one of the most important equations in studying elastic wave propagation. The quantity $\rho_0 C_0$ is usually called the wave impedance or the **sound impedance**, and is one of the most important parameters describing the mechanical properties of materials under impact loading. Table 2.1 collects the elastic longitudinal wave speed C_0 and the wave impedance $\rho_0 C_0$ of several common materials (Kolsky, 1953).

Note that, as shown in Fig. 2.3, the constant value region AOX on (X, t) plane is mapped onto the (v, ε) plane as one point O, and the simple wave region AOt on (X, t) plane is mapped onto (v, ε) plane as a line segment Oa. In other words, any characteristic line in the simple wave region on (X, t) plane, representing a non-zero (un-trivial) disturbance, is mapped as a point on (v, ε) plane. An important conclusion can be drawn from the above discussions: **A simple wave region is always neighboring a constant wave region**.

The above discussions are based on the prescribed velocity boundary condition. If the boundary condition is prescribed as a strain condition:

$$\varepsilon(0, t) = \varepsilon_0(\tau) \quad t \geq 0$$

or, as a stress condition:

$$\sigma(0, t) = \sigma_0(\tau) \quad t \geq 0$$

we can obtain the similar results.

Table 2.1. The elastic longitudinal wave velocity C_0 and the wave impedance $\rho_0 C_0$ of several common materials.

	Steel	Copper	Aluminum	Glass	Rubber
ρ_0 (10^3 kg/m^3 = g/cm^3)	7.8	8.9	2.7	2.5	0.93
E (GPa = 10^{10} dyne/cm^2)	210	120	70	70	2.0×10^{-3}
C_0 (km/s)	5.19	3.67	5.09	5.30	0.046
$\rho_0 C_0$ (MPa/m/s = 10^6 kg/m^2/s)	40.5	32.7	13.7	13.3	42.8×10^{-3}

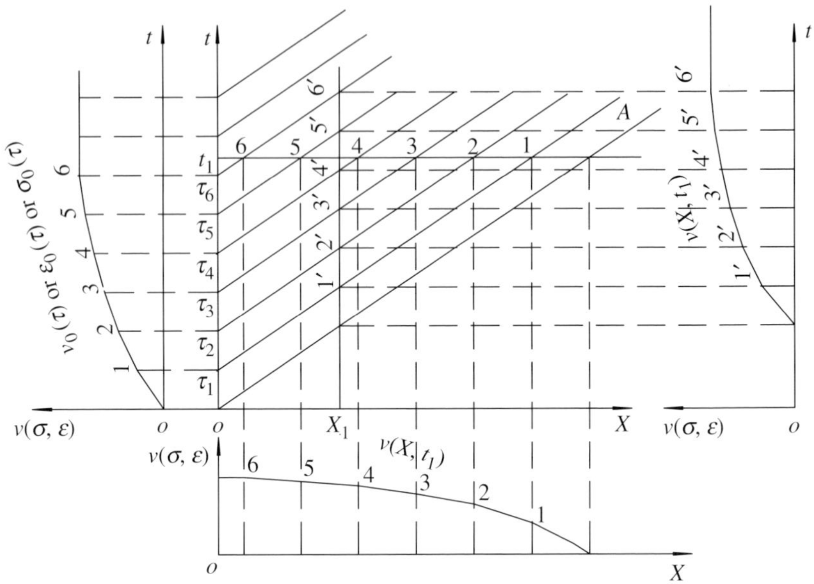

Fig. 2.4. The wave profiles (waveform curves) at any time, or the wave histories (temporal curves) on any cross section of a bar can be determined by using the method of characteristics.

Based on the method of characteristics, a simple and convenient drawing method can be used to determine the stress (or strain, or particle velocity) distributions at any time, or the stress (or strain, or material velocity) history curves on any cross section. As shown in Fig. 2.4, we can draw a horizontal line $t = t_1$ on (X, t) plane. The line intersects with each characteristic line at points 1, 2, 3, 4, 5, and 6. Since along each characteristic line, v (or ε, or σ) equals the known boundary value $v_0(t)$ (or $\varepsilon_0(t)$, or $\sigma_0(t)$) depending on the type of the boundary condition, therefore the velocity distribution at the time $t = t_1$, and the corresponding strain distribution and stress distribution can be determined, which are called the **waveform curves** as shown in the bottom of Fig. 2.4. Similarly, we can draw a vertical line $X = X_1$ on (X, t) plane, which intersects with the characteristic lines at points $1', 2', 3', 4', 5'$, and $6'$. The velocity, strain, and stress at each point (representing different times) can be solved; hence we obtain the **temporal curves** at the cross section X_1, as shown in the right part of Fig. 2.4. Using the waveform curves at a sequence of times, or the temporal curves at different points, we can imaginarily describe the propagation process of stress waves. The waveforms of the linear elastic waves are not changing during propagation, because the wave velocity C_0 is a constant.

2.4.2 *Elastic–plastic loading waves*

Consider an initially stationary and unstressed long bar that is subjected to an impact loading at its end. By the simple wave relationship $v = -(\sigma/\rho_0 C_0)$, (2.33), we know that the stress magnitude of the stress wave is proportional to the impact velocity. If the dynamic uni-axial yielding stress of the material is denoted as Y, then once the impact velocity v exceeds the so-called **yielding velocity** v_Y, namely,

$$|v| > v_Y = \frac{Y}{\rho_0 C_0} \tag{2.36}$$

the material enters into the plastic deformation region and plastic waves will propagate along the bar.

We now use the method of characteristics to solve the problem of elastic–plastic wave propagation. As the wave speed $C = \sqrt{(1/\rho_0)\,(d\sigma/d\varepsilon)}$ is a function of strain ε, the characteristic Eq. (2.23) and the compatibility Eq. (2.24)

$$dX = \pm C dt$$

$$dv = \pm C d\varepsilon$$

do not represent straight lines on (X, t) plane and (v, ε) plane. However, by introducing a variable

$$\varphi = \int_0^{\varepsilon} C d\varepsilon = \int_0^{\sigma} \frac{d\sigma}{\rho_0 C} \tag{2.37}$$

the compatibility equations on the characteristic lines, either Eq. (2.24) or Eq. (2.26), can be written as the following unified form:

$$dv = \pm d\varphi \tag{2.38}$$

which means that on (v, φ) plane the characteristic lines are two groups of orthogonal straight lines that intersect with the coordinate axes by $\pm 45°$:

$$\begin{cases} v - \varphi = R_1 \\ v + \varphi = R_2 \end{cases} \tag{2.38$'$}$$

Obviously, if the relationship $\sigma = \sigma(\varepsilon)$ is known, the function $C = C(\varepsilon)$ and $\varphi = \varphi(\varepsilon)$ are also known functions (Fig. 2.5).

To solve the problem of elastic–plastic wave propagation in a semi-infinite bar with the initial-boundary conditions (2.31), the procedures previously applied for solving the elastic wave problem are still available, namely, the problem is reduced to solve a Cauchy problem in *AOX* region and a Picard problem in *AOt* region (Fig. 2.5). The elastic wave parts in the constant value region *AOX* and in the simple wave region *AOt* are exactly identical with that discussed previously (Fig. 2.4). The plastic wave part corresponding to the boundary condition $|v_0(\tau)| \geq v_Y$, however, should be handled separately. Since all negative characteristic lines will intersect with X-axis, the zero disturbance initial condition (2.31a) eventually leads the Riemann invariables R_2 in Eq. (2.38) to be zero. Therefore in the plastic simple wave region following conditions exist:

$$v = -\varphi = -\int_0^{\varepsilon} C d\varepsilon = -\int_0^{\sigma} \frac{d\sigma}{\rho_0 C} \tag{2.39}$$

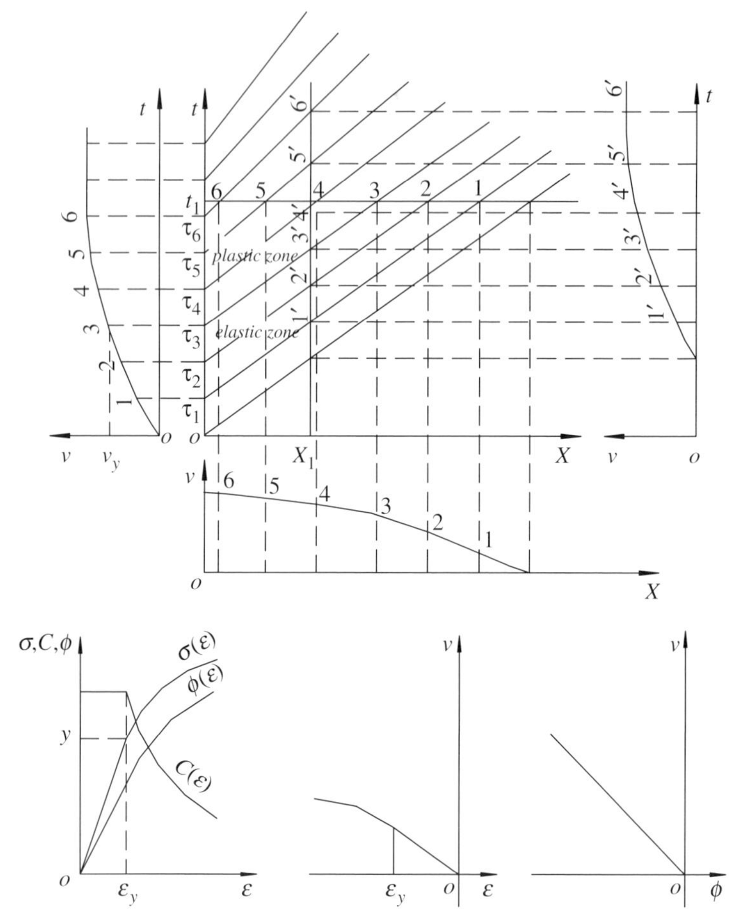

Fig. 2.5. Up: Characteristics lines in a simple wave region on (X,t) plane; Low left: Variances of stress σ, wave velocity C, and variable ϕ with strain; Low Middle: States of simple wave region on (v,ε) plane; Low Right: States of simple wave region on (ϕ,ε) plane.

and along the positive characteristic lines the Riemann invariable R_2 can be determined via boundary condition (2.31b), as:

$$v = -\varphi = \frac{R_1}{2} = v_0(\tau)$$

namely, along a positive characteristic line the particle velocity v, strain ε, and stress σ are invariant, and so $C(\varepsilon)$ is invariant either. However, on different positive characteristic lines, the values of C are different. As a result, in simple plastic wave region the positive characteristic lines are a group of straight lines with different slopes (Fig. 2.5):

$$X = C(\varepsilon)(t - \tau)$$

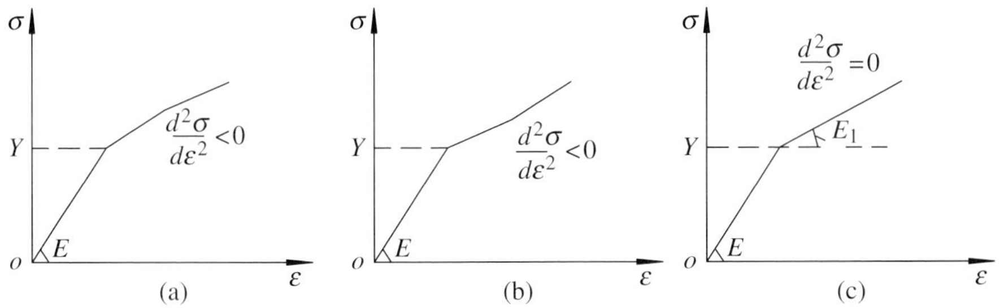

Fig. 2.6. Stress–strain curves showing different hardening properties: (a) decreasing hardening; (b) increasing hardening; and (c) linear hardening.

where the quantity C physically represents the propagation velocity of the plastic wave.

According to Eq. (2.15) the plastic wave speed C is determined by the material density ρ_0 and the tangent modulus of the material dynamic stress–strain curve $(d\sigma/d\varepsilon_0)$. Depending on the different strain-hardening characters of the stress–strain curve $\sigma = \sigma(\varepsilon)$ there exist three different cases as shown in Fig. 2.6, wherein Fig. 2.6(a) represents decreasingly hardening character, (b) represents increasingly hardening character, and (c) represents linearly hardening character.

Case (a). In this case the material tangent modulus decreases with the increase of strain (**decreasing hardening material**), namely the $\sigma \sim \varepsilon$ curve is convex upwards $(d^2\sigma/d\varepsilon^2 < 0)$. The plastic wave speed decreases with the increase of strain. This means that in loading process the propagation velocity of a higher magnitude plastic disturbance is less than that of a lower magnitude plastic disturbance ahead. Therefore in a process of wave propagation, the waveform becomes flatter and flatter (called the **divergent wave**).

Case (b). In this case the material tangent modulus increases with the increase of strain (**increasing hardening material**), namely the $\sigma \sim \varepsilon$ curve is concave upwards $(d^2\sigma/d\varepsilon^2 > 0)$. The plastic wave velocity increases with the increase of strain. This means that during a loading process the propagation velocity of a higher magnitude plastic disturbance is higher than that of a lower magnitude plastic disturbance ahead. Therefore, the waveform becomes more and more steep with stress wave propagation (called the **convergent wave**). Finally jumps of the particle velocity, stress and strain across a wave front will form, and the plastic stress wave becomes a **shock wave**. Such shock wave problem will be discussed further in Section 2.6.

In a special case, the material tangent modulus can be approximated as a constant E_1 $(d^2\sigma/d\varepsilon^2 = 0)$. This kind of material is called the **linear hardening material**, and E_1 is called the linear hardening modulus. Then the plastic wave velocity is constant $C_1 = \sqrt{E_1/\rho_0}$. In general $E_1 < E$ – for example, $E_1/E = 0.003 - 0.010$ for steel, $E_l/E = 0.05 - 0.1$ for dry soil (Nowaski, 1978) – therefore the propagation of plastic waves is much slower than that of the elastic wave.

The problem solved in Fig. 2.5 corresponds to the decreasing hardening material. On the (X, t) plane the characteristic lines in elastic region are parallel straight lines, and the positive characteristic lines in the plastic simple wave region are divergent straight lines. The shape of the elastic parts of the temporal curve (before point $3'$) and the waveform curve (before point 3) is invariable. Nevertheless the shape of the plastic parts is divergent, and becomes more and more flat during propagation. The plastic waveform is elongated. If the material is linear hardening, on the (X, t) plane the positive characteristic lines in the plastic simple wave region is another group of parallel straight lines, whose slope differs from that in elastic region. In this case, the elastic waveform and the plastic waveform are invariable during propagation, but the interval between them becomes larger and larger. The readers can show this by geometric drawing.

In Fig. 2.5 the corresponding v–ε diagram and v–φ diagram are also plotted. Same as previously discussed, on these diagrams the constant value region is mapped to a point, and the simple wave region to a line segment. It is notable that on the (v, ε) plane the plastic simple wave region is mapped to a *curve* segment, which is described by the plastic simple wave relationship (2.39). However, after introducing a variable φ on the (v, φ) plane the region is again mapped as a segment of straight line.

The simple wave relationship (2.39) is satisfied everywhere within a simple wave region. When applied to the impact end of the bar, this condition leads to the relationship between particle velocity and strain or stress at the contact boundary. Further it can be deduced that corresponding to the material's ultimate strength σ_b, the **critical impact velocity** v_c is expressed as:

$$v_c = -\int_0^{\sigma_b} \frac{d\sigma}{\rho_0 C} \tag{2.40}$$

Failure will occur if the particle velocity at the impact end of the bar reaches this critical value.

2.5 Governing Equations of Longitudinal Waves in Bars in Spatial Coordinates

Now we use Euler method to study the elastic–plastic wave propagation. In this method we consider a specified region in the space, called the **controlled volume**. In general the conservation laws of mass, momentum, and energy are applied to a fixed mass of the material. So when applying these laws to a controlled volume we must account the changes of the physical quantities within the controlled volume, and the flow of the quantities across the surfaces of this volume (the **controlled surfaces**). All physical quantities involved in the wave propagation are functions of independent Euler variables: the spatial coordinate x and the time t.

For the current one-dimensional uniaxial stress wave problem, we consider a controlled volume between x and $x + dx$, as illustrated in Fig. 2.7. Assuming that during the wave propagation any cross section of the bar is planar with uniform distributions of all physical quantities, the mathematical one-dimensional spatial problem is posed below.

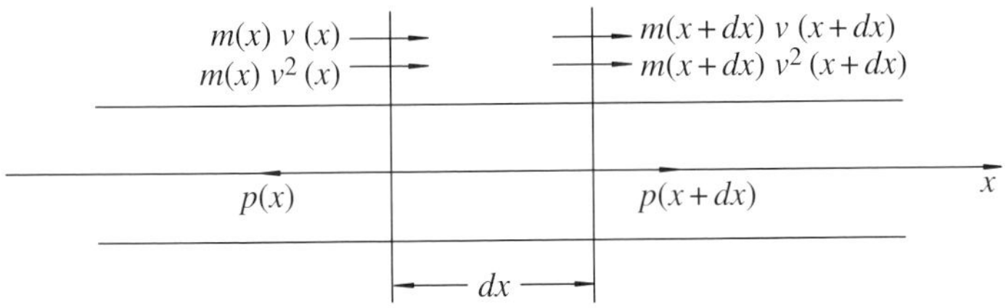

Fig. 2.7. An infinitesimal controlled volume of bar with uniform cross section in a Euler coordinate system.

The mass occupied by a spatial segment of dx is:

$$M = \rho A dx = \rho_0 A_0 dX$$

where ρ and A are the actual density and cross-sectional area of the bar, ρ_0 and A_0 are the initial density and cross-sectional area of the bar before deformation, dX is the length of the current bar segment dx before deformation. The above equation represents the mass conservation of a bar segment before and after deformation. From the definition of strain $dx = (1 + \varepsilon)dX$, the line density (defined as mass per unit length) of the bar is:

$$m = \rho A = \frac{\rho_0 A_0}{1 + \varepsilon} \tag{2.41}$$

According to the mass conservation of the controlled volume, the mass rate within a spatial segment dx equals to the difference between the mass flowing into the volume and the mass flowing out of the volume:

$$\frac{\partial (m dx)}{\partial t} = m(x)v(x) - m(x + dx)v(x + dx)$$

Simultaneously, the momentum rate within dx equals to the momentum flow difference into and out of the volume, plus the difference of the external forces:

$$\frac{\partial (m v dx)}{\partial t} = m(x)v^2(x) - m(x + dx)v^2(x + dx) - P(x) + P(x + dx)$$

Here $P = \sigma A_0$ is the total force applied on a cross section. The above two equations can be simplified as the following forms:

$$\frac{\partial \varepsilon}{\partial t} + v\frac{\partial \varepsilon}{\partial x} - (1 + \varepsilon)\frac{\partial v}{\partial x} = 0 \tag{2.42}$$

$$\frac{\rho_0}{1 + \varepsilon}\left(\frac{\partial v}{\partial t} + v\frac{\partial v}{\partial x}\right) = \frac{\partial \sigma}{\partial x} \tag{2.43}$$

which are the **continuous equation** and the **dynamic equation** expressed by Euler coordinates.

Here again we assume that stress is a function of strain:

$$\sigma = \sigma(\varepsilon)$$

and denote the quantity $(1/\rho_0)(d\sigma/d\varepsilon)$ as C^2, (2.15), the Eq. (2.43) can be rewritten as:

$$(1+\varepsilon)C^2\frac{\partial \varepsilon}{\partial x} - \frac{\partial v}{\partial t} - v\frac{\partial v}{\partial x} = 0 \tag{2.44}$$

Equations (2.42) and (2.44) are a group of the first-order partial differential equations of the unknown functions v and ε. These equations are similar to the propagation equations of finite magnitude plane wave in ideal compressible fluid, which were treated by Earnshow and Reimann (see Courant and Friedrichs, 1948). The continuous plastic wave is therefore called **Riemann wave** in some occasions. We now use the method of characteristics to solve these equations. A linear combination of these equations is:

$$L\frac{\partial \varepsilon}{\partial t} + \left[Lv + M(1+\varepsilon)C^2\right]\frac{\partial \varepsilon}{\partial x} - M\frac{\partial v}{\partial t} - [L(1+\varepsilon) + Mv]\frac{\partial v}{\partial x} = 0$$

here L and M are undetermined coefficients. According to the characteristic line definition, in case of a characteristic line this combination should contain only the directional derivatives along the characteristic line, namely:

$$\frac{dx}{dt} = \frac{Lv + M(1+\varepsilon)C^2}{L} = \frac{Mv + L(1+\varepsilon)}{M}$$

From the second equation we obtain $L = \pm MC$, therefore the characteristic lines in Euler system must satisfy:

$$dx = [v \pm (1+\varepsilon)C]dt \tag{2.45}$$

and the corresponding compatibility equations are:

$$dv = \pm C d\varepsilon \tag{2.46}$$

The corresponding equations in Lagrange system are Eqs. (2.23) and (2.24), respectively.

Equation (2.45) implies a relationship between Euler wave velocity $c(=dx/dt)$ and Lagrange wave velocity C, which is just the previously deduced Eq. (2.11). In the current discussion the spatial location of particles before deformation ($t = 0$) are taken as the material coordinate. When a wave front propagates in a material coordinate system with velocity C, considering the deformation of the material coordinate itself, in the spatial coordinate system the *relative* wave speed *with respect to the particle ahead of the wave front* should be $(1+\varepsilon)C$. This quantity is equivalent to the local sound speed in fluid mechanics. Consider, again, that the particle itself is moving with velocity v, so that in the spatial coordinate system the *absolute* spatial wave velocity should be $[v \pm (1+\varepsilon)C]$. This is the physical meaning of Eqs. (2.11) or (2.45). In these equations, (+) corresponds to rightward waves and (−) corresponds to leftward waves.

Equation (2.46) is exactly the same as Eq. (2.24) in a material coordinate system, as it ought to be, since the compatible relationships represent the continuous condition, momentum conservation, and material constitutive relation, all these should not be dependent on what kind of coordinate system is chosen.

As a matter of fact, the fundamental equations in Lagrange system (2.12) and (2.16), and those in Euler system (2.42) and (2.44), are convertible by simply using the coordinate conversion:

$$\left(\frac{\partial}{\partial t}\right)_X = \left(\frac{\partial}{\partial t}\right)_x + v\left(\frac{\partial}{\partial x}\right)_t$$

$$\left(\frac{\partial}{\partial X}\right)_t = \frac{\partial x}{\partial X}\left(\frac{\partial}{\partial x}\right)_t = (1+\varepsilon)\left(\frac{\partial}{\partial x}\right)_t$$

Although the appearances of these equations are different in different coordinate systems, they represent the same physical essence.

In observing stress wave propagation, if the sensor is fixed in the observing space, the measured wave velocity is a Euler speed; if the sensor is attached to the specimen such as the bar, then the measured wave velocity is a Lagrange speed. In case the particle velocity and the strain ahead of the wave front are zero, both measurements give the same value.

2.6 Strong Discontinuity and Weak Discontinuity; Shock Waves and Continuous Waves

On a wave front the particle displacement u must be continuous to satisfy the continuity condition. However, the derivatives of u can be discontinuous. Mathematically this kind of interface with "derivative discontinuity" is called a **singular interface**. If the first-order derivatives, namely the particle velocity $v(=\partial u/\partial t)$ and the strain $\varepsilon(=\partial u/\partial x)$ have a jump across a wave front, this interface (wave front) is called a **first-order singular interface**, or a **strong discontinuity**. For example, in an increasing hardening material, as the plastic wave velocity of higher magnitude disturbance is faster than that of a lower magnitude disturbance, a strong discontinuity will finally form. This kind of stress wave is called a **shock wave**. If u and its first-order derivatives are continuous but the second-order derivative(s) such as the acceleration $a(=\partial v/\partial t=\partial^2 u/\partial t^2)$ is discontinuous, the interface is called a **second-order singular interface**. This kind of stress wave is frequently called an **acceleration wave**. And so forth there exist even higher singular interfaces. All second or higher order singular interfaces are called **weak discontinuities**. For example, in a decreasing hardening material the plastic wave of higher magnitude disturbance is slower than that of lower magnitude disturbance, so that the plastic waves in this material are weak discontinuous. This kind of waves have continuous waveforms, and therefore are called the **continuous waves**.

In the previous discussions on stress wave propagation in a semi-infinite bar, as shown in Figs. 2.4 and 2.5, we assumed that the impact velocity at the end of the bar increases with time, namely the boundary loading is the so-called gradually increasing loading. Mathematically the function $v_0(\tau)$ in the boundary condition (2.31b) is a continuous function with respect to τ. Assume $v_0(\tau)$ in Fig. 2.4 is a constant when $\tau > \tau_6$. As τ_6 approaches to zero, the boundary loading is equivalent to a suddenly applied constant velocity impact. The boundary condition is strong discontinuous in this case.

Whether the stress wave propagates in the form of strong discontinuity or weak discontinuity is determined by the material's stress–strain relationship and the boundary conditions. We will discuss several different cases as follows (Fig. 2.8):

(a) In the case of linear elastic materials (Fig. 2.8a): If the boundary condition is a weak discontinuous function of time, the elastic stress wave is a weak discontinuous wave. If the boundary condition is of a strong discontinuity, the elastic stress wave is a strong discontinuous wave. In other words, the type of stress wave is determined by the boundary condition. Actually, if the boundary function $v_0(\tau)$ in Fig. 2.4 is constant for $\tau \geq \tau_6$, the region $6\tau_6 t$ on X–t plane is

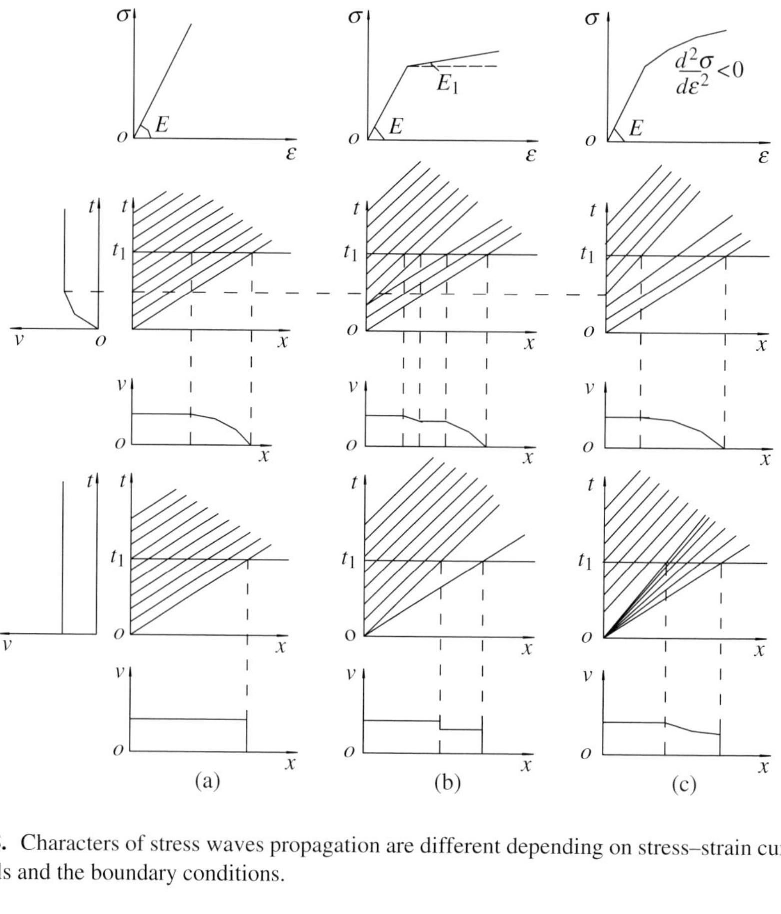

Fig. 2.8. Characters of stress waves propagation are different depending on stress–strain curves of materials and the boundary conditions.

a constant value region. As τ_6 approaches to zero, all the previous positive characteristic lines starting from the points between 0 and τ_6 on t axis overlap on one line *OA*. The two regions across *OA* are constant value regions. Across the *OA* line the quantities v, ε, and σ jump. Therefore the characteristic line *OA* is the propagation locus of a strong discontinuity. The propagation velocity of this strong discontinuity is also C_0. Summarizing the above discussions, for a linear wave system, the type of the boundary discontinuity determines the type of propagation. This is one of the important properties of the linear hyperbolic partial differential system.

(b) In the case of linear elastic-linear hardening plastic materials (Fig. 2.8b): If the boundary condition is a continuous function of time, both the elastic stress wave and the plastic stress wave propagate with invariable waveforms. The intervals between these two waveforms, however, become larger and larger. Here we have assumed that the plastic hardening modulus E_1 is less than the elastic modulus E. If the boundary loading is a suddenly applied velocity (the strong discontinuity), two strong discontinuous waves are created (the **double-wave structure**), the first faster wave propagates as the characteristic line with slope C_0, across which an elastic jump arises. The second slower wave propagates as the characteristic line with slope C_1, across which a plastic jump arises. The distance between these two abrupt wave fronts (shock waves) increases during wave propagation.

(c) In case of linear elastic-decreasing hardening plastic materials (Fig. 2.8c): If the boundary condition is a continuous function of time, as discussed in Fig. 2.5, the weak discontinuous elastic–plastic waves are generated. Now consider the boundary loading being a suddenly applied constant velocity. This loading is equivalent to the problem of Fig. 2.5 when the boundary velocity $v_0(\tau)$ is constant as $\tau \geq \tau_6$ and τ_6 approaches to zero. Therefore the previous parallel characteristic lines in the elastic wave region overlap as one line *OA*, representing a strong discontinuous elastic wave. The previous divergent characteristic lines in the plastic region now start from one point O (which is a singular point), forming a so-called **centered-wave** structure. In the current situation, the elastic part of the strong discontinuous boundary disturbance, propagates as a strong discontinuous wave, and the plastic part propagates as a series of divergent continuous waves, which are of weak discontinuity. The waveform becomes more and more flat during propagation. Therefore, for material of $d^2\sigma/d\varepsilon^2 < 0$, even if a strong discontinuity exists in the initial condition or boundary condition, this discontinuity disappears when waves start propagating. The strong discontinuity does not proagate in the form of strong discontinuous wave.

(d) In the case of linear elastic-increasing hardening plastic materials (see Fig. 2.6b). As pointed out previously, for plastic waves:

$$\frac{dC}{d\varepsilon} = \frac{1}{2\rho_0 C}\frac{d^2\sigma}{d\varepsilon^2} > 0$$

namely the velocity of a high magnitude plastic disturbance is larger than that of a low magnitude plastic disturbance. Therefore the plastic waves are so-called

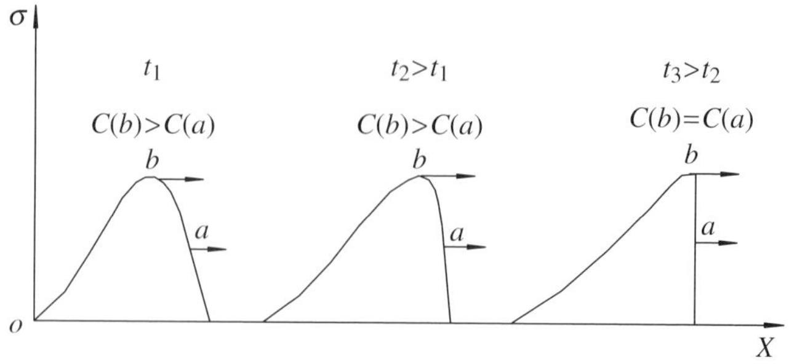

Fig. 2.9. Formation of a shock wave in material with increasing hardening plasticity.

convergent waves. The waveform becomes steeper and steeper during propagation, and finally the stress, strain, and particle velocity almost jump instantaneously, forming a shock wave (Fig. 2.9). Note that this strong discontinuous (shock) wave differs from the above discussed (in a and b) shock waves, which are generated by the suddenly applied boundary loading. Actually, the strong discontinuous waves in (a) and (b) originate from the strong discontinuous boundary disturbance, and thermodynamically no extra entropy production exists in wave propagation. The current strong discontinuous wave, however, is created by the convergent property of the plastic waves, whatever the boundary conditions are. In the subsequent discussions (Section 2.7) we will see that this shock wave accompanies extra entropy production.

Although in case (d) the shock wave formation originates from the specific nonlinear constitutive property of the material, the shock wave's formation time and site are also dependent on the initial or boundary conditions. Take the rightward simple wave case as an example. Assume at time $t = 0$ the initial waveform be expressed as:

$$\sigma(X, 0) = f(\xi)$$

where ξ is the X locations of a stress disturbance σ at $t = 0$. For rightward waves, the location of the same disturbance at time t is $X = \xi + C(\sigma)t$, therefore the waveform at time t is expressed by:

$$\sigma = f[X - C(\sigma)t]$$

Using function F, the inverse function of f, the location of disturbance σ at time t is:

$$X = F(\sigma) + C(\sigma)t$$

Condition of the shock wave formation is:

$$\frac{\partial\sigma}{\partial X} = \infty \text{ or } \frac{\partial X}{\partial\sigma} = 0 \tag{2.47}$$

from which the time of shock wave formation t_0, and the corresponding shock wave location X_0 are deduced as:

$$t_0 = \left\{ -\frac{1}{C'(\sigma) f'} \right\}_{\min} = \left\{ -\frac{F'(\sigma)}{C'(\sigma)} \right\}_{\min}$$

$$X_0 = F(\sigma) - \frac{C(\sigma) F'(\sigma)}{C'(\sigma)} \tag{2.48}$$

where $\{a\}_{\min}$ represents the minimum of argument a. Actually any stress disturbance corresponds to a time that satisfies condition (2.47), the earliest one corresponds to the time of shock wave formation. If the initial condition already includes a strong discontinuity, then we have $f'(\xi) = \infty$, and $t_0 = 0$.

Similar approach is applied to the problem with a given boundary condition:

$$\sigma(0, t) = g(\tau)$$

For rightward stress wave propagation, as $X = C(\sigma)(t - \tau)$, we have:

$$\sigma = g\left[t - \frac{X}{C(\sigma)} \right]$$

By representing G as the inverse function of g, the above equation is rewritten as:

$$t = G(\sigma) + \frac{X}{C(\sigma)}$$

Condition of the shock wave formation is:

$$\frac{\partial \sigma}{\partial t} = \infty \text{ or } \frac{\partial t}{\partial \sigma} = 0$$

from which the time of shock wave formation t_0, and the corresponding shock wave location X_0 are deduced as:

$$X_0 = \left\{ \frac{C^2}{g' C'} \right\}_{\min} = \left\{ \frac{C^2 G'}{C'} \right\}_{\min}$$

$$t_0 = G(\sigma) + \frac{C G'}{C'} \tag{2.49}$$

If the boundary condition already contains a strong discontinuity, then $g' = \infty$, $X_0 = 0$, namely the suddenly applied boundary loading propagates as a strong discontinuous wave from the very beginning.

It is noteworthy that for a suddenly applied boundary loading, for example, the case of $v(0,t) = v_1$, the problem of the elastic–plastic wave propagation in a semi-infinite bar has a **self-similar solution** (see Appendix 3). The first plastic wave theory proposed by von Karman (1942) is for this kind of problem where the material is of decreasing hardening plasticity. Such a problem does not contain a characteristic length or time. Applying dimensional analysis to the governing Eqs. (2.12) and (2.16) where ε and v are unknown variables, from π theorem the strain ε should be expressed as the following non-dimensional form:

$$\varepsilon = g\left(\frac{C(\varepsilon)}{v_0}, \frac{X}{v_0 t}\right)$$

In other words, the variables X and t are not independent but appear in a combined form of X/t. Introducing a variable $\beta = X/t$, the strain is a function of β:

$$\varepsilon = f\left(\frac{X}{t}\right) = f(\beta)$$

Accordingly, the displacement u is:

$$u = \int_{\infty}^{X} \frac{\partial u}{\partial X} dX = \int_{\infty}^{X} f(\beta) dX = t \int_{\infty}^{\beta} f(\beta) d\beta$$

and the second-order partial derivatives of u with respect to X and t are:

$$\begin{aligned} \frac{\partial^2 u}{\partial X^2} &= \frac{1}{t} f'(\beta) \\ \frac{\partial^2 u}{\partial t^2} &= \frac{\beta^2}{t} f'(\beta) \end{aligned}$$

Substituting this into the governing Eq. (2.18), we have

$$\left(\beta^2 - C^2\right) f'(\beta) = 0$$

then two solutions may exist (Fig. 2.10):

$$\begin{aligned} f'(\beta) &= 0 \\ \beta^2 &= C^2 \end{aligned}$$

The first solution renders $\varepsilon = f(\beta) = const.$ that corresponds to a constant value region. The second solution renders $X/t = C(\varepsilon)$ that corresponds to a centered wave region. As long as X/t is constant, the strain ε is invariable, regardless of what the value of X or t is.

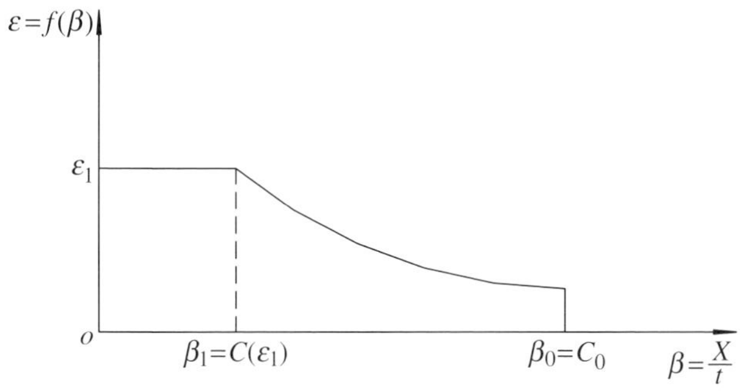

Fig. 2.10. Self-similar solution for the elastic-plastic wave propagation in a semi-infinite bar under constant velocity impact.

From the boundary condition

$$u = u(0, t) = v_1 t = t \int_{\infty}^{0} f(\beta) d\beta$$

we obtain

$$v_1 = -\int_{0}^{\infty} f(\beta) d\beta = -\int_{0}^{\infty} \varepsilon(\beta)\, d\beta$$

which is the area integration under the curve shown in Fig. 2.10, and can be rewritten as

$$v_1 = -\int_{0}^{\varepsilon_1} \beta(\varepsilon) d\varepsilon = -\int_{0}^{\varepsilon_1} C(\varepsilon) d\varepsilon$$

The above equation is the application of Eq. (2.39) at the end of the bar.

2.7 Conservation Conditions across Wave Front, Rankine–Hugoniot Relations

As a planar wave propagates as a singular surface in an infinite media, the physical quantities across the wave front must satisfy certain conditions. In this section we will discuss these conditions, in other words, the compatibility equations on the wave front.

Consider a planar wave front propagating along the positive X axis (rightward wave) with the material wave velocity $\mathscr{D} = dX/dt$, here X denotes Lagrange location of the wave front at time t. Suppose we stand on the wave front to observe a physical quantity $\psi(X, t)$. The total rate of ψ with respect to time, namely the wave derivative (2.10) is:

$$\frac{d\psi}{dt} = \frac{\partial \psi}{\partial t} + \mathscr{D} \frac{\partial \psi}{\partial X} \tag{2.50}$$

Denote the values of ψ just before and after the wave front by ψ^+ and ψ^-, the difference between them as $[\psi]$

$$[\psi] = \psi^- - \psi^+ \tag{2.51}$$

Obviously, if ψ is continuous across the wave front $[\psi] = 0$; otherwise $[\psi] \neq 0$. The quantity $[\psi]$ therefore represents the jump of ψ across the wave front. Taking the wave derivative of ψ^+ and ψ^- by (2.50), the difference between them is:

$$\frac{d}{dt}[\psi] = \left[\frac{\partial \psi}{\partial t}\right] + \mathscr{D}\left[\frac{\partial \psi}{\partial X}\right] \tag{2.52}$$

If across the wave front ψ is continuous, but its first-order derivatives are not (namely the wave front is a first-order singular surface), we have

$$\left[\frac{\partial \psi}{\partial t}\right] = -\mathscr{D}\left[\frac{\partial \psi}{\partial X}\right] \tag{2.53}$$

which is the famous **Maxwell theorem**.

We use $\partial\psi/\partial t$, $\partial\psi/\partial X$ instead of the ψ variable in Eq. (2.52), obtaining:

$$\frac{d}{dt}\left[\frac{\partial \psi}{\partial t}\right] = \left[\frac{\partial^2 \psi}{\partial t^2}\right] + \mathscr{D}\left[\frac{\partial^2 \psi}{\partial t \partial X}\right]$$

$$\frac{d}{dt}\left[\frac{\partial \psi}{\partial X}\right] = \left[\frac{\partial^2 \psi}{\partial t \partial X}\right] + \mathscr{D}\left[\frac{\partial^2 \psi}{\partial X^2}\right]$$

Therefore if across the wave front ψ and all of its first-order derivatives are continuous, but its second-order derivatives are not (namely the wave front is a second-order singular surface), we have

$$\left[\frac{\partial^2 \psi}{\partial t^2}\right] = -\mathscr{D}\left[\frac{\partial^2 \psi}{\partial X \partial t}\right] = \mathscr{D}^2\left[\frac{\partial^2 \psi}{\partial X^2}\right] \tag{2.54}$$

The Eqs. (2.52)–(2.54) are the **kinematic compatibility conditions on wave front**, corresponding to cases when ψ itself, its first-order derivatives, and its second-order derivatives are discontinuous, respectively. Similar deductions can be derived on cases of higher-order singular interfaces. For cases when the wave front is leftward, we can simply substitute the quantity $\mathscr{D}$ by $-\mathscr{D}$ and all equations are true.

Take the quantity ψ as the particle displacement $u(X, t)$. By the condition of continuity, the displacement must be continuous across a wave front, namely $[u] = 0$. In case of a shock wave front, by the compatibility condition of a first-order discontinuity (2.53) we have:

$$[v] = -\mathscr{D}[\varepsilon] \tag{2.55}$$

In case of an acceleration wave front, by the compatibility condition of a second-order discontinuity (2.54) we have:

$$\left[\frac{\partial^2 u}{\partial t^2}\right] = -\mathscr{D}\left[\frac{\partial^2 u}{\partial X \partial t}\right] = \mathscr{D}^2\left[\frac{\partial^2 u}{\partial X^2}\right]$$

or expressed by v and ε:

$$\left[\frac{\partial v}{\partial t}\right] = -\mathscr{D}\left[\frac{\partial \varepsilon}{\partial t}\right] = -\mathscr{D}\left[\frac{\partial v}{\partial X}\right] = \mathscr{D}^2\left[\frac{\partial \varepsilon}{\partial X}\right] \tag{2.56}$$

Equations (2.55) and (2.56) are the **kinematic compatibility conditions** across a **shock wave front** and an **acceleration wave front**, respectively. These conditions reflect physically the **condition of mass conservation**.

Now we consider another type of compatibility conditions based on the law of dynamics.

Consider a strong discontinuous wave front. At time t the front is located at AB (Fig. 2.11), in a time dt it arrives at location $A'B'$. The distance between these locations are connected to the wave velocity $\mathscr{D}$ by $dX = \mathscr{D}dt$. By the momentum conservation of the region $ABA'B'$:

$$\left(\sigma^+ - \sigma^-\right)A_0 dt = \rho_0 A_0 dX(v^- - v^+)$$

which can be simplified as:

$$[\sigma] = -\rho_0 \mathscr{D}\,[v] \tag{2.57}$$

In case the wave front is weak continuous, since $[v] = 0$, $[\sigma] = 0$, we must consider the relationships between the derivatives of v and σ. As an example, for an acceleration wave we can apply the differential momentum conservation Eq. (2.13) to the materials before and after the wave front; the difference of these two equations leads to:

$$\left[\frac{\partial \sigma}{\partial X}\right] = \rho_0\left[\frac{\partial v}{\partial t}\right] \tag{2.58}$$

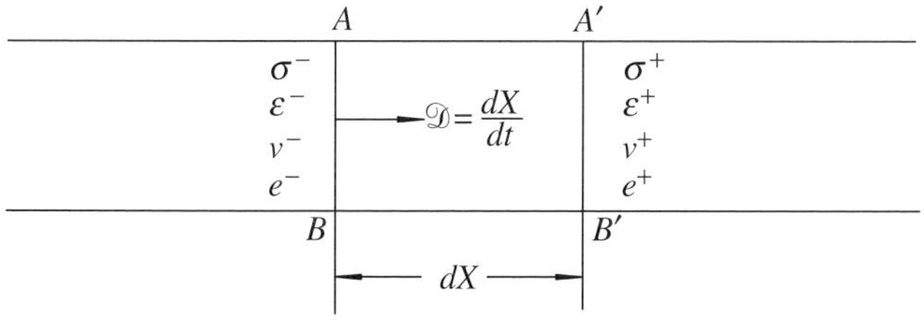

Fig. 2.11. Propagation distance of shock (strong discontinuous) wave front in time dt is dX $(= \mathscr{D}dt)$.

Equations (2.57) and (2.58) are the **dynamic compatibility conditions** across a **shock wave front** and an **acceleration wave front**, respectively, in a continuous medium. These conditions reflect physically the **condition of momentum conservation**.

For a shock wave front, from the kinematic Eq. (2.55) and the dynamic Eq. (2.57) we can eliminate the term $[v]$, giving:

$$[\sigma] = \rho_0 \mathscr{D}^2 [\varepsilon]$$

from which the shock wave velocity $\mathscr{D}$ is related to the stress jump $[\sigma]$ and the strain jump $[\varepsilon]$ across the wave front, as:

$$\mathscr{D} = \sqrt{\frac{1}{\rho_0} \frac{[\sigma]}{[\varepsilon]}} \tag{2.59}$$

For an acceleration wave front, from the kinematic Eq. (2.56) and the dynamic Eq. (2.58) we can eliminate the term $[\partial v/\partial t]$, giving:

$$\left[\frac{\partial \sigma}{\partial X}\right] = \rho_0 \mathscr{D}^2 \left[\frac{\partial \varepsilon}{\partial X}\right]$$

from which the acceleration wave velocity $\mathscr{D}$ is related to the stress gradient jump $[\partial\sigma/\partial X]$ and the strain gradient jump $[\partial\varepsilon/\partial X]$, by:

$$\mathscr{D} = \sqrt{\frac{1}{\rho_0} \frac{\left[\dfrac{\partial \sigma}{\partial X}\right]}{\left[\dfrac{\partial \varepsilon}{\partial X}\right]}} \tag{2.60}$$

Note that in the deductions of the kinematic and the dynamic compatibility equations, the material properties are not involved. Therefore, these equations are true for the plane waves in any continuum. Nevertheless, the determination of the wave speed $\mathscr{D}$ is related to the constitutive relationship of the materials. In general, the relation between $[\sigma]$ and $[\varepsilon]$ and the relations between $[\partial\sigma/\partial X]$ and $[\partial\varepsilon/\partial X]$ are different, therefore the velocities of a shock wave and an acceleration wave are different.

Under the framework of the rate-independent theory, we assume that there exists an exclusive dynamic stress–strain relationship, namely $\sigma = \sigma(\varepsilon)$. By this condition we have:

$$\left[\frac{\partial \sigma}{\partial X}\right] = \frac{d\sigma}{d\varepsilon} \left[\frac{\partial \varepsilon}{\partial X}\right]$$

so that the Eq. (2.60) is reduced to (2.15). This means that the velocity of a continuous wave is determined by the tangent slope of the $\sigma \sim \varepsilon$ curve. This conclusion has been obtained previously in elastic–plastic wave analysis. On the other side, Eq. (2.59) means that the velocity of a shock wave is determined by the secant slope of the $\sigma \sim \varepsilon$ curve, namely the slope of the straight-line linking the pre-shock state and post-shock states. Similar to the shock wave discussions in fluid dynamics (Readers are referred to Courant and Friedrichs, 1948), this secant line is called **Rayleigh line** or **shock secant**.

Take the elastic-increasing hardening plastic material for example. The plastic shock wave velocity is determined by the slope of Rayleigh line *AB* linking the yield point *A* (the initial state prior to shock) and *B* (the final state after shock), as shown in Fig. 2.12. As long as this slope is less than the elastic modulus, the plastic shock velocity $\mathscr{D}_p$ is always slower than the elastic wave speed C_0, resulting in a double-wave structure in the media. Conversely, once $\mathscr{D}_p \geq C_0$, a single elastic–plastic shock wave is created. If the stress and strain are linearly related, $[\sigma]/[\varepsilon] = d\sigma/d\varepsilon = const.$, and the shock wave speed is identical with the continuous wave speed.

Now we discuss the energy conservation on a shock wave front. Consider the region $ABA'B'$ in Fig. 2.11, denoting the material's internal energy in a unit mass by e, the energy conservation of this region is:

$$\left(\sigma^+ v^+ - \sigma^- v^-\right) A_0 dt = \left(e^- - e^+\right) \rho_0 A_0 dX + \frac{1}{2}\left\{\left(v^-\right)^2 - \left(v^+\right)^2\right\} \rho_0 A_0 dX$$

After simplification, the equation is:

$$[\sigma v] = -\rho_0 \mathscr{D}[e] - \frac{1}{2}\rho_0 \mathscr{D}\left[v^2\right] \tag{2.61}$$

which is the **energy conservation equation** across a **shock wave front**.

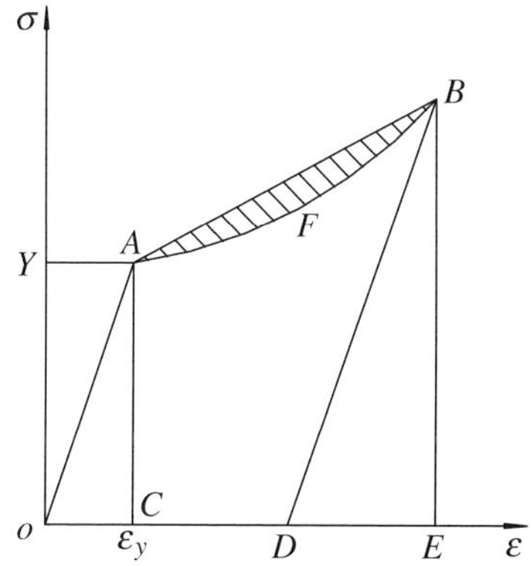

Fig. 2.12. For the "linear elastic-increasing hardening plastic" material, the shock wave velocity is determined by the slope of the shock secant *AB*.

Introducing the material's internal energy in a unit volume, $\mathscr{E} = \rho_0 e$, and substituting (2.55) into (2.57), we notice that:

$$
\begin{aligned}
[\sigma v] + \frac{1}{2}\rho_0 \mathscr{D}\left[v^2\right] &= [\sigma v] + \frac{1}{2}\rho_0 \mathscr{D}\,[v]\left(v^- + v^+\right) = [\sigma v] - \frac{1}{2}[\sigma]\left(v^- + v^+\right) \\
&= \sigma^- v^- - \sigma^+ v^+ - \frac{1}{2}\{\sigma^- v^- + \sigma^- v^+ - \sigma^+ v^- - \sigma^+ v^+\} \\
&= \frac{1}{2}\{(\sigma^- + \sigma^+)v^- - (\sigma^- + \sigma^+)v^+\} = \frac{1}{2}(\sigma^- + \sigma^+)[v] \\
&= -\frac{\mathscr{D}}{2}\left(\sigma^- + \sigma^+\right)[\varepsilon]
\end{aligned}
$$

after some mathematical manipulations we obtain another form of the energy conservation condition:

$$
\mathscr{E}^- - \mathscr{E}^+ = \tfrac{1}{2}\left(\sigma^- + \sigma^+\right)\left(\varepsilon^- - \varepsilon^+\right) \tag{2.62}
$$

For the linear elastic, increasing-hardening plastic material shown in Fig. 2.12, after a suddenly applied loading, two stress waves are generated: an elastic precursor wave followed by a plastic shock wave. Assume that the bar is originally at rest and stress-free, its initial state before the elastic precursor wave is represented by the point O in the figure. After the elastic wave passes, the material's state is represented by the dynamic yielding point A, therefore:

$$
\sigma^- = Y, \quad \varepsilon^- = \varepsilon_y, \quad \mathscr{E}^- = \tfrac{1}{2} Y \varepsilon_y
$$

This means that across the wave front of the elastic precursor, the internal energy jump $[E]$ is represented by the area of the triangle *OAC* below elastic path *OA*. This quantity is exactly equal to the elastic strain energy. The state prior to the plastic shock wave is at point A, and the state after the shock is at point B. The location of B is determined by the boundary condition of the sudden loading. The energy conservation Eq. (2.62) means that across the shock front the internal energy jump $[\mathscr{E}]$ is represented by the trapezoidal area *ABEC* below the shock secant *AB*. This area is bigger than the plastic strain energy $\int_{\varepsilon_A}^{\varepsilon_B} \sigma\, d\varepsilon$ by an area of region AFB (shaded in the figure). Expression of this part of energy is:

$$
\Delta Q = \frac{1}{2}(\sigma_B + \sigma_A)(\varepsilon_B - \varepsilon_A) - \int_{\varepsilon_A}^{\varepsilon_B} \sigma\, d\varepsilon = \mathscr{E}_B - \int_{\varepsilon_A}^{\varepsilon_B} \sigma\, d\varepsilon
$$

which corresponds to the thermal energy that has been dissipated during the formation of the shock wave. Indeed the very large velocity gradient across a shock wave front activates the internal friction of the solid – which has been neglected in a strain-rate independent theory – and results in irreversible energy dissipation. Therefore, the state change across a shock wave front by nature is an adiabatic process but not an isentropic process. This thermodynamic process is frequently called the **shock adiabatic process**. Certainly we should point out, the plastic deformation process itself is also an irreversible process and the plastic wave is thus not isentropic. Among the plastic deformation energy $\int_0^{\varepsilon_B} \sigma\, d\varepsilon$, only the part equivalent to the triangular area *BDE* is elastically recoverable, and the remains is transferred to the thermal dissipation during the plastic deformation. In case of the plastic shock wave, an extra entropy-increase corresponding to ΔQ is produced. Since the energy dissipation is always positive, there must be

$$\Delta Q \geq 0$$

For the linear hardening material $d^2\sigma/d\varepsilon^2 = 0$, the extra energy dissipation $\Delta Q = 0$ and the internal energy increase equals the plastic deformation energy. No extra entropy increase is produced in a shock wave process for such materials.

Across a shock wave front, the mass conservation Eq. (2.55), the momentum conservation Eq. (2.57), and the energy conservation Eq. (2.62) are collectively called the **shock jump condition** or the **Rankine–Hugoniot relationship**. The equations listed here are Lagrange forms for an elastic–plastic bar.

Given the initial state of the material (all quantities with superscript + are known), the shock jump condition and the material's constitutive relationship provide four equations linking any two of the five unknown quantities σ^-, ε^-, v^-, e^-, and $\mathscr{D}$. These equations are called the **shock adiabatic lines** or the **Hugoniot lines**. They describe the loci of any possible final equilibrium states arrived by shock jumping from the known initial state. These relationships, nevertheless, do not represent all the state points which material experiences in an actual shock process. For example, a shock adiabatic $\sigma \sim \varepsilon$ curve is not material's constitutive $\sigma \sim \varepsilon$ curve in adiabatic condition. Note that the rate-independent theory assumes the existence of a unique dynamic constitutive $\sigma \sim \varepsilon$ curve, and therefore, neglects the difference between a shock adiabatic $\sigma \sim \varepsilon$ relationship and a constitutive $\sigma \sim \varepsilon$ relationship. This assumption made it possible for us to avoid using the energy conservation equation, and from only the mass conservation Eq. (2.55), the momentum conservation Eq. (2.57) and the material's constitutive relationship can we determine the quantities σ^-, ε^-, v^-, and $\mathscr{D}$ with the given initial-boundary conditions. In a more strict theory, the state equation of solid should be used instead of Eq. (2.14), and the energy conservation equation should be included. We will discuss this topic further in Chapter 7 when discussing the one-dimensional plane strain shock wave problem (see Section **7.7**).

As the jumps across a shock wave front decrease from finite values to infinitely small values, the three conservation Eqs. (2.55), (2.57), and (2.62) become the conservative conditions across a weak discontinuous wave front:

$$\begin{cases} dv = \mp C d\varepsilon & \text{(2.63a)} \\ d\sigma = \mp \rho_0 C dv & \text{(2.63b)} \\ de = \sigma d\varepsilon / \rho_0 & \text{(2.63c)} \end{cases}$$

The minus (−) in the first two equations correspond to rightward wave, and the plus (+) to leftward wave. These signs are opposite to that of characteristic compatibility relationships (2.24) and (2.26). The reason is: Eq. (2.63) represents the relationships between state variables ahead of and behind a wave front, namely the state variables **across wave front**;

conversely the characteristic compatibility equations represent the relationships between state variables **following the characteristic lines**. As a disturbance propagates along a rightward characteristic line, it is across a series of leftward characteristic lines namely the leftward wave fronts, therefore the signs in two types of relationships differ by a negative sign.

In studying the stress wave propagation, we can start from the governing equations, as has been done in Sections **2.1–2.5**; or we can work to establish the conservative conditions on wave fronts, as the work in this section shows. These two approaches are interlinked and interchangeable. In the subsequent discussions we may use either or both approaches.

2.8 Dispersion Effects Induced by the Transverse Inertia

In the previous discussions the cross section of the bar is assumed to be planar during deformation, and the axial stress is assumed to be uniformly distributed on the cross section. This approximate theory is called the primary theory or the engineering theory, as it neglects the inertia effects of particle transversal movement, or the contribution of the transversal expansions or contractions of the bar to the kinetic energy.

Within the range of elastic waves, we now discuss the effects of the transversal inertia. This study will help us understand the limitations of the primary theory, and make clear under what conditions such approximate theory is applicable.

Under a uni-axial stress $\sigma_X(X,t)$, the axial strain of the bar is:

$$\varepsilon_X = \frac{\partial u_X}{\partial X} = \frac{\sigma_X(X,t)}{E}$$

Besides this component, the bar also has transversal strain due to Poisson effect:

$$\varepsilon_Y = \frac{\partial u_Y}{\partial Y} = -\nu\varepsilon_X(X,t)$$
$$\varepsilon_Z = \frac{\partial u_Z}{\partial Z} = -\nu\varepsilon_X(X,t)$$

In the above equations, u_X, u_Y, and u_Z are the displacement components in X, Y, and Z axes, respectively, ν is the Poisson ratio. Since the axial stress σ_X is a function of X and t, so is the axial strain ε_X. Integrating the above equations lead to:

$$u_Y = -\nu Y\varepsilon_X = -\nu Y\frac{\partial u_X(X,t)}{\partial X}$$
$$u_Z = -\nu Z\varepsilon_X = -\nu Z\frac{\partial u_X(X,t)}{\partial X} \quad (2.64)$$

where the center of the cross section is taken as the origin point of Y and Z axes. Therefore the particle velocity v_Y and v_Z, and the particle acceleration a_Y and a_Z are respectively:

$$\begin{aligned}
v_Y &= \frac{\partial u_Y}{\partial t} = -\nu Y \frac{\partial \varepsilon_X}{\partial t} = -\nu Y \frac{\partial v_X}{\partial X} \\
v_Z &= \frac{\partial u_Z}{\partial t} = -\nu Z \frac{\partial \varepsilon_X}{\partial t} = -\nu Z \frac{\partial v_X}{\partial X} \\
a_Y &= \frac{\partial v_Y}{\partial t} = -\nu Y \frac{\partial^2 \varepsilon_X}{\partial t^2} = -\nu Y \frac{\partial^2 v_X}{\partial t \partial X} = -\nu Y \frac{\partial a_X}{\partial X} \\
a_Z &= \frac{\partial v_Z}{\partial t} = -\nu Z \frac{\partial^2 \varepsilon_X}{\partial t^2} = -\nu Z \frac{\partial^2 v_X}{\partial t \partial X} = -\nu Z \frac{\partial a_X}{\partial X}
\end{aligned} \tag{2.65}$$

The above equations show that on a cross section there exist non-uniformly distributed transversal particle displacements, velocities, and accelerations. Accordingly, there must be non-uniformly distributed transversal stresses, resulting in the warping of the cross section. Therefore attributed to the particle transversal movement of the bar, the stress state of the bar is not uni-axial, and the cross section is no more planar. Strictly it should be treated as a three-dimensional problem, or at least an axial-symmetric (for cylindrical bars or tubes) two-dimensional problem.

From the energy viewpoint, neglecting the particle transversal inertia means that the kinetic energy related to the particle transversal motion has been neglected. The average kinetic energy density (energy per unit volume) can be calculated from Eq. (2.65) as:

$$\frac{1}{A_0 dX} \int_{A_0} \frac{1}{2} \rho_0 \left(v_Y^2 + v_Z^2 \right) dX dY dZ = \frac{1}{2} \rho_0 \nu^2 r_g^2 \left(\frac{\partial \varepsilon_X}{\partial t} \right)^2$$

where r_g is the rotational radius of the cross section with respect to X axis:

$$r_g = \frac{1}{A_0} \int_{A_0} \left(Y^2 + Z^2 \right) dY dZ$$

The energy conservation Eqs. (2.61) and (2.63c), deduced from the primary theory in the previous section, account only the longitudinal kinetic energy and neglect this transversal kinetic energy.

To investigate the effects of this energy, we consider again the bar segment shown in Fig. 2.1. As the involved quantities are all those with respect to the X axis, we omit the subscript X for convenience. The bar segment dX is subjected to a balanced force $A_0 \sigma$ and an unbalanced force $A_0(\partial \sigma / \partial X) dX$. The latter is related to the longitudinal inertia of the segment. In unit time the work done by this net force equals the increase rate of the longitudinal kinetic energy of the bar:

$$A_0 \frac{\partial \sigma}{\partial X} dX \cdot v = \frac{\partial}{\partial t} \left(\frac{1}{2} \rho_0 A_0 dX \, v^2 \right)$$

The above equation, after rearrangement, leads to the motion equation of the bar segment (2.13):

$$\rho_0 \frac{\partial v}{\partial t} = \frac{\partial \sigma}{\partial X}$$

Therefore in primary theory, the work conducted by the balanced force is completely converted into the internal energy of the system, i.e. the strain energy during elastic wave motion [readers can refer to Eq. (2.63)]. In the current analysis the transversal motion must be accounted. The work done by the balanced external force, therefore, is converted into the longitudinal strain energy and the transversal kinetic energy. Approximately in unit time and unit volume, the energy equilibrium equation is:

$$\sigma \frac{\partial \varepsilon}{\partial t} = \frac{\partial}{\partial t}\left(\frac{1}{2}E\varepsilon^2\right) + \frac{\partial}{\partial t}\left(\frac{1}{2}\rho_0 \nu^2 r_g^2 \left(\frac{\partial \varepsilon}{\partial t}\right)^2\right)$$

from which we obtain:

$$\sigma = E\varepsilon + \rho_0 \nu^2 r_g^2 \frac{\partial^2 \varepsilon}{\partial t^2} \tag{2.66}$$

When neglecting the transversal kinetic energy, the above equation becomes the one-dimensional Hooke's law. Therefore, Eq. (2.66) can be viewed as a rate-dependent stress–strain relationship instead of Hooke's law when the bar's transversal inertia is taken into account. Since the transversal inertia term is proportional to the second-order time derivative of strain, this term becomes important only when the second-order derivative is significant.

By substituting (2.66) into the motion Eq. (2.13), we obtain the following wave equation in terms of the displacement:

$$\frac{\partial^2 u}{\partial t^2} - \nu^2 r_g^2 \frac{\partial^4 u}{\partial^2 X \partial t^2} = \frac{E}{\rho_0}\frac{\partial^2 u}{\partial X^2} = C_0 \frac{\partial^2 u}{\partial X^2} \tag{2.67}$$

Comparing this equation with Eq. (2.18) in elementary theory, we see that the only difference here is the second term in the left-hand side of equation. Because of this term, the elastic longitudinal wave no longer propagates in the bar with a constant velocity C_0. Instead, the harmonic waves with different frequency f (or wavelength λ) will propagate with different wave velocity (**phase velocity**). Actually, substituting the harmonic wave solution

$$u(X,t) = u_0 \exp\{i(\omega t - kt)\}$$

into Eq. (2.67), we obtain:

$$\omega^2 + \nu^2 r_g^2 \omega^2 k^2 = C_0^2 k^2$$

where ω is the angular frequency ($\omega = 2\pi f$), and k is the wave number ($k = 2\pi/\lambda$). From this equation, the phase velocity C ($C = \omega/k = f\lambda$) of a harmonic wave with the angular frequency ω is:

$$C^2 = C_0^2 - \nu^2 r_g^2 \omega^2 = \frac{C_0^2}{1 + \nu^2 r_g^2 k^2} \tag{2.68a}$$

In case $\nu^2 r_g^2 k^2 \ll 1$, we have the following approximation:

$$\frac{C}{C_0} \approx 1 - \frac{1}{2}\left(\nu r_g k\right)^2 = 1 - 2\nu^2\pi^2\left(\frac{r_g}{\lambda}\right)^2 \tag{2.68b}$$

For a solid cylindrical bar with radius a, $r_g = a/\sqrt{2}$, leading to:

$$\frac{C}{C_0} \approx 1 - \nu^2\pi^2\left(\frac{a}{\lambda}\right)^2 \tag{2.68c}$$

The above equation is the so-called Rayleigh approximate solution that includes the modification of the transversal inertia effect (Rayleigh, Lord, 1887). This approximate solution gives quite accurate results in the range $a/\lambda \leq 0.7$.

From the above equations, it can be seen that high frequency waves (short waves) propagate slower, and low frequency waves (long waves) propagate faster. In case of linear elastic waves, a wave with arbitrary waveform can be regarded as a superposition of a series of harmonic components by Fourier theorem. Since different components propagate with different velocity, the waveform is likely to change during propagation. In other words, the wave will **disperse** as it propagates. This **dispersion phenomenon** originates from the transversal inertial effect. It differs from the **nonlinear constitutive dispersion** discussed previously, which originates from the nonlinearity of the material's stress–strain relationship. On the other side, the current dispersion also differs from the **constitutive viscous dispersion** (will be discussed in Chapter 6), which originates from the viscous effect of the material. Because this dispersion is closely related to the geometry of the bar (such as the bar's radius), it is sometimes called the **geometric dispersion**.

In experiments related to stress wave propagation, for example, in a Hopkinson pressure bar test (will be discussed in Chapter 3), the measured stress waves more or less exhibit the geometric dispersion. The local oscillations of a real stress wave shown in Fig. 2.13 resulted from this dispersion effect. Specifically, a strong discontinuous stress wave contains many high frequency harmonic components which are prone to dispersion. As a result, the abrupt front of a strong discontinuous stress wave is no longer kept as a jump during wave propagation. Actually most so-called shock waves propagating along a bar only have a relatively steep wave front.

The recent two-dimensional (axi-symmetric) numerical simulations on the elastic wave propagation in a cylindrical bar (e.g., Liu and Hu 2000a; Wang and Wang, 2005; etc.)

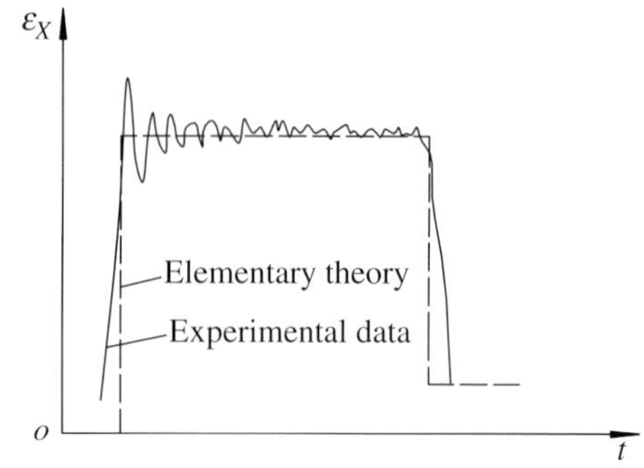

Fig. 2.13. A typical waveform recorded during a Hopkinson pressure bar test.

have exposed the following phenomena that are connected to the transversal inertial effect of a one-dimensional bar:

1. *The non-uniform stress distribution on cross section.* Different from the elementary theory that assumes uniform stress distribution on a bar cross section, the transversal inertia effect introduces non-uniform, two-dimensional stress state. Consider a steel cylindrical bar of diameter $D = 2R = 37$ mm, the material's parameters are: Young's modulus $E = 200$ GPa, density $\rho = 7.8 \times 10^3$ kg/m^3, Poisson ratio $\nu = 0.3$. At the end of the bar $X = 0$ a trapezoidal impulse is applied. The loading parameters are: magnitude $\sigma_0 = 800$ MPa, total loading time 120 μs including the rising and decreasing time 10 μs respectively. Figure 2.14(a) plots the history curves of the non-dimensional axial stresses at three points ($r/R = 0$, 0.5, and 1) on a cross section $X_1 = 0.5D$. Along the radial direction, the axial stress decreases from center to surface. At the axial center of the bar the axial stress has the maximum magnitude (the material is in an approximately one-dimensional strain state). At the surface of the bar the axial stress has the minimum magnitude (the material in an approximately one-dimensional stress state). The axial stress magnitude at location $0.5R$ is intermediate. Nevertheless, as the stress pulse propagates, the axial stress distribution on a cross section gradually becomes uniform, as Fig. 2.14(b) shows, though the stress histories still retain significant high frequency oscillations.

2. *Stress wave oscillations.* According to Eq. (2.68c), the phase velocity of a harmonic wave is dependent on the ratio of the bar diameter to wavelength. As the bar diameter increases, or the wavelength decreases, the waveform oscillations shown in Fig. 2.13 become more significant. Figure 2.15(a–d) plot the stress wave profiles on cross sections $X = 0$, 100, 200, 300, 400, and 500 mm of steel bars with diameters 5, 14.5, 37, and 74 mm, respectively. The prescribed boundary condition is same as that stated above.

   As can be seen, for a given bar diameter the oscillations of a waveform increase as the wave propagates; on the other hand the oscillations increase significantly as the bar diameter increases. Indeed, the trapezoidal wave with 10 μs rising time propagates along the 5 mm diameter bar with negligible oscillations, while the same

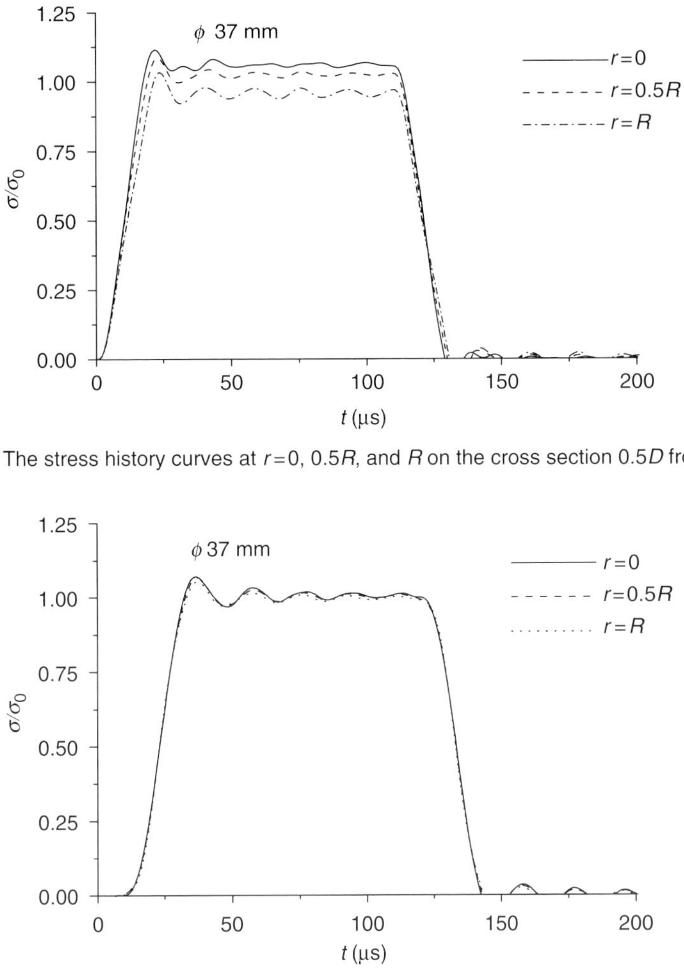

(a) The stress history curves at $r=0$, $0.5R$, and R on the cross section $0.5D$ from end.

(b) Stress histories at same radial locations on cross section $2D$ from end.

Fig. 2.14. Axial stress waveforms in different radial locations of a 37 mm diameter steel bar.

wave propagates along the 74 mm bar with significant distortions. Such waveform oscillations may pose a severe problem to the experiments of split Hopkinson pressure bar (will be discussed in Section 3.8). If a thick bar is used in the experiments, adequate measures must be taken to improve the experimental accuracy.

3. *Increases of wave front rising time.* From Fig. 2.15 it is seen that due to the transversal inertia effect, the waveform of a stress pulse flattens during propagation. One result is that the rising time in the front of the stress pulse t_s (the duration when stress pulse increases from starting point to peak) increases with the propagation distance X. This effect is more marked as the bar diameter increases. Figure 2.16 plots the t_s *vs.* X curves corresponding to different bar diameters. It is understandable that as

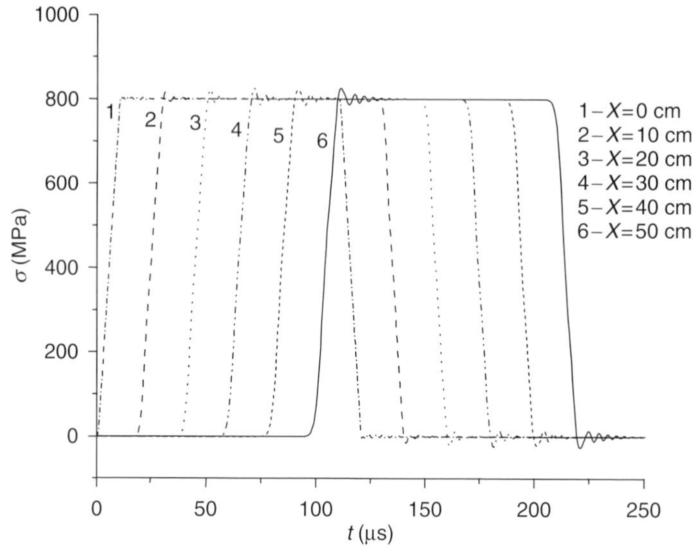

(a) Stress waveforms in 5 mm diameter bar.

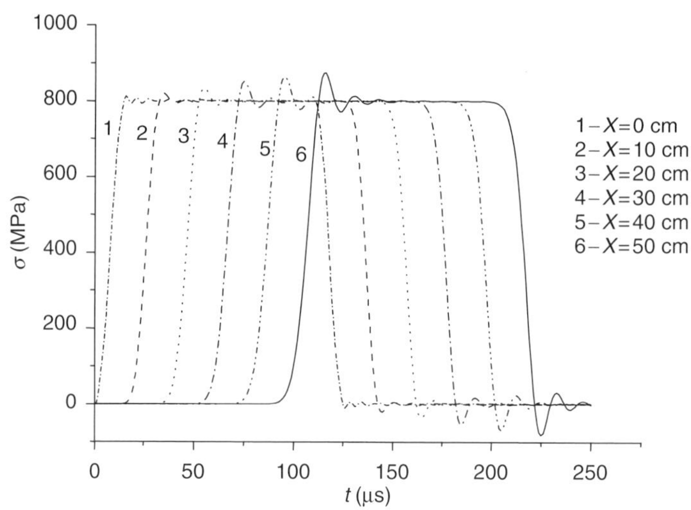

(b) Stress waveforms in 14.5 mm diameter bar.

Fig. 2.15. Stress waveforms propagate in bars of different diameters.

a thicker bar has more transversal inertia, the rising time increase in a thick bar is more significant than that in a thin bar. The rising time increase is larger in the early propagation stage, and tends to level up as time goes on.

4. *Attenuations of stress peaks during propagation.* Another phenomenon of the geometric dispersion is: the peak level of a stress pulse decreases during its propagation.

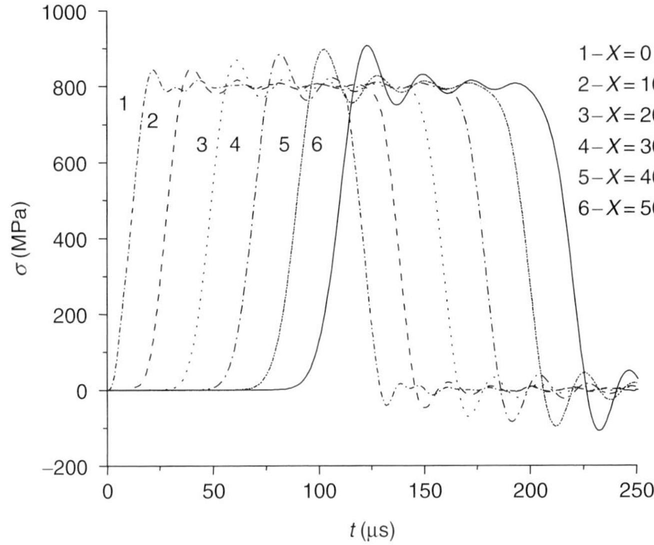

(c) Stress waveforms in 37 mm diameter bar.

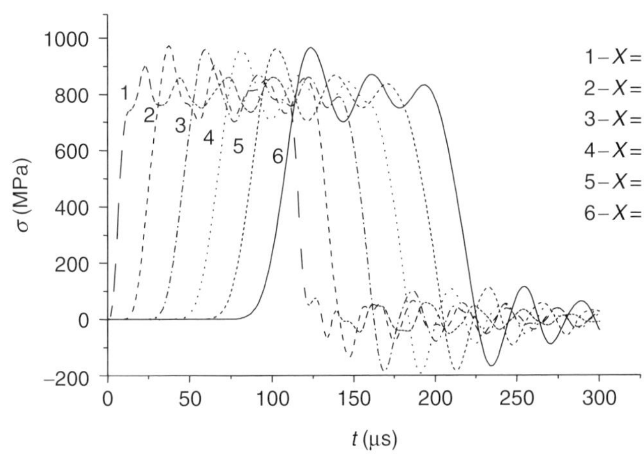

(d) Stress waveforms in 74 mm diameter bar.

Fig. 2.15. Continued.

As a trapezoidal stress pulse results in wave oscillations so that the attenuation of wave peak is difficult to quantify, we consider the propagation process of a triangular stress pulse to illustrate this peak attenuation phenomenon. The triangular stress pulse applied at the bar end has a peak magnitude $\sigma_0 = 800$ MPa. The rising time and the descending time are 150 μs. Figure 2.17 plots the waveforms of the stress pulse after propagating different distances along a 37 mm diameter bar. Obvious peak stress

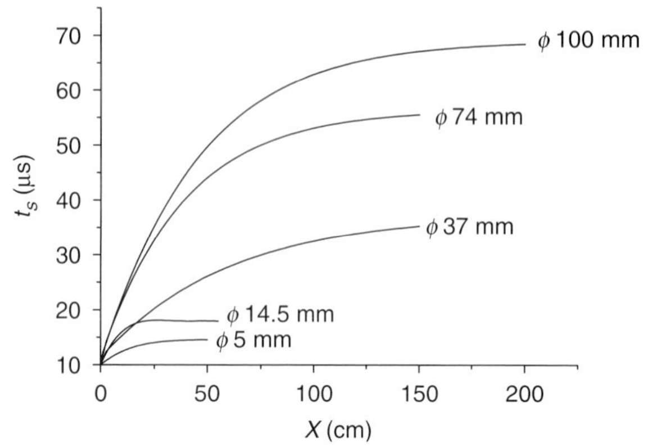

Fig. 2.16. The rising times of stress pulse front (t_s) as functions of propagation distance (X). Results corresponding to different bars.

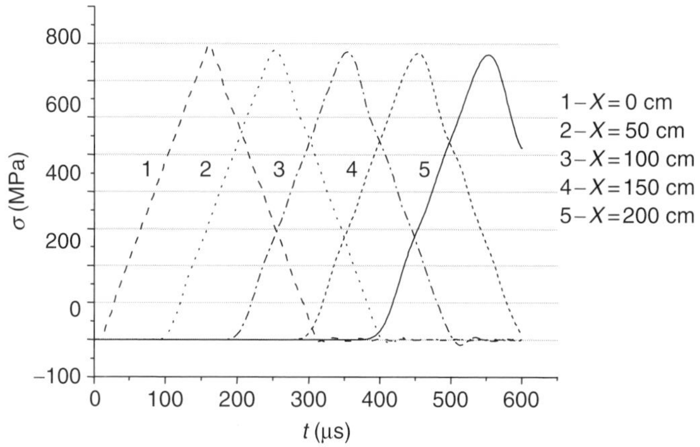

Fig. 2.17. Waveforms of triangular stress pulse propagating in a 37 mm diameter steel bar.

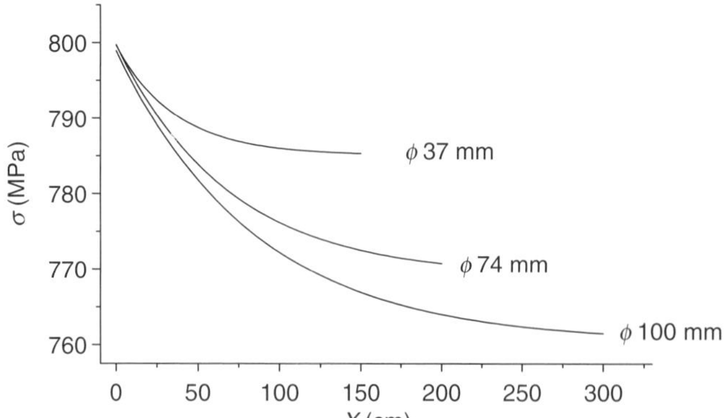

Fig. 2.18. Attenuations of peaks of a triangular stress pulse propagating in different bars.

attenuation phenomenon is seen. Figure 2.18 plots the peak stress values as functions of propagation distance X in 37, 74, and 100 mm diameter bars. Obviously a larger bar diameter results in more significant attenuation. This is consistent with the fact that a thicker bar has more marked effect of transversal inertia.

In summarization, if the transversal dimension of the bar is much less than the wavelength of a stress pulse, then the transversal kinetic energy of the bar is much small than the longitudinal kinetic energy. In this case the elementary one-dimensional stress wave theory provides sufficiently accurate results. If the bar's transversal dimension is comparable with the wavelength, geometric dispersions resulting from transversal inertia is should be taken into account. Further study shows, for stress waves within the range $a/l \leq 0.7$, the Rayleigh approximate solution (2.68) gives quite good transversal inertia modification. However, for the waves that have even shorter wavelengths, the more exact but complicated Pochhammer–Chree solutions should be taken into consideration (the readers can refer the books of Kolsky, 1953 or Miklowitz, 1978).

2.9 Torsion Waves in Cylindrical Bars

The discussions up to now are for uni-axial longitudinal stress waves. In concluding this chapter we briefly discuss the propagation of a simple kind of transverse wave, namely the elastic torsion wave.

Lagrange coordinate system is used in the current analysis. Under the planar cross section assumption[1], we investigate the pure torsion motion of a cylinder with uniform cross section. The cylinder is subjected to a torque M, and twists an angle φ (Fig. 2.19). The angular velocity and the twist per unit length, denoted by ω and θ respectively, are the first-order derivatives of φ:

$$\omega = \frac{\partial \varphi}{\partial t}$$
$$\theta = \frac{\partial \varphi}{\partial X} \tag{2.69}$$

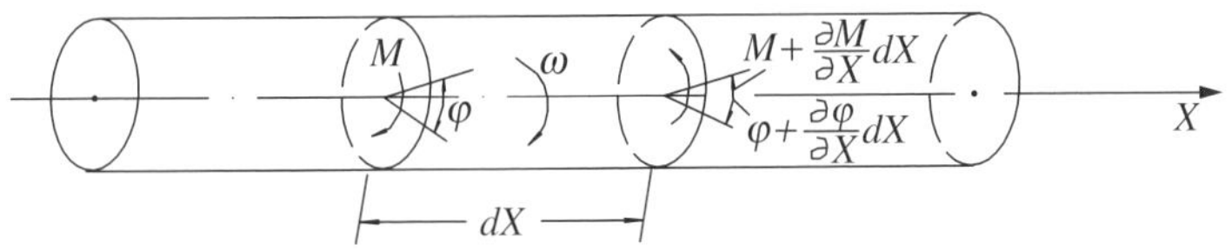

Fig. 2.19. Pure torsion movement of a cylindrical bar with uniform cross sections.

[1]According to the elasticity theory, the planar cross-section assumption is true only to the cylinder bar or tube with the circular cross sections. For the bar with other shape of cross section, the cross section warps when the bar is twisted.

The geometric compatibility equation between ω and θ, and the conservation equation of the angular momentum in a segment dX are:

$$\frac{\partial \omega}{\partial X} = \frac{\partial \theta}{\partial t} \tag{2.70}$$

$$I\frac{\partial \omega}{\partial t} = \frac{\partial M}{\partial X} \tag{2.71}$$

where I is the rotational inertia of a unit length cylinder with respect to its axis X:

$$I = \int r^2 dm = \rho_0 \int r^2 dA \tag{2.72}$$

and m is the line density ($dm = \rho_0 dA$). In case the material is linear elastic, the relationship between the shear stress τ and shear strain γ is:

$$\tau = G\gamma \tag{2.73}$$

where G is the shear modulus. As the cross sections of the cylinder is not deformed during twist, we have:

$$\gamma = r\theta \tag{2.74}$$

$$M = \int \tau r dA \tag{2.75}$$

where r is distance of a point on cross section to the twisting axis and A is the cross sectional area of the cylinder. Substituting (2.73, 2.74) into (2.75) and considering (2.72) gives:

$$M = \int \tau r ds = G\theta \int r^2 ds = \frac{GI\theta}{\rho_0} \tag{2.76}$$

Substituting (2.76) into the angular momentum conservation Eq. (2.71) leads to:

$$\frac{\partial \omega}{\partial t} = C_T^2 \frac{\partial \theta}{\partial X} \tag{2.77}$$

where we have introduced a quantity:

$$C_T = \sqrt{\frac{G}{\rho_0}} \tag{2.78}$$

Equations (2.70) and (2.77) constitute a hyperbolic first-order partial differential system having the unknown functions of the angular velocity ω and the relative angular twist θ. Formally, these equations are identical to the longitudinal wave Eqs. (2.12) and (2.16).

Substituting (2.69) into (2.77) gives a second-order hyperbolic PDE having the unknown function of the twist angle φ:

$$\frac{\partial^2 \varphi}{\partial t^2} - C_T^2 \frac{\partial^2 \varphi}{\partial X^2} = 0 \tag{2.79}$$

Obviously the quantity C_T is the wave velocity of the elastic torsion wave. As $G = E/[2(1+\nu)]$, and the value of Poisson ratio ν is generally between 0 and 0.5, the torsion wave velocity is less than the longitudinal wave velocity in the bar. The ratio of $C_0/C_T = \sqrt{2(1+\nu)}$ is between $\sqrt{2}$ and $\sqrt{3}$. Table 2.2 collects the C_T values of some common materials (Kolsky, 1953). These data can be compared to those listed in Table 2.1.

In case of a thin-walled-thickness cylindrical tube with average radius a, the shear stress is uniformly distributed on the tube's cross section. Equations (2.70) and (2.77) can be rewritten as:

$$\begin{aligned} \frac{\partial (a\omega)}{\partial X} &= \frac{1}{\rho_0 C_T^2} \frac{\partial \tau}{\partial t} \\ \frac{\partial (a\omega)}{\partial t} &= \frac{1}{\rho_0} \frac{\partial \tau}{\partial X} \end{aligned} \tag{2.80}$$

which in form are identical to the longitudinal wave Eqs. (2.17) and (2.13). The strategy of using the method of characteristics to solve these equations is similar to that discussed previously in Eqs. (2.24) and (2.26). It is easy to prove that along the characteristic lines:

$$dX = \pm C_T dt \tag{2.81a}$$

the angular velocity ω and the relative twist angle θ satisfy the compatibility equation:

$$d\omega = \pm C_T d\theta \tag{2.81b}$$

or represented by the shear stress τ and the tangent twist speed $(a\omega)$, the characteristic compatibility equation is:

$$d\tau = \pm \rho_0 C_T d(a\omega) \tag{2.81c}$$

Table 2.2. Shear modulus and the elastic torsion wave velocity of common materials.

	Steel	Copper	Aluminum	Glass	Rubber
G (GPa = 10^{10} dyne/cm^2)	81	45	26	28	7.0×10^{-4}
C_T (km/s)	3.22	2.25	3.10	3.35	0.027

For a **simple torsion wave** propagating in a tube that is initially at rest and stress-free, behind the wave front the following simple torsion wave relationships exist:

$$\omega = \mp C_T \theta$$

$$\tau = \mp \rho_0 C_T a \omega$$

where (−) corresponds to the rightward wave and (+) corresponds to the leftward wave.

Note that when a tube is subjected to pure torsion deformation, the direction that inclines to the tube axis X by 45°, namely the X' direction shown in Fig. 2.20, is the principle stress direction. Therefore under pure torsion loading, the spiral fiber that is 45° inclined to the tube axis is under a tensile stress of value τ in X' direction, and a compressive stress of value $-\tau$ in the direction normal to X'. As Johnson (1972) pointed out, a torsion wave propagating with the velocity C_T along the tube axis (the X axis) can be viewed as a longitudinal wave propagating with the velocity C_0' along the 45° spiral fiber direction (the X' axis), accordingly:

$$C_T = C_0' \cos 45^\circ = \frac{C_0'}{\sqrt{2}} \tag{2.82}$$

In X' direction, the stress–strain relationship accounting for the transversal compression effect is:

$$\varepsilon_{X'} = \frac{1}{E}(\tau + \nu\tau) = \frac{1+\nu}{E}\tau$$

from which the longitudinal wave velocity, expressed by Eq. (2.15), is deduced as:

$$C_0' = \sqrt{\frac{E}{\rho_0(1+\nu)}}$$

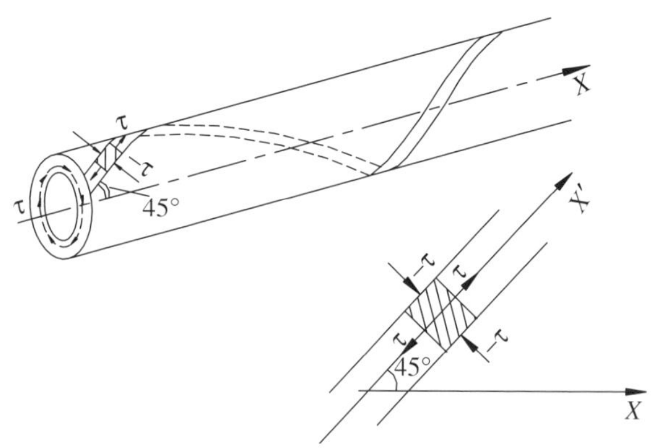

Fig. 2.20. Propagation of torsion waves along the axis of a thin elastic tube is equivalent to that of longitudinal waves along the spiral fibers on the tube surface 45° inclined to the tube axis.

Substituting the above result into (2.82), and noticing $G = E/2(1 + \nu)$, the torsion wave velocity C_T is:

$$C_T = \frac{C_0'}{\sqrt{2}} = \sqrt{\frac{E}{2\rho_0(1+\nu)}} = \sqrt{\frac{G}{\rho_0}}$$

This expression is exactly identical with Eq. (2.78).

It is worthwhile to note that although the theory of the longitudinal wave propagation in a bar, based on invariable plane cross-section assumption, is an approximate theory; the torsion wave theory based on the same assumption leads solutions that are identical with Pochhammer's elastodynamic exact solution (Kolsky, 1953). All harmonic torsion waves of different frequencies propagate with the same phase velocity $C_T = \sqrt{G/\rho_0}$, and no frequency dispersion exists during torsion wave propagation.

As a matter of fact, the elastic torsion wave velocity C_T is the isometric wave velocity, or the distortional wave velocity, in an infinite media. Sometimes this velocity is called the shear wave velocity. In Chapter 11 we will further discuss such wave velocity.

CHAPTER 3

Interaction of Elastic Longitudinal Waves

3.1 Coaxial Collision of Two Elastic Bars

Before discussing the interaction between two elastic waves, we first discuss the longitudinal coaxial collision of two elastic bars. This is helpful to understand the interaction between two elastic waves.

Consider two elastic bars B_1 and B_2 with the same section sizes but different sound impedance $(\rho_0 C_0)_1$ and $(\rho_0 C_0)_2$, respectively. Before their collision, the initial stresses in both bars is zero, while the initial particle velocities are v_1 and v_2 respectively, and assuming $v_1 < v_2$, as schematically shown in Fig. 3.1(a). On the σ–v plane, such initial states correspond to the points 1 and 2, respectively as shown in Fig. 3.1(d). When a longitudinal coaxial collision takes place, originating from the collision interface, a rightward strong discontinuous elastic wave propagates in bar B_1 and simultaneously a leftward strong discontinuous elastic wave propagates in bar B_2 as shown in Fig. 3.1(c). At the collision interface, the particle velocities of the two bars should be identical (the continuity condition), as well as the stresses of the two bars should be identical (the action force and reaction force equality condition). Based on these, and moreover according to the momentum conservation condition across a strong discontinuous interface [Eq. (2.57)], we obtain

$$\sigma = -(\rho_0 C_0)_1 (v - v_1) = (\rho_0 C_0)_2 (v - v_2)$$

From this, the particle velocity v and the stress σ in the two bars after collision can be determined respectively:

$$\left.\begin{aligned} v &= \frac{(\rho_0 C_0)_1 v_1 + (\rho_0 C_0)_2 v_2}{(\rho_0 C_0)_1 + (\rho_0 C_0)_2} \\ \sigma &= -\frac{v_2 - v_1}{\dfrac{1}{(\rho_0 C_0)_1} + \dfrac{1}{(\rho_0 C_0)_2}} \end{aligned}\right\} \tag{3.1}$$

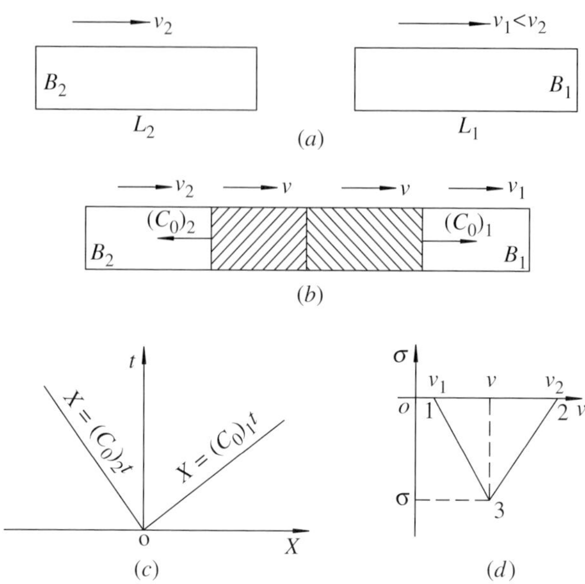

Fig. 3.1. Longitudinal coaxial collision of two elastic bars.

On the σ–v plane, this state corresponds to the intersection point 3 of the leftward characteristics passing point 1 with slope $-(\rho_0 C_0)_1$ and the rightward characteristics passing point 2 with slope $(\rho_0 C_0)_2$ as shown in Fig. 3.1(d).

If the material of the two bars are identical, or the sound impedance are identical, namely if $(\rho_0 C_0)_1 = (\rho_0 C_0)_2 = \rho_0 C_0$, then $v = (1/2)(v_1 + v_2)$ while $\sigma = -(1/2)\rho_0 C_0 (v_2 - v_1)$. Further if $v_1 = v_2$, then $v = 0$ and $\sigma = \rho_0 C_0 v_2$. This is equivalent to such a situation where bar B_2 impacts onto a rigid wall at a velocity v_2. This result coincides with that of a rigid wall condition where $(\rho_0 C_0)_1 \to \infty$ and $v_1 = 0$ are applied. On the other hand, If $(\rho_0 C_0)_2 \to \infty$, then $v \to v_2$ and $\sigma \to -(\rho_0 C_0)_1 (v_2 - v_1)$, which corresponds to such a situation that a rigid bar B_2 impacts onto an elastic bar B_1. The above results are valid from the beginning of collision up to the time when the stress waves arrive at the other end of the bars. Once the stress waves arrive at the other end of the bars, the situation will be different since the wave reflection occurs (refer to Section 3.4).

3.2 Interaction of Two Elastic Longitudinal Waves

Consider an elastic bar which is initially at rest and in natural state and is suddenly subjected to a constant impact load at both ends as shown in Fig. 3.2(a). Then, originating from the two ends, two strong discontinuous elastic waves propagate toward each other (called the primary waves), which are represented by the rightward characteristics *OA* and the leftward characteristics *LA* on the *X*–*t* plane as shown in Fig. 3.2(e). The bar part swept by the leftward primary wave front (namely across the characteristics *LA*) is in the state of $v_1, \varepsilon_1, \sigma_1$. According to Eq. (2.57) and reversing the sign for leftward waves, there exists

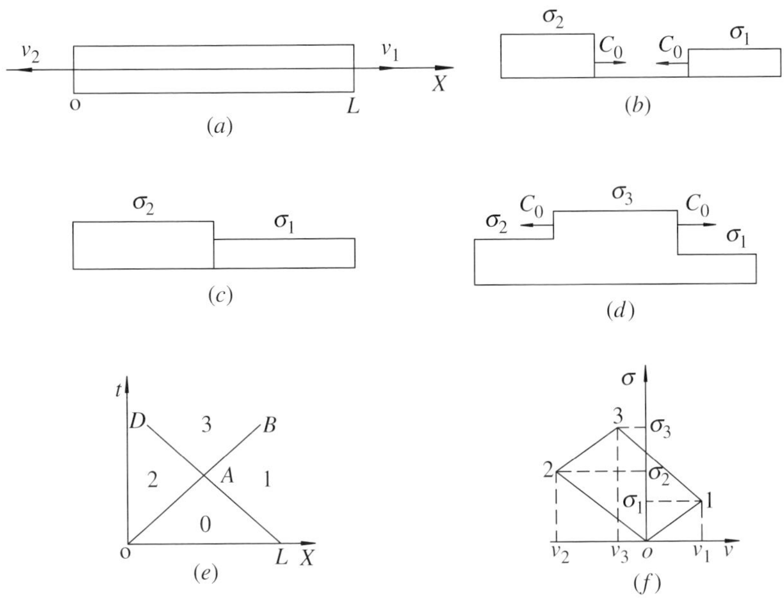

Fig. 3.2. Interaction of two head-on propagating strong discontinuous elastic waves.

the relation $\sigma_1 = \rho_0 C_0 v_1$. The bar part swept by the rightward primary wave (namely across the characteristics *OA*) is in the state of $v_2, \varepsilon_2, \sigma_2$. According to Eq. (2.57) there exists the relation $\sigma_2 = -\rho_0 C_0 v_2$. On the σ–v plane, these correspond to the state-jump from point 0 to point 1 and point 2, respectively as shown in Fig. 3.2(f). All these are entirely the same as what we discussed for the elastic waves in semi-infinite bars. At the instant that two waves meet at an interface as shown in Fig. 3.2(c), the particle velocity of the bar on the right side of the interface is v_1, while the particle velocity of the bar on the left side of the interface is v_2, just similar to a coaxial collision of two elastic bars shown in Fig. 3.1. So an interaction of two waves in a bar is equivalent to an internal collision within the bar, sometimes called the **internal collision**. As a result of such an internal collision, from the interface where the two primary waves meet, two internal reflective waves are generated, namely the rightward secondary wave *AB* and the leftward secondary wave *AD*, and propagate towards the two ends of the bar respectively, as shown in Fig. 3.2(d and e).

The state of the bar swept by the rightward secondary wave front (namely across the characteristics *AB*) changes suddenly from state $v_1, \varepsilon_1, \sigma_1$ to the state $v'_3, \varepsilon'_3, \sigma'_3$, and according to the compatibility condition [Eq. (2.57)] there is,

$$\sigma'_3 - \sigma_1 = -\rho_0 C_0 (v'_3 - v_1)$$

On the σ–v plane, this corresponds to a state-jump from point 1 to point 3′ in the direction coincident to the direction across the characteristics *AB* on the X–t plane.

The state of the bar swept by the leftward secondary wave front changes suddenly from state $v_2, \varepsilon_2, \sigma_2$, to state $v_3'', \varepsilon_3'', \sigma_3''$, and according to Eq. (2.57), there is,

$$\sigma_3'' - \sigma_2 = \rho_0 C_0 (v_3'' - v_2)$$

On the σ–v plane, this corresponds to a state-jump from point 2 to point 3″ in the direction coincident to the direction across the characteristics AD on the X–t plane. But, according to the particle-velocity continuity condition and the stress equality condition at the internal collision interface, there should be $v_3' = v_3'' = v_3$, and $\sigma_3' = \sigma_3'' = \sigma_3$. On the σ–v plane, this corresponds to the determination of the intersection point 3 of the leftward characteristics passing the point 1 and the rightward characteristics passing the point 2. Thus the problem is solvable. It should be noticed that the Eq. (2.57) is deduced for the rightward waves, so it is needed to replace $\mathscr{D}$ with $-\mathscr{D}$ (reversing sign) for the leftward waves, and the initial states ahead of the primary wave front and the secondary wave front are different. Thereby, we obtain

$$\underbrace{\sigma_3 = -\rho_0 C_0 v_2 + \rho_0 C_0 (v_3 - v_2)}_{\substack{\text{(rightward primary)} + \text{(leftward secondary)} \\ \text{left part of bar}}} = \underbrace{\rho_0 C_0 v_1 - \rho_0 C_0 (v_3 - v_1)}_{\substack{\text{(leftward primary)} + \text{(rightward secondary)} \\ \text{right part of bar}}}$$

Then, the particle velocity v_3 and the stress σ_3 in the bar experiencing the interaction of two elastic waves can be respectively determined as follows [Fig. 3.2(f)]:

$$\left.\begin{aligned} v_3 &= v_1 + v_2 \\ \sigma_3 &= \sigma_1 + \sigma_2 \end{aligned}\right\} \tag{3.2}$$

It is clear that the final state of the bar experiencing the interaction of two elastic waves can be obtained by a superposition of the results respectively induced by each elastic wave propagation. This is because the governing equations of elastic waves are linear so that the **principle of superposition** must be valid.

Bear in mind, the basic principles and methods to treat the interaction of stress waves used here will be repeatedly used in the following sections. For whatever be the type of waves like strong discontinuous waves or the weak discontinuous waves, the loading waves or the unloading waves, the elastic waves, the elasto-plastic wave or the hydrodynamic waves, these principles are still valid.

3.3 Reflection of Elastic Longitudinal Waves at Fixed End and Free End

When a stress wave propagating in a bar arrives at the other end of the bar, a wave reflection occurs. The reflective wave varies depending on the actual boundary condition. With respect to the incident wave, the boundary condition is a new disturbance to the state behind the incident wave front, and the propagation of this new disturbance is displayed as a reflective wave. The state behind a reflective wave depends on the total result of the incident wave together with the reflective wave, satisfying the given boundary condition. For the elastic waves, the total result of the incident wave together with the reflective wave

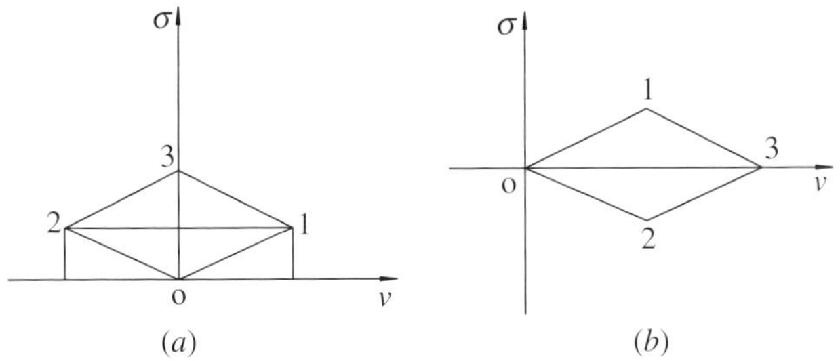

Fig. 3.3. Reflection of an elastic wave at a fixed end.

can be determined based on the principle of superposition. Thereby, the reflection of an elastic wave at the fixed end and at the free end can be viewed as special examples of the general discussions on the interaction of two elastic waves conducted in the previous section.

As a special case of that shown in Fig. 3.2, let $v_2 = -v_1$, then we have $v_3 = 0$ and $\sigma_3 = 2\sigma_1$ as shown in Fig. 3.3(a), namely the final particle velocity is zero but the final stress is doubled. This satisfies the boundary condition of particle velocity at a fixed end. Thus, this special case corresponds to the **reflection at a fixed end (a rigid wall)** of a normally incident elastic wave. Therefore, if the bar end is imagined as a mirror, when a normally incident elastic wave arrives at a fixed end, the reflective wave is just a positive image of the incident wave.

Based on the reflection of an elastic wave at a fixed end, the well-known drop-weight tests for the impact tension of steel string performed by Hopkinson (1872) can be explained. From the Hopkinson's tests as schematically shown in Fig. 3.4, it was shown that when a steel string is subjected to an impact tension by a dropping weight, it breaks not at the loaded end A but at the fixed end B, and moreover whether the string breaks or not is mainly dependent on the height of the weight (or the impact velocity), but almost independent of the mass of weight. This is because that for a given material (namely the sound impedance $\rho_0 C_0$ is known), the amplitude of the incidence stress wave generated

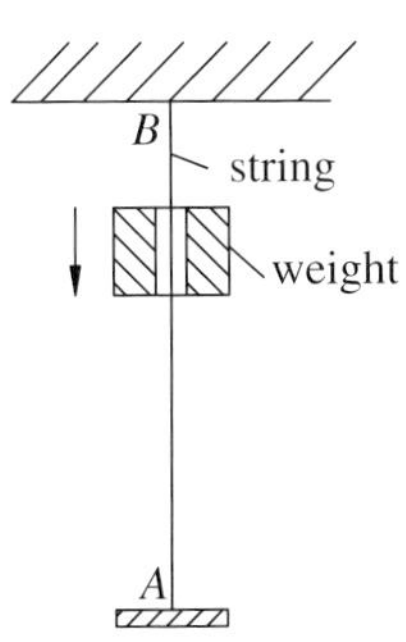

Fig. 3.4. Drop tensile test.

at point A only depends on the impact velocity v_0 regardless of the mass of weight, $\sigma = \rho_0 C_0 v_0$ [see Eq. (2.33)], and the stress at the fixed end B reaches two times of the incidence stress σ due to the wave reflection at a fixed end, $2\sigma = 2\rho_0 C_0 v_0$. A further analysis shows that the dropping mass will influence the profile of the incident wave, which will be discussed in the next section, refer to Eq. (3.5).

As another special case of that shown in Fig. 3.2, let $v_2 = v_1$, then we have $v_3 = 2v_1, \sigma_3 = 0$ as shown in Fig. 3.3(b), namely the final stress is zero while the particle velocity is doubled. This satisfies the boundary condition of stress at a free end. Thus, this special case corresponds to the **reflection at a free end (free surface)** of a normal incident elastic wave. Therefore, if the bar end is imagined as a mirror, when a normally incident elastic wave is reflected at a free end, the reflective wave is just an inverted image of the incident wave.

3.4 Coaxial Collision of Two Elastic Bars with Finite Length

In the previous discussion for the coaxial collision of two elastic bars shown in Fig. 3.1, we temporarily only focused on the primary elastic waves generated by the collision, and neither the wave reflection nor the wave interaction was involved. Having the knowledge of the reflection of elastic waves at the fixed end and the free end discussed above, now we can further discuss the coaxial collision of two elastic bars with finite lengths.

Assume that in Fig. 3.1 the length of bar B_2 is L_2 and of bar B_1 is L_1, and $L_2 < L_1$, namely, we consider a problem where a short bar B_2 impacts onto a long bar B_1. For convenience, let the long bar B_1 be initially at rest, namely $v_1 = 0$. The primary elastic waves generated at the beginning of collision ($t = 0$) can be computed according to Eq. (3.1). Subsequently, at time $t = L_2/(C_0)_2$, the leftward wave is first reflected at the free end, and at time $t = 2L_2/(C_0)_2$, this rightward reflected wave returns to the collision interface. The subsequent situations depend on the ratio of wave impedances (sound impedances) of these two bars, and can be categorized into three cases:

(a) $(\rho_0 C_0)_2/(\rho_0 C_0)_1 = 1$ [Fig. 3.5(a)].

In this case, since wave impedances of the two bars are identical, the rightward unloading wave reflected from the free end propagates through the collision interface without any additional disturbance, just as propagating in the same bar, although the wave velocities in the two bar can be different, namely there may be $(C_0)_2 \neq (C_0)_1$. Therefore, at time $t = 2L_2/(C_0)_2$, the contact interface of two bars is unloaded to $v = 0$ and $\sigma = 0$. This indicates the termination of collision (no interaction between the two bars). After this, a stress pulse consisting of a strong discontinuous loading wave front and a strong discontinuous unloading wave front still propagates in the long bar. In the case of $(C_0)_2 = (C_0)_1$, the pulse length λ is two times the length of the short bar, namely $\lambda = 2L_2$. By changing the length of the short bar, we can obtain stress pulses with various pulse lengths. This is a typical method to control the length of stress pulses in stress wave experiments.

In the special case of $L_2/(C_0)_2 = L_1/(C_0)_1$, the primary elastic waves generated by the collision simultaneously arrive at each free end, and the reflected elastic

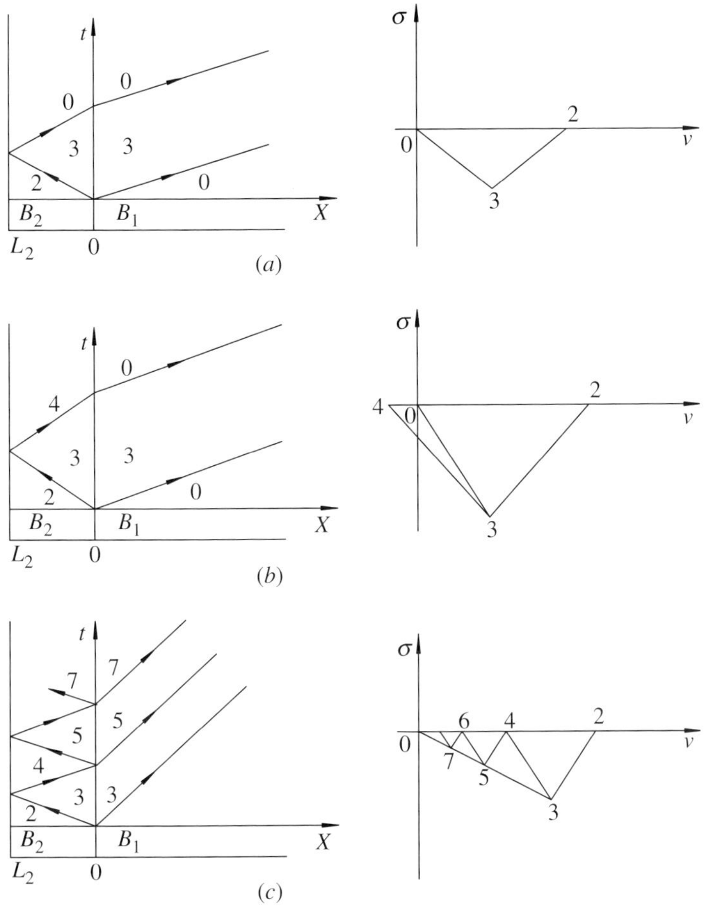

Fig. 3.5. Coaxial collision between a short elastic bar and a long elastic bar.

unloading waves also simultaneously return to the collision interface. At time $t = 2L_2/(C_0)_2$, the collision is finished, the whole bar B_2 is at rest ($v = 0$ and $\sigma = 0$), and bar B_1 is moving as a whole ($v = v_2$ and $\sigma = 0$). We advise the readers to plot these states simultaneously on the X–t plane and the σ–v plane by themselves. Note that before the collision the particle velocity of bar B_2 is v_2 while that of bar B_1 is 0, but after the collision the particle velocity of bar B_2 is 0 while that of bar B_1 is v_2. It is clear that through the collision, the momentum and kinetic energy of bar B_2 is entirely transferred to bar B_1 (readers can compare this process with the elastic collision of two homogenous balls with the same mass in the textbook of physics). It should be noticed that in the case of $L_2 < L_1$, although the whole bar B_2 is at rest, bar B_1 is not in a state like a rigid body motion (namely, all its particles move at a same velocity), but in a state of back-and-forth propagating stress waves.

(b) $(\rho_0 C_0)_2/(\rho_0 C_0)_1 < 1$ [Fig. 3.5(b)]

In this case, when the leftward incident wave in the short bar B_2 is reflected at the free end, the stress decreases to zero and the particle velocity changes to negative, as shown by point 4 on the σ–v plane in Fig. 3.5(b), since the increment of particle velocity $(v_4 - v_2)$ of a reflective wave at a free end is twice the increment of particle velocity $(v_3 - v_2)$ of an incident wave. When this reflective unloading wave returns to the collision interface, namely at time $t = 2L_2/(C_0)_2$, the whole short bar B_2 is at the state of $v = v_4$ (negative!) and $\sigma = 0$, while for the long bar B_1 both the stress and the particle velocity at the contact interface are unloaded to zero. At this time, the velocities of the two bars are different. As $v_4 < 0$, the short bar B_2 rebounds at velocity v_4 and is no longer in contact with the long bar B_1. Thus, the collision is finished. After that, a stress pulse with a length of $\lambda = 2L_2(C_0)_1/(C_0)_2$ still propagates in the long bar B_1.

In the special case of $L_2/(C_0)_2 = L_1/(C_0)_1$, at time $t = 2L_2/(C_0)_2$, the whole bar B_2 rebounds at velocity v_4, and the whole bar B_1 moves at velocity $2v_3$. At the end of collision, only a part of the initial momentum and kinetic energy of bar B_2 is transferred to bar B_1. The readers are suggested to scheme the corresponding figure on the X–t plane and the σ–v plane on their own.

If $(\rho_0 C_0)_1 \to \infty$, this corresponds to an elastic short bar impacting onto a rigid target. Obviously, at time $t = 2L_2/(C_0)_2$, the short bar rebounds at a velocity, the absolute value of which is the same as the impact velocity v_2 but in the opposite direction $(v_4 = -v_2)$. The readers are suggested to plot the corresponding figure on the X–t plane and the σ–v plane themselves.

(c) $(\rho_0 C_0)_2/(\rho_0 C_0)_1 > 1$ [Fig. 3.5(c)]

In this case, when the leftward primary incident wave in the short bar B_2 is reflected at the free end, the stress decreases to zero, and the particle velocity also decreases, though it is still positive $[v_4 - v_2 = 2(v_3 - v_2)]$, as shown by point 4 on the σ–v plane in Fig. 3.5(c). When this reflected unloading wave returns to the collision interface, namely at time $t = 2L_2/(C_0)_2$, the whole short bar B_2 is in the state of $v = v_4$ (positive) and $\sigma = 0$. However, for the long bar B_1, if the stress at the collision interface decreases to zero, then the particle velocity should also decrease to zero. Since now $v_4 > 0$, it means that the collision has not yet ended, and now it is equivalent to such a situation where the bar B_2 moving with velocity v_4 impacts onto the bar B_1 at rest, and then the secondary waves are generated. The state of such secondary waves is represented by point 5 on the σ–v plane. Based on this analogy, before the unloading wave reflected from the free end of the long bar returns to the collision interface, the collision never ends, but the amplitude of the incident waves propagating in the bar B_1 decreases successively. The readers are suggested to plot by themselves, the stress wave profile on the σ–X plane for the long bar B_1 at a time t (after the wave in the short bar has been reflected back-and-forth several times) and discuss the relationship between such step-shaped wave profile and the short bar length L_2.

In the special case of $L_2/(C_0)_2 = L_1/(C_0)_1$, at time $t = 2L_2/(C_0)_2$, bar B_2 moves as a whole at velocity v_4 $(0 < v_4 < v_2)$, while the bar B_1 moves as a whole at a much larger velocity $2v_3(>v_2)$, then the collision is finished. After the end of collision, only a part of the initial momentum and kinetic energy of bar B_2 is transferred to the bar B_1. The readers are suggested to plot the corresponding figure on the X–t plane and the σ–v plane themselves.

If $(\rho_0 C_0)_2 \to \infty$, this corresponds to a situation where a rigid body impacts onto an elastic bar. If the mass of the rigid body is denoted by M and the reflection at the other end of the elastic bar is disregarded, then according to the particle velocity continuity condition and the stress equality condition at the collision interface, we can write the following formulas for the elastic bar and the rigid body respectively

$$\sigma = -\rho_0 C_0 v, \tag{3.3}$$

$$\sigma = \frac{M}{A}\frac{dv}{dt} \tag{3.4}$$

where A is the section area of the elastic bar. From these equations, we obtain

$$v = v_0 \exp\left(-\frac{\rho_0 C_0 A}{M}t\right) = v_0 \exp\left(-\frac{M_t}{M}\right) \tag{3.5a}$$

where v_0 is the initial impact velocity, M_t $(= \rho_0 C_0 A t)$ denotes the mass in the bar swept by the elastic wave front at time t. This formula indicates that the particle velocity profile is displayed as a strong discontinuous wave front followed by an exponentially attenuating wave tail.

If we introduce the initial impact stress $\sigma_0 = -\rho_0 C_0 v_0$ according to Eq. (3.3), then corresponding to Eq. (3.5a) we obtain

$$\sigma = \sigma_0 \exp\left(-\frac{\rho_0 C_0 A}{M}t\right) = \sigma_0 \exp\left(-\frac{M_t}{M}\right) \tag{3.5b}$$

which indicates that stress σ attenuates with t exponentially. The readers are suggested to compare this result with the time-histories profile σ–t at a certain X of the long bar in Fig. 3.5(c).

If $M_t \ll M$, then $v \to v_0$, then it is reduced to an approximate constant velocity collision. Recalling the Hopkinson's impact tension tests on steel strings (Fig. 3.4), obviously, a more complete analysis is presented here.

From the above discussion, we can further understand that, unlike the static theory of solid mechanics, the theory of stress waves focuses on the unsteady and non-uniform motion of media varying with time and coordinates. We can also further understand the coupling between loads and media and the important influence of the mechanical constitutive behavior of materials on the propagation of stress waves.

Although in the above discussions, the collisions of only two elastic bars were taken as examples, the principles mentioned are also suitable to analyze the coaxial collision of three or more elastic bars.

For example, consider three identical elastic bars B_1, B_2, and B_3, the bars B_2 and B_1 are originally at rest and in contact with each other, while bar B_3 moves at a velocity v_0 and

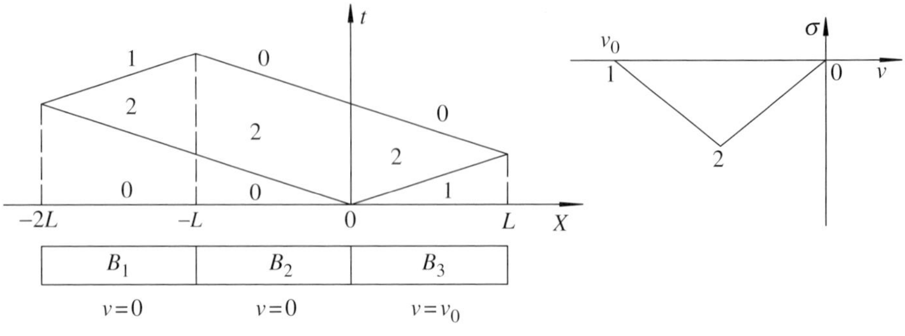

Fig. 3.6. Coaxial collision of three identical bars.

impacts onto the bar B_2 as shown in Fig. 3.6. It is easy to prove that the collision ends at time $t = 3L/C_0$, and at this time bar B_1 flies out as a whole at a velocity v_0, while both the bars B_2 and B_3 are now at rest. The initial momentum and kinetic energy of bar B_3 are entirely transferred to bar B_1 through bar B_2.

3.5 Reflection and Transmission of Elastic Longitudinal Waves at the Interface of Two Different Bars

In the previous discussion on the collision of two finite length bars with different wave impedances, when the reflected wave in the short bar returns from the free end to the collision interface, we have actually dealt with the transmission and reflection of elastic wave at the interface between two different materials. In the following, we will discuss the general cases of this sort.

Consider a situation when an elastic wave propagates from one medium (the related parameters are denoted by the subscript 1) to another (the related parameters are denoted by the subscript 2) and the propagating direction of the wave is perpendicular to the interface of the two media, namely a normal incidence. When an incident elastic wave disturbance $\Delta\sigma_I$ arrives at the interface, new disturbances not only with respect to the first medium but also to the second medium, are generated, namely a reflective disturbance $\Delta\sigma_R$ and a transmission disturbance $\Delta\sigma_T$ will propagate into the two media respectively, as shown in Fig. 3.7. So long as the two media experiencing the reflection–transmission are still in contact at their interface (namely they are not separated and can undergo both compression and tension), according to the continuity condition and the Newton's third law, the particle velocities on both sides and the stresses on both sides should be identical respectively.

$$\Delta v_I + \Delta v_R = \Delta v_T \tag{3.6}$$

$$\Delta\sigma_I + \Delta\sigma_R = \Delta\sigma_T \tag{3.7}$$

where the subscripts *I*, *R*, and *T* denote the parameters relevant to the incident wave, reflective wave, and transmission wave.

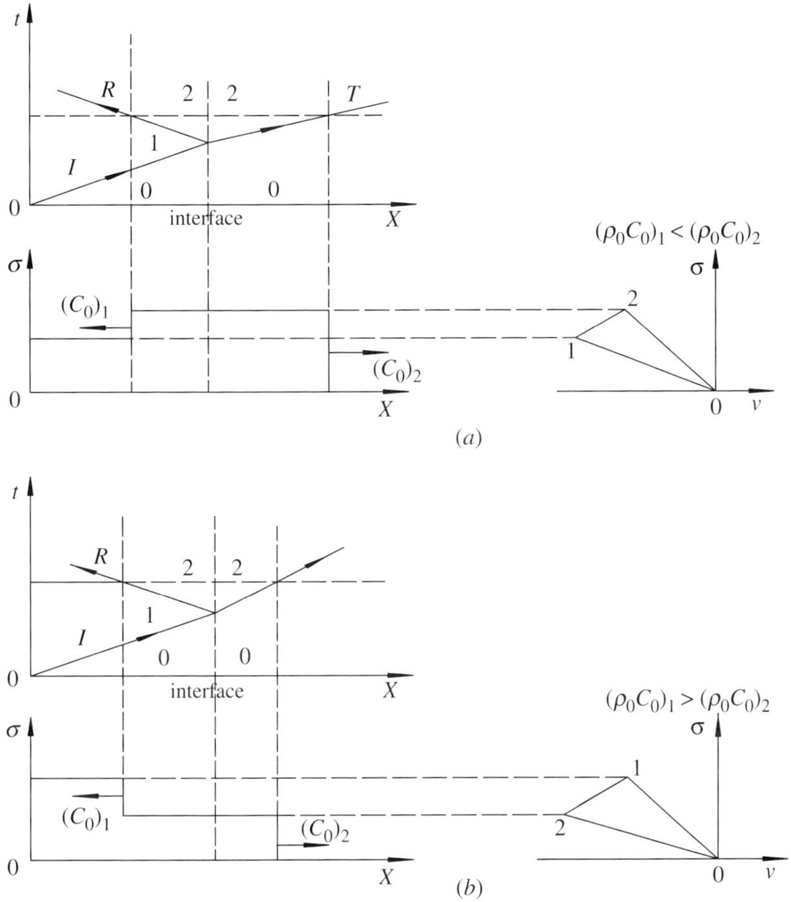

Fig. 3.7. Wave reflection and transmission in bars of different materials.

Based on the momentum conservation condition across wave front [Eq. (2.57)], Eq. (3.6) can be written as

$$\frac{\Delta\sigma_I}{(\rho_0 C_0)_1} - \frac{\Delta\sigma_R}{(\rho_0 C_0)_1} = \frac{\Delta\sigma_T}{(\rho_0 C_0)_2} \tag{3.8}$$

Solving the simultaneous Eqs. (3.8) and (3.7), we obtain

$$\left.\begin{aligned} \Delta\sigma_R &= F \cdot \Delta\sigma_I \\ \Delta v_R &= -F \cdot \Delta v_I \end{aligned}\right\} \tag{3.9}$$

$$\left.\begin{aligned} \Delta\sigma_T &= T \cdot \Delta\sigma_I \\ \Delta v_T &= nT \cdot \Delta v_I \end{aligned}\right\} \tag{3.10}$$

where

$$\left.\begin{aligned} n &= \frac{(\rho_0 C_0)_1}{(\rho_0 C_0)_2} \\ F &= \frac{1-n}{1+n} \\ T &= \frac{2}{1+n} \end{aligned}\right\} \tag{3.11}$$

where n is the ratio of wave impedances of two media, F and T are called the **reflection coefficient** and **transmission coefficient** respectively, which depend on the ratio of wave impedances n. It is clear that

$$1 + F = T$$

On the σ–v plane, the leftward σ–v characteristics 0–1 passing point 0 in the first media corresponds to the incident disturbance $\Delta\sigma_I$; the rightward σ–v characteristics 1–2 passing point 1 in the first medium corresponds to the reflection disturbance $\Delta\sigma_R$; while the leftward σ–v characteristics 0–2 passing point 0 in the second medium corresponds to the transmission disturbance $\Delta\sigma_T$. The above approach shows how to determine the state of the interface after reflection and transmission, namely σ_2 and v_2, from the intersection point 2 of the characteristics 1–2 and 0–2, as shown in Fig. 3.7. It should be emphasized that, either from the angle of concept or in the actual analyses of wave reflection and transmission, the reflection disturbance $\Delta\sigma_R$ itself should not be confused with the stress state σ_2 after the reflection.

It should be noted that T is always positive and consequently the sign of the transmission wave is always the same as that of the incident wave, while the sign of F depends on the relative value of the wave impedances of two media. There are two different cases which will be discussed below:

(a) If $n < 1$, namely $(\rho_0 C_0)_1 < (\rho_0 C_0)_2$, then $F > 0$. In such a case, the reflective stress disturbance and the incident stress disturbance have the same sign (**reflection loading**), while the stress of transmitted wave is larger than that of the incident wave ($T > 1$). This is the so-called wave transmission from a "softer" material into a "harder" material, as shown in Fig. 3.7(a).

   In the special case of $(\rho_0 C_0)_2 \to \infty (n \to 0)$, which corresponds to the reflection of elastic waves at the rigid wall (fixed end), there is $T = 2$, $F = 1$.

(b) If $n > 1$, namely $(\rho_0 C_0)_1 > (\rho_0 C_0)_2$, then $F < 0$. In such a case, the reflective stress disturbance and the incident stress disturbance have opposite signs (**reflection unloading**), while the stress of transmitted wave is smaller than that of incident wave ($T < 1$). This is the so-called wave transmission from a "harder" material into a "softer" material, as shown in Fig. 3.7(b). Thereby, we can explain the cushion effect of "softer" materials.

   In the special case of $(\rho_0 C_0)_2 \to 0$ (corresponding to $n \to \infty$), which corresponds to the reflection of an elastic wave at the free surface (free end), there is $T = 0$, $F = -1$.

It should be noticed, for two different media, although their ρ_0 and C_0 are different, so long as their wave impedances are the same, namely $(\rho_0 C_0)_1 = (\rho_0 C_0)_2$ (corresponding to $n = 1$), there is no reflection ($F = 0$) at the interface of two media, which is named the **impedance matching**. In the cases where no wave reflection is expected or required, the requirement of impedance matching should be satisfied in the selection of materials. For instance, in the experimental study of stress wave propagation, if the wave impedance of gauges matches with that of the tested materials, the unwanted disturbances on the measured wave signals can be avoided.

3.6 Reflection and Transmission of Elastic Waves in Bars with Varying Cross Sections

In a bar with different cross sections, when a stress wave passes through a sudden varying section, a wave reflection and transmission occurs, as shown in Fig. 3.8. The stress distribution in the bar before the reflection is shown in Fig. 3.8(a), and the stress distribution after the reflection is shown in Fig. 3.8(b). In such a case, the requirement of stress equality across the interface [Eq. (3.7)] should be replaced by the requirement of total action force equality across the interface, while the requirement of particle velocity continuity across the interface [Eq. (3.8)] is still valid, so we have

$$A_1(\Delta\sigma_I + \Delta\sigma_R) = A_2(\Delta\sigma_T),$$

$$\frac{\Delta\sigma_I}{(\rho_0 C_0)_1} - \frac{\Delta\sigma_R}{(\rho_0 C_0)_1} = \frac{\Delta\sigma_T}{(\rho_0 C_0)_2}$$

Solving these simultaneous equations, we obtain

$$\left.\begin{aligned} \Delta\sigma_R &= F(\Delta\sigma_I) \\ \Delta v_R &= -F(\Delta v_I) \end{aligned}\right\} \tag{3.12}$$

$$\left.\begin{aligned} \Delta\sigma_T &= T(\Delta\sigma_I)A_1/A_2 \\ \Delta v_T &= nT(\Delta v_I) \end{aligned}\right\} \tag{3.13}$$

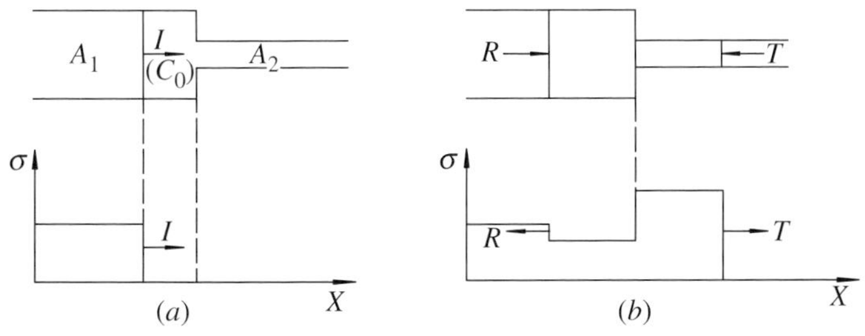

Fig. 3.8. Reflection and transmission of stress waves in a bar with sudden varying cross sections.

$$\left.\begin{aligned} n &= \frac{(\rho_0 C_0 A)_1}{(\rho_0 C_0 A)_2} \\ F &= \frac{1-n}{1+n} \\ T &= \frac{2}{1+n} \end{aligned}\right\} \tag{3.14}$$

where $(\rho_0 C_0 A)$ is called the **generalized wave impedance**.

If the wave impedances on the two sides of an interface are identical and the wave reflection and transmission are caused only by a section-area discontinuity of the bar, such as the case of a stepped bar, then there is $n = A_1/A_2$. Introducing $T_A = nT$, then $\Delta\sigma_T = T_A \Delta\sigma_I$. Since n and T are always positive, and the same for T_A, consequently, the signs of transmission wave and reflective wave are always the same; while the sign of F depends on the relative value of A_1 and A_2.

When a stress wave propagates from a bar segment with a small section into a bar segment with a large section ($A_1 < A_2$ or $n < 1$), the signs of the reflected stress and the incident stress are identical (reflection loading). But, due to $T_A = 2n/\,(1+n) < 1$, the transmission stress is less than the incident stress.

When a stress wave propagates from a bar segment with a large section into a bar segment with a small section ($A_1 > A_2$ or $n>1$), the signs of the reflected stress and the incident stress are opposite (reflection unloading). But the transmission stress is larger than the incident stress ($T_A > 1$).

This is different from the reflection and transmission caused only by the different wave impedance ($\rho_0 C_0$). Based on these, it is clear that, if the larger end of a stepped bar is subjected to an impact loading, then the smaller segment acts as a **wave trap**. However, when $A_2/A_1 \to 0$ ($n \to \infty$), $T_A \to 2$. Thereby, for a multi-stepped bar, when a stress wave passes an interface of section-area discontinuity, the magnification limit of a transmission stress is 2 for each step.

A taper or conical bar can be approximately viewed as a series of stepped bars with strong discontinuous interfaces of section-area. As shown in Fig. 3.9, when a compressive pulse propagates rightward from the bottom of a large section to the top of a small section, each time the stress wave passes through a strong discontinuous interface, the rightward transmitted stress disturbance increases $T_A = [2A_n/(A_n + A_{n+1})]$ times and simultaneously a leftward tension unloading disturbance is reflected at an interface. When this leftward tension unloading disturbance propagates from the small section to the large section, a rightward intensified tension disturbance will be reflected. From these, it can be imagined that with the propagation of a compressive pulse from the large bottom to the small top of the stepped bar, the compressive stress intensity at the pulse front will successively increase, while a tension zone following the compression zone of the pulse will be developed. The tensile stress intensity at the pulse tail zone will also successively increase. Once the tensile stress is large enough, it will result in a tensile break at a certain section near the top and the related segment will fly off. Such a dynamic fracture may take place even before the compressive pulse arrives at the top, so it is different from another form of dynamic fracture, the so-called spallation, which will be discussed in Section 3.9.

An example with calculated data is also shown in Fig. 3.9 (Johnson, 1972). A conical bar is approximated by a stepped bar consisting of several segments with equal length l. The diameter ratios of each segment to the minimum diameter (segment F) are 2, 3, 4, 5, and so on, or the section-area ratios are 4, 9, 16, 25, and so on. Assume that a rectangle compressive pulse of unit intensity ($\sigma_0 = -1$) with pulse length of l propagates rightward from the segment A to the segment B at a wave velocity of C_0. Introducing $T(= l/C_0)$

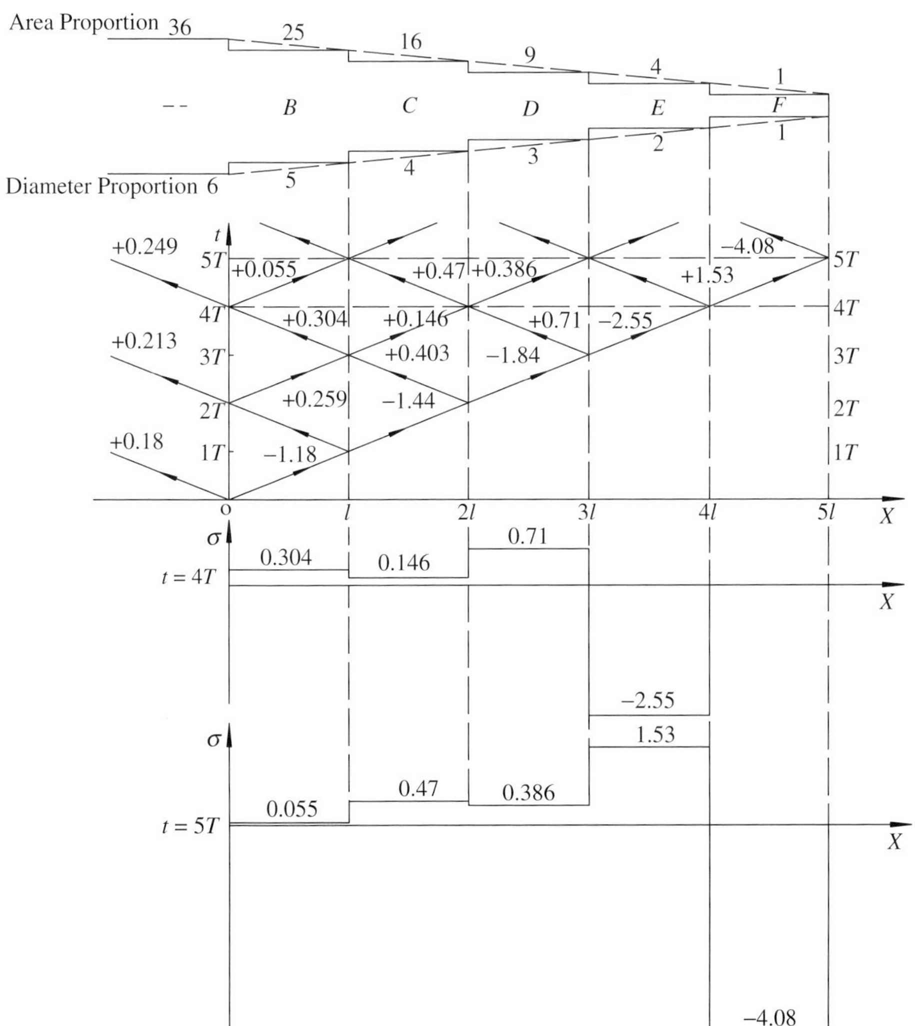

Fig. 3.9. Reflection and transmission of an elastic stress pulse in a stepped bar.

to denote the time duration required for propagating a distance l, then the computational results of the wave propagation process before the stress disturbance reaches the top of the bar, including the stress distribution profiles at $t = 4T$ and $t = 5T$, are shown in the bottom part of Fig. 3.9. It can be seen that, at time $t = 4T$, the intensity of the compressive pulse arriving at the segment E already reaches $\sigma/\sigma_0 = 2.55$, and the tension stress following it already reaches $\sigma/\sigma_0 = -0.71$. At time $t = 5T$, the intensity of the compressive pulse arriving at F further increases to $\sigma/\sigma_0 = 4.08$, and the tension stress following it further increases to $\sigma/\sigma_0 = -1.53$, namely, the tension stress induced by the section-area decrease has exceeded the intensity of incident compressive pulse.

Indeed, after introducing the generalized wave impedance, a longitudinal wave propagating in a conical bar can be approximated by a 1D stress longitudinal plane wave. For an elastic conical bar, assuming that its section area $A(X)$ is a function of coordinate X, its governing equations consist of the continuity equation, the motion equation, and the Hooke's constitutive relation as shown below:

$$\left.\begin{aligned} &\frac{\partial \varepsilon}{\partial t} = \frac{\partial v}{\partial X} \\ &\rho_0 A(X)\frac{\partial v}{\partial t} = \frac{\partial(\sigma A)}{\partial X} = A\frac{\partial \sigma}{\partial X} + \sigma\frac{\partial A}{\partial X} \\ &\sigma = E\varepsilon \end{aligned}\right\} \tag{3.15a}$$

Eliminating ε in these equations and introducing the dimensionless parameters below,

$$\overline{X} = \frac{X}{L}, \bar{t} = \frac{t}{t_L}, \overline{\sigma} = \frac{\sigma}{\sigma_0}, \bar{v} = \frac{v}{v_0}$$

where L is the characteristic length of the conical bar, σ_0 the maximum stress of the incident wave, $t_L = L/C_0$ the characteristic time, $v_0 = \sigma_0/(\rho_0 C_0)$ the characteristic particle velocity, then Eq. (3.15) can be rewritten as the following first-order differential equations with respect to the dimensionless parameter $\overline{\sigma}$ and $\bar{v}$:

$$\left.\begin{aligned} &\frac{\partial \bar{v}}{\partial \bar{t}} - \frac{\partial \overline{\sigma}}{\partial \overline{X}} - \overline{\sigma}\frac{d \ln A}{d\overline{X}} = 0 \\ &\frac{\partial \bar{v}}{\partial \overline{X}} - \frac{\partial \overline{\sigma}}{\partial \bar{t}} = 0 \end{aligned}\right\} \tag{3.15b}$$

According to the theory of characteristics discussed in Chapter 2, it is easy to deduce that the differential equation of characteristics and the differential equation of characteristic compatibility relationships are respectively

$$\left.\begin{aligned} &d\overline{X} \pm d\bar{t} = 0, \\ &d\overline{\sigma} + \overline{\sigma} d \ln A \pm d\bar{v} = 0 \end{aligned}\right\} \tag{3.15c}$$

where the negative sign corresponds to the rightward waves and the positive sign corresponds to the leftward waves. It is clear that, the cone shape $A(X)$ influences the stress wave propagation through the item $\ln A(X)$ in the characteristic compatibility relationship, which results in the distortion of wave profile.

Once the $A(X)$ has been given, by using the characteristics numerical method, the problem is solvable. For example, consider an elastic straight conical bar, its radius and area of the large end are R_L and A_L, respectively; and its radius and area of the small end are R_S and A_S, respectively. If we introduce the dimensionless area $\overline{A} = A/A_L$ and the dimensionless radius $\overline{R} = R_L/R_S$, then $A(X)$ can be expressed in a dimensionless form as below

$$\overline{A} = \frac{A(X)}{A_L} = \left[1 - \left(1 - \frac{R_S}{R_L}\right)\frac{X}{L}\right]^2 = \left[1 - \left(1 - \frac{1}{\overline{R}}\right)\overline{X}\right]^2$$

For a straight conical bar with dimensionless radius $\overline{R} = R_L/R_S = 3$, if its large end is subjected to a triangle stress pulse with a dimensionless pulse length $\bar{\lambda}$ (namely the dimensionless pulse duration $\bar{\tau}$) defined as

$$\bar{\tau} = \frac{\tau}{t_L} = \frac{\tau C_0}{L} = \frac{\lambda}{L} = \bar{\lambda}$$

then the characteristics numerical results of the transmission stress wave at the small end are shown in Fig. 3.10 for $\bar{\tau} = 0.5, 1, 2, 3, 4$, and 6 (Wang and Hu, 1988). It is clear that the conical bar plays a role of a stress amplifier, when a stress wave propagates from the large end to the small end. On the contrary, the conical bar plays a role of a stress attenuator, when a stress wave propagates from the small end to the large end. It can also be seen from Fig. 3.10 that, with the decreasing pulse length of incident stress pulse at the large end, the magnification of dimensionless stress $\overline{\sigma} = \sigma_S/\sigma_0$ at the small end increases, but its upper limit is $\sigma_S/\sigma_L = R_L/R_s$, which can be obtained from Eq. (3.15c) and

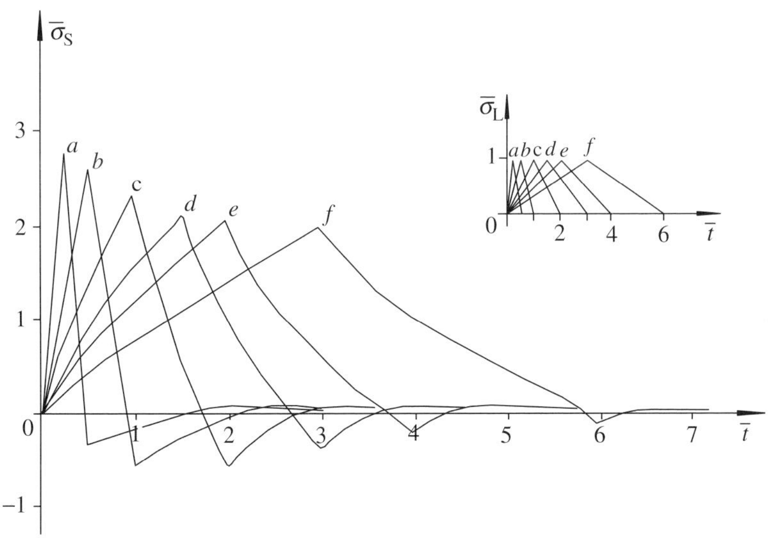

Fig. 3.10. Triangle stress pulses with various pulse lengths propagate in a straight conical bar, showing the stress magnification effect.

the dynamic compatibility condition across a strong discontinuous wave front (Eq. 2.57), namely $[\overline{\sigma}] = \pm[\overline{v}]$ in dimensionless form.

This feature of a conical bar, in modulating the shape and intensity of stress wave, can be applied in stress wave working and profile-modulation in the split Hopkinson pressure bar technique, etc. (Refer to Wang and Hu, 1988; Liu and Hu, 2000).

3.7 Hopkinson Pressure Bar and Flying Piece

Utilizing the property that the compressive pulse is reflected as a tension pulse at the free end of a bar, Hopkinson (1914) presented a crafty test to measure the pressure–time waveform in a bar when one end of the bar is loaded by an explosion or shot by a bullet. The equipment used by him is usually named the Hopkinson pressure bars (HPBs), as schematically shown in Fig. 3.11. Its main component is an elastic cylindrical steel bar, which is horizontally hung by strings and can swing in the perpendicular plane. One end of the bar is subjected to a sudden pressure pulse, which is produced by either an explosion or an impact of a bullet; on the other end the bar connects with a short cylinder (called **timer** or **flying piece**). The connected surfaces of both the bar and the timer have been grinded as smooth as possible. Then the two objects connect together by a little grease or magnetic force so that the connection does not influence the propagation of stress wave on the one hand, but cannot bear tension on the other hand.

Thereby, when a pressure pulse is reflected as a tension pulse at the free end of the timer, a net tension stress can be formed at the bar–timer interface due to the superposition of the incident pressure pulse and the reflected tension pulse, and then the timer flies off with the momentum passing into it. The momentum of the timer can be measured by using a **ballistic pendulum**, as schematically shown in Fig. 3.11, while the momentum kept in the bar can be determined by measuring the swing amplitude of the bar.

Obviously, if the length of the timer is equal to or larger than half of the pulse length of the incident pressure pulse, the entire momentum of the pressure pulse is passed to the timer; consequently, the bar is at rest when the timer flies off. Thereby, by changing the length of

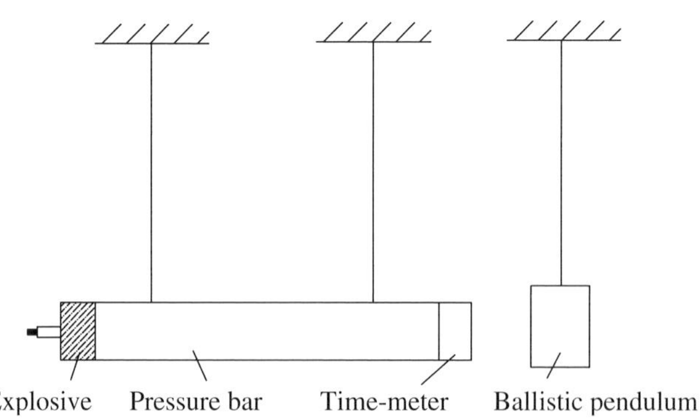

Fig. 3.11. Schematic of Hopkinson pressure bar.

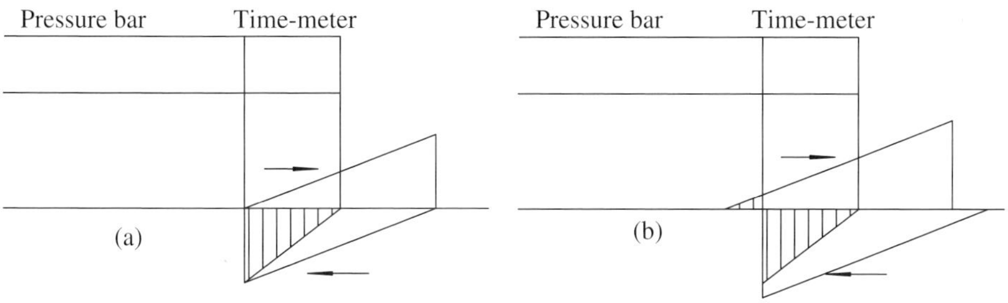

Fig. 3.12. Measuring waveform of a pressure pulse by using timers of different lengths.

the timer, finding the minimum length l_0 of the timer when the bar is at rest while the timer flies off, we obtain the pulse length $\lambda = 2l_0$, or the pulse duration $\tau = \lambda/C_0 = 2l_0/C_0$ of the incident pressure pulse. Figure 3.12(a) shows the superposition of the incident wave and the reflected wave in such a case. If the length of the timer is smaller than l_0, then only a part of the momentum of the pressure pulse is passed to the timer, as shown in Fig. 3.12(b). Note that the momentum m corresponds to the area under the pressure–time curve of stress wave, namely

$$m = A \int_0^T \sigma(t)dt$$

where A is the section area of the timer (or the bar), T is the time duration that a stress wave propagates back-and-forth in the timer. Therefore, by using a series of timers of different length, the waveform of a pressure pulse can be obtained approximately.

Certainly, the precision of the waveform of a pressure pulse measured by using timers of different length is not as high as that of the σ–t relationship measured directly, and the connection between the bar and the timer can also make errors. In principle, the maximum pressure measured cannot exceed the yield strength of the bar material. Moreover, the bar diameter should be much smaller than the pulse length, otherwise, the assumption that transverse inertia is negligible, required in the theory of 1D stress waves cannot be satisfied, in which case there will be a stress dispersion. These limit the application of Hopkinson pressure bar. However, before the establishment of the modern technology used to measure the pressure pulse $\sigma(t)$ directly, this experimental method was very helpful. Up till now, various kinds of improved Hopkinson pressure bars, especially the Split Hopkinson Pressure Bar (**SHPB**) are still being applied and developed, which will be discussed in the next section.

On the other hand, the flying-plate technology based on the timer flying-off principle has been widely applied to measure the particle velocity caused by the wave reflection at the free surface, or applied as a high-velocity impact technique. For example, the velocity of a flying-plate driven by explosion can reach to up 5 mm/μs. The pressure generated under such high velocity reaches up to 200 GPa, which is much higher than the velocity induced by the contact explosion. Thereby, the flying-plate technology is of significance in the research of state equation of materials under high pressure.

3.8 Split Hopkinson Pressure Bar

Kolsky (1949) first modified the Hopkinson pressure bar discussed in the previous section to study the dynamic material behavior at high strain rates and their mathematical expressions, namely, the dynamic constitutive relations of materials. He established an experimental technology, called the **split Hopkinson pressure bar (SHPB)** or **Kolsky bar**.

It should be pointed out that, to distinguish from the research on the static behavior of materials, in the research on the dynamic behavior of materials at high strain rates, in general, two dynamic effects are considered, namely, the inertia effect in structures and the strain rate effect in materials. The key to the problem is how to distinguish these two effects. The main difficulty is that these two effects are usually interconnected, interdependent, or intercoupled, which make the problem much more complicated. In fact, on the one hand, no wave propagation in structures can be analyzed without knowing the corresponding constitutive relation of material, and consequently the basic characteristics of wave propagation inevitably depend on the mechanical behavior of materials, as discussed earlier. On the other hand, in the study of dynamic constitutive relationships of materials at high strain rates, the stress wave propagation either in the specimens or in the experimental equipments should not be neglected. Thus, we have to handle a hot potato – a kind of recurrence of "whether hen before egg or egg before hen?"

How to solve this circular argument? In the view of the research of dynamic constitutive behavior of materials, there are two approaches at present. The first one uses specimens of simple structures that are easy to analyze with the theory of stress wave propagation. With the known explosion or impact loading (namely the initial and boundary conditions are known), experimentally measure the dynamic information (such as the wave signals) or the residual information (such as the residual strain distribution) of the wave propagation. Based on these, we can deduce the dynamic constitutive relationship of materials. In principle, this approach is to determine the dynamic constitutive relationship of materials from the information of stress wave propagation, namely, to solve the so-called "second kind of reverse problem". The relevant discussion will be conducted in the following sections.

The key to the second approach is to try to uncouple the stress wave effect and the strain rate effect in experiments. Experiments using the split Hopkinson pressure bar (SHPB) that will be discussed in this section are the typical examples of this approach and has been extensively used. Its schematic is shown in Fig. 3.13, where the strike bar (bullet), input bar (incident bar), and output bar (transmission bar) is generally of identical diameter and material, namely the elastic modulus E, wave velocity C_0, and wave impedance $\rho_0 C_0$ are identical, and are required to keep in an elastic one-dimensional stress state. During an experiment, a short specimen is sandwiched between the long input bar and the long output bar. When a strike bar of length L_0 driven by an air-gun hits the input bar at velocity v^*, an incident pulse $\sigma_I(t)$ is generated. The amplitude of the incident pulse can be adjusted by controlling the impact velocity v^*, while its duration ($= 2L_0/C$) can be adjusted by controlling the length L_0 of the strike bar. The short specimen is deformed in high strain rate under the action of incident pulse. Meanwhile, a reflected pulse $\sigma_R(t)$ propagates back to the input bar and a transmission pulse $\sigma_R(t)$ propagates to the output bar.

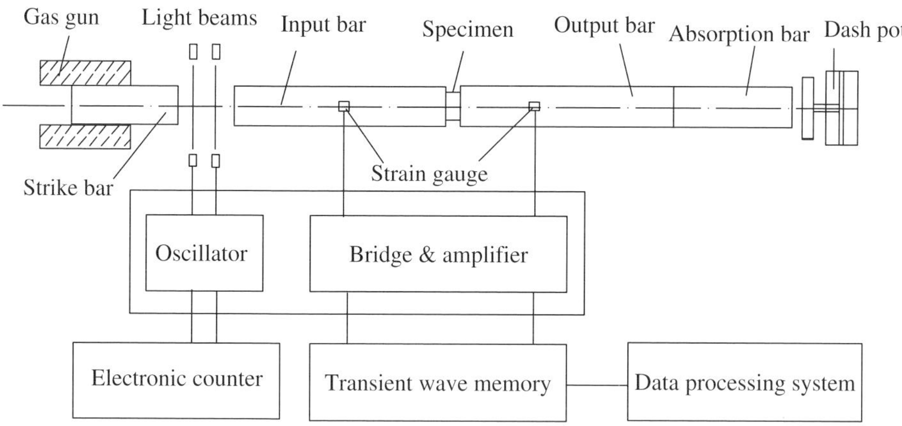

Fig. 3.13. Schematic of the split Hopkinson pressure bar (SHPB) set up.

These two pulses reflect the dynamic mechanical behavior of the specimen. The signals of these two pulses are measured and recorded by a system mainly consisting of two strain gauges mounted on the surfaces of the two bars respectively, an instrument for super-dynamic strain measurement and a transient wave memory, etc. while the impact velocity is measured by a system consisting of a pair of parallel light beams and photoelectric cell, an amplification circuit, an electronic time counter, etc. The absorption bar works like the flying plate (timer) as shown in Fig. 3.11. When the transmitted pulse is reflected at the free end of the absorption bar, the bar flies off with all the momentum passed to it so that it keeps the output bar at rest. All the transmitted energy is finally dissipated by a dash-pot, as shown in Fig. 3.13.

It should be emphasized that the SHPB technique is established on two basic assumptions, namely (1) one-dimensional (1D) stress state in bars, and (2) uniform distribution of stress–strain along the length of the short bar.

We will examine the first assumption. Under the condition that the 1D stress assumption is satisfied, once the stress $\sigma(X_1, t)$ and particle velocity $v(X_1, t)$ at the interface X_1 between the specimen and the input bar (Fig. 3.14) as well as the stress $\sigma(X_2, t)$ and the particle velocity $v(X_2, t)$ at the interface X_2 between the specimen and the output bar (Fig. 3.14) are measured, the average stress $\sigma_s(t)$, strain rate $\dot{\varepsilon}_s(t)$ and strain $\varepsilon_s(t)$ of the specimen

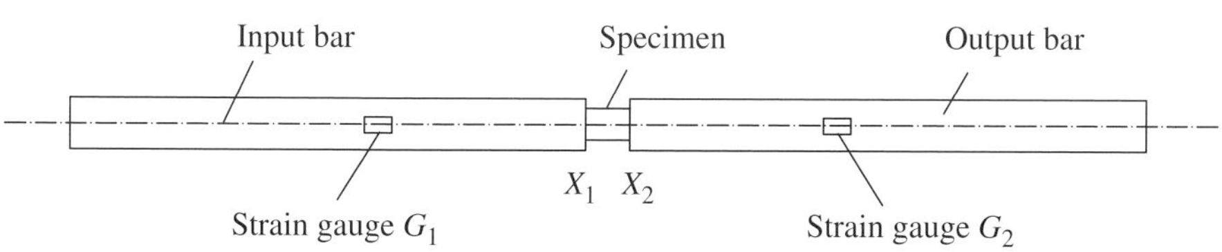

Fig. 3.14. Layout of input bar, specimen, and output bar.

can then be determined according to the equations below,

$$\sigma_s(t) = \frac{A}{2A_s}\left[\sigma(X_1,t) + \sigma(X_2,t)\right] = \frac{A}{2A_s}\left[\sigma_I(X_1,t) + \sigma_R(X_1,t) + \sigma_T(X_2,t)\right] \tag{3.16a}$$

$$\dot{\varepsilon}_s(t) = \frac{v(X_2,t) - v(X_1,t)}{l_s} = \frac{v_T(X_2,t) - v_I(X_1,t) - v_R(X_1,t)}{l_s} \tag{3.16b}$$

$$\varepsilon_s(t) = \int_0^t \dot{\varepsilon}_s(t)dt = \frac{1}{l_s}\int_0^t \left[v_T(X_2,t) - v_I(X_1,t) - v_R(X_1,t)\right]dt \tag{3.16c}$$

where A is the section area of pressure bar, A_s the section area of specimen, and l_s the length of specimen. In the case of **elastic pressure bar**, from the previous discussion on the 1D elastic waves in this chapter, it has been established that there exists the following **linear relationship** between the strain and the stress as well as the particle velocity,

$$\left.\begin{aligned} &\sigma_1 = \sigma(X_1,t) = \sigma_I(X_1,t) + \sigma_R(X_1,t) = E\left[\varepsilon_I(X_1,t) + \varepsilon_R(X_1,t)\right] \\ &\sigma_2 = \sigma(X_2,t) = \sigma_T(X_2,t) = E\varepsilon_T(X_2,t) \\ &v_1 = v(X_1,t) = v_I(X_1,t) + v_R(X_1,t) = C_0\left[\varepsilon_I(X_1,t) - \varepsilon_R(X_1,t)\right] \\ &v_2 = v(X_2,t) = v_T(X_2,t) = C_0\varepsilon_T(X_2,t); \end{aligned}\right\} \tag{3.17}$$

Then, the problem is converted to how to measure the incident strain wave $\varepsilon_I(X_1,t)$ and the reflected strain wave $\varepsilon_R(X_1,t)$ at the interface X_1 and the transmitted strain wave $\varepsilon_T(X_2,t)$ at the interface X_2. However, so long as the pressure bar is in elastic state, all waveforms at different positions are the same. In other words, by using the characteristic that there is no wave profile distortion when an elastic wave propagates in a long bar in the 1D stress state, the incident strain wave $\varepsilon_I(X_1,t)$ and the reflected strain wave $\varepsilon_R(X_1,t)$ at the interface X_1 can be replaced by the incident strain wave $\varepsilon_I(X_1,t)$ and the reflected strain wave $\varepsilon_R(X_1,t)$ measured by the gauge G_1 mounted at X_{G1} on the input bar, and the transmitted strain wave $\varepsilon_T(X_2,t)$ at the interface X_2 can be replaced by the transmitted strain $\varepsilon_T(X_{G2},t)$ measured by the gauge G_2 mounted at X_{G2} on the output bar. Therefore, we can determine the dynamic stress $\sigma_s(t)$ and strain $\varepsilon_s(t)$ of the specimen through the signals measured by the gauge G_1 and the gauge G_2.

$$\sigma_s(t) = \frac{EA}{A_s}\varepsilon_T(X_{G2},t) = \frac{EA}{A_s}\left[\varepsilon_I(X_{G1},t) + \varepsilon_R(X_{G1},t)\right] \tag{3.18a}$$

$$\varepsilon_s(t) = -\frac{2C_0}{l_s}\int_0^t \varepsilon_R(X_{G1},t)dt = \frac{2C_0}{l_s}\int_0^t \left[\varepsilon_I(X_{G1},t) - \varepsilon_T(X_{G2},t)\right]dt \tag{3.18b}$$

It should be pointed out that when a reflected wave generated at the interface X_1 and a transmitted wave generated at the interface X_2 propagate in the input bar and the output bar respectively, simultaneously there are stress waves propagating back-and-forth in the specimen between the interface X_1 and the interface X_2. We can imagine that if the specimen is short enough, the stress–strain is quickly distributed uniformly along its length and then the stress wave effect in it can be neglected. This is the second assumption that

SHPB experiments are based on. According to this assumption of uniform stress–strain distribution, there is $\sigma_1 = \sigma_2$; or combining the theory of 1D stress waves, there is

$$\sigma_I + \sigma_R = \sigma_T, \qquad \varepsilon_I + \varepsilon_R = \varepsilon_T \tag{3.19}$$

Then, if any two of the incident strain wave $\varepsilon_I(X_1, t)$, the reflected strain wave $\varepsilon_R(X_1, t)$ and the transmission strain wave $\varepsilon_T(X_2, t)$ have been measured, we can determine the dynamic stress $\sigma_s(t)$ and strain $\varepsilon_s(t)$ according to Eq. (3.18). After eliminating the time parameter t, we obtain a dynamic stress–strain curve, σ_s–ε_s, of the specimen material at high strain rate.

From the above discussion, it is known that the art of SHPB technology lies in that it **uncouples the stress wave effect and the strain rate effect**. On the one hand, with respect to the input bar and the output bar which play the roles of both impact loading and dynamic measuring, since they are always in elastic state, only the stress wave propagation should be considered while the strain rate effect can be neglected. So long as the bar diameter is small enough, the theory of 1D stress waves can be applied to analyze these experiments. On the other hand, if the specimen sandwiched between the input bar and the output bar is short enough so that the specimen can be viewed as in the state of uniform deformation, then, only the strain rate effect should be considered for the specimen while the stress wave effect can be neglected. In fact, if the specimen is so short that the duration for a stress wave propagating between the two ends of the specimen is much shorter than the loading duration due to the incident wave, then the specimen can be approximately regarded as in a state of uniform deformation. Thus, the stress wave effect and the strain rate effect in the pressure bars and the specimen are cleverly uncoupled respectively, and consequently the rate-dependent mechanical response of specimen material can be obtained by analyzing the signals of stress wave propagating in elastic bars. For the specimen, this corresponds to conducting a “quasi-static” test, but at high strain rates; for the pressure bar, this corresponds to solving an inverse problem, namely, to deduce the constitutive response of the adjacent specimen from the signals of wave propagation.

Certainly, the assumption of 1D stress waves in bars and the assumption of uniform distribution of stress–strain along the specimen length (or henceforth the **uniformity assumption** for short) are also the constraints to assure the validity and reliability of SHPB experiments. Considering how to satisfy the assumption of 1D stress waves in bars, the key is whether the transverse inertia effects can be neglected; readers can refer to Section 2.8. In the following, we will examine the factors influencing the uniformity assumption based on the theory of elastic wave propagation in bars.

Obviously, once the specimen is loaded by an incident pulse, the earlier the uniformity assumption is satisfied, the more reliable the experimental results will be. Obviously, it would be perfect if the uniformity assumption could be satisfied when the specimen is still in an elastic state. Therefore, in the analysis below, we assume that the input bar–specimen–output bar in Fig. 3.14 are all in elastic state and denote the generalized elastic wave impedance by $(\rho CA)_B$ and $(\rho CA)_S$ for the pressure bar and specimen respectively. Without loss of generality, we also assume that the section areas of pressure bar and specimen are identical. As discussed in Section 3.5, the wave reflection and transmission in a system of input bar–specimen–output bar mainly depend on the ratio of $(\rho C)_B$ and ρC_S.

Thus, based on the analysis of elastic waves discussed earlier in this chapter, the process of reflection and transmission of elastic waves in the input bar–specimen–output bar system, and the state of stress σ and particle velocity v in each step, can be determined and graphically displayed on the X–t plane and σ–v plane, as shown in Fig. 3.15(a and b). In these figures, it has been assumed that the strike bar hits the input bar at velocity v_0, and so a strong discontinuous elastic wave (with a rectangle wave front) is generated, the stress amplitude of which is $\sigma_A = -(\rho C)_B v_0/2$.

According to Eqs. (3.9)–(3.11), when an incident wave σ_A propagates with the elastic wave velocity $(C)_B$ in the input bar and arrives at the interface X_1, the primary transmission and reflection occur. The transmitted wave passes to the specimen at the elastic wave velocity $(C)_S$ and induces a strong discontinuous stress disturbance

$$\Delta\sigma_1 = \sigma_1 - 0 = T_{B-S}\sigma_A$$

where the transmission coefficient T_{B-S} can be calculated according to Eq. (3.11)

$$T_{B-S} = \frac{2}{1 + n_{B-S}}, \qquad n_{B-S} = \frac{(\rho C)_B}{(\rho C)_S}$$

where the subscript "B–S" of the transmission coefficient T and the ratio of wave impedance n indicates that the stress waves propagate from bar B to the specimen S.

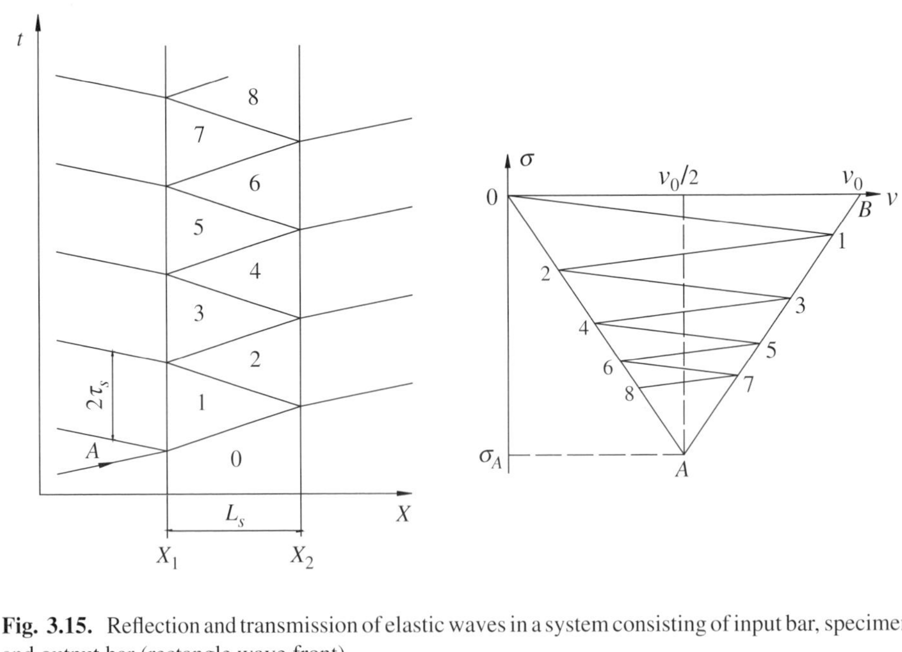

Fig. 3.15. Reflection and transmission of elastic waves in a system consisting of input bar, specimen, and output bar (rectangle wave front).

At time $\tau_S = L_S/(C)_S$, where L_S is the length of the specimen, the secondary transmission and reflection occur. According to Eq. (3.9), the reflected wave propagating back to the specimen induces a strong discontinuous stress disturbance

$$\Delta\sigma_2 = \sigma_2 - \sigma_1 = F_{S-B}\Delta\sigma_1$$

where the reflection coefficient F_{S-B} can be calculated according to Eq. (3.11)

$$F_{S-B} = \frac{1 - n_{S-B}}{1 + n_{S-B}}$$

where the subscript "S–B" of the reflection coefficient F indicates that the stress waves propagate from the specimen S to bar B. Noticing that n_{B-S} and n_{S-B} are reciprocal, and rewriting n_{S-B} as β, it can be easily proved that there exist the following relationship between T_{B-S} and F_{S-B} and can be written respectively as:

$$\left.\begin{aligned} T_{B-S} &= \frac{2}{1 + n_{B-S}} = \frac{2\beta}{1+\beta} \\ F_{S-B} &= \frac{1-\beta}{1+\beta} \\ 1 - F_{S-B} &= 1 - \frac{1-\beta}{1+\beta} = \frac{2\beta}{1+\beta} = T_{B-S} \end{aligned}\right\} \tag{3.20}$$

When the reflected wave returns to the interface X_1, the third transmission and reflection occur. The strong discontinuous stress disturbance induced in the specimen is

$$\Delta\sigma_3 = \sigma_3 - \sigma_2 = F_{S-B}\Delta\sigma_2 = F_{S-B}^2\Delta\sigma_1$$

On this analogy, when the kth transmission and reflection occur, a strong discontinuous stress disturbance induced in the specimen is

$$\Delta\sigma_k = \sigma_k - \sigma_{k-1} = F_{S-B}\Delta\sigma_{k-1} = F_{S-B}^{k-1}\Delta\sigma_1 \tag{3.21}$$

while the final state of stress σ_k in the zone k (Fig. 3.15) after kth transmission and reflection is

$$\sigma_k = \sum_{i=1}^{k}\Delta\sigma_i = \left(1 + F_{S-B} + F_{S-B}^2 + F_{S-B}^3 + \cdots + F_{S-B}^{k-1}\right)\Delta\sigma_1 \tag{3.22a}$$

By using the expansion of a binomial below

$$1 - x^k = (1-x)\left(1 + x + x^2 + x^3 + \cdots + x^{k-1}\right)$$

and considering the relationship between T_{B-S} and F_{S-B} given by Eq. (3.20) and the expression of β, Eq. (3.22a) can finally be written as

$$\sigma_k = \frac{1 - F_{S-B}^k}{1 - F_{S-B}} \Delta\sigma_1 = \frac{1 - F_{S-B}^k}{1 - F_{S-B}} T_{B-S}\sigma_A = \left(1 - F_{S-B}^k\right)\sigma_A = \left[1 - \left(\frac{1-\beta}{1+\beta}\right)^k\right]\sigma_A \tag{3.22b}$$

This indicates that after k-times back-and-forth transmissions and reflections in the specimen, the stress σ_k in the specimen depends on both the number of transmission–reflection times and the ratio of impedances of specimen and pressure bar. Note, the number of times k indeed equals the non-dimensional time $\bar{t} = t/\tau_S = t(C)_S/L_S$.

For a given β, when k is an even number (refer to Fig. 3.15), Eq. (3.22) shows how the transmission stress at the interface X_2 varies with the number of transmission–reflection times k or the dimensionless time $\bar{t} = t/\tau_S = t(C)_S/L_S$), while when k is an odd number, Eq. (3.22) shows how the reflective stress at the interface X_1 varies with the number of transmission–reflection times k or the dimensionless time $\bar{t}$.

As an example, when $\beta = 1/10$, according to Eq. (3.22), the variation of dimensionless stress σ/σ_A with k (or $\bar{t}$) at the interface X_1 and interface X_2 are shown in Fig. 3.16(a and b). Both of them gradually approach 1, which indicates that the stresses along the specimen length gradually realize a uniform distribution, namely it is a process depending on both β and k.

Noticing that Eq. (3.21) just shows the stress difference between that at the interface X_1 and at the interface X_2 (refer to Fig. 3.15), a dimensionless stress difference (relative stresses) for the two ends of the specimen can be defined as

$$\alpha_k = \frac{\Delta\sigma_k}{\sigma_k} \tag{3.23}$$

For a rectangular strong discontinuous incident wave, substituting Eqs. (3.21) and (3.22) into Eq. (3.23), we obtain

$$\alpha_k = \frac{\Delta\sigma_k}{\sigma_k} = \frac{F_{S-B}^{k-1}}{\dfrac{1 - F_{S-B}^k}{1 - F_{S-B}}} = \frac{\left(\dfrac{1-\beta}{1+\beta}\right)^{k-1}\left(1 - \dfrac{1-\beta}{1+\beta}\right)}{1 - \left(\dfrac{1-\beta}{1+\beta}\right)^k} = \frac{2\beta\,(1-\beta)^{k-1}}{(1+\beta)^k - (1-\beta)^k} \tag{3.24}$$

This formula depicts analytically how the relative stress difference for the two ends varies with the impedance ratio β and the number of transmission–reflection times k. For different values of β (= 1/2, 1/4, 1/6, 1/10, 1/25, 1/100), the variation of α_k with k calculated according to Eq. (3.24) is shown in Fig. 3.17. It is clear that the α_k increases with decreasing β for a given k, or in other words, with the decrease of the impedance ratio β, stress waves must be reflected more times in the specimen in order to satisfy the requirement of the uniformity assumption.

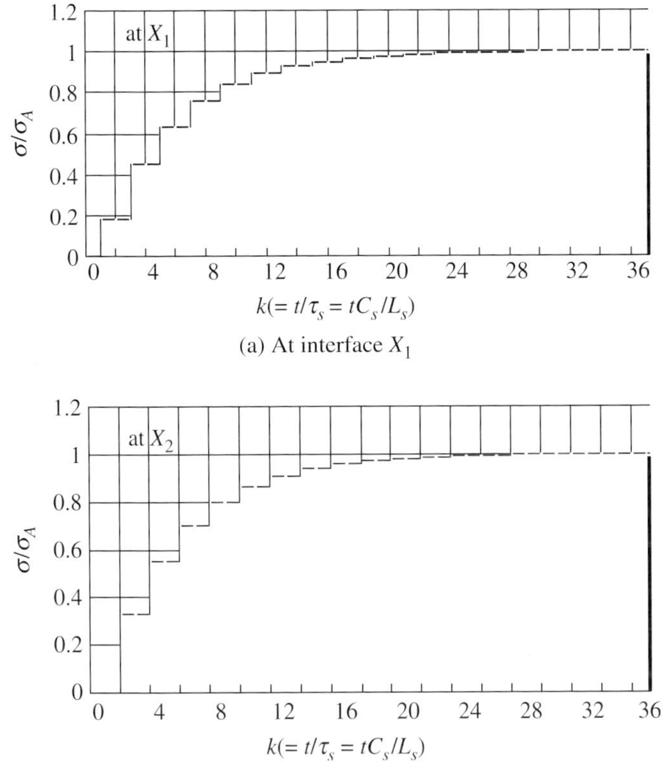

Fig. 3.16. Variation of the dimensionless stress σ/σ_A with k (or $\bar{t}$) at the interface X_1 and the interface X_2, when $\beta = 1/10$.

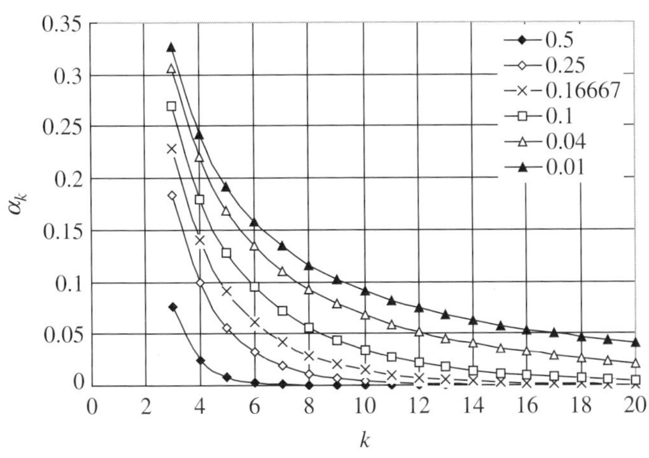

Fig. 3.17. Variation of the dimensionless stress difference α_k with the wave impedance ratio β and the number of reflection times k (for a rectangular incident wave front).

As suggested by Zhou *et al* (1992a) and Ravichandran and Subhash (1994), if the stress (strain) distribution along the bar can be regarded as approximately satisfying the uniformity assumption when $\alpha_k \leq 5\%$, then it can be seen from Fig. 3.17 that in the case of $\beta = 1/2$ the minimum number of reflective times is $k_{\min} = 4$, while in the case of $\beta = 1/100$, the minimum number of reflective times increases to $k_{\min} = 18$.

The above results are obtained for the rectangular strong discontinuous incident wave. However, in most SPHB experiments, the incident wave with a trapezoidal front, namely with a certain "rise time" t_r, are usually observed. For such forms of waves, the analysis methods similar to those used in the above discussion are still valid, although the problem becomes more complicated. Assuming that the rise time of the trapezoidal wave front t_r just equals the time an elastic wave propagates around the specimen, namely $t_r = 2\tau_S = 2L_S/(C)_S$, Yang and Shim (2005) obtained an analytical solution for the case after an elastic wave propagates back-and-forth one time in the specimen ($k > 2$), as

$$\alpha_k = \frac{2\beta^2(1-\beta)^{k-2}}{(1+\beta)^k - (1-\beta)^{k-2}}, \quad k > 2 \tag{3.25}$$

For different values of β ($= 1/2, 1/4, 1/6, 1/10, 1/25, 1/100$), the variation of α_k with k calculated according to Eq. (3.25) is shown in Fig. 3.18. It is clear that, contrary to the case for a rectangular incident wave (Fig. 3.17), the α_k–k curves now move downwards with decreasing β, namely for a given k the α_k decreases with decreasing β. It is worthwhile to notice that in the range of β discussed in this example, the minimum number of reflection times for satisfying the uniformity assumption is only 3–4.

Let us now consider a linear ramp wave, namely an incident wave with a longer rise time. Assume that the rise time t_r linearly increases with time, namely, it can be depicted by $\sigma_I(t) = \sigma^* t/\tau_S = \sigma^*(C)_S t/L_S$, where σ^* denotes the stress when $t = \tau_S = L_S/(C)_S$.

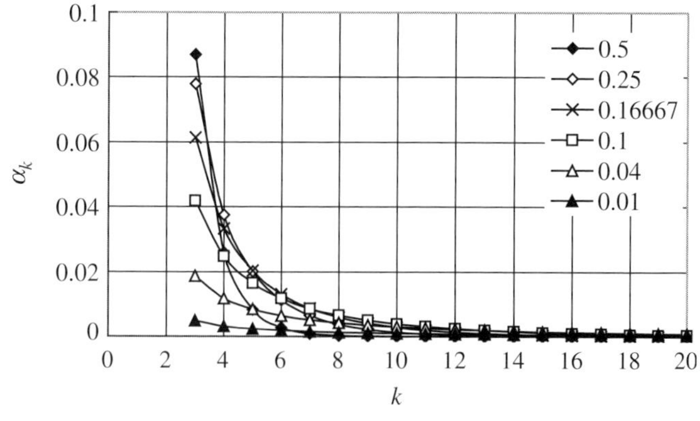

Fig. 3.18. Variation of the dimensionless stress difference α_k with the wave impedance ratio β and the numbers of reflection times k (for a trapezoidal incident wave front).

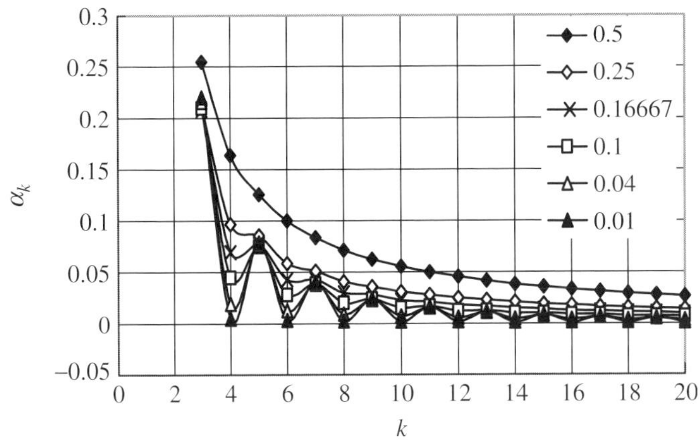

Fig. 3.19. Variation of the dimensionless stress difference α_k with the wave impedance ratio β and the numbers of reflection times k (for a linear ramp incident wave front).

Yang and Shim (2005) presented an analytical solution as below,

$$\alpha_k = \frac{2\beta^2\left[1-\left(-\dfrac{1-\beta}{1+\beta}\right)^k\right]}{2k\beta - 1 + \left(\dfrac{1-\beta}{1+\beta}\right)^k} \tag{3.26}$$

For different values of β $(= 1/2, 1/4, 1/6, 1/10, 1/25, 1/100)$, the variation of α_k with k calculated according to Eq. (3.26) is shown in Fig. 3.19. It is clear that the α_k–k curves for the linear ramp wave also move downwards with decreasing β, but are accompanied by a marked oscillation of the curve. In addition, in the range of β discussed in this example, the minimum number of reflection times for satisfying the uniformity assumption in the present case is larger than that for the trapezoidal wave.

For the same β, the comparison of α_k–k curves between the rectangular wave, the trapezoidal wave, and the linear ramp wave is shown in Fig. 3.20.

It can be seen from Fig. 3.20 that, when $\beta = 0.5$, there is almost no difference between the rectangular wave and the trapezoidal wave, and both are better than the linear ramp wave with respect to satisfying the uniformity assumption; but with the decrease of β, the α_k–k curves of the trapezoidal wave and the linear ramp wave move downwards, but the α_k–k curves of the rectangular wave move upwards, consequently the situation changes. Therefore, when $\beta = 0.1$, the rectangular wave becomes the worst one while the trapezoidal wave becomes the best one with respect to satisfying the uniformity assumption.

From the above discussion on the propagation of stress waves in the system consisting of input bar, specimen, and output bar, both the impedance ratio β and the waveform,

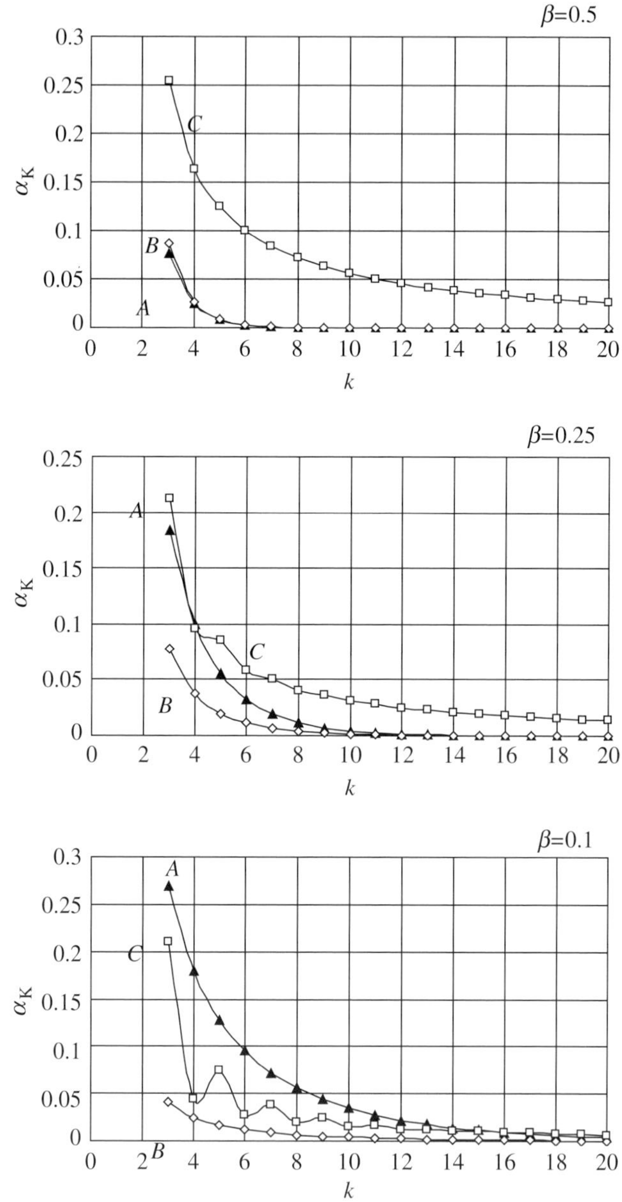

Fig. 3.20. Comparison of α_k–k curves of the rectangle, trapezoidal, and linear ramp wave front (denoted by curve *A*, *B*, and *C* respectively) under an identical β.

particularly the rise time of the incident wave front, markedly influence the minimum number of reflection times $k_{\min}$ for satisfying the uniformity assumption.

3.9 Dynamic Fracture Induced by Reflective Unloading Waves

When a compressive pulse is reflected at the free surface of a bar (or a plane) as a tension pulse, the tension stress in a certain zone near the free surface may be quite large. Once a certain criterion of dynamic fracture (spalling criterion) is satisfied, the material in such a zone will be cracked. When the crack is large enough, a whole cracked piece will fly off with the momentum in it. Such a type of **dynamic fracture** near the back surface of the bar (or plane) caused by the reflection of compressive pulse at the free surface is named **spalling** or **scabbing**. The fly-off crack piece is called **spalling piece** or **scab**. Figure 3.21(a) schemes the back spalling of a thick steel plant caused by a contact explosion.

In the above situation, once a spalling takes place, a new free surface is also formed. The continuously incident compressive pulse will be reflected at this new free surface; consequently, a second spalling could be formed, if the spalling criterion is satisfied. On this analogy a multiple spalling may be formed in a certain condition, and a succession of multilayer scabs will be generated. Figure 3.21(b) schemes a multi-spalling at the free end of a cement bar caused by a contact explosion at another end.

It is necessary to emphasize that a compressive pulse always consists of a compressive loading front and an unloading tail following it. Most engineering materials can undergo quite intense compressive stress waves without fracture but cannot undergo tension stress waves of the same stress amplitude. The occurrence of spalling is attributed to a tension stress that is formed by the reflection of compressive pulse at the free surfaces and this tension stress is large enough to satisfy the dynamic fracture criterion of materials, while the occurrence of tension stress is attributed to the interaction of the reflective tension wave which is formed when an incident compressive wave is reflected at free surface, and the unloading tail of the incident compressive wave; or briefly, the tension stress is caused by the **interaction of the reflected unloading wave and the incident unloading wave**. Therefore, the intensity and shape of incident compressive pulse greatly influence whether the spalling occurs, where it happens, and how many scabs are formed, etc. Indeed, the formation of tension stress is only a prerequisite, whether the spalling occurs depends on whether the tension stress satisfies the **dynamic fracture criterion** of materials.

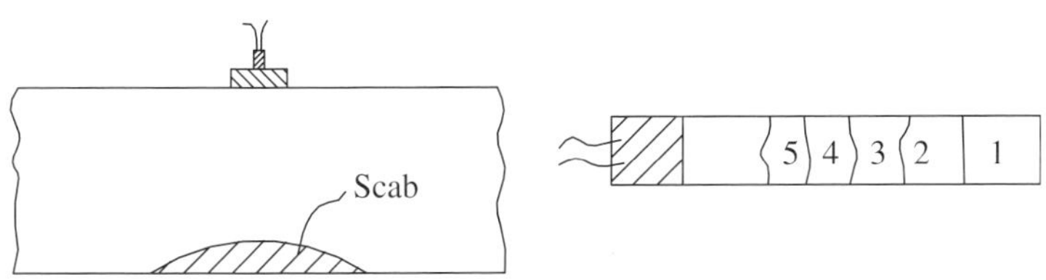

(a) (b)

Fig. 3.21. Spalling caused by the reflection of compressive pulse at free surface.

The first and simplest dynamic fracture criterion of materials is the **maximum tension stress instantaneous fracture criterion**. According to this criterion, once a tension stress σ reaches or exceeds the critical tension stress of materials, σ_c,

$$\sigma \geq \sigma_c \tag{3.27}$$

a spalling occurs instantaneously, where σ_c is the parameter representing the material behavior resisting to dynamic tension fracture, called the **dynamic fracture strength** of materials. This criterion was formerly a generalization of the maximum normal stress criterion in the theory of static strength to the dynamic situation, namely, the fracture is regarded to occur instantaneously when the criterion is satisfied. So it belongs to the rate/time-independent theory of fracture. However, the σ_c is measured in dynamic tests and is generally much higher than the static strength of materials σ_b; in this view, the rate/time-dependency has been taken into consideration in this criterion.

Based on the theory of 1D elastic waves, according to the criterion of instantaneous fracture at maximum tension stress, how the spalling takes place under the compressive pulses with various profiles will be discussed below.

Figure 3.22 schemes the wave profiles at five typical times when an incident rectangular pulse of length λ is reflected at a free surface: (a) an incident rectangular pulse approaches the free surface; (b) when one-fourth of the incident pulse has been reflected, the sum of incident compressive stress and reflected tension stress is zero in the distance range of $\lambda/4$ from the free surface; (c) when half of the incident pulse has been reflected, the stress superposed is just zero, but the particle velocity in the distance range of $\lambda/2$ from the free surface is twice the particle velocity of the incident wave; after this, the tension stress will be generated due to the interaction between the incident unloading wave and the reflected unloading wave; (d) when three-fourth of the incident pulse has been reflected, a tension zone with a length of $\lambda/2$ is formed, while the stress in the distance range of $\lambda/4$ from the free surface still is zero; (e) when the reflection has ended, the rightward

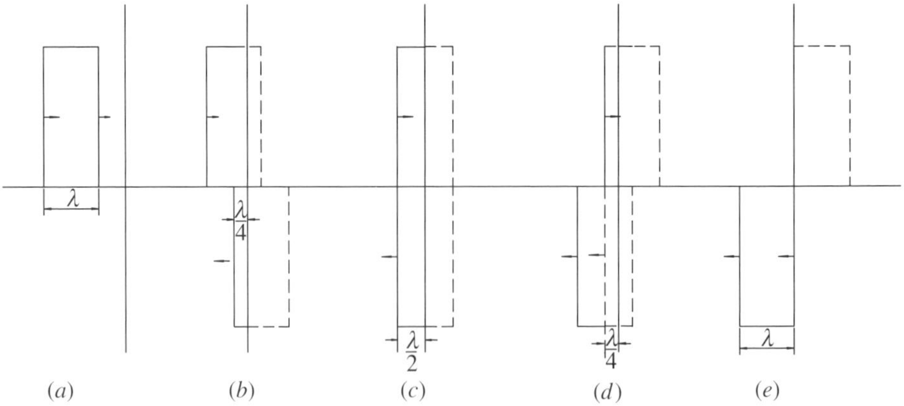

Fig. 3.22. Wave profiles at five typical times when an incident rectangle pulse of length λ is reflected at a free surface.

incident compressive pulse has been totally reflected to a leftward tension pulse. Therefore, for a rectangle pulse, so long as the stress amplitude of the pulse $|\sigma_m|$ is larger than the dynamic fracture strength σ_c, namely, $|\sigma_m| > \sigma_c$, then just after half of the incident wave is reflected at the free surface, namely at time $t = \lambda/(2C_0)$, spalling occurs; the thickness of the spalling piece $\delta = \lambda/2$. Since the spalling piece flies off with the total momentum of incident compressive pulse, on the one hand, the velocity of the flying-off spalling piece v_f can be calculated which is twice the particle velocity of incident wave $[\sigma_m/(\rho_0 C_0)]$:

$$v_f = \frac{\sigma_m \dfrac{\lambda}{C_0}}{\rho_0 \delta} = \frac{2\sigma_m}{\rho_0 C_0}$$

On the other hand, for a rectangle pulse, no matter how large the amplitude of incident pulse is, multiple spalling will not occur.

For a saw-toothed compressive stress pulse with a linearly attenuating tail, the situations are different. When it is reflected at the free surface, the wave profiles at five typical times are schemed in Fig. 3.23. It can be seen that even at the beginning of reflection, a net

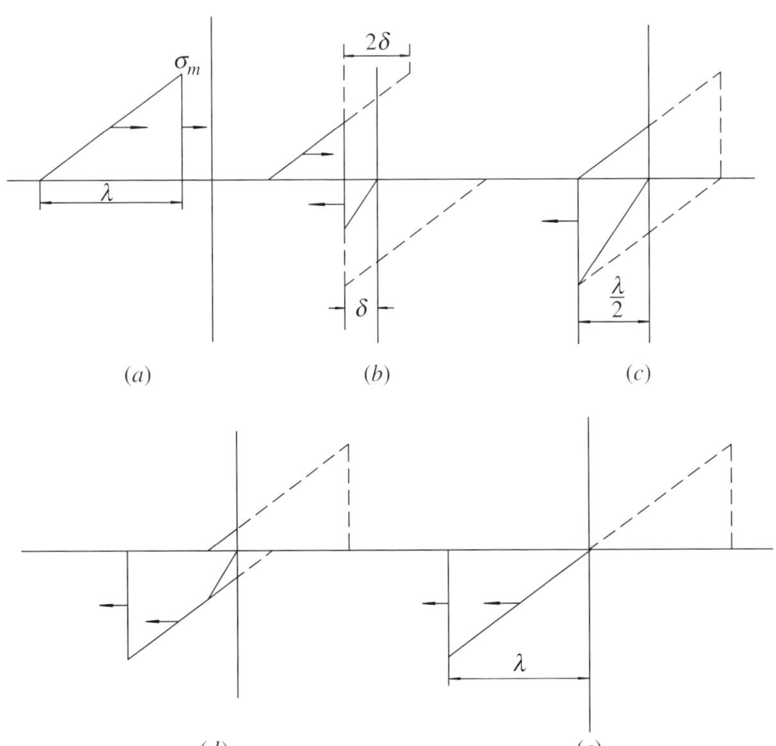

Fig. 3.23. Wave profiles at five typical times when an incident saw-toothed pulse of length λ is reflected at a free surface.

tensile stress zone is simultaneously formed due to the interaction of reflected unloading wave and incident unloading wave. The net tensile stress at the front is larger than other parts of the reflected wave, and it continuously increases with the process of reflection. When half of the incident wave is reflected, it reaches its maximum. However, before this, once Eq. (3.27) is satisfied, spalling occurs.

When a stress pulse is expressed as a function of time, namely it is represented by a time-histories curve $\sigma(t$ at an arbitrary point X, assume that the time when the wave front arrives at the point is the start of time, $t = 0$, if the stress at a distance of δ from the free surface satisfies the instantaneous fracture criterion based on maximum tensile normal stress [Eq. (3.27)], then spalling occurs and obviously there should be

$$\sigma(0) - \sigma\left(\frac{2\delta}{C_0}\right) = \sigma_c \tag{3.28}$$

For the saw-toothed stress pulse with linearly attenuating tail as shown in Fig. 3.23, the time-histories profile $\sigma(t)$ can be expressed as

$$\sigma = \sigma_m\left(1 - \frac{C_0 t}{\lambda}\right) \tag{3.29}$$

where σ_m is the peak value of the stress pulse. According to Eq. (3.28), the thickness of the first spalling piece δ_1 can be determined as,

$$\delta_1 = \frac{\lambda}{2} \cdot \frac{\sigma_c}{\sigma_m}$$

The time t_1 (after the beginning of reflection) that spalling occurs is

$$t_1 = \frac{\delta_1}{C_0} = \frac{\lambda}{2C_0}\frac{\sigma_c}{\sigma_m}$$

The velocity of the spalling piece can be approximately calculated, according to that the momentum of the spalling piece equals the impulse of the part of the pulse entrapped in the spalling piece, namely

$$v_f = \frac{1}{\rho_0 \delta}\int_0^{2\delta/C_0} \sigma(t)dt$$

Substituting Eq. (3.29) into the above equation, we obtain

$$v_f = \frac{2\sigma_m - \sigma_c}{\rho_0 C_0}$$

If $|\sigma_m| = \sigma_c$, similar to the case of a rectangular pulse, then spalling occurs at time $t = \lambda/(2C_0)$ after the beginning of reflection, the thickness of spalling piece $\delta = \lambda/2$, all the pulse impulse is entrapped in the spalling piece and multiple spalling is impossible to occur. But the flying-off velocity of the spalling piece is less than that for an incident

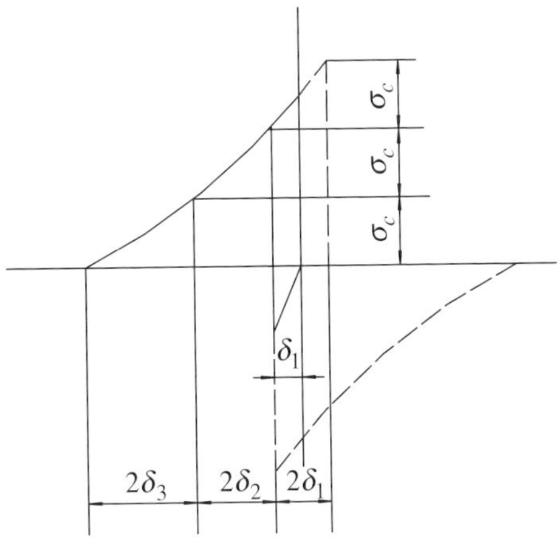

Fig. 3.24. Multiple spalling for a saw-toothed pulse with exponentially attenuating tail.

rectangular pulse. If $|\sigma_m| > \sigma_c$, then a multiple spalling is possible to occur. The readers are suggested to prove that if $n\sigma_c \leq |\sigma_m| < (n+1)\sigma_c$, then the number of spalling is n and the thickness of all spalling pieces are identical, while the flying-off velocity of each spalling piece will successively decrease.

On the analogy of the above analysis, it is easy to illustrate that for a saw-toothed pulse with exponentially attenuating tail as shown in Fig. 3.24, the thickness of each spalling piece will successively increase when multiple spalling occurs.

The above discussion is an approximate analysis based on the theory of 1D elastic waves, practical problems are much complicated. Even so, it can be seen that, because of the interaction of stress waves, dynamic fracture can occur not only in the loading condition but also in the unloading condition; the latter is a kind of the so-called **unloading failure** due to stress waves.

It should also be pointed out that, except for perfect crystals, any fracture or rupture of engineering materials does not occur instantaneously, but is a final result of a time/rate-dependent process developing at a finite velocity; especially at high strain rates, displayed as the so-called **fracture-delay**, the decrease of critical stress with increasing duration of loading, the increase of flow stress with increasing strain rate and so on. This shows that the material break depends on not only the magnitude of applied stress but also on the loading duration or the stress–strain rate. Therefore, in principle an instantaneous fracture criterion of material like Eq. (3.27) should be replaced by a time- or rate-dependent dynamic fracture criterion. This research topic has attracted a lot of attention. One class of time/rate-dependent fracture criterion of material is the damage accumulation criterion in various forms, in which the evolution process of internal damage of material has been considered.

For instance, a time-dependent spalling criterion was presented by Tuler and Butcher (1968), which states that when the stress $\sigma(t)$ as a function of time t at an arbitrary point satisfies the following equation, spalling occurs

$$\int_0^t (\sigma - \sigma_0)^\alpha dt = K \tag{3.30}$$

where α, K, and σ_0 are all material parameters, σ_0 denotes the minimum stress (threshold stress) for spalling. When $\sigma(t) < \sigma_0$, dynamic fracture does not occur however long the duration of the stress pulse is. When $\alpha = 1$, Eq. (3.30) is reduced to the so-called **impulse criterion**, which indicates that dynamic fracture occurs only when the impulse reaches a critical value.

Finally, it should be pointed out that, similar to the back spalling near the free surface of a plate or bar, caused by the tensile stress induced by the interaction between an incident unloading wave and an unloading wave reflected from a free surface, if a compressive

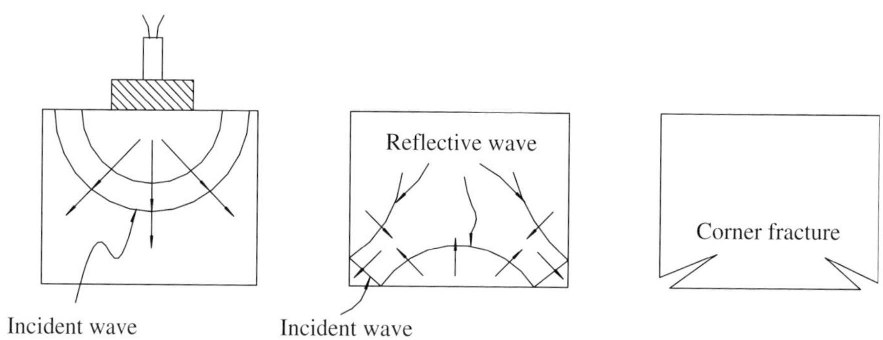

Fig. 3.25. Corner spalling caused by the net tensile stress induced by two unloading waves reflected at the two free surfaces respectively of a corner.

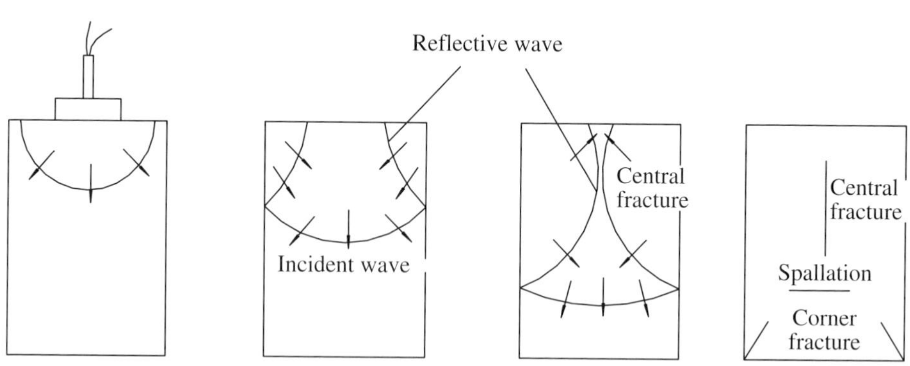

Fig. 3.26. Central spalling occurs due to the interaction of two unloading waves generated by the reflection of a compressive wave at two free surfaces.

wave propagates to a corner enclosed by two free surfaces at a certain angle[1], then net tensile stress may be induced by the interaction between the two unloading waves reflected from these two free surface respectively and then cause a spalling at the corner, which is called the corner spalling, as approximately schemed in Fig. 3.25.

If two unloading tensile waves which are generated by the reflection of an incident compressive wave at two free surfaces respectively, meet at the center of a solid body, the so-called **central spalling** may occur, if the net tensile stress satisfies the spalling criterion, as shown in Fig. 3.26. From the view of fracture mechanism, either the ordinary spalling, or the corner spalling and central spalling are all relevant to the tensile stress inducted by the reflection of stress waves, so all of them can be called **reflective tensile fracture**.

[1]Here, the reflection of stress wave at a free surface is in fact an oblique reflection, which will be discussed in Chapter 11. When a longitudinal wave is incident obliquely, there are both oblique transverse wave and oblique longitudinal wave reflected; and consequently the real case is much more complicated than that discussed here.

CHAPTER 4

Interaction of Elastic–Plastic Longitudinal Waves in Bars

When we study the problems of wave interaction in the case of linear elastic stress waves, as discussed in Chapter 3, there is no need to distinguish whether such a wave interaction is a loading process or an unloading process. This is because for linear elastic material, the stress–strain relationship is linear, identical for both the loading and unloading cases, representing a reversible thermal–mechanical process. Therefore, there is no essential distinction between a linear elastic loading wave and a linear elastic unloading wave; both propagate at speed $C_0 \, (= E/\rho_0)^{1/2}$. Nevertheless, when we start to discuss the interaction between elastic–plastic waves, the problem becomes much more complicated. First, the stress–strain curves are not linear, thus the superimposition principle is no longer held good. On the other hand, the stress–strain curves of an elastic–plastic material are different in the loading and unloading cases, representing irreversible thermal–mechanical processes. As a result, it is necessary to distinguish whether the stress wave interaction is a loading case or an unloading case, and then to treat the respective case differently.

In the following sections, we will first discuss the loading cases of two elastic–plastic wave interaction, and then the unloading cases.

4.1 Interaction of Two Longitudinal Elastic–Plastic Loading Waves in Bars

The interaction between two loading elastic–plastic waves can be categorized as two types: the **head-on loading** and the **pursuing loading**. The former type is the interaction process when two loading waves propagate in the opposite direction and interact with each other when they meet together. Such waves must have the same signs (namely, both are compressive waves or tensile waves), to guarantee that the interaction will load the material further. If the two stress waves have different signs, the interaction will involve unloading process. The latter type of interaction is the process when two loading stress waves propagate in the same direction, and the back one pursues the front one. This process only happens in the material with increasing strain-hardening coefficient, so that the plastic wave velocity increases with the plastic deformation. Such a pursuing loading interaction is actually the process of shock wave formation and growing, which has been qualitatively discussed in the earlier chapter.

In this section we will discuss the **head-on loading** case in detail. It does not matter whether material is of increasing strain-hardening or decreasing strain-hardening. When the reader gets familiar with the basic approaches of the analysis method, it will be easy to treat the **pursuing loading** case.

The general analytical method is, in principle, the same as discussed previously for elastic wave interaction. The interaction between two stress waves is considered as an internal impact process. On the boundaries of interaction between any two stress waves, the equality of material velocities (condition of material continuation) and the equality of stresses (principle of reaction force) should be satisfied. Moreover, the momentum-conservation condition across a wave front, such as Eq. 2.57 for strong-discontinuous wave front or Eq. 2.63 for weak-discontinuous wave front, should be satisfied on both sides of the interaction boundary, respectively. The problem then can be solved from these equations. Different from the linear elastic case, the wave speed $\mathscr{D}$ for strong-discontinuous shock wave, or $C(\varepsilon)$ for weak-discontinuous plastic wave, is no longer constant. Consequently, the principle of superimposition does not hold good any more. Therefore the problem becomes more complex.

4.1.1 Head-on loading interaction of two strong-discontinuous stress waves

First we assume that the material is of linear strain-hardening. In this case both the elastic wave speed $C_0\ (=\sqrt{E/\rho_0})$ and the plastic wave speed $C_1\ (=\sqrt{E_1/\rho_0})$ are constants. The problem is slightly simpler.

Consider a bar with uniform cross section and length l, which is originally in a rest, stress-free condition. At time $t=0$, its two ends are suddenly loaded by a constant-speed impact loading. At the right end of the bar ($X=l$), the boundary condition is $v_3>0$. At the left end of the bar ($X=0$), the boundary condition is $v_4<0$. Therefore, along the bar two strong-discontinuous elastic–plastic tensile waves are propagating, as illustrated in Fig. 4.1. This is a typical case of the interaction between two strong-discontinuous elastic–plastic waves propagating in the opposite directions, namely a head-on loading case.

Before the two waves meet, the elastic–plastic wave propagation is the same as that propagating in an infinite bar, as discussed in the previous chapter. Thus the stress and velocity states of the zones marked as 0, 1, 2, 3, and 4 on the X–t plane in Fig. 4.1 are all known. Let the subscript of the state variables be the same as their zone number, we have:

$$\sigma_0 = v_0 = 0$$

$$\sigma_1 = \sigma_2 = Y$$

$$v_1 = -v_2 = v_Y = \frac{Y}{\rho_0 C_0}$$

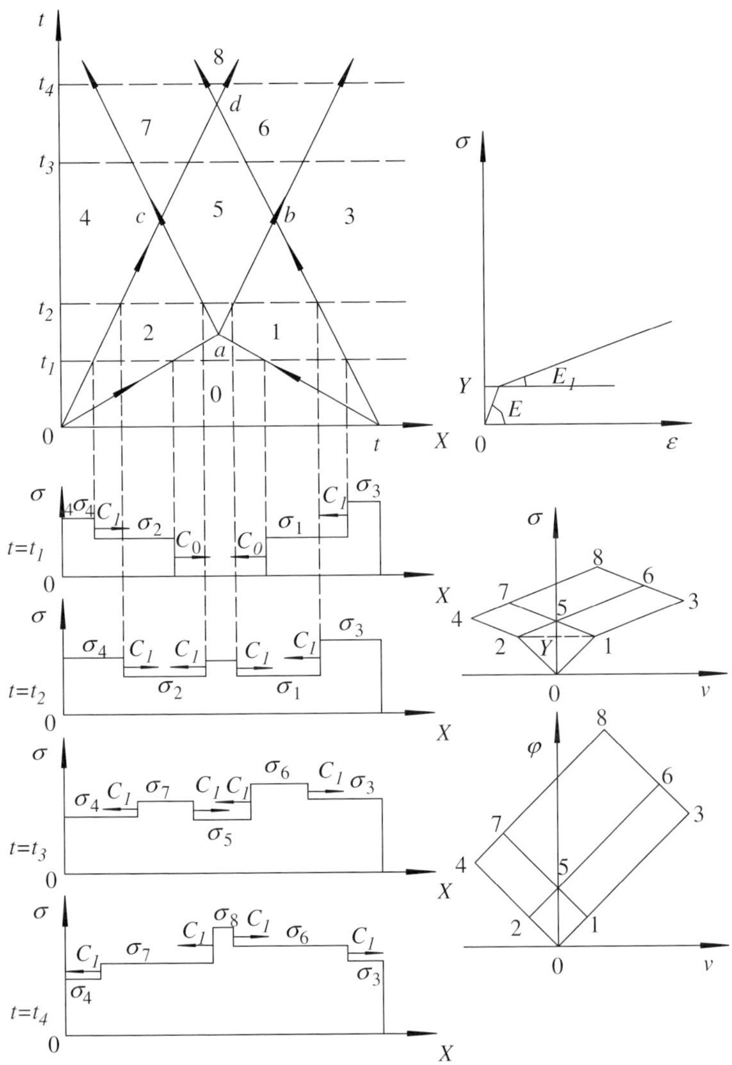

Fig. 4.1. An uniform cross-sectioned bar is subjected to sudden constant velocity pulling at both ends: (up left) different zones on X–t plane; (up right) elastic, linear hardening plastic stress–strain curve; (low left) stress profiles at different times marked in X–t plane; (low right) the corresponding σ–v states and φ–v states of different zones in X–t plane.

$$\sigma_3 = \sigma_1 + \rho_0 C_1 (v_3 - v_1)$$

$$\sigma_4 = \sigma_2 - \rho_0 C_1 (v_4 - v_2)$$

In the following, we will discuss the situation after the two waves meet. As the elastic wave is faster than the plastic wave ($C_0 > C_1$), the two elastic precursor waves first meet at the time t_a at the point a. Since the stress level of both elastic precursor waves has reached the elastic limit Y, the reflected loading waves after the interaction must be plastic waves. According to the momentum conservative condition at the wave front Eq. (2.57),

$$[\sigma] = -\rho_0 \mathscr{D}[v]$$

on the right side of the internal impact point a, we have:

$$v_5' = v_1 - \frac{\sigma_5' - \sigma_1}{\rho_0 C_1} = v_Y - \frac{\sigma_5' - Y}{\rho_0 C_1}$$

Similarly, on the left side of the internal impact point, we have:

$$v_5'' = v_2 + \frac{\sigma_5'' - \sigma_2}{\rho_0 C_1} = -v_Y + \frac{\sigma_5'' - Y}{\rho_0 C_1}$$

Moreover, the material velocity and the stress should be the same on this boundary, namely:

$$v_5' = v_5'' = v_5$$
$$\sigma_5' = \sigma_5'' = \sigma_5$$

From the above four equations we get:

$$v_5 = 0, \quad \sigma_5 = Y + \rho_0 C_1 v_Y = \left(1 + \frac{C_1}{C_0}\right) Y$$

After this, the rightward internal reflective plastic wave *ab* will meet the leftward incident plastic wave *lb* at the point *b*, and the leftward internal reflective plastic wave *ac* will meet the rightward incident wave *oc* at the point *c*. These wave-interactions all result in further plastic loading. The secondary internal reflective plastic waves, *bd* and *cd*, will finally meet at the point *d*. Only by this time the complete interaction between the two elastic–plastic loading waves finishes. The states of zone 6, 7, and 8 can be determined, respectively, by the approach similar to that applied to determine the state of zone 5. That is, each time when two waves meet, two stress-velocity relationships can be obtained according to the momentum conservative conditions [Eq. (2.57)] at both sides of the meeting point. Then, by using the condition that both the particle velocity and the stress should be equal at the meeting interface, the problem can be solved.

Figure 4.1 gives the wave profiles of stress distribution at four typical times: t_1, t_2, t_3, and t_4, respectively. These wave profiles vividly depict the course of the loading interaction between two strong-discontinuous elastic–plastic waves. In addition to the σ–v diagram, the corresponding φ–v diagram is also given in Fig. 4.1. By introducing the quantity φ [Eq. (2.37)], the nonlinear compatibility relationship along the characteristics lines is now linearlized.

4.1.2 Head-on loading interaction of two weak-discontinuous stress waves

For the decreasingly hardening plastic material ($d^2\sigma/d\varepsilon^2 < 0$), the plastic wave speed $C(\varepsilon) = \sqrt{(1/\rho)(d\sigma/d\varepsilon)}$ is not a constant but decreases with increasing stress. Therefore the plastic wave propagates in the form of continuous wave. If we view the continuous wave as a series of discontinuous incremental waves, then the loading interaction between two continuous elastic–plastic waves can be imagined as the interactions of many discontinuous incremental waves. Each time when two incremental waves meet, both the rightward and the leftward internal reflective wave result to a further loading, and in the meantime the wave speed decreases further, just as the situation previously discussed at the point a in Fig. 4.1. Therefore, it is imaginable that in X–t plane, the characteristics lines, which represent a series of interaction between two simple continuous elastic–plastic waves, are a set of curves with decreasing, absolute value of its slope $\left(\left|\frac{dX}{dt}\right|\right)$.

Now, the problem is solved in the same way as that used to solve the discontinuous wave interaction. The only difference is that the momentum conservative condition for continuous waves, Eq. (2.63), should be used instead of the momentum conservative condition for discontinuous waves, Eq. (2.57).

Take a decreasingly hardening bar with finite length as an example. The initial state of the bar is: velocity $v_a(>0)$, stress $\sigma_a(>0)$. Assume that the right end of the bar is loaded by a gradually increasing impact velocity up to $v_b(v_b > v_a)$ and then is held on at a constant velocity; while the left end of the bar is loaded by a gradually increasing impact velocity up to $v_c(v_c < v_a)$ and then is held on at a constant velocity. Therefore, two weak-discontinuous elastic–plastic tensile waves are propagating within the bar in the opposite direction. Before they meet, both waves are simple waves. The stress-velocity conditions after the wave fronts are respectively:

$$\left.\begin{aligned} v_b &= v_a + \int_{\sigma_a}^{\sigma_b} \frac{d\sigma}{\rho_0 C} \\ v_c &= v_a - \int_{\sigma_a}^{\sigma_c} \frac{d\sigma}{\rho_0 C} \end{aligned}\right\} \tag{4.1}$$

Assume that the leftward simple wave and the rightward simple wave begin to meet at the point a, and the interactions finish at the point d, as shown in Fig. 4.2. The state variables v_d and σ_d can be determined as follows. According to Eq. (2.63)

$$d\sigma = \mp \rho_0 C dv$$

at the right side of the bar we have:

$$v_d' = v_b - \int_{\sigma_b}^{\sigma_d'} \frac{d\sigma}{\rho_0 C}$$

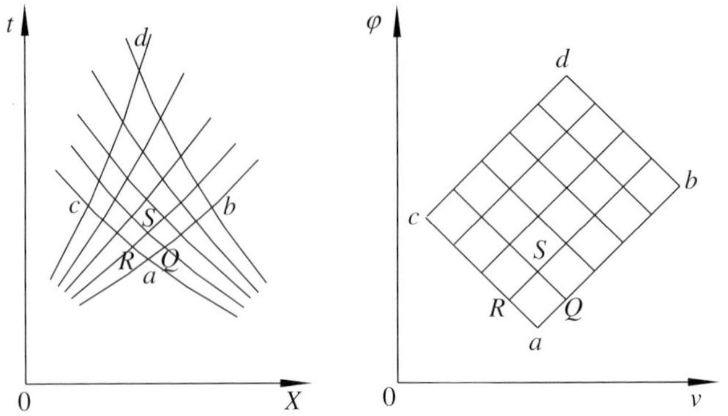

Fig. 4.2. Interaction of two sets of weak-discontinuous elastic-plastic tensile waves propagating in a bar in the opposite direction: (left) X–t figure showing different zones; (right) the physical states of different zones on φ–v plane.

while at the left side of the bar we have:

$$v''_d = v_c + \int_{\sigma_c}^{\sigma''_d} \frac{d\sigma}{\rho_0 C}$$

Moreover, at the interface (i.e. at the point d), the continuation conditions for stress and velocity should be held: $v'_d = v''_d = v_d, \sigma'_d = \sigma''_d = \sigma_d$. Thus, the state variables v_d and σ_d are solved.

Introducing a substitution variable φ [defined in Eq. (2.37)], the two stress-velocity conditions listed above are rewritten as:

$$\left.\begin{aligned} v_d &= v_b + \varphi_b - \varphi_d \\ v_d &= v_c - \varphi_c + \varphi_d \end{aligned}\right\} \tag{4.2}$$

while Eq. (4.1) is rewritten as:

$$\left.\begin{aligned} v_b &= v_a + \varphi_b - \varphi_a \\ v_c &= v_a - \varphi_c + \varphi_a \end{aligned}\right\} \tag{4.3}$$

The variable φ_d can be eliminated from the two equations of Eq. 4.2; and similarly the variable φ_a can be eliminated from the two equations of Eq. 4.3. Then two equations that contain the variable $(\phi_b - \phi_c)$ are obtained. Eliminating this variable again we get an equation that only contains the variable v_d. By similar approach, we can get an equation that only contains the variable φ_d. The final forms of these two equations are:

$$\left.\begin{aligned} v_d &= v_c + v_b - v_a \\ \varphi_d &= \varphi_c + \varphi_b - \varphi_a \end{aligned}\right\} \tag{4.4}$$

which can be further rewritten as:

$$\left.\begin{aligned}(v_d - v_a) &= (v_c - v_a) + (v_b - v_a)\\ (\varphi_d - \varphi_a) &= (\varphi_c - \varphi_a) + (\varphi_b - \varphi_a)\end{aligned}\right\}$$

As can be seen, the above equations are actually a generalization of Eq. (3.2) from the case of elastic waves to the case of elastic–plastic waves. The only difference is that the variable σ is substituted by the variable φ. In other words, if we use the variable φ instead of the old variable σ, then the superposition principle is still applicable. In fact, Eq. (3.2) is only a special case of Eq. (4.4) under the conditions that $v_a = \varphi_a = \sigma_a = 0$ and $C(\varepsilon) = C_0$.

The above discussions are applicable not only to the determination of the state variables in zone d, where the interactions of two weak-discontinuous elastic–plastic waves finished completely. They can also be applied to determine the state variables at any point of the region *abcd*, representing a specified point of the interaction process. Actually, the location of point d can be determined only after the state variables and the locations of all points within the interaction zone have been known. The reason is that in case of elastic–plastic wave interaction, the tangent slope of a characteristic line at any point of X–t plane, namely the characteristic line direction, is a function of the state variable at this point $[dX/dt = \pm C(\varepsilon)]$. Unless the state variables at the points along the characteristic lines *bd* and *cd* have been determined, the locations of the characteristic lines and consequently their intersection point d are impossible to be determined. In the previous discussions, we only assume the existence of the point d, but its location has not been determined yet. Compared to a linear wave problem, this reflects the complexity of a nonlinear wave problem. Nevertheless, since the state variables of any point along the characteristic lines *ab* and *ac*, and therefore its location, are known, we can use the finite differential approach, where the differential algorithm is performed along the characteristic lines, to solve the problem approximately. In practice, according to the accuracy requirement, the characteristic lines *ab* and *ac* can be divided into m and n segments respectively (see Fig. 4.2). The interaction region is then divided by the leftward characteristic lines starting at the segmental points along *ab* line and the rightward characteristic lines starting at the segmental points along *ac* line into $m \times n$ meshes. Within each mesh, the state variables, such as the particle velocity and the stress–strain, can be regarded as uniform. Therefore, the slope of the characteristic line segment *QS* can be approximately determined by the state of point Q, and similarly the slope of a characteristic line segment *RS* can be approximately determined by the state of point R. In this way, the algebraic equations:

$$X_S - X_Q = -C(\varepsilon_Q)(t_S - t_Q)$$

$$X_S - X_R = +C(\varepsilon_R)(t_S - t_R)$$

are used instead of the differential equations along the characteristic lines. Therefore the location of point S, which is the intersection of the line segments *QS* and *RS*, is determined (X_S, t_S). The state variables of point S can be calculated from the state variables of points a, Q, and R, according to Eq. (4.4):

$$v_S = v_Q + v_R - v_a$$

$$\varphi_S = \varphi_Q + \varphi_R - \varphi_a$$

Repeating the same approach we can determine the locations and the state variables at all mesh points.

In a problem like the previous one, if the state variables v and σ (or ε) along two characteristic lines of the different sets are given, then within the area bounded by these two characteristic lines and another two characteristic lines starting from the other ends of these two characteristics lines, there exists a single-valued solution. Such a type of problem is called a Darboux problem, or a characteristic line boundary value problem. Therefore, the problem to solve the loading interaction between two weak-discontinuous elastic–plastic waves is actually attributed to solve a Darboux problem.

Now we have discussed three types of solution-determined problem: the Cauchy problem (the initial value problem), the Picard problem (the mixed initial-boundary value problem), and the Darboux problem (the characteristic line boundary value problem).

4.2 Reflection of Longitudinal Elastic–Plastic Loading Waves at a Fixed End

For the problem shown in Fig. 4.2, if the two head-on elastic–plastic waves have the same stress magnitude, i.e. $\sigma_c = \sigma_b$ ($\varphi_c = \varphi_b$), then from Eqs. (4.2) and (4.3) we have:

$$v_d = v_a$$

This is equivalent to the case when an elastic–plastic wave reflects from a rigid wall (a special case is $v_a = 0$). After wave reflection, the change of the φ value $(\mathscr{D}\varphi)_R$ is the two times of the incident φ value change $(\mathscr{D}\varphi)_I$, namely:

$$(\Delta\varphi)_R = \varphi_d - \varphi_a = 2(\varphi_b - \varphi_a) = 2(\Delta\varphi)_I \tag{4.5}$$

We know that the stress level doubles when an elastic wave reflects from a rigid wall. Actually this is only a special case of Eq. (4.5) when $C = C_0$.

As a typical example of elastic–plastic wave interaction including fixed end reflection, we consider a decreasingly hardening bar with finite length. The left end of the bar ($X = 0$) is fixed. At time $t = 0$, the right end of the bar ($X = L$) is suddenly loaded by a constant velocity v^* (>0). The elastic–plastic waves will be reflected back-and-forth between the fixed end and the impact end, and gradually the wave intensity increases. This process is shown in Fig. 4.3 (von Karman *et al.*, 1942), wherein the characteristic lines are shown not only in the (X, t) plane and the (ε, t) plane, but also in (ϕ, v) plane. As stated previously, in the (ϕ, v) plane, all characteristic lines are straight lines with an angle $\pm 45°$ to the axes of coordinates.

Before the stress wave reaches the fixed end ($t < L/C_0$), the situation is the same as that in an infinite bar loaded by a sudden constant velocity impact. A beam of central simple waves, pioneered by a strong-discontinuous elastic wave, emit from the impact end. Assume that the bar is at a stress-free, rest state before impact (corresponding to the point 0 on the state plane). Across the precursor elastic wave front LA, the stress, strain, particle velocity, and φ value jump from zero to Y (the elastic limit), $\varepsilon_Y (= Y/E)$, $v_Y (= Y/\rho_0 C_0)$,

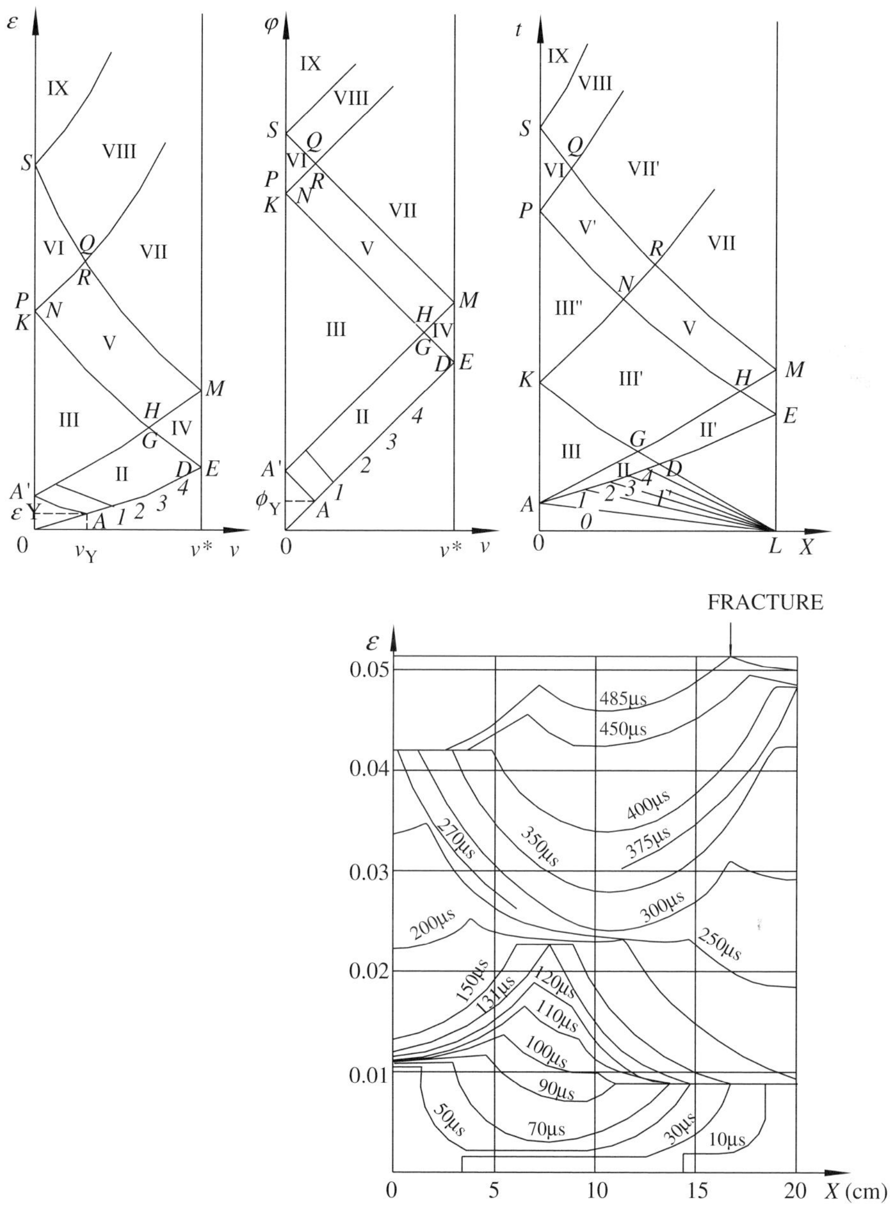

Fig. 4.3. Wave propagation in a finite bar of decreasingly-hardening material, the left end of bar ($X = 0$) is fixed, and the right end of the bar $X = L$ is suddenly loaded by a constant velocity v^* (>0).

and $\phi_Y(= Y/\rho_0 C_0)$, respectively. Accordingly, on the state plane, the state point jumps from the point 0 to the point A (the characteristic line *LA* on the physical plane corresponds to the characteristic line *OA* on the state plane). After this discontinuous elastic wave front, the continuous plastic waves come. On the physical plane (X–t plane) it is the simple wave region I' bounded by the curves *LA* and *LD*. On the state plane (v–ε plane) it corresponds to the line segment AD. The point D is determined by the boundary condition $v = v^*$. The constant value region I'' on the X–t plane corresponds to the point E on the state plane. The point E is identical to the point D.

At the time $t = L/C_0$, the elastic wave with the stress level of Y reflects at the fixed end. As mentioned before, this reflection wave must be a kind of loading plastic wave. Since the material is decreasing hardening as assumed, the incident strong-discontinuous elastic wave in fact reflects into a beam of weak-discontinuous plastic wave. On state plane this reflection wave region is represented by the line segment AA'. The state variable at point A' is determined by the boundary condition at the fixed end [Eq. (4.5)], namely, $\varphi_A = 2\varphi_Y$. Therefore at the fixed end of the bar ($X = 0$), at time $t = L/C_0$, a beam of central plastic wave should be reflected. Nevertheless, from the very beginning of reflection, those rightward reflection waves meet the leftward incident plastic waves. Therefore solving the region II on X–t plane is reduced to a problem that two weak-discontinuous plastic wave head-on interact and load each other further. As aforementioned, this is a solution-determined Darboux problem. In fact, the slope of the characteristic line at an arbitrary point P along the reflective characteristic line *AD* on X–t plane, with regard to the absolute value, should be equal to the slope of the characteristic line *LP*, namely,

$$\frac{dX}{dt} = \frac{L - X}{t}$$

Integrating the above function, and using the boundary condition of $X = 0, t = L/C_0$, we can find that the characteristic line *AD* is a segment of a hyperbolic curve expressed by:

$$t(L - X) = L^2/C_0$$

Therefore, the locations of the characteristic line *AD* and the state variables at these locations are known. Correspondingly, on the state plane, the boundary characteristic lines of Region II, the line *AD* and the line AA' in different sets, are known. Thus the region II can be solved as a characteristic line boundary value problem (Darboux problem).

After the interaction between the reflective plastic wave tail (*AG*) and the incident plastic wave tail (*LDG*) is completed, the reflective plastic waves will no longer interact with any other head-on plastic waves and then propagate in the constant value zone I″. Therefore, the region II′ on X–t plane is a simple wave zone. From this, it is easy to understand that a simple wave zone is always neighboring a constant value zone.

Since the region II has been solved, one boundary of the region III – the characteristic line *AG* is known. Another boundary of the region III, the time axis AK, which is not a characteristic line, is prescribed by the boundary condition $v = 0$ (the fixed end boundary condition). Therefore, the solution of the region III is attributed to solve as a Picard problem (mixed boundary problem).

On the analogy of these approaches, the solutions of the other regions on the X–t plane can be attributed to solve either as a Darboux problem, or as a Picard problem, until the whole field is solved. In the X–t plane shown in Fig. 4.3, three types of regions exist as listed below:

(a) The constant value regions, which are marked by the superscript ($''$). For such regions, the positive and negative characteristic lines on the X–t plane are all straight lines, while on the state plane they are correspondingly reduced to a point.

(b) The simple wave regions, which are marked by the superscript ($'$). They are always neighboring a constant value region. On the state plane such a region is reduced to a line segment. If this line segment is positive, then on the X–t plane the negative characteristic lines are straight lines, while the other set of the characteristic lines are curves.

(c) The complex wave regions, which are not marked with any superscript. For such regions, a one-to-one area correspondence exists on the physical plane and the state plane. If one boundary of such a region is the t axis, then this region is attributed to solve as a mixed boundary problem. The regions III, IV, VI, VII, and IX are all of this type. Otherwise the problem is attributed to solve as a characteristic line boundary problem. The regions II, V, VIII are all of this type.

In principle, the exact solution can be found for this problem. However, in practice, because of the complicated function of $\sigma = \sigma(\varepsilon)$, the problem is usually solved by a numerical approach. An example solved by the characteristics numerical approach is given in Fig. 4.3(d), showing the calculated strain wave profiles along the bar at different times. The calculation indicates that at $t = 485$ μs, the fracture happens at a point near the impact end of the bar, because the ultimate tensile strength is reached. Note that a secondary fracture will happen at another point near the fixed end before the unloading wave due to the first fracture arrives at the fixed end of the bar.

4.3 Governing Equations and Characteristic Lines of Unloading Waves

We now start to discuss the unloading problem of elastic–plastic stress wave interaction. It is well known that the stress–strain curves in the loading process and in the unloading process are the same for an elastic (or even for a nonlinear elastic) material. However, the stress–strain relationship of an elastic–plastic material differs for the loading and the unloading processes, and consequently the governing equations for the loading waves and the unloading waves in an elastic–plastic material are also different.

From the phenomenological point of view, the stress wave propagation in an elastic–plastic material can be treated as the same as the stress wave propagation in a nonlinear elastic material, if only the loading process is concerned. In fact, without knowing the unloading stress–strain curve of a material, it is impossible to distinguish an elastic–plastic material from a nonlinearly elastic material. In other words, in a loading process, the stress–strain curve for an elastic–plastic material can be mathematically regarded as that for a nonlinear elastic material. Consequently, the loading wave propagation for the two materials can be mathematically regarded as the same. Only when the unloading happens, the two materials behave in different ways: the elastic–plastic material exhibits irreversible plastic deformation, while the elastic material exhibits reversible elastic deformation. In this sense, the essential features of plastic wave propagation, which distinguishes from the elastic wave propagation, are mainly reflected in the unloading process.

Therefore, when we analyze an elastic–plastic wave problem including both the loading and unloading processes, it is necessary to identify what the state is for any arbitrary material particle at any time. Is it in a loading process or in an unloading process? In other words, on the X–t plane, it is necessary to distinguish between the plastic loading zone and the unloading zone, and to determine the unknown boundary between these two zones. Note that the location of such a boundary, X, changes with the propagation and interaction of stress waves, and so changes with time t. Thus, the problem can now be expressed as: to determine the propagation of the boundary $X = f(t)$ between the plastic loading zone and the unloading zone, and meanwhile to solve the plastic loading zone and the unloading zone, which are bounded by the $X = f(t)$, by different governing equations. Mathematically, it is a kind of problem with undetermined internal boundaries. This makes the problem much more complicated.

The governing equations in a plastic loading zone have been stated previously [see Eq. (2.18)]. Now, we need to discuss the governing equations in an unloading zone. Similar to the case for loading waves, the governing equations for unloading waves should be composed of the kinematics equation, the dynamics equation and the constitutive equation. The basic assumptions applied are the same as for the loading waves in bars too. In the following, we still use Lagrange variables to describe those governing equations.

Consider the unloading stress–strain curve of an elastic–plastic material that has experienced plastic loading. If before unloading, the material is in a state with stress σ_m and strain ε_m, as shown in Fig. 4.4. No matter whether the material is unloaded or reloaded again after previous unloading, provided that the stress does not go over σ_m, we assume that the unloading stress–strain curve is a linear line, and its slope is same as the initial elastic modulus. In other words, the unloading and reloading (below σ_m) curve obeys the same Hooke's law for the elastic loading (Fig. 4.4). This is the so-called **elastic unloading hypothesis**. Note that when the material is unloaded from σ_m and then reloaded again, the yield limit of the material is raised to σ_m, representing the so-called work-hardening effect or the strain-hardening effect.

Using an overbar to indicate the variables after unloading, the elastic unloading stress–strain relationship for material, in the one-dimensional stress state, is:

$$\bar{\sigma} = \sigma_m + E(\bar{\varepsilon} - \varepsilon_m) \tag{4.6}$$

The total strain ε can be viewed as the sum of the elastic strain ε^e and the plastic strain ε^p. The elastic strain ε^e is recoverable, and satisfies Hooke's law, $\varepsilon^e = \sigma/E$. The plastic strain ε^p is unrecoverable.

Note that, generally speaking, during the unloading process, the stress value σ_m may be different for different points along the bar. However, once the unloading process begins, as long as plastic re-yielding does not occur again, the value of σ_m will not change with time.

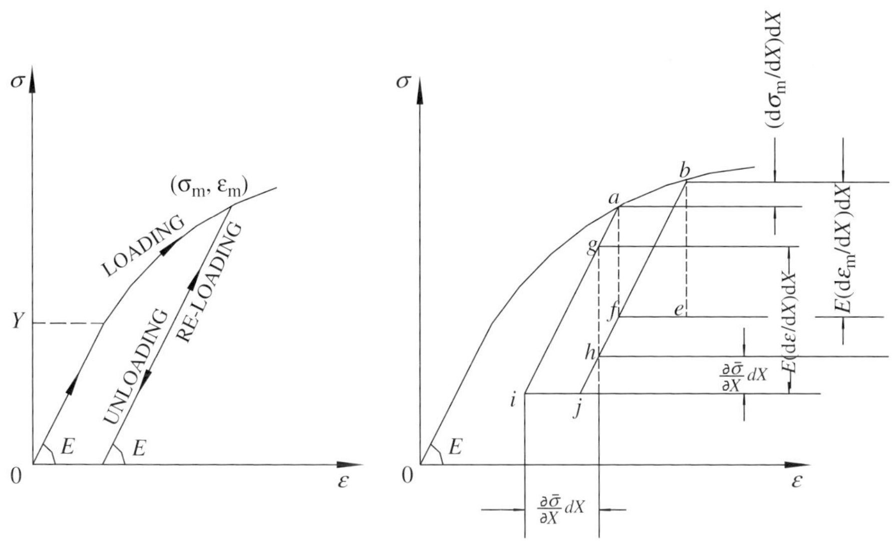

Fig. 4.4. Elastic-plastic loading and elastic unloading stress–strain relationship.

Therefore, in an unloading region, σ_m and ε_m are functions of X and are independent of the variable t. We then get:

$$\frac{\partial \bar{\sigma}}{\partial t} = E\frac{\partial \bar{\varepsilon}}{\partial t} = \rho_0 C_0^2 \frac{\partial \bar{\varepsilon}}{\partial t}$$

$$\frac{\partial \bar{\sigma}}{\partial X} = E\frac{\partial \bar{\varepsilon}}{\partial X} + \frac{d}{dX}(\sigma_m - E\varepsilon_m) = \rho_0 C_0^2 \frac{\partial \bar{\varepsilon}}{\partial X} + \frac{d\sigma_m}{dX} - \rho_0 C_0^2 \frac{d\varepsilon_m}{dX}$$

If at time t, the stresses before unloading, σ_m, are denoted by the symbol a and b in Fig. 4.4(b) for the point X and $X + dX$, respectively, the unloading stress $\bar{\sigma}$ for the point X and $X + dX$ are denoted by the symbol i and h in Fig. 4.4(b), respectively, then each term in the above equation for $\partial\bar{\sigma}/\partial X$ is illustrated in the same figure, showing its physical meaning respectively.

The kinematics equation and the dynamics equation for the unloading waves are the same as those for the loading waves, namely Eqs. (2.12) and (2.13). Those equations, together with the unloading stress–strain relationship, constitute the governing equations of the unloading waves:

$$\left.\begin{aligned} \frac{\partial \bar{\varepsilon}}{\partial t} &= \frac{\partial \bar{v}}{\partial X} \\ \rho_0 \frac{\partial \bar{v}}{\partial t} &= \frac{\partial \bar{\sigma}}{\partial X} \\ \bar{\sigma} &= \sigma_m + E(\bar{\varepsilon} - \varepsilon_m) \end{aligned}\right\} \tag{4.7}$$

We can eliminate the variable $\bar{\sigma}$ from the above equations, getting two first order partial differential equations with regard to the variables $\bar{\varepsilon}$ and $\bar{v}$:

$$\left.\begin{aligned}\frac{\partial \bar{v}}{\partial X} &= \frac{\partial \bar{\varepsilon}}{\partial t}\\ \frac{\partial \bar{v}}{\partial t} &= C_0^2\frac{\partial \bar{\varepsilon}}{\partial X} + \frac{1}{\rho_0}\frac{d\sigma_m}{dX} - C_0^2\frac{d\varepsilon_m}{dX}\end{aligned}\right\} \tag{4.8}$$

Or, we can eliminate the variable $\bar{\varepsilon}$ from the Eq. (4.7), getting two first order, homogeneous partial differential equations with regard to the variables $\bar{\sigma}$ and $\bar{v}$:

$$\left.\begin{aligned}\frac{\partial \bar{\sigma}}{\partial t} &= \rho_0 C_0^2\frac{\partial \bar{v}}{\partial X}\\ \frac{\partial \bar{\sigma}}{\partial X} &= \rho_0\frac{\partial \bar{v}}{\partial t}\end{aligned}\right\} \tag{4.9}$$

Moreover, we can get an equivalent second order partial differential equation with regard to the unknown displacement $\bar{u}$:

$$\frac{\partial^2 \bar{u}}{\partial t^2} = C_0^2\frac{\partial^2 \bar{u}}{\partial X^2} + \frac{1}{\rho_0}\frac{d\sigma_m}{dX} - C_0^2\frac{d\varepsilon_m}{dX} \tag{4.10}$$

The characteristic line method can be used to solve these equations. Using the procedures similar to that stated in Section 2.1, the characteristic line equations and the compatibility equations along the characteristic lines can be obtained without any difficulty. For example, corresponding to Eq. (4.8), we have:

$$\left.\begin{aligned}dX &= \pm C_0 dt\\ d\bar{v} &= \pm\left(C_0 d\bar{\varepsilon} + \frac{1}{\rho_0 C_0}d\sigma_m - C_0 d\varepsilon_m\right)\end{aligned}\right\} \tag{4.11}$$

And corresponding to Eq. (4.9), we have:

$$\left.\begin{aligned}dX &= \pm C_0 dt\\ d\bar{v} &= \pm\frac{1}{\rho_0 C_0}d\bar{\sigma}\end{aligned}\right\} \tag{4.12}$$

The above equations are equivalent to each other. Among them the governing equations with regard to the variables $\bar{\sigma}$ and $\bar{v}$ [Eq. (4.9)], and their corresponding characteristic line equations [Eq. (4.12)], are completely identical to those for the elastic waves. It is also more convenient to use this set of equations to solve the problem with wave interactions, as will be shown below. Therefore we tend to use this set of equations in the following discussions.

From Eq. (4.11) or Eq. (4.12) we also know that: **the unloading disturbance propagates with the elastic wave speed** C_0. This is an inevitable outcome resulted from the **elastic unloading hypothesis**. It should be emphasized that the propagation of unloading disturbance and the propagation of the boundary of plastic-loading/elastic-unloading are two different concepts. The former is the **unloading wave**, which propagates along the elastic characteristic line. Under the elastic unloading hypothesis, the speed of the unloading wave is always C_0. On the other hand, the latter is the spread or propagation of the so-called **elastic-plastic boundary**. Generally the elastic–plastic boundary propagates neither along the elastic characteristic line nor along the plastic characteristic line. Its propagation speed can only be determined by a detailed analysis of each case. Only under some specific conditions, the propagation of an elastic–plastic boundary coincides with the propagation of an unloading wave.

On the X–t plane, if the curve $X = f(t)$ presents the elastic–plastic boundary during an unloading process, which will be shortly called **unloading boundary** below, for convenience, then the slope of this curve:

$$\overline{C} = \left.\frac{dX}{dt}\right|_{X=f(t)} = f'(t) \tag{4.13}$$

represents the propagation speed of this boundary. Note that there were literatures where this elastic–plastic boundary propagation had been called "unloading wave", or "Rakhmatulin wave". Since it may induce confusion in the basic concept, it will not be accepted in our discussion.

Thus, for an elastic–plastic wave problem including both the plastic loading and the elastic unloading processes, the problem is reduced to: under the prescribed initial condition and a boundary condition, simultaneously solve the partial differential equations in an elastic–plastic loading zone (Eq. (2.18) or other equivalences) and the partial differential equations in an elastic unloading zone (Eq. (4.10) or other equivalences), and meanwhile on the boundary of these two zones, $X = f(t)$, the conditions of the displacement continuation and the momentum conservation must be satisfied. The main difficulty lies in that generally the elastic–plastic boundary $X = f(t)$ is not known before the whole problem has been solved.

In the next few sections, we will discuss the problem of unloading wave propagation, first in a semi-infinite bar. The end of the bar is loaded to the elastic–plastic state, and then is unloaded. Because the unloading disturbance propagates faster than the plastic loading disturbance ($C_0 > C$), the unloading disturbances will pursue the proceeding plastic loading disturbances and then interact each other. This kind of wave interaction is called a **pursuing unloading** problem. Afterwards, we will discuss the problem of unloading wave propagation in a finite bar, wherein an unloading disturbance propagates from one end, and interacts with a head-on plastic loading disturbance propagated from the other end, resulting in a head-on unloading. Such kind of wave interaction is called a **head-on unloading** problem. Finally, we will discuss the general properties of elastic–plastic boundary propagation.

4.4 Pursuing Unloading by Strong-Discontinuous Unloading Disturbances

4.4.1 Sudden unloading of strong-discontinuous loading wave in a linear hardening bar

Consider a semi-infinite bar made of linear hardening material. The bar is in a natural, rest state before loading is applied. At time $t = 0$, a sudden constant value impact loading σ^* is applied on the end of the bar. The impact loading is large enough to cause the bar to deform plastically. After time t_1 the applied load is suddenly released to 0. This problem is equivalent to the case that two bars impact suddenly and then bounce back suddenly. In such a case, both the loading disturbance and the unloading disturbance propagate within the bar in the form of strong-discontinuity. Since the unloading disturbance is faster than the plastic loading disturbance, it will pursue and interact with the loading disturbance.

When $t < t_1$, namely, for the region 1 and region 2 on the X–t plane shown in Fig. 4.5(a), it is equivalent to solving the elastic–plastic loading wave propagation in a semi-infinite bar as has been discussed previously, so it is easy to get:

$$\sigma_1 = -Y, \varepsilon_1 = -\varepsilon_Y = -\frac{Y}{E}, v_1 = \frac{Y}{\rho_0 C_0} = v_Y$$

$$\sigma_2 = \sigma^*, \varepsilon_2 = \frac{\sigma^*}{E_1} + \left(\frac{1}{E_1} - \frac{1}{E}\right)Y$$

$$v_2 = \left(\frac{1}{\rho_0 C_0} - \frac{1}{\rho_0 C_1}\right)Y - \frac{\sigma^*}{\rho_0 C_1} = v^*$$

where Y and ε_Y are the yield stress and the yield strain of the material (absolute values). $C_0 = \sqrt{E/\rho_0}$, $C_1 = \sqrt{E_1/\rho_0}$, and E_1 is the linear hardening modulus [Fig. 4.5(c)].

At $t = t_1$, the plastic wave front arrives at the location $l_1 = C_1 t_1$, while the stress at the impact end is suddenly decreased to 0, which means that the end of the bar becomes a free end. So a strong-discontinuous unloading disturbance is propagating in the plastic constant value region 2, with speed $C_0 (> C_1)$[1]. This unloading wave pursues the forerunning plastic loading disturbance, which propagates with the velocity C_1. The state of the unloading region 3 can be determined by the prescribed unloading boundary condition, the unloading stress–strain relationship [Eq. (4.6)], and the dynamic compatibility condition across a strong-discontinuous wave front [Eq. (2.57)], namely:

$$\left.\begin{aligned} &\bar{\sigma}_3 = 0 \\ &\bar{\varepsilon}_3 = \varepsilon_2 - \frac{\sigma_2}{E} = \bar{\varepsilon}^* \\ &\bar{v}_3 = v_2 + \frac{\sigma_2}{\rho_0 C_0} = v^* + \frac{\sigma^*}{\rho_0 C_0} = \bar{v}^* \end{aligned}\right\} \quad (4.14)$$

[1]As discussed in Section 4.3, under the elastic unloading hypothesis, the unloading disturbance propagates by the elastic wave velocity C_0. This is true no matter whether the unloading disturbance is weak-discontinuous or strong-discontinuous. Actually when Eq. (4.6) is substituted into Eq. (2.59) we get $\mathscr{D} = C_0$.

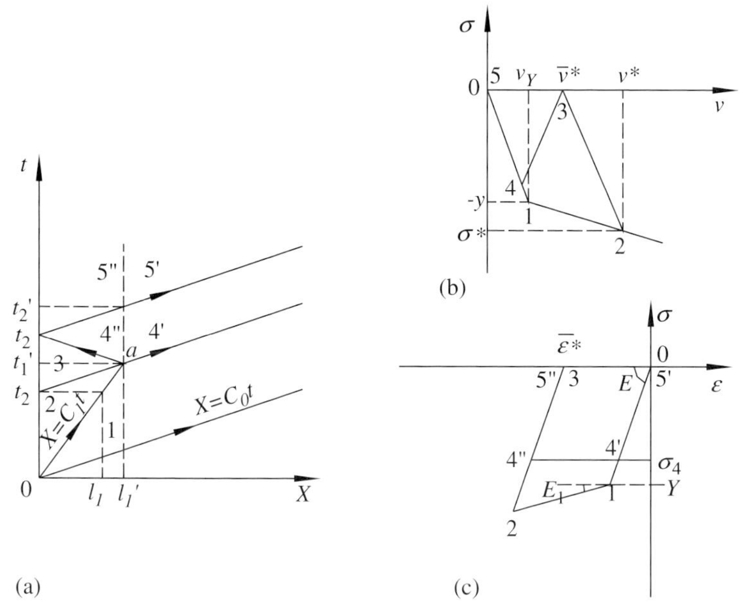

Fig. 4.5. Sudden unloading of strong discontinuous loading wave in a linear-hardening bar ($|\bar{v}^*| < 2|v_Y|$).

As can be seen that although the stress $\bar{\sigma}_3$ is unloaded to zero, the material still has **residual strain** $\bar{\varepsilon}^*$ and the **residual particle velocity** $\bar{v}^*$, which are shown in Fig. 4.5(c and b), respectively. This feature reflects the irreversible property of plastic deformation. Note that the concept of residual velocity is important in understanding the internal impact phenomenon when an unloading disturbance has caught up with a plastic loading disturbance. The internal impact phenomenon will be discussed below.

Assume that at time $t = t'$ the unloading disturbance catches up with the plastic loading disturbance, the value of t' can be determined by the condition that on the X–t plane the line $t_1 a$ intersects with the line oa, which gives:

$$C_0(t_1' - t_1) = C_1 t_1'$$

Then we have:

$$t_1' = \frac{C_0 t_1}{C_0 - C_1} = \frac{t_1}{1 - C_1/C_0} \tag{4.15a}$$

and the location where the two waves interact with each other is:

$$l_1' = C_1 t_1' = \frac{C_1 t_1}{1 - C_1/C_0} = \frac{l_1}{1 - C_1/C_0} \tag{4.15b}$$

Therefore, while $X < l_1'$, $t < t_1'$, the unloading disturbance propagates in the plastic constant value zone 2, and behind the strong-discontinuous front of the propagating unloading disturbance, the stress $\bar{\sigma}_3$ is unloaded to zero, but the particle velocity $\bar{v}_3$ is not unloaded to zero, although this residual velocity $\bar{v}_3 = \bar{v}^*$ is uniformly distributed along the bar segment $X \le l_1'$. However, once the unloading disturbance catches up with the forerunning plastic loading disturbance, let us imagine that the unloading disturbance now propagates in the bar segment $X > l_1$, which is the elastic constant value zone 1, so that behind the unloading wave front both the stress and the particle velocity are now unloaded to 0. Thus, across the bar section $X = l_1$, a particle velocity jump $(\bar{v}_3 - 0)$ occurs. In other words, an internal impact should take place at $X = l_1$. Therefore, when the unloading disturbance catches up with the plastic loading disturbance, the interaction between those two waves will induce the so-called **secondary stress waves** or the **internal reflective waves**.

After the interaction of the unloading wave and the plastic loading wave, the concrete state variables will be determined depending on the magnitude of residual velocity difference $(\bar{v}_3 - 0)$ across the section point $X = l_1'$. It is similar to a case that a bar traveling with the speed $\bar{v}_3$ longitudinally impacts on a bar at rest. The states of the bars just after the impact will be determined depending on the level of $\bar{v}_3$. From our previous discussions about the elastic–plastic bar impact [e.g. refer to Eq. (3.1)], if the condition

$$|\bar{v}_3| < |2v_Y| \tag{4.16}$$

is satisfied, then the level of the secondary stress wave is less than the yield limit Y. On the contrary, when $|\bar{v}_3| > |2v_Y|$, the secondary stress wave will include plastic wave.

Figure 4.5 shows the case of $|\bar{v}_3| < |2v_Y|$. The state variables in region 4 can be determined as follows. At the left side and the right side of the section point $X = l_1'$, according to the dynamic compatibility condition across a strong-discontinuous wave front [Eq. (2.57)], we have:

$$\sigma_4'' = \bar{\sigma}_3 + \rho_0 C_0(\bar{v}_4'' - \bar{v}_3) = \rho_0 C_0(\bar{v}_4'' - \bar{v}_3)$$

$$\sigma_4' = \sigma_1 - \rho_0 C_0(\bar{v}_4' - v_1) = -\rho_0 C_0 \bar{v}_4'$$

Moreover, the stress and particle velocity should be continuous at the point $X = l_1'$:

$$\bar{\sigma}_4'' = \bar{\sigma}_4' = \sigma_4, \; \bar{v}_4'' = \bar{v}_4' = v_4$$

Therefore the following results can be obtained:

$$\bar{v}_4 = \frac{\bar{v}_3}{2}$$

$$\bar{\sigma}_4 = -\rho_0 C_0 \frac{\bar{v}_3}{2}$$

However, from Fig. 4.5 we note that:

$$\bar{\varepsilon}_4' = \frac{\bar{\sigma}_4}{E}$$

$$\bar{\varepsilon}_4'' = \bar{\varepsilon}_3 + \frac{\bar{\sigma}_4}{E} = \bar{\varepsilon}_3 + \bar{\varepsilon}_4'$$

Therefore, at the point $X = l_1'$, the strain is discontinuous. This strong-discontinuity of strain will retain at the point $X = l_1'$, unless the so-called secondary plastic loading happens in the future. Such kind of discontinuity is called a **stationary discontinuous interface**. A stationary strain-discontinuous interface, originating from the interaction between a strong-discontinuous unloading disturbance and a strong-discontinuous plastic-loading disturbance, reflects the strain history difference at the two sides of the interaction point. The difference of the plastic strain history causes the difference of the stress–strain relationships. Therefore, once a stationary strain-discontinuous interface appears in a uniform bar, the mechanical properties (e.g. at least the yield stress) of the material will be different at two sides of the interface, and the bar is no longer a uniform bar. Under certain conditions like those shown in Fig. 4.5, suppose a stress wave with the stress level between σ_1 and σ_2 (i.e. $Y < |\sigma| < |\sigma_2|$) propagates in the bar, when this stress wave passes through the stationary strain-discontinuous interface l_1', transmitting and reflecting phenomena will appear, just like a wave passing through an interface of different media. Note that the elastic wave resistance, $\rho_0 C_0$, is the same on two sides of the stationary strain-discontinuous interface. So the propagation of an elastic wave with level $|\sigma| < Y$, or an unloading wave, will not be influenced by the stationary strain-discontinuous interface. For example, in the case shown in Fig. 4.5, the leftward reflective wave $l_1' t_2$ originating at the interface l_1' will be reflected again from the free end at time $t = t_2$, the rightward reflective unloading wave $t_2 b$ will pass through the interface l_1' without any interference.

For the problem illustrated in Fig. 4.5, the distributions of the stress, strain, and the particle velocity at several typical times are shown in Fig. 4.6. As can be seen, once the unloading process appears in the propagation of plastic wave, the stress wave profile, the strain wave profile and the particle velocity wave profile become different from each other.

It should be noted that in general the propagation of unloading boundary should be distinguished from the propagation of unloading disturbance itself. However, in the special case discussed here, when the propagating locus of unloading boundary $t_1 a$ (Fig. 4.5) just coincides with the propagating locus of strong-discontinuous unloading disturbance itself, then the propagating velocity of unloading boundary $\overline{C} = C_0$.

In the case shown in Figs. 4.5 and 4.6, the problem is relatively simple, because the boundary impact loading is only high enough to result in a residual particle velocity as small as ($\bar{v}^* < 2v_Y$), and consequently the secondary stress wave σ_4 created by the internal impact is less than the yield limit ($|\sigma_4| < Y$), which means that once the unloading wave catches up with the forerunning plastic loading wave and interacts each other, the plastic loading wave "disappears" completely.

If the impact loading σ^*, or correspondingly v^*, is high enough to result in a residual particle velocity $\bar{v}_4^* > 2v_Y$, the situation becomes more complex. In this case, the secondary stress wave originating from the internal impact has a stress level beyond the yield limit Y. Therefore the secondary stress wave propagating to the right side of the stationary strain-discontinuous interface l_1' is a strong-discontinuous plastic wave, although the stress level is lower than the initial plastic wave. This means that the initial plastic wave, after interacting with the unloading disturbance, is **weakened**. On the other side, the secondary stress wave propagating to the left side of the stationary strain-discontinuous interface l_1' is an elastic wave propagating in the unloading region, because its stress level is lower

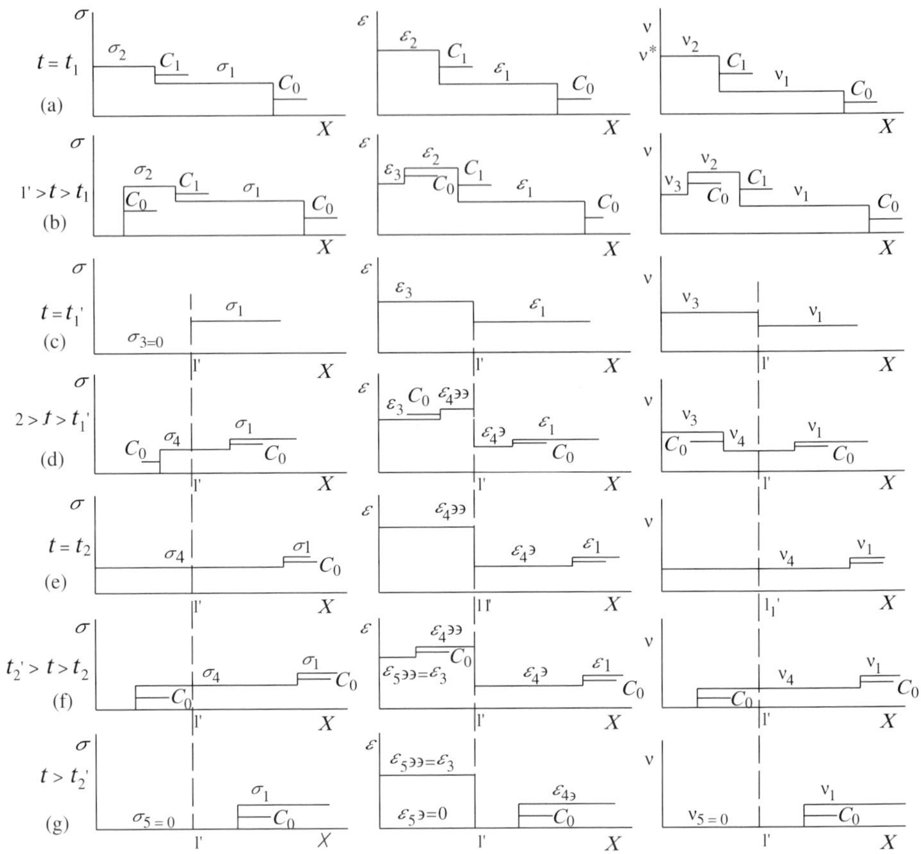

Fig. 4.6. The distributions of the σ, ε, and v at several typical times in Fig. 4.5.

than the initial plastic wave. At $t = t_2$ this leftward wave reaches the bar end (which is now a free end), and is reflected again as the secondary unloading disturbance. At the point $X = l_2'$, this strong-discontinuous, secondary unloading disturbance catches up with the already weakened, forerunning plastic loading wave. This time a second stationary strong-discontinuous strain interface is formed at l_2'. The plastic loading wave propagating to the right side of the interface l_2' is weakened further.

On the analogy of this unloading process, in each unloading cycle a stationary strong-discontinuous strain interface is formed and the rightward plastic loading wave is weakened, until the forerunning plastic wave disappears, that is, unloaded to an elastic wave. Clearly, higher the initial impact loading σ^* (or v^*) is, more will be the number of stationary strong-discontinuous strain interfaces. Figure 4.7 shows an example where three stationary strong-discontinuous strain interfaces are formed. The detailed explanation for the processes of wave propagation and interaction in Fig. 4.7 is omitted here. Readers can easily understand it by referring to the previous discussions for the case in Fig. 4.5. It is

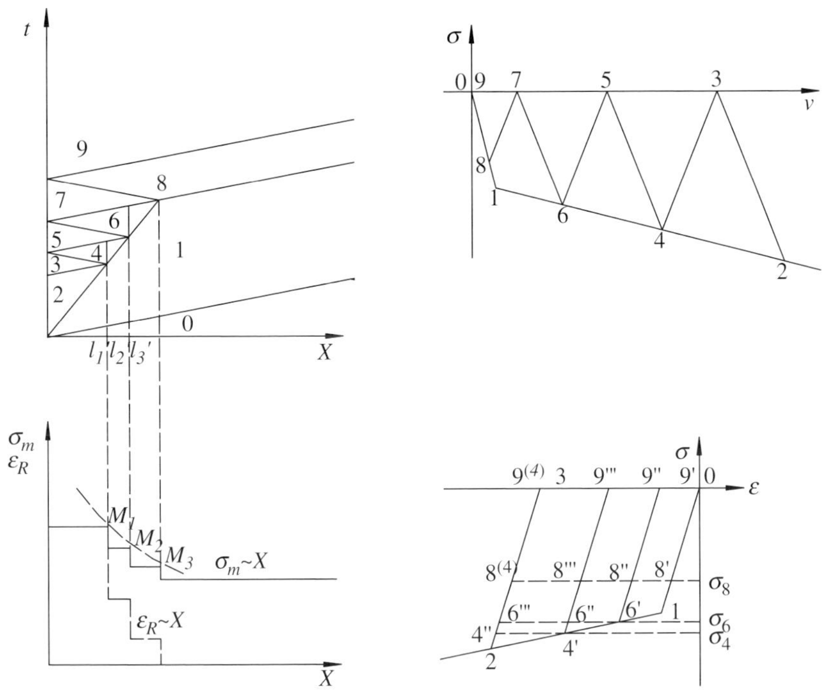

Fig. 4.7. Sudden unloading of strong discontinuous loading wave in a linear-hardening bar ($|\bar{v}^*| > 2|v_Y|$).

recommended that the readers go through the deduction to acquire a better understanding. Besides, it can be easily proved by the readers that if the following dimensionless parameter β (>1) is introduced,

$$\beta = \frac{C_0 + C_1}{C_0 - C_1} = \frac{1 + C_1/C_0}{1 - C_1/C_0} > 1 \tag{4.17}$$

then the length of the $(n+1)$th stationary strain-discontinuous interfaces, l'_{n+1}, is β times the length of the nth stationary strain-discontinuous interfaces, l'_n; and, the maximum plastic stress experienced at point l'_{n+1}, $\sigma_{2(n+1)}$, is $1/\beta$ times the maximum plastic stress at point l'_n, σ_{2n}, that is:

$$\left.\begin{aligned} l'_{n+1} &= \beta l'_n \\ \sigma_{2(n+1)} &= \frac{\sigma_{2n}}{\beta} \end{aligned}\right\} \quad (n = 1, 2, 3, \ldots)$$

This means that when a stationary strain-discontinuity forms in each unloading cycle, the maximum stress decreases by a factor of $1/\beta$, while the distribution length of the maximum stress increases by a factor of β. Thus the points $M_1, M_2, M_3, \ldots$ on the plot of maximum stress distribution in Fig. 4.7 fall upon a hyperbolic curve expressed by:

$$\sigma_2 l_1' = \sigma_4 l_2' = \ldots \sigma_{2n} l_2' = \sigma_{2(n+1)} l_{(n+1)}' = const.$$

In the same plot the distribution of the residual strain after impact, ε_R, is also shown. As can be seen, the residual deformation is concentrated to a small region near the impact end of the bar, and decreases step by step. The number of the steps is the number of the stationary strain-discontinuities. This result is completely different from the case under static loading.

4.4.2 Sudden unloading of continuous loading waves in a linear hardening bar

In the previous discussions, both the unloading disturbance and the plastic loading disturbance are strong-discontinuous waves. Now we consider a slightly complex case where the unloading disturbance is a strong-discontinuous wave but the forerunning plastic loading disturbances are a series of continuous waves. We need to deal with the wave interactions between the strong-discontinuous unloading disturbance and **a series of** continuous plastic loading disturbances.

To help understand the image of such a kind of wave interaction process, we start the discussion on the basis of the example shown in Fig. 4.5 for a semi-infinite bar made of linear hardening material, but with a boundary condition of gradually increasing impact loading. That is, at time $t = 0$, the end of the bar is loaded by an increasing external load, which is linearly changed with time, and at time $t = T_0$ the external load is suddenly released [Fig. 4.8(a)]. In such a situation, the unloading disturbance propagates as a strong-discontinuous wave, while the plastic loading disturbances propagate as a series of weak-discontinuous waves but with the constant speed C_1. As in our previous discussions, such continuous plastic waves can be regarded as a series of incremental waves. Similarly, the linearly increasing boundary load can be regarded as a series of step-like incremental boundary conditions [Fig. 4.8(b)]. Therefore, the interactions between a strong-discontinuous unloading disturbance and a series of weak-discontinuous plastic disturbances can be viewed as the interactions between a strong-discontinuous unloading disturbance and *a series of* small strong-discontinuous plastic loading disturbances in succession. Since the interaction between a strong-discontinuous unloading disturbance and a strong-discontinuous plastic-loading disturbance has been dealt with previously, the problem can be solved step by step as follows.

From the time $t = T_0$, the strong-discontinuous unloading disturbance propagates with wave speed C_0 in the constant-value zone 5. Behind the unloading wave front, the stress drops from σ^* to zero, and the particle velocity drops from v^* to the residual velocity $\bar{v}^* = v^* + \sigma^*/(\rho_0 C_0)$. At the time $t = T_1$, the unloading disturbance catches up with the first small-discontinuous plastic wave at the point $X = L_1$ and interact with each other. This interaction process can be understood as follows: since a particle velocity jump $(\Delta v)_1$ and a stress jump $(\Delta\sigma)_1$ exist across this small-discontinuous interface, if both stresses are unloaded to zero, then a residual particle velocity jump $(\Delta\bar{v})_1$ is produced:

$$(\Delta\bar{v})_1 = (\Delta v)_1 + \frac{(\Delta\sigma)_1}{\rho_0 C_0} = -\frac{(\Delta\sigma)_1}{\rho_0 C_1} + \frac{(\Delta\sigma)_1}{\rho_0 C_0} = \frac{(\Delta\sigma)_1}{\rho_0}\left(\frac{1}{C_0} - \frac{1}{C_1}\right) \tag{4.18}$$

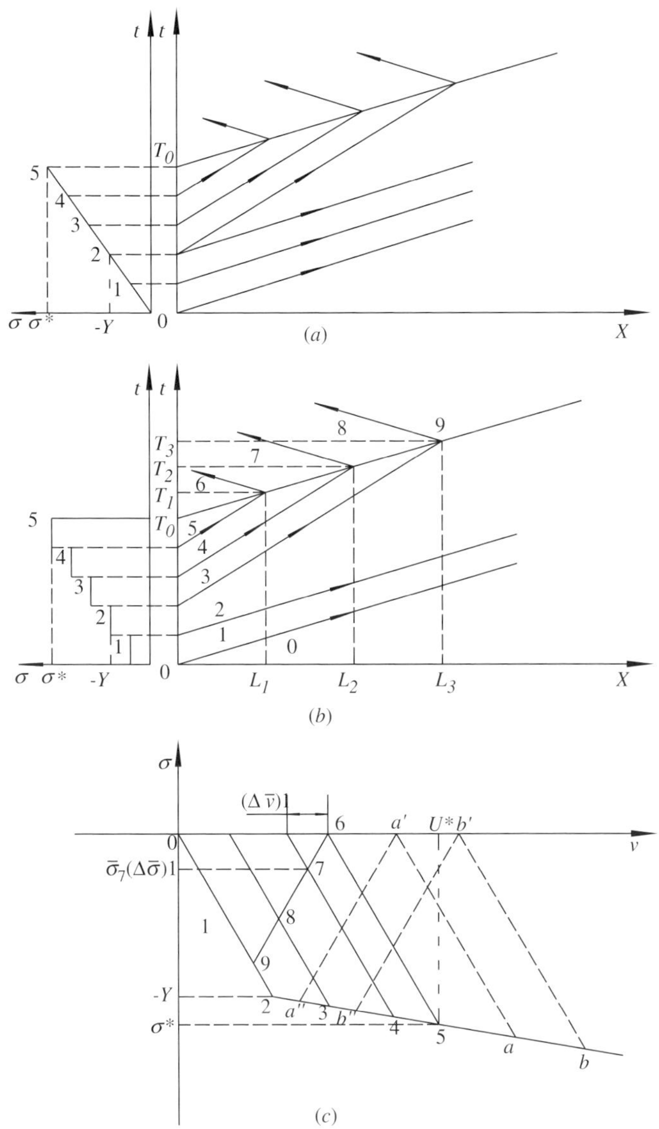

Fig. 4.8. Sudden unloading of continuous loading waves in a linear hardening bar.

Therefore, internal impact occurs. The stress jump $(\Delta\bar{\sigma})_1$ induced by this internal impact is within the unloading region, thus can be calculated for the elastic waves as:

$$(\Delta\bar{\sigma})_1 = -\rho_0 C_0 \frac{(\Delta\bar{v})_1}{2} = \frac{(\Delta\sigma)_1}{2}\left(\frac{C_0}{C_1} - 1\right) = -\frac{1}{2}[(\Delta\sigma)_1 + \rho_0 C_0 (\Delta v)_1] \qquad (4.19)$$

This is corresponding to the point 7 in the σ–v plane as shown in Fig. 4.8(c).

After the first internal impact, on the one hand, the intensity of the plastic wave is weakened by the unloading disturbance, namely, its stress magnitude drops from $|\sigma_5|$ to $|\sigma_4| = |\sigma_5 - (\Delta\sigma)_1| = |\sigma^* - (\Delta\sigma)_1|$. On the other hand, the intensity of the unloading disturbance is weakened by the plastic wave, namely, its stress magnitude drops from $|\sigma_5 - \bar{\sigma}_6| = |\sigma^*|$ to $|\sigma_4 - \bar{\sigma}_7| = |\sigma^* - (\Delta\sigma)_1 - (\Delta\bar{\sigma})_1| = |\sigma^* - [(\Delta\sigma)_1 + (\Delta\bar{\sigma})_1]|$.

Thus, the intensity of the strong-discontinuous unloading disturbance decreases by $|(\Delta\sigma)_1 + (\Delta\bar{\sigma})_1|$.

At the time $t = T_2$, the unloading disturbance catches up with the next small-discontinuous plastic wave and the second internal impact happens at the point $X = L_2$. Similar to Eqs. (4.18) and (4.19), we have:

$$(\Delta\bar{v})_2 = \frac{(\Delta\sigma)_2}{\rho_0}\left(\frac{1}{C_0} - \frac{1}{C_1}\right)$$

$$(\Delta\bar{\sigma})_2 = -\frac{1}{2}[(\Delta\sigma)_2 + \rho_0 C_0 (\Delta v)_2]$$

After the internal impact, the intensity of the secondary stress wave in unloading zone is raised from $|\bar{\sigma}_7|$ to $|\bar{\sigma}_8| = |\bar{\sigma}_7 + (\Delta\bar{\sigma})_2| = |(\Delta\bar{\sigma})_1 + (\Delta\bar{\sigma})_2|$, while the intensity of plastic wave decreases from $|\sigma_4|$ to $|\sigma_3| = |\sigma_4 - (\Delta\sigma)_2|$. Therefore the strong-discontinuous unloading disturbance is weakened to the level:

$$|\sigma_3 - \bar{\sigma}_8| = |\sigma^* - \{(\Delta\sigma)_1 + (\Delta\sigma)_2 + (\Delta\bar{\sigma})_1 + (\Delta\bar{\sigma})_2\}|$$

Compared to the initial stress level σ^*, the intensity of the unloading disturbance drops by

$$|(\Delta\sigma)_1 + (\Delta\sigma)_2 + (\Delta\bar{\sigma})_1 + (\Delta\bar{\sigma})_2|$$

On the analogy of those procedures, when the strong-discontinuous unloading disturbance pursues the ith small-discontinuous plastic wave and the corresponding internal impact occurs, the difference of the residual particle velocity:

$$(\Delta\bar{v})_i = (\Delta v)_i + \frac{(\Delta\sigma)_i}{\rho_0 C_0} = \frac{(\Delta\sigma)_i}{\rho_0}\left(\frac{1}{C_0} - \frac{1}{C_1}\right) \tag{4.20}$$

will induce a stress jump:

$$(\Delta\bar{\sigma})_i = \frac{(\Delta\sigma)_i}{2}\left(\frac{C_0}{C_1} - 1\right) = -\frac{1}{2}[(\Delta\sigma)_i + \rho_0 C_0 (\Delta v)_i] \tag{4.21}$$

so that the secondary stress in the unloading zone is reloaded to a level of:

$$\begin{aligned}\bar{\sigma} &= \sum_{k=1}^{i} (\Delta\bar{\sigma})_k = \sum_{k=1}^{i} \frac{(\Delta\sigma)_k}{2}\left(\frac{C_0}{C_1} - 1\right) = \frac{1}{2}\left(\frac{C_0}{C_1} - 1\right)(\sigma^* - \sigma) \\ &= \frac{1}{2}(\sigma + \rho_0 C_0 v) - \frac{1}{2}(\sigma^* + \rho_0 C_0 v^*)\end{aligned} \tag{4.22}$$

while the strong-discontinuous unloading disturbance is weakened to $|\sigma - \bar{\sigma}|$:

$$
\begin{aligned}
|\sigma - \bar{\sigma}| &= \sigma^* - \sum_{k=1}^{i} [(\Delta\sigma)_k + (\Delta\bar{\sigma})_k] = \sigma^* - \sum_{k=1}^{i} \frac{(\Delta\sigma)_k}{2}\left(\frac{C_0}{C_1} + 1\right) \\
&= \frac{1}{2}\left(1 + \frac{C_0}{C_1}\right)\sigma + \frac{1}{2}\left(1 - \frac{C_0}{C_1}\right)\sigma^* \\
&= \frac{1}{2}(\sigma - \rho_0 C_0 v) + \frac{1}{2}(\sigma^* + \rho_0 C_0 v^*)
\end{aligned} \tag{4.23}
$$

On the σ–v plane shown in Fig. 4.8(c), the intensity of the secondary stress wave, $|\sigma|$, is raised with the internal impact in succession, as illustrated by that the state point moves correspondingly from point 6 to point 7, 8, ... in succession along a positive characteristic line. Obviously, once $|\sigma - \bar{\sigma}| = 0$, the unloading strong-discontinuity vanishes. The point 6 in Fig. 4.8(c) represents the case that the initial residual velocity v^* is less than $2v_Y$ [Eq. (4.16)]. In such a case, the strong-discontinuous unloading disturbance can pass through the whole plastic zone and is satisfied the condition of $|\sigma_2 - \bar{\sigma}| = Y - |\bar{\sigma}| > 0$. In other words, having interacted with all the plastic loading disturbances and causing the plastic region to be unloaded, the unloading disturbance still retains a strong-discontinuity. The final unloading disturbance becomes a strong-discontinuous unloading wave propagating in the elastic loading zone (the point 9). In this case, the locus of the unloading boundary propagation coincides with the locus of the unloading disturbance propagation.

However, if a stronger load such as $|v^*| > 2v_Y$ impacts the bar, then after the strong-discontinuous unloading disturbance interacts with only a part of the plastic loading disturbances, its intensity $|\sigma_2 - \bar{\sigma}|$ will decrease to zero. Namely, the unloading disturbance will be "absorbed" somewhere in the plastic loading zone and cannot pass through the whole plastic zone. It means that the first phase of unloading is finished. An example is shown below:

Suppose that the point a in Fig. 4.8(c) presents the loading state of the plastic loading zone just before the sudden unloading. The initial unloading state is represented by the point a' ($\bar{v}_{a'} > 2v_Y$). Along with the interactions between the strong-discontinuous unloading disturbance and the plastic loading disturbances, the plastic wave intensity $|\sigma|$ decreases along the line aa'', while the secondary stress wave intensity $|\bar{\sigma}|$ increases along the line $a'a''$. At the point a'', the intensity of the unloading disturbance $|\sigma_2 - \bar{\sigma}|$ finally decreases to zero, implying that the unloading disturbance is now "absorbed" by the plastic zone. The plastic waves are only weakened but not yet vanished, and thus they continue to propagate rightward with the maximum stress level represented by the point a''.

Similarly, if the initial loading state is represented by the point b, then the initial unloading state is represented by the point b', and the strong-discontinuous unloading disturbance vanishes at the point b''. From now on, the plastic waves continue to propagate rightwards with the intensity of $|\sigma| \leq |\sigma_b''|$, and will be unloaded again only when the secondary unloading waves arrive. In fact, at the first phase of unloading, the leftwards internal reflective waves will be reflected again at the free end ($X = 0$) of the bar, forming a set of rightwards "secondary unloading waves". Once those secondary unloading waves pursue the forerunning already-weakened plastic loading waves, they will be unloaded secondary, the so-called secondary phase of unloading. Obviously, if the initial plastic waves are strong enough, it is possible that the plastic waves be unloaded by the "third phase unloading", the "fourth phase unloading", and so on.

In the above discussions, when the incremental value between the small-discontinuous plastic loading waves approaches to infinitely small, those plastic disturbances are actually reduced to continuous waves. Accordingly, when the finite jump value is substituted by the infinite small increment, and the summation algorithm is substituted by integral, the above equations should be still effective. In this way, the discussions toward the problem shown in Fig. 4.8(b) are applied to the problem shown in Fig. 4.8(a). Actually, the X–t plot and the σ–v plot are identical for both problems.

The above results can be extended conveniently to the problem where the material is decreasingly hardening plastic, as first analyzed by White and Griffis for the unloading waves (White and Griffis, 1942, 1948). In such cases, the plastic wave speed is no longer a constant value, but a function of strain or stress [Eq. (2.15)], $C = \sqrt{(1/\rho_0)(d\sigma/d\varepsilon)} = C(\sigma)$, then the Eqs. (4.20)–(4.23) are converted to the following forms, respectively:

$$d\bar{v} = dv + \frac{d\sigma}{\rho_0 C_0} = \left(\frac{1}{C_0} - \frac{1}{C}\right)\frac{d\sigma}{\rho_0} \tag{4.24}$$

$$d\bar{\sigma} = -\frac{1}{2}(d\sigma + \rho_0 C_0 dv) = \left(\frac{C_0}{C} - 1\right)\frac{d\sigma}{2} \tag{4.25}$$

$$\bar{\sigma} = \frac{1}{2}\int_{\sigma}^{\sigma^*}\left(\frac{C_0}{C} - 1\right)d\sigma = \frac{1}{2}(\sigma + \rho_0 C_0 v) - \frac{1}{2}(\sigma^* + \rho_0 C_0 v^*) \tag{4.26}$$

$$\begin{aligned} \sigma - \bar{\sigma} &= \frac{1}{2}\int_0^{\sigma}\left(1 + \frac{C_0}{C}\right)d\sigma + \frac{1}{2}\int_0^{\sigma^*}\left(1 - \frac{C_0}{C}\right)d\sigma \\ &= \frac{1}{2}(\sigma - \rho_0 C_0 v) + \frac{1}{2}(\sigma^* + \rho_0 C_0 v^*) \end{aligned} \tag{4.27}$$

As can be seen, the final forms for $\bar{\sigma}$ and $(\sigma - \bar{\sigma})$ do not change.

4.4.3 *Sudden unloading of centered plastic loading waves*

In the following, we consider an example where a strong-discontinuous unloading disturbance pursues a set of weak-discontinuous plastic loading disturbances in a semi-infinite bar, which is made of decreasingly hardening material. The bar is in a natural, rest state before the impact load is applied. At the time $t = 0$, a sudden constant-value impact loading σ^* is applied on the end of the bar ($X = 0$), while at the time $t = T$ the applied load is suddenly released to 0. The associated X–t plot and the σ–v plot are shown in Fig. 4.9 (Bohnenblust, 1942b). Some key points are emphasized below, while the details of the analysis are omitted herein.

(1) When the unloading process begins, the strong-discontinuous unloading disturbance propagates with the wave speed C_0, catching up with the forerunning plastic waves that propagate with the velocity $C(=\sqrt{(1/\rho_0)(d\sigma/d\varepsilon)})$. So long as the unloading wave still keeps a strong-discontinuous, after having interacted with the plastic waves, the propagating locus of the unloading disturbance is the same as the propagating locus of the unloading boundary, namely, the propagating locus of the elastic–plastic boundary $\overline{C} = C_0$. Therefore, on the X–t plane, the location of this boundary segment A_0A_2 can be determined.

(2) To the right side of the boundary A_0A_2, it is the plastic loading zone. According to the previous discussions about elastic–plastic loading waves, the state variables σ and v in this zone can be determined. Particularly, under the situation of the simple waves propagating in a semi-infinite bar, there exists the following simple wave relationship in the whole plastic zone:

$$v = -\int_0^{\sigma} \frac{d\sigma}{\rho_0 C_0} \tag{4.28}$$

To the left side of the boundary A_0A_2, it is the unloading zone, where the state variables need to be calculated. Note the strong-discontinuous feature of this boundary, the state variables $\bar{\sigma}$ and $\bar{v}$ in the unloading zone should satisfy the momentum-conservation equation for a discontinuous wave [Eq. (2.57)], which is:

$$\bar{v} + \frac{\bar{\sigma}}{\rho_0 C_0} = v + \frac{\sigma}{\rho_0 C_0} \tag{4.29}$$

At the same time, the boundary A_0A_2 coincides with the rightward characteristic line in the unloading region that starts from the point A_0. Therefore the state variables $\bar{\sigma}$ and $\bar{v}$ in the unloading zone should satisfy the compatibility equation along this characteristic line [Eq. (4.12)]. Moreover, since the state variables at the point A_0 after unloading have been known as:

$$\bar{\sigma}(A_0) = 0, \quad \bar{v}(A_0) = \bar{v}^* = v^* + \frac{\sigma^*}{\rho_0 C_0}$$

we can get:

$$\bar{v} - \frac{\bar{\sigma}}{\rho_0 C_0} = v^* + \frac{\sigma^*}{\rho_0 C_0} \tag{4.30}$$

Thus the variables $\bar{\sigma}$ and $\bar{v}$ can be simultaneously solved from Eqs. (4.29) and (4.30). For example, we can subtract Eq. (4.30) from Eq. (4.29), and the resulting equation is substituted with Eq. (4.28). This renders:

$$\bar{\sigma} = \frac{1}{2}(\sigma + \rho_0 C_0 v) - \frac{1}{2}(\sigma^* + \rho_0 C_0 v^*) = \frac{1}{2}\int_{\sigma}^{\sigma^*} \left(\frac{C_0}{C} - 1\right) d\sigma$$

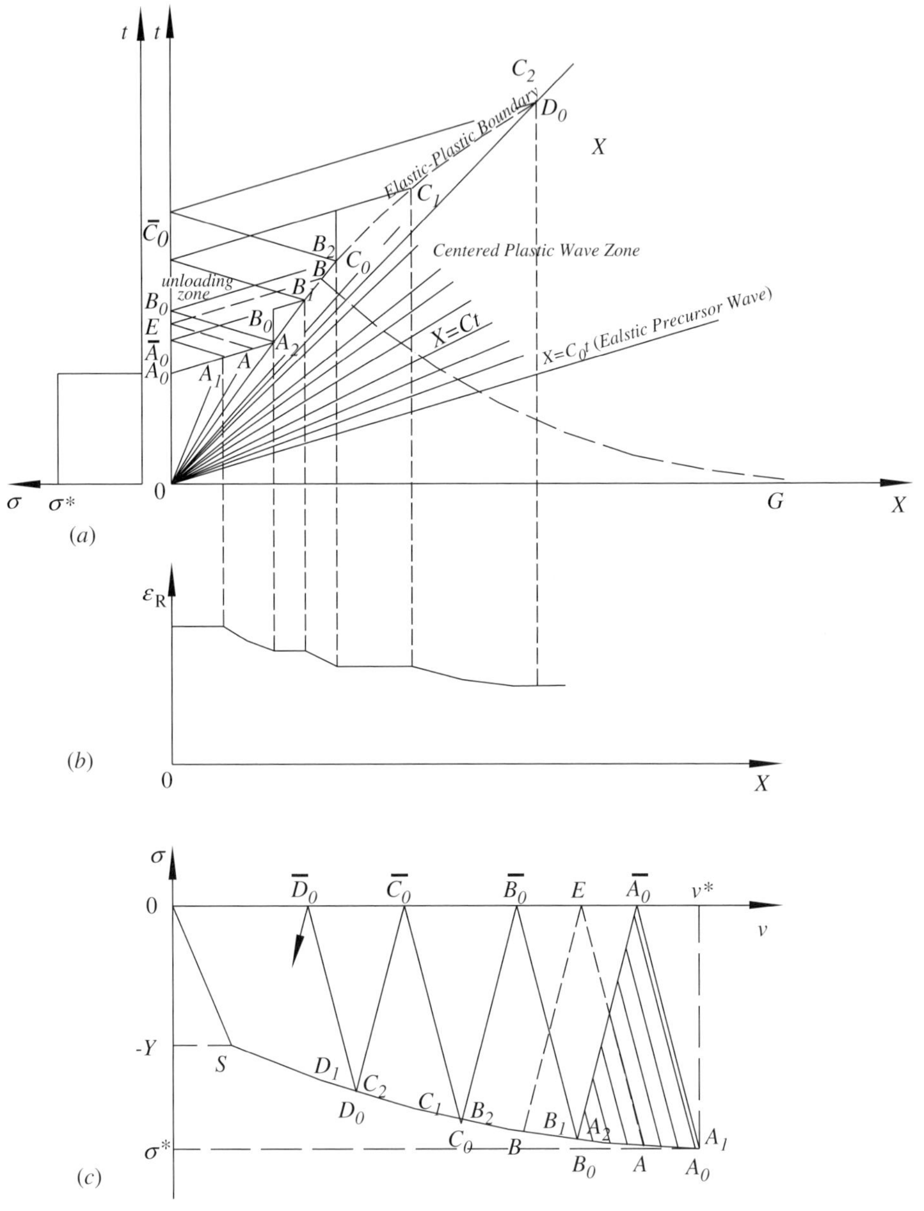

Fig. 4.9. Sudden unloading of centered plastic loading waves.

Furthermore, the stress jump along the unloading boundary can be determined:

$$\sigma - \bar{\sigma} = \frac{1}{2}(\sigma - \rho_0 C_0 v) + \frac{1}{2}(\sigma^* + \rho_0 C_0 v^*)$$
$$= \frac{1}{2}\int_0^{\sigma}\left(1 + \frac{C_0}{C}\right)d\sigma + \frac{1}{2}\int_0^{\sigma^*}\left(1 - \frac{C_0}{C}\right)d\sigma$$

Note that the above two equations are just the same as that have been obtained previously when we discussed the internal impact between an unloading disturbance and a plastic loading disturbance [Eqs. (4.26) and (4.27)]. Here the equations are deduced by the characteristics method.

(3) On the plastic loading side of the elastic–plastic boundary, the lowest stress level is the plastic yield limit Y. If the condition $|\sigma - \bar{\sigma}| > 0$ is satisfied even though the loading stress level σ is less than Y, then the strong-discontinuous unloading disturbance can pass through the whole plastic zone and does not vanish. According to Eq. (4.27), the condition that the strong discontinuous unloading wave will not be absorbed by the plastic loading zone, is:

$$v^* + \frac{\sigma^*}{\rho_0 C_0} = \bar{v}^* < 2v_Y$$

Note that this formula has also been obtained previously [Eq. (4.16)]. In the contrary situation, if the strong-discontinuity vanishes at a certain point, say the point A_2, then at this point we must have: $\sigma(A_2) - \bar{\sigma}(A_2) = 0$. According to Eq. (4.27) we get:

$$v(A_2) - \frac{\sigma(A_2)}{\rho_0 C_0} = v(A_0) + \frac{\sigma(A_0)}{\rho_0 C_0} \tag{4.31a}$$

or

$$\int_0^{\sigma(A_2)}\left(1 + \frac{C_0}{C}\right)d\sigma = \int_0^{\sigma^*}\left(\frac{C_0}{C} - 1\right)d\sigma \tag{4.31b}$$

From this equation we can determine the value of $\sigma(A_2)$. Therefore the intersection point A_2 of the straight line OA_2 whose slope is $C(\sigma(A_2))$ and the straight line A_0A_2 whose slope is C_0 can be determined. As a result, all information about the elastic–plastic boundary A_0A_2 in the first phase of unloading is known. This includes the location of A_0A_2 on the X–t plane, and the corresponding state map of A_0A_2 on the σ–v plane. Thus, the state variables at the zones in both sides of A_0A_2, namely σ, v, $\bar{\sigma}$, and $\bar{v}$, can all be determined.

(4) With regard to the unloading zone, since the variables $\bar{\sigma}$ and $\bar{v}$ along the characteristic line A_0A_2 have been determined and the boundary condition governs that $\bar{\sigma}|_{x=0} = 0$ along the time axis, the unloading zone problem is reduced to a mixed boundary (Picard) problem in the case of $|v^*| < |2v_Y|$ (the point A_2 is infinitely far away). However, in the case that $|v^*| > |2v_Y|$, only in the triangle region $A_0A_2\bar{B}_0$,

the problem is reduced to a mixed boundary problem, where the point $\overline{B}_0$ is the intersection of the time axis with the leftward characteristic line emitted from the point A_2. Within this region, along the leftward characteristic lines what propagate are the influences of interactions between the unloading disturbance and the plastic loading disturbances, while along the rightward characteristic lines what propagate are the influences of boundary disturbances. For example, at an arbitrary point E on t-axis the particle velocity $\bar{v}_E$ can be calculated by the compatibility condition along the leftward characteristic line AE and the boundary condition $\bar{\sigma}_E = 0$. As the state variables at the point A ($\bar{\sigma}_A$ and $\bar{v}_A$) have been known, the variable $\bar{v}_E$ is:

$$\bar{v}_E = \bar{v}_A + \frac{\bar{\sigma}_A}{\rho_0 C_0} \tag{4.32}$$

(5) If the strong-discontinuous unloading disturbance is absorbed by the plastic loading zone and vanishes at the point A_2, then after this point the elastic–plastic boundary is reduced to a weak-discontinuous unloading boundary. Across this boundary, the stress and the particle velocity should satisfy the continuity conditions:

$$\bar{\sigma} = \sigma, \quad \bar{v} = v, \text{ along the weak-discontinuous boundary} \tag{4.33}$$

However, the locus of this boundary is not known in advance and this is just what we need to solve. To determine the location of this boundary, now consider an arbitrary point on the boundary, say the point B. In the plastic loading zone, the leftward characteristic line passing through the point B intersects X-axis at the point G. According to the compatibility condition along BG [Eq. (2.26)] and the zero initial condition, or according to the simple wave relationship Eq. (4.28), we have:

$$v_B + \int_0^{\sigma_B} \frac{d\sigma}{\rho_0 C} = 0 \tag{4.34}$$

On the other hand, in the unloading zone, a rightward characteristic line passing through point B intersects t-axis at the point E. According to the compatibility condition along BE [Eq. (4.12)], and taking account of Eq. (4.32), we have:

$$\bar{v}_B - \frac{\bar{\sigma}_B}{\rho_0 C_0} = \bar{v}_E = \bar{v}_A + \frac{\bar{\sigma}_A}{\rho_0 C_0} \tag{4.35a}$$

Since the state variables at the point A are known, the above two equations, and Eq. (4.33) applied to the point B, totally provide four equations for the unknown variables v_b, σ_B, $\bar{v}_B$, and $\bar{\sigma}_B$. Alternatively, substituting Eqs. (4.33) and (4.29) into Eq. (4.35a), we get:

$$v_B - \frac{\sigma_B}{\rho_0 C_0} = v_A + \frac{\sigma_A}{\rho_0 C_0} \tag{4.35b}$$

Therefore from Eqs. (4.34) and (4.35b) we can solve for the unknown variables v_b and σ_B. The rightward characteristic line OB in the plastic loading region has a slope $C(\sigma_B)$. The location of point B can be determined by the intersection of this line with the rightward characteristic line EB in the unloading region, while the point E is the point where the leftward characteristic line AE intersects with the t-axis. Therefore, the state of the arbitrary point B on the weak-discontinuous unloading boundary during the second unloading phase is completely determined by the state of the corresponding point A on the unloading boundary during the first unloading phase. In other words, when A is known, B can be determined. The particle velocity and the stress at the points A and B are related by Eq. (4.35a) or Eq. (4.35b). These relationships are called the **conjugate relationship**. The point pair on the unloading boundary like A and B are called the **conjugate points**.

(6) Since the point B is an arbitrary point, the conjugate relationship is held on through the whole unloading boundary. Therefore, from the already determined boundary A_0A_2 in the first phase of unloading, the conjugate boundary segment B_0B_2 can be obtained. Particularly, by applying Eq. (4.35b) to the point A_0 and comparing it with Eq. (4.31a), it is clear that the starting point of B_0B_2 (the point B_0) is just the finishing point of A_0A_2 (the point A_2), and consequently, corresponding to the constant-value part A_0A_1, the boundary segment B_0B_1 is also a constant-value part. So the boundary segment B_0B_1 must coincide with the characteristic line OB_1 in the plastic simple wave region. Since B_0B_2 has been determined from A_0A_2, we can use the conjugate relationship to determine the boundary segment C_0C_2 from B_0B_2. By the analogy of this approach, the whole unloading boundary can be solved until all plastic waves vanish.

(7) With regard to the unloading region, since the stress $\bar{\sigma}$ and the particle velocity $\bar{v}$ along the whole unloading boundary are known, as well as the boundary stress $\bar{\sigma}$ or boundary velocity $\bar{v}$ along the t-axis has been prescribed by the boundary condition, the solution within the whole unloading region can be reduced to one of the different kinds of solution-determinable boundary value problems, like a mixed boundary problem, or a characteristic line boundary problem, etc. Thus the problem can be solved accordingly.

(8) On the σ–v plane [Fig. 4.9(c)], the line OS corresponds to the strong-discontinuous elastic wave, and the curve SA_0 corresponds to the centered plastic wave [Eq. (4.28)]. In the whole plastic zone, including the plastic side of the unloading boundary, the Eq. (4.28) is satisfied. So the curve SA_0 is the mapping of the plastic side along the whole unloading boundary. As for the strong-discontinuous unloading boundary A_0A_2, the states on the unloading side start from the point $\overline{A}_0$ and change along the line $\overline{A}_0A_2$ up to the point A_2, where $\sigma - \bar{\sigma} = 0$, namely the unloading strong-discontinuity vanishes. Therefore the line $\overline{A}_0A_2$ is the mapping of the unloading side along the strong-discontinuous unloading boundary A_0A_2. As for the weak-discontinuous unloading boundary, since the continuity conditions for σ and v [Eq. (4.33)] are satisfied, the line SB_0 is precisely the mapping of the whole weak-discontinuous unloading boundary. It is thus clear that although the location of the unloading boundary on the X–t plane is not yet known, their mapping on the σ–v plane can be determined immediately. Besides, an arbitrary point A and its conjugate point B on the unloading boundary can be easily determined by the characteristic lines AE and EB on the σ–v plane. The readers are suggested to deduce

the conjugate relationship [Eq. (4.35)] from the geometric relationship of the curved triangle *AEB*.

(9) From this example, it is thus clear that if the unloading disturbance is in the form of a strong-discontinuity from the very beginning, or, if the unloading disturbance propagates in a constant-value plastic region from the very beginning regardless of whether it is a strong-discontinuity or a weak-discontinuity, then the locus of the unloading disturbance propagation coincides with the locus of unloading boundary, $\overline{C} = C_0$. However, in general, the propagating locus of the weak-discontinuous unloading boundary is different from the propagating locus of the unloading disturbance.

(10) The distribution of residual strain ε_R along the bar is shown in Fig. 4.9(b). Each constant-value plastic zone on the X–t plane is corresponding to a constant-strain platform on the ε_R–X plot. Comparing with the strain distribution given in Fig. 4.7, the stationary strong-discontinuous strain interface in Fig. 4.7 is spread into a "transition region", wherein the residual strain gradually changes from one step level to the next step level. This is because that the plastic loading waves herein is a continuous wave, so that the residual strain distribution is weak-discontinuous and thus no stationary strain-discontinuity appears.

Now we summarize the above discussions. The pursuing unloading problem where an unloading disturbance pursues and interacts with the forerunning plastic loading disturbance in the semi-infinite bar has been analyzed. We have shown that when the stress–strain relationship of material, as well as the initial-boundary conditions are known, this problem can be solved so long as the unloading disturbance from the very beginning is a strong-discontinuity, such as the case of sudden unloading, no matter whether the forerunning plastic loading disturbance is a strong- or a weak-discontinuity. This problem is classified to solve a kind of **direct problem**. It is solvable because first of all, the sudden unloading particularity can be utilized. In the case of sudden unloading, the initial part of the unloading boundary is a strong-discontinuous boundary, which coincides with the propagating locus of the unloading disturbance. Therefore, this initial part of the unloading boundary, the strong-discontinuous part, is determinable. Second, once this strong-discontinuous unloading boundary vanishes, the conjugate relationship can be utilized to determine the successive parts of the weak-discontinuous unloading boundary piece by piece. Finally, after the whole unloading boundary is determined, the rest of the problems within the unloading region are reduced to solving either a mixed boundary value problem, or a characteristic boundary value problem, both of which are determinable.

4.5 Pursuing Unloading by Weak-Discontinuous Unloading Disturbances

Now we consider a semi-infinite bar made of decreasingly hardening material. The bar is in a natural, rest state before the impact load is applied. The boundary loading on the end of the bar ($X = 0$) is prescribed as: $\sigma(0, t) = \sigma_0(t)$, where the function of $\sigma_0(t)$ is graphically shown in Fig. 4.10, namely, first gradually increasing in the duration $0 < t < t_0$, then gradually decreasing down to zero in the duration $t_0 < t < T_0$, and finally keeping zero

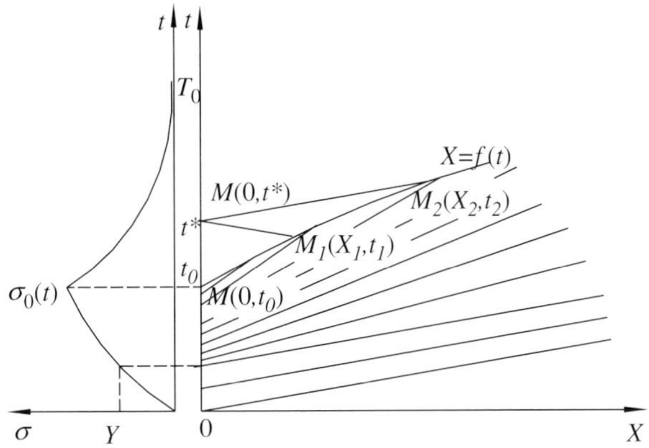

Fig. 4.10. Pursuing unloading of weak-discontinuous loading disturbances by weak-discontinuous unloading disturbances.

(i.e. the stress-free end condition) for $t \geq T_0$. This is a kind of boundary loading what we frequently encounter when studying the dynamic response of structures under explosive loading.

Now the problem we will solve is such a problem where the after- running weak-discontinuous unloading disturbances pursue the forerunning weak-discontinuous plastic-loading disturbance. The interaction of these two kinds of disturbances results in an unloading process and consequently a weak-discontinuous unloading boundary. On this unloading boundary $X = f(t)$ the particle velocity, stress, and strain are all continuous:

$$\begin{cases} \bar{v}(X,t)|_{X=f(t)} = v(X,t)|_{X=f(t)} = v_m(X) \\ \bar{\sigma}(X,t)|_{X=f(t)} = \sigma(X,t)|_{X=f(t)} = \sigma_m(X) \\ \bar{\varepsilon}(X,t)|_{X=f(t)} = \varepsilon(X,t)|_{X=f(t)} = \varepsilon_m(X) \end{cases} \tag{4.36}$$

However, the derivative of the above variables may be discontinuous across the unloading boundary $X = f(t)$. The complexity lies in that, in general, the whole weak-discontinuous unloading boundary $X = f(t)$ is completely unknown at the present stage.

Consider an arbitrary point $M_2(X_2, t_2)$ on the boundary $X = f(t)$. In the unloading zone, the rightward characteristic line passing through the point M_2, the line MM_2, intersects the t-axis at the point $M(0, t^*)$, while the leftward characteristic line passing through the point M intersects the unloading boundary $X = f(t)$ at the point $M_1(X_1, t_1)$. According to the compatibility equation [Eq. (4.12)] along the characteristic line MM_2, we have the following relationship that links the state variables at points M and M_2:

$$\rho_0 C_0 \bar{v}(X_2, t_2) - \bar{\sigma}(X_2, t_2) = \rho_0 C_0 \bar{v}(0, t^*) - \bar{\sigma}(0, t^*)$$

And, from the compatibility equation along the characteristic line MM_1, we have the following relationship that links the state variables at points M and M_1:

$$\rho_0 C_0 \bar{v}(X_1, t_1) + \bar{\sigma}(X_1, t_1) = \rho_0 C_0 \bar{v}(0, t^*) + \bar{\sigma}(0, t^*)$$

Eliminating the variable $\bar{v}(0, t^*)$ from the above two equations, and using the continuity relationship of Eq. (4.36), we obtain:

$$\rho_0 C_0 v_m(X_2) - \sigma_m(X_2) = \rho_0 C_0 v_m(X_1) + \sigma_m(X_1) - 2\sigma_0(t^*) \tag{4.37a}$$

This is an extension of the conjugate relationship given in Eq. (4.35) to the case that the unloading stress at bar end $\sigma_0(0, t)$ is not zero. Note that as the weak-discontinuous unloading boundary is now propagating in a simple wave region, the simple wave relationship [Eq. (2.63)] must be satisfied. After introducing the wave speed ratio μ:

$$\mu(\sigma) = \frac{C_0}{C(\sigma)} \geq 1$$

The conjugate relationship Eq. (4.37a) is rewritten as:

$$\int_0^{\sigma_m(X_2)} (\mu + 1) d\sigma = \int_0^{\sigma_m(X_1)} (\mu - 1) d\sigma + 2\sigma_0(t^*) \tag{4.37b}$$

which is an extension of Eq. (4.31b) given in the previous section.

From Eq. (4.37), it is clear that any point M_2 on the weak-discontinuous boundary $X = f(t)$ is linked to its conjugate point M_1 through the unloading boundary condition $\sigma_0(t^*)$. This is the most important relationship for studying the weak-discontinuous boundary. By this basic equation, if we have known a point M_1 on $X = f(t)$, then the conjugate point M_2 on $X = f(t)$ can be calculated. By the same argument, if a small part of $X = f(t)$, say the segment M_0M_1 is known, then the rest of the unloading boundary can be calculated piece by piece. However, in general, even a very short segment of the unloading boundary is unknown in advance. Thus, the key is how to determine an initial short segment of the weak-discontinuous boundary $X = f(t)$. This will be further discussed in Section 4.7.

4.5.1 Continuously unloading of centered plastic waves

If the end of the bar is loaded suddenly and immediately unloaded gradually, then the unloading problem is relatively easy to treat. In this case, since the plastic zone is a simple, centered wave region as shown in Fig. 4.11, at any point on the weak-discontinuous unloading boundary, we have the following relationship:

$$\frac{X}{t} = C[\sigma_m(X)] \quad on\ X = f(t) \tag{4.38a}$$

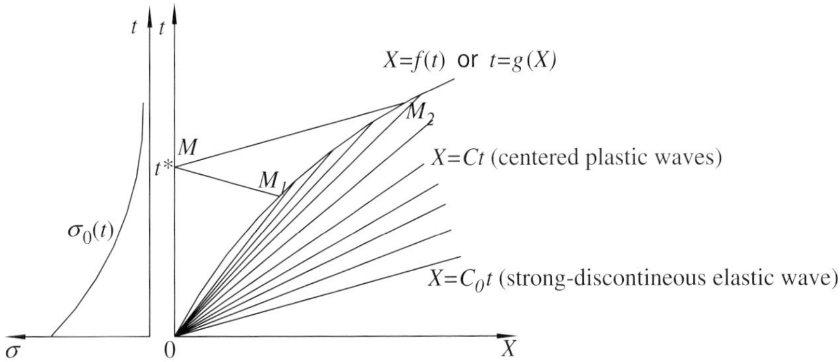

Fig. 4.11. Continuously unloading of centered plastic waves.

Using the geometric relationship between the characteristic lines MM_1 and MM_2 in the unloading zone, we obtain:

$$X_1 = \frac{C_0 t^*}{\mu[\sigma_m(X_1)] + 1}, \quad X_2 = \frac{C_0 t^*}{\mu[\sigma_m(X_2)] - 1} \tag{4.38b}$$

If the stress–strain relationship of the material is known, and if the residual strain distribution along the bar $\varepsilon_R(X)$ has also been experimentally measured, which is equivalent to that, the function $\sigma_m(X)$ and the function $\varepsilon_m(X)$ can be determined according to the unloading stress–strain relationship [Eq. (4.6)]:

$$\varepsilon_m(X) - \frac{\sigma_m(X)}{E} = \varepsilon_R(X)$$

meanwhile the unloading boundary location $X = f(t)$ can be determined from Eq. (4.38), then the loading character at the bar end $\sigma_0(t)$ can be solved from Eq. (4.37). Such kind of problem is classified as the so-called **inverse problem**, namely, to determine the boundary condition from the known stress–strain relationship of material and the known dynamic response of the bar. It is clear from the above example that if the boundary loading is suddenly applied then the inverse problem is solvable.

As to the **direct problem** for the same example discussed, it can be briefly stated as: when the stress–strain relationship of the material and the initial-boundary conditions are known, the demand is to determine the dynamic response of the bar, such as the stress, strain, and the particle velocity as functions of X and t, or the residual strain distribution $\varepsilon_R(X)$. Although the accurate solution is almost unable to be found in the case of the present example, nevertheless the approximate solution is obtainable, as will be discussed in the following.

First, a **progressive approximation method** is introduced to solve the direct problem for the case shown in Fig. 4.11.

Under the boundary condition of suddenly applied loading, Eq. (4.37b) can be rewritten as below by using Eq. (4.38):

$$\int_0^{\sigma_m(\nu X)}[\mu(\sigma)+1]d\sigma=\int_0^{\sigma_m(X)}[\mu(\sigma)-1]d\sigma+2\sigma_0\left[\frac{(\mu[\sigma_m(X)]+1)X}{C_0}\right] \tag{4.39a}$$

where

$$\nu=\frac{\mu[\sigma_m(X)]+1}{\mu[\sigma_m(\nu X)]-1} \tag{4.39b}$$

In case that $C_0\to\infty$, $\nu\to 1$, we have:

$$\sigma_m(X)=2\sigma_0\left(\frac{X}{C}\right) \tag{4.40}$$

This is equivalent to approximately regarding the unloading part of the bar as a rigid body, of which the inertial effect is neglected. By this equation we can determine from the boundary condition $\sigma_0(t)$, the maximum plastic stress distribution along the bar $\sigma_m(X)$ just before unloading. Simultaneously we can use Eq. (4.38a) to determine the unloading boundary $X=f(t)$, or its inverse function $t=g(X)$, namely:

$$t=g(X)=\frac{X}{C[\sigma_m(X)]} \tag{4.41}$$

Obviously the solution thus obtained does not satisfy the elastic unloading Eq. (4.39a). Nevertheless, we can use this solution as the 0th order approximate solution, written as $\sigma_m^{(0)}(X)$ and $g^{(0)}(X)$. This 0th order solution is substituted into the right side of Eq. (4.39a) again to obtain the 1st order solution, $\sigma_m^{(1)}(X)$ and $g^{(1)}(X)=X\Big/\left\{C\left[\sigma_m^{(1)}(X)\right]\right\}$, from the left side of Eq. (4.39a). Similarly, substitute the 1st order solutions into the right side of Eq. (4.39a), then the left side of Eq. (4.39a) provides the 2nd order solutions: $\sigma_m^{(2)}(X)$ and $g^{(2)}(X)=X\Big/\left\{C\left[\sigma_m^{(2)}(X)\right]\right\}$. Repeating this approach, the iterative formula for the nth order approximate solution can be written as:

$$\int_0^{\sigma_m^{(n)}(\nu X)}[\mu(\sigma)+1]d\sigma=\int_0^{\sigma_m^{(n-1)}(X)}[\mu(\sigma)-1]d\sigma+2\sigma_0\left[\frac{\left(\mu\left[\sigma_m^{(n-1)}(X)\right]+1\right)X}{C_0}\right] \tag{4.42}$$

In substance, the above approximate approach is like the previous approach where the point M_2 is calculated from the point M_1 by the conjugate relationship [Eq. (4.37)]. Now the 0th order solutions (equivalent to the point M_1) are used to obtain the 1st order solutions (equivalent to the point M_2) through the conjugate Eq. (4.39). Again the 1st order solutions (now equivalent to point M_1) provide the 2nd order solutions (now equivalent to point M_2), and so on. By each iterative operation, the solutions are corrected once again through the boundary condition at the bar end $\sigma_0(t)$ and approach the exact solutions. The choice of the 0th order solution does not matter much to the final results, but can affect the number of iterations to get the accurate solutions. In the above discussion, Eq. (4.40) was taken as the 0th order solution, which was first suggested by Буравцев (Buravtsev) (1970). However, the progressive approximation algorithm used by Буравцев subjects to the limitation that $1/\mu \ll 1$, which is no longer needed in the present algorithm modified by the present author.

4.5.2 Attenuation of shock waves in linearly hardening materials

In the previous example, if the material is linearly hardening plastic instead of decreasingly hardening plastic, then rather than being a set of continuous waves, the plastic wave will be a strong-discontinuous shock wave propagating with constant velocity $C_1 = \sqrt{E_1/\rho}$. The unloading process is a pursuing interaction between weak-discontinuous unloading disturbances and a strong-discontinuous plastic loading disturbance propagating with constant velocity C_1.

Obviously, in this situation, the propagating locus of the unloading boundary coincides with the propagating locus of the strong-discontinuous plastic loading disturbance, namely $\overline{C} = C_1$. On the X–t plane, the unloading boundary is expressed as:

$$X = f(t) = \overline{C}t = C_1 t$$

This result can also be obtained from the previous problem shown in Fig. 4.10. In the case of linearly hardening material, first the rightward characteristic lines in the plastic loading region in Fig. 4.10 become a set of parallel lines with constant slope C_1, then let $t_0 \to 0$ (representing the boundary condition of sudden loading), the original problem is reduced to the present problem, as shown in Fig. 4.12. Note that the strong-discontinuity

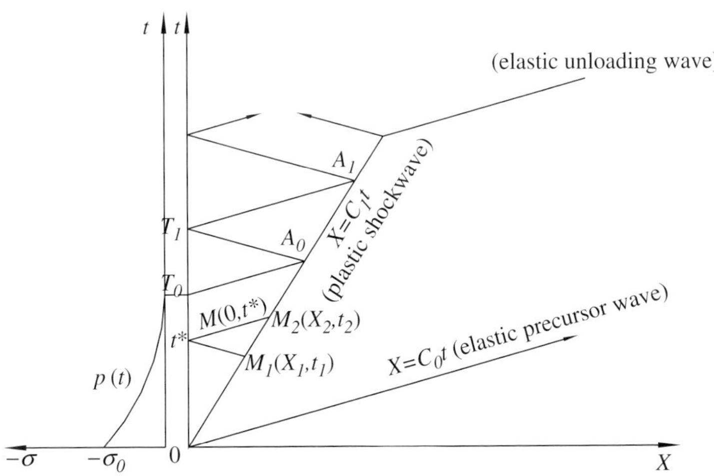

Fig. 4.12. Pursuing unloading of constant-velocity propagating strong-discontinuous plastic loading disturbances by weak-discontinuous unloading disturbances.

along $X = C_1 t$ is attributed to the plastic suddenly loading process. The unloading process is, in nature, still a weak-discontinuity.

Because of the linearly hardening plastic behavior, the conjugate relationship [Eq. (4.37)] is rewritten as:

$$(\mu_1 + 1)\sigma_m(X_2) = (\mu_1 - 1)\sigma_m(X_1) + 2\sigma_0(t^*)$$
$$\mu_1 = \frac{C_0}{C_1} = \sqrt{\frac{E}{E_1}} \tag{4.43}$$

which, by taking account of Eq. (4.38), can be further reduced to:

$$\frac{1}{2}(\mu_1 + 1)\sigma_m\left[\frac{C_0 t}{\mu_1 + 1}\right] - \frac{1}{2}(\mu_1 + 1)\sigma_m\left[\frac{C_0 t}{\mu_1 - 1}\right] = -\sigma_0(t) = p(t) \tag{4.44}$$

Assume that the function of boundary unloading pressure, $p(t)$, can be expanded in a power series [Рахматулин (Rakhmatulin), 1945b]:

$$p(t) = \sum_{n=0}^{\infty} p_n t^n \tag{4.45a}$$

Then from Eq. (4.44) we know that $\sigma_m(X)$ should have the following form:

$$\sigma_m(X) = \sum_{n=0}^{\infty} b_n X^n \tag{4.45b}$$

where

$$b_n = \frac{2p_n}{(\mu_1 - 1)\left(\dfrac{C_0}{\mu_1 + 1}\right)^n - (\mu_1 + 1)\left(\dfrac{C_0}{\mu_1 - 1}\right)^n}$$

Accordingly, the functions of strain $\varepsilon_m(X)$ and the residual strain distribution $\varepsilon_R(X)$ are expressed as:

$$\varepsilon_m(X) = \frac{\sigma_m(X)}{E_1} + \left(1 - \mu_1^2\right)\varepsilon_Y = \frac{1}{E_1}\sum_{n=0}^{\infty} b_n X^n + \left(1 - \mu_1^2\right)\varepsilon_Y \tag{4.46}$$

$$\varepsilon_R(X) = \varepsilon_m(X) - \frac{\sigma_m(X)}{E} = \left(\frac{1}{E_1} - \frac{1}{E}\right)\sum_{n=0}^{\infty} b_n X^n + \left(1 - \mu_1^2\right)\varepsilon_Y \tag{4.47}$$

Therefore the direct problem is solvable. This approach is called a **power series approximate method**.

The length of plastic deformation l can be determined from Eq. (4.47) by the condition $\varepsilon_R(l) = 0$. For example, suppose that the boundary unloading pressure, $p(t)$, has a form:

$$p(t) = p_m \left\{1 - \left(\frac{t}{T_0}\right)^n\right\} \tag{4.48}$$

where p_m is the peak pressure applied on the bar end when $t = 0$. T_0 is the time when boundary load is unloaded to zero. We can obtain

$$l = \frac{C_1 T_0}{1 - \dfrac{1}{\mu_1^2}} \left\{\frac{\mu_1}{2}\left[\left(1 + \frac{1}{\mu_1}\right)^{n+1} - \left(1 - \frac{1}{\mu_1}\right)^{n+1}\right]\left(1 - \frac{Y}{p_m}\right)\right\}^{\frac{1}{n}} \tag{4.49}$$

Generally $1/\mu = (C_1/C_0) < 1$, when we omit the small term $1/\mu^2$ and those of the higher order, Eq. (4.49) is reduced to

$$l = C_1 T_0 \left\{(n+1)\left(1 - \frac{Y}{p_m}\right)\right\}^{\frac{1}{n}}$$

It is thus clear that the length of plastic deformation l increases with the plastic wave velocity of the linearly hardening material, C_1, the maximum boundary-loading p_m, and the boundary loading duration T_0, while it is independent of the material elastic modulus E.

Suppose that the boundary loading is linearly decreasing with time. This means that the index n of Eq. (4.48) is 1. Or equivalently, from Eq. (4.45a) we take

$$p_0 = p_m, \quad p_1 = -\frac{p_m}{T_0}, \quad p_i = 0\ (i = 2, 3, \ldots)$$

Then from Eq. (4.45b) we have:

$$b_0 = -p_m, \quad b_1 = \frac{\mu_1^2 - 1}{2\mu_1}\frac{p_m}{C_0 T_0}, \quad b_i = 0\ (i = 2, 3, \ldots)$$

Therefore we have

$$\sigma_m(X) = -p_m\left(1 - \frac{\mu_1^2 - 1}{2\mu_1}\frac{X}{C_0 T_0}\right) \tag{4.50}$$

From Eq. (4.49) we get

$$l = \frac{2C_1 T_0\left(1 - \dfrac{Y}{p_m}\right)}{1 - \dfrac{1}{\mu_1^2}}$$

Of course, the same problem can be solved by the **progressive approximation method** based on Eq. (4.39). In case of linearly hardening material, Eq. (4.39) is reduced to:

$$\sigma_m(\nu_1 X) = \frac{1}{\nu_1}\sigma_m(X) + \frac{2}{\mu_1+1}\sigma_0\left[\frac{(\mu_1+1)X}{C_0}\right] \tag{4.51a}$$

or

$$\sigma_m(X) = \lambda_1\sigma_m(\lambda_1 X) + \frac{2}{\mu_1+1}\sigma_0\left[\frac{(\mu_1+1)\lambda_1 X}{C_0}\right] \tag{4.51b}$$

where the parameters ν_1 and λ_1 are known constants:

$$\nu_1 = \frac{\mu_1+1}{\mu_1-1} = \frac{1}{\lambda_1}$$

Take the solution when $C_0 \to \infty$ as the 0th order approximation, namely from Eq. (4.48) when $n = 1$, we have:

$$\sigma_m^{(0)}(X) = -p_m\left(1 - \frac{X}{C_1 T_0}\right)$$

Substitute this 0th order approximation into the right side of Eq. (4.51), the left side of Eq. (4.51) provides the 1st order approximate solution:

$$\begin{aligned}\sigma_m^{(1)}(X) &= -\lambda_1 p_m\left(1 - \frac{\lambda_1 X}{C_1 T_0}\right) - \frac{2}{\mu_1+1}p_m\left(1 - \frac{(\mu_1+1)\lambda_1 X}{C_0 T_0}\right)\\ &= -p_m\left[1 - \left(2\lambda_1 + \lambda_1^2\mu_1\right)\frac{X}{C_0 T_0}\right]\end{aligned}$$

Similarly, substitute this 1st order solution into the right side of Eq. (4.51), the left side of Eq. (4.51) provides the 2nd order approximate solution:

$$\begin{aligned}\sigma_m^{(2)}(X) &= \lambda_1\sigma_m^{(1)}(X) + \frac{2}{\mu_1+1}\sigma_0\left[\frac{(\mu_1+1)\lambda_1 X}{C_0}\right]\\ &= -\lambda_1 p_m\left[1 - \left(2\lambda_1 + \lambda_1^2\mu_1\right)\frac{X}{C_0 T_0}\right] - \frac{2}{\mu_1+1}p_m\left(1 - \frac{(\mu_1+1)\lambda_1 X}{C_0 T_0}\right)\\ &= -p_m\left\{1 - \left[2\lambda_1\left(1+\lambda_1^2\right) + \lambda_1^4\mu_1\right]\frac{X}{C_0 T_0}\right\}\end{aligned}$$

Repeating this approach, the nth order approximate solution can be expressed as:

$$\begin{aligned}\sigma_m^{(n)}(X) &= -p_m\left\{1 - \left[2\lambda_1\left(1+\lambda_1^2+\lambda_1^4+\cdots+\lambda_1^{2(n-1)}\right) + \lambda_1^{2n}\mu_1\right]\frac{X}{C_0 T_0}\right\}\\ &= -p_m\left\{1 - \left[\frac{2\lambda_1(1-\lambda_1^{2n})}{1-\lambda_1^2} + \lambda_1^{2n}\mu_1\right]\frac{X}{C_0 T_0}\right\}\end{aligned} \tag{4.52}$$

Table 4.1. The B values calculated by different equations for $\mu_1 = 2$.

Calculation equation	Exact solution [Eq. (4.50)]	Approximate solution of Eq. (4.52)				
		0th order	1st order	2nd order	3rd order	4th order
B value	$3/4 = 0.750$	2.00	0.889	0.765	0.752	0.750

As $\lambda_1 < 1$, when $n \to \infty$ we get:

$$\lim_{n \to \infty} \sigma_m^{(n)}(X) = -p_m \left\{ 1 - \frac{\mu_1^2 - 1}{2\mu_1} \frac{X}{C_0 T_0} \right\}$$

which is exactly the same as the formula in Eq. (4.50). As can be seen, by two different methods we get the same results.

In the nondimensional form, the approximate solution given by Eq. (4.52) and the exact solution given by Eq. (4.50) can be unified as the following linear functions:

$$\left(\frac{\sigma_m(X)}{-p_m} \right) = 1 - B\left(\frac{X}{C_0 T_0} \right) \tag{4.53}$$

The differences between the different order approximate solution and the exact solution are expressed by the coefficient B, the slope of this linear function. Assume $\mu_1 = 2$, the B value calculated by Eqs. (4.50) and (4.52), respectively, are listed in Table 4.1. As can be seen, comparing the B values calculated up to three significant digits, the 4th order approximation gives the same results as the exact solution. Of course, the speed of divergence is also influenced by the proper choice of the 0th order approximate form.

It should be pointed out that the above solutions are applicable to the OA_0 segment in Fig. 4.12 corresponding to the unloading stage when $|p(t)| > 0$. At the time when $t \geq T_0$, the boundary load becomes zero, $p(t) = 0$. The solution at $t \geq T_0$ can be calculated from the OA_0 by the conjugate relationship [Eq. (4.35)]. Note that at the time $t = T_0$, the first order derivative of $p(t)$ is discontinuous, the zigzag curve $T_0A_0T_1A_1\ldots$ represents the propagating locus of this weak-discontinuity.

4.6 Shock Wave Attenuation Due to Pursuing Unloading

In the previous section we have discussed the pursuing unloading problem where a weak-discontinuous unloading disturbance pursues and interacts with a strong-discontinuous plastic wave (plastic shock wave), but the material is elastic–plastic with linearly hardening character so that the speed of plastic shock wave is constant. If the material is increasingly hardening plastic, the speed of plastic shock wave is no longer constant but changes with the intensity of shock wave. The problem can be treated in the similar way as discussed previously, except that the varied shock wave velocity, $\mathcal{D}$, should be taken into account.

In such a case, since the unloading boundary still coincides with the propagating locus of the shock wave, the problem to determine the unloading boundary is equivalent to the problem to determine the propagating locus of the shock wave which is interacted with the tail-following unloading disturbance. Alternatively, it is equivalent to investigating the attenuation of a shock wave caused by the pursuing unloading disturbance. The following example shows an approach to solve this problem.

Let us consider a semi-infinite bar made of increasingly hardening material. The bar is at a natural, rest state before the impact loading is applied. At the time $t = 0$ the end of the bar $(X = 0)$ is applied a sudden load $\sigma_m(0)$ and then is gradually released to zero, as shown in Fig. 4.13. The present problem differs from the previous problem (Fig. 4.12) in that the strong-discontinuous plastic loading wave (i.e. plastic shock wave) no longer propagates with a constant speed. When the intensity of the shock wave decreases, its speed $\mathscr{D}$ also decreases. The unloading boundary, which coincides with the shock wave locus, is an unknown curve expressed by function $X = f(t)$ or the inverse function $t = g(X)$.

For the convenience of discussion, we assume that the stress–strain curve of the material can be simplified as shown in Fig. 4.13 (the left diagram). This means that the linear elastic loading part of the stress–strain curve can be omitted. Soil is a typical material example that exhibits this type of stress–strain relationship. The above assumption does not change the basic features of our problem. Adding a linear elastic loading part to the σ–ε curve only means that a strong-discontinuous elastic precursor wave runs before the plastic shock wave.

The key of this problem is how to determine the attenuation properties and thus the propagating locus of the plastic shock wave. With the present stress–strain relationship and the zero initial conditions, the shock wave velocity is:

$$\mathscr{D} = f'(t) = \frac{1}{g'(t)} = \sqrt{\frac{1}{\rho_0}\frac{\sigma_m}{\varepsilon_m}} \tag{4.54}$$

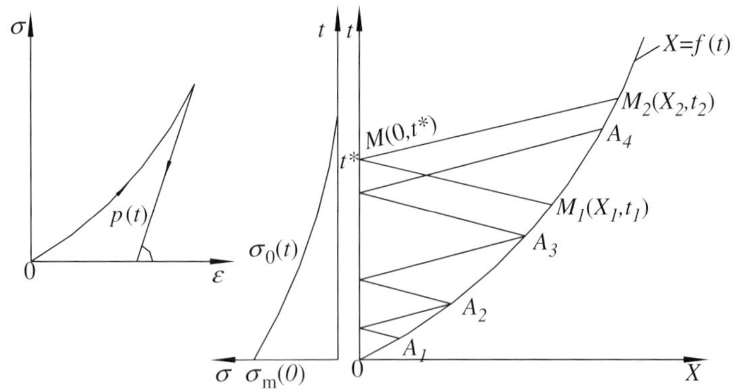

Fig. 4.13. An increasingly-hardening semi-infinite bar subjected to a suddenly applied loading and subsequently a gradually unloading.

and the momentum conservation equation across the shock wave front is reduced to

$$\sigma_m = -\rho_0 \mathscr{D} v_m \tag{4.55}$$

In the unloading zone, an arbitrary point M_1 on the unloading boundary is related to the conjugate point M_2 on the same boundary, via the characteristic lines M_1M and MM_2, as shown in Fig. 4.13. Referring to Eq. (4.37a), the conjugate relationship between the point M_1 and the point M_2 is

$$\rho_0 C_0 v_m(X_2) - \sigma_m(X_2) = \rho_0 C_0 v_m(X_1) + \sigma_m(X_1) - 2\sigma_0(t^*)$$

Using Eq. (4.55), the above equation is reduced to:

$$\left\{\frac{C_0}{\mathscr{D}[\sigma_m(X_2)]} + 1\right\}\sigma_m(X_2) = \left\{\frac{C_0}{\mathscr{D}[\sigma_m(X_1)]} - 1\right\}\sigma_m(X_1) + 2\sigma_0(t^*) \tag{4.56}$$

By introducing a dimensionless parameter μ_D representing the velocity ratio of C_0 to $\mathscr{D}$

$$\mu_D = \frac{C_0}{\mathscr{D}(\sigma)} \tag{4.57}$$

the conjugate relationship can be rewritten as:

$$\{\mu_D[\sigma_m(X_2)] + 1\}\sigma_m(X_2) = \{\mu_D[\sigma_m(X_1)] - 1\}\sigma_m(X_1) + 2\sigma_0(t^*) \tag{4.58}$$

Since the σ–ε relationship is given, the functions $\mathscr{D}(\sigma)$ and thus $\mu_D(\sigma)$ are also known.

On the other side, from the geometric relationship of characteristic lines M_1M and MM_2, we can deduce the coordinate relationships between the point X_1, the point X_2 and the point t^*:

$$\left.\begin{aligned} t^* &= t_1 + \frac{X_1}{C_0} = g(X_1) + \frac{X_1}{C_0} = \{\mu_S(X_1) + 1\}\frac{X_1}{C_0} \\ t^* &= t_2 - \frac{X_2}{C_0} = g(X_2) - \frac{X_2}{C_0} = \{\mu_S(X_2) - 1\}\frac{X_2}{C_0} \\ X_1 &= \frac{\mu_S(X_2) - 1}{\mu_S(X_1) + 1}\cdot X_2 = \lambda_S X_2 \end{aligned}\right\} \tag{4.59}$$

where the quantities $\mu_S(X_1)$ and $\mu_S(X_2)$ defined at points X_1, X_2, respectively, represent the ratio of C_0 to the slope of straight line OM_1 or OM_2

$$\mu_S(X) = \frac{C_0 g(X)}{X} \tag{4.60}$$

and the quantity λ_S is defined as:

$$\lambda_S(X) = \frac{\mu_S(X) - 1}{\mu_S(\lambda_S X) + 1} \tag{4.61}$$

Substitute the above equations into Eq. (4.58), we get:

$$\{\mu_D[\sigma_m(X)]+1\}\sigma_m(X)=\{\mu_D[\sigma_m(\lambda_S X)]-1\}\sigma_m(\lambda_S X)+2\sigma_0\left[\frac{(\mu_S(X)-1)X}{C_0}\right] \tag{4.62}$$

If $t = g(X)$ is known, then the functions $\mu_S(X)$ and accordingly $\lambda_S(X)$ are all known. Therefore, based on Eq. (4.58), we can use the **progressive approximation method** to determine the unloading boundary $t = g(X)$.

Imagine that the intensity of shock wave during the unloading process is approximately equal to the boundary loading, which is equivalent to the situation that $C_0 \to \infty$, and regard this situation as the 0th order approximate solution, we have

$$\sigma_m^{(0)}(t)=\sigma_0(t) \tag{4.63}$$

Since the stress–strain relationship of the material has been known, the function $\mathscr{D}(t)$ can be determined from Eq. (4.54). By a further integration, the 0th order locus of the shock wave can be obtained

$$X=f^{(0)}(t)=\int_0^t \mathscr{D}(t)dt \tag{4.64}$$

Equivalently, its inverse function $t = g^{(0)}(X)$ is also obtained. Substituting it into Eq. (4.63) we get the function $\sigma_m^{(0)}(X)$, and accordingly the functions $\mu_S^{(0)}(X)$ and $\lambda_S^{(0)}(X)$, where the $\lambda_S^{(0)}(X)$ is actually determined by an iteration manipulation of Eq. (4.61).

The 0th order approximate solutions are substituted into the right side of Eq. (4.58). The 1st order approximation of the function $\sigma_m(X)$, namely $\sigma_m^{(1)}(X)$ can be got from the left side of Eq. (4.58) through an iteration manipulation:

$$\left\{\mu_D\left[\sigma_m^{(1)}(X)\right]+1\right\}\sigma_m^{(1)}(X)=\left\{\mu_D\left[\sigma_m^{(0)}\left(\lambda_S^{(0)}X\right)\right]-1\right\}\sigma_m^{(0)}\left(\lambda_S^{(0)}X\right)+2\sigma_0\left[\frac{\left(\mu_S^{(0)}(X)-1\right)X}{C_0}\right] \tag{4.65}$$

By Eq. (4.54) the 1st order locus of shock wave, $t = g^{(1)}(X)$, and accordingly the functions $\mu_S^{(1)}(X)$ and $\lambda_S^{(1)}(X)$ can be determined too. Similarly the 1st order approximate solutions are substituted into the right side of Eq. (4.62) and the left side of Eq. (4.62) gives the 2nd order approximations. This iteration process is repeated until the solutions satisfy the accuracy requirement.

Once the shock wave locus $OA_1A_2\ldots$ is determined, the intensity of the shock wave $\sigma_m(X)$ is also obtained, which reflects the attenuation of the shock wave during the pursuing unloading process. The whole unloading zone can be solved as a Picard problem. Thus this problem is completely solved.

4.7 Propagating Properties of Elastic–Plastic Boundaries in Semi-Infinite Bars

From the discussions up till now, we know that in general the propagating locus of an elastic–plastic boundary differs from that of an unloading disturbance. The exceptions are the cases that either the elastic unloading disturbance or the plastic loading disturbance is a strong-discontinuity. The elastic–plastic boundary is not a wave motion of mechanical disturbance itself. It may either move forward, or move backward, depending on the interaction of waves. The propagation velocity of an elastic–plastic boundary, $\overline{C}$, generally varies during the interaction process of the elastic–plastic waves. An elastic–plastic boundary can be a loading boundary, which develops with time from an elastic zone to a plastic zone, or an unloading boundary, which develops with time from a plastic zone to an elastic zone.

In case of a semi-infinite bar, because no wave propagates from the other end, the elastic–plastic loading wave is a simple wave. Therefore, the loading boundary can be easily determined. The difficulty is to determine the unloading boundary. In the following, we will use the characteristic line method to investigate the general propagating features of a weak-discontinuous unloading boundary in a semi-infinite bar. Particularly, we will discuss what are the major factors which influence the propagating speed of the boundary $\overline{C}$.

4.7.1 An analysis by the characteristic line method

Suppose that the weak-discontinuous unloading boundary is expressed by the function $X = f(t)$ or its inverse function $t = g(X)$. Now consider an arbitrary point $M_1(X_1, t_1)$ and its neighboring point $M_2(X_2, t_2)$ on the unloading boundary. Those two points are connected by the characteristic lines to a third point $M(X, t)$ in the unloading zone. The characteristic line MM_1 in the unloading zone is leftwards, and the characteristic line MM_2 in the unloading zone is rightwards, as shown in Fig. 4.14. In the plastic loading zone, the leftward characteristic line passing through point M_1 interacts with the rightward characteristic line passing through point M_2 at the point N.

From the compatibility relationship along the characteristic line MM_1:

$$\rho_0 C_0 \bar{v}(M) + \bar{\sigma}(M) = \rho_0 C_0 v_m(M_1) + \sigma_m(M_1)$$

and from the compatibility relationship along the characteristic line MM_2:

$$\rho_0 C_0 \bar{v}(M) - \bar{\sigma}(M) = \rho_0 C_0 v_m(M_2) - \sigma_m(M_2)$$

by eliminating the quantity $\bar{v}(M)$, and taking account of that in the semi-infinite bar there exists a simple wave relationship between v_m and σ_m [Eq. (4.28)], finally we have:

$$\bar{\sigma}(X, t) = \frac{1}{2}\int_0^{\sigma_m(X_1)}\left(1 - \frac{C_0}{C}\right)d\sigma + \frac{1}{2}\int_0^{\sigma_m(X_2)}\left(1 + \frac{C_0}{C}\right)d\sigma \qquad (4.66)$$

The above equation is actually an extension of the conjugate relationship [Eq. (4.37b)] to the case that the point M is not located on the t-axis.

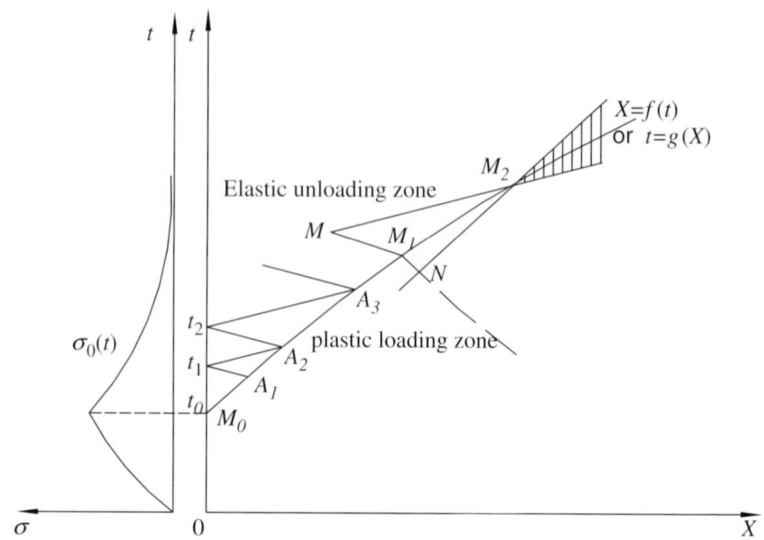

Fig. 4.14. An analysis of general propagating features of a weak-discontinuous unloading boundary in a semi-infinite bar by the characteristics method.

The expressions of the characteristic lines MM_1 and MM_2 are:

$$\left.\begin{aligned} MM_1:&\quad X_1 = X + C_0\{t - g(X_1)\}\\ MM_2:&\quad X_2 = X - C_0\{t - g(X_2)\}\end{aligned}\right\} \tag{4.67a}$$

Taking account of, that X and t are independent variables and $g' = 1/\overline{C}$, differentiating X_1 and X_2 in Eq. (4.67) with respect to variable t, we have:

$$\left.\begin{aligned} \frac{\partial X_1}{\partial t} &= C_0\left(1 - g'\frac{\partial X_1}{\partial t}\right) = \frac{C_0}{1 + \dfrac{C_0}{\overline{C}}}\\ \frac{\partial X_2}{\partial t} &= C_0\left(g'\frac{\partial X_2}{\partial t} - 1\right) = \frac{C_0}{\dfrac{C_0}{\overline{C}} - 1}\end{aligned}\right\} \tag{4.67b}$$

Therefore from Eq. (4.66) we get:

$$\begin{aligned} \frac{\partial \sigma}{\partial t} &= \frac{1}{2}\left(1 - \frac{C_0}{C}\right)\frac{d\sigma_m}{dX_1}\frac{\partial X_1}{\partial t} + \frac{1}{2}\left(1 + \frac{C_0}{C}\right)\frac{d\sigma_m}{dX_2}\frac{\partial X_2}{\partial t}\\ &= \frac{1}{2}\left(1 - \frac{C_0}{C}\right)\cdot\frac{C_0}{\dfrac{C_0}{\overline{C}} + 1}\frac{d\sigma_m}{dX_1} + \frac{1}{2}\left(1 + \frac{C_0}{C}\right)\frac{C_0}{\dfrac{C_0}{\overline{C}} - 1}\frac{d\sigma_m}{dX_2}\end{aligned} \tag{4.68}$$

When $M_2 \to M_1$, $M \to M_1$, the relationship between the quantities $\partial\bar{\sigma}/\partial t$ and $d\sigma_m/dX$ at an arbitrary point M_1 on the weak-discontinuous unloading boundary $t = g(X)$ is obtained

$$\frac{\partial\bar{\sigma}}{\partial t} = \frac{1}{2}\left\{\frac{1-\dfrac{C_0}{C}}{\dfrac{1}{\overline{\overline{C}}}+\dfrac{1}{C_0}} + \frac{1+\dfrac{C_0}{C}}{\dfrac{1}{\overline{\overline{C}}}-\dfrac{1}{C_0}}\right\}\frac{d\sigma_m}{dX_2} \tag{4.69}$$

On the other hand, in the plastic loading simple wave zone, along the characteristic line NM_2 we have:

$$\left.\begin{aligned} \sigma(N) &= \sigma(M_2) \\ \frac{\partial\sigma}{\partial t} &= \frac{d\sigma_m}{dX_2}\frac{\partial X_2}{\partial t} \end{aligned}\right\} \tag{4.70}$$

while from the expression of the characteristic line NM_2 we have:

$$\begin{aligned} X_2 &= X + C\{g(X_2) - t\} \\ \frac{\partial X_2}{\partial t} &= C\left(g'\frac{\partial X_2}{\partial t} - 1\right) + (g(X_2) - t)\frac{dC}{d\sigma}\frac{d\sigma_m}{dX_2}\frac{\partial X_2}{\partial t} \\ &= \frac{C}{\dfrac{C}{\overline{\overline{C}}} - 1 + (g(X_2) - t)\dfrac{dC}{dX_2}\dfrac{d\sigma_m}{dX_2}} \end{aligned} \tag{4.71}$$

Substitute Eq. (4.71) into Eq. (4.70), and let $M_2 \to M_1$, we get the relationship between the quantities $\partial\sigma/\partial t$ and $d\sigma_m/dt$ at an arbitrary point M_1 on the boundary $t = g(X)$:

$$\frac{\partial\sigma}{\partial t} = \frac{\dfrac{d\sigma_m}{dX}}{\dfrac{1}{\overline{\overline{C}}} - \dfrac{1}{C}} \tag{4.72}$$

Since both C_0 and C are known quantities, from Eqs. (4.69) and (4.72) we can eliminate the quantity $d\sigma_m/dX$ and thus obtain the following relationship between the propagation velocity of the unloading boundary $\overline{\overline{C}}$ and the quantities $\partial\sigma/\partial t$ and $\partial\bar{\sigma}/\partial t$ at an arbitrary point M_1 on the unloading boundary $t = g(X)$:

$$\frac{\dfrac{\partial\sigma}{\partial t}}{\dfrac{\partial\bar{\sigma}}{\partial t}} = \frac{\left(C_0^2 - \overline{\overline{C}}^2\right)C^2}{\left(C^2 - \overline{\overline{C}}^2\right)C^2} = \frac{1 - \left(\dfrac{\overline{\overline{C}}}{C}\right)^2}{1 - \left(\dfrac{\overline{\overline{C}}}{C}\right)^2} \tag{4.73a}$$

Alternatively, the quantity $\overline{C}$ can be solved in the explicit form

$$\overline{C} = \sqrt{\frac{\dfrac{\partial \sigma}{\partial t} - \dfrac{\partial \bar{\sigma}}{\partial t}}{\dfrac{1}{C^2}\dfrac{\partial \sigma}{\partial t} - \dfrac{1}{C_0}\dfrac{\partial \bar{\sigma}}{\partial t}}} = \sqrt{\frac{C^2\dfrac{\partial \bar{\varepsilon}}{\partial t} - C^2\dfrac{\partial \varepsilon}{\partial t}}{\dfrac{\partial \bar{\varepsilon}}{\partial t} - \dfrac{\partial \varepsilon}{\partial t}}} \tag{4.73b}$$

In the present situation, $t = g(X)$ expresses the unloading boundary, across which the stress state changes from plastic loading region to the elastic unloading region. As long as the quantities $\partial\sigma/\partial t$ and $\partial\bar{\sigma}/\partial t$ are not simultaneously zero, these quantities must have different signs, namely $(\partial\bar{\sigma}/\partial t)/(\partial\sigma/\partial t) \leq 0$ or $(\partial\sigma/\partial t)/(\partial\bar{\sigma}/\partial t) \leq 0$. Thus from Eq. (4.73a) we have:

$$C_0 \geq \overline{C} \geq C \tag{4.74}$$

This means that in a semi-infinite bar the moving speed of an unloading boundary $\overline{C}$ is between the elastic wave velocity C_0 and the plastic wave velocity C. For example, in Fig. 4.14, the unloading boundary passing through the point M_2 must fall within the shaded area shown in the figure.

Under the following four specific situations, as shown in Fig. 4.15, $\overline{C}$ either reaches its upper limit C_0, or reaches its lower limit C:

$$\left.\begin{array}{llll}
\text{(a)} & \dfrac{\partial \sigma}{\partial t} = 0, & \dfrac{\partial \bar{\sigma}}{\partial t} \neq 0: & \overline{C} = C_0 \\[2ex]
\text{(b)} & \dfrac{\partial \sigma}{\partial t} \neq \infty, & \dfrac{\partial \bar{\sigma}}{\partial t} = -\infty: & \overline{C} = C_0 \\[2ex]
\text{(c)} & \dfrac{\partial \sigma}{\partial t} \neq 0, & \dfrac{\partial \bar{\sigma}}{\partial t} = 0: & \overline{C} = C \\[2ex]
\text{(d)} & \dfrac{\partial \sigma}{\partial t} = \infty, & \dfrac{\partial \bar{\sigma}}{\partial t} \neq -\infty: & \overline{C} = C
\end{array}\right\} \tag{4.75}$$

Note that the Case (b) is in accordance with the situation when a strong-discontinuous unloading disturbance pursues a weak-discontinuous plastic wave ($\overline{C} = C_0$); the Case (d) is in accordance with the situation when a weak-discontinuous unloading disturbance pursues a strong-discontinuous plastic wave ($\overline{C} = C$), while the Case (a) coincides with the situation that the unloading disturbance propagates in the constant-value plastic zone ($\overline{C} = C_0$).

In the previous discussions, we have assumed that the derivatives $\partial\sigma/\partial t$ and $\partial\bar{\sigma}/\partial t$ are not simultaneously zero, which means that the unloading boundary is a first order weak-discontinuity with respect to the stress σ and particle velocity v. In case that $\partial\sigma/\partial t = \partial\bar{\sigma}/\partial t = 0$, the right side of Eq. (4.73a) represents an indefinite quantity. In order to understand the features of the unloading boundary, we must further consider the second order derivatives of stress: $\partial^2\sigma/\partial t^2$, $\partial^2\bar{\sigma}/\partial t^2$, etc.

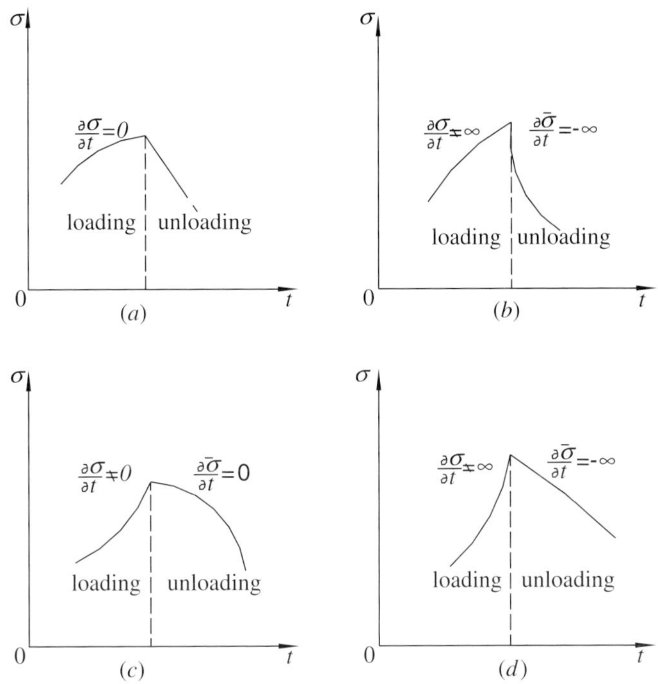

Fig. 4.15. Four specific situations of unloading boundary propagation in a semi-infinite bar.

In the unloading region, from Eq. (4.68), we have the following relationship at the point M:

$$\begin{aligned}\frac{\partial^2\bar{\sigma}}{\partial t^2} &= \frac{1}{2}\left(1-\frac{C_0}{C}\right)\frac{d^2\sigma_m}{dX_1^2}\left(\frac{\partial X_1}{\partial t}\right)^2 + \frac{1}{2}\left(1+\frac{C_0}{C}\right)\frac{d^2\sigma_m}{dX_2^2}\left(\frac{\partial X_2}{\partial t}\right)^2 \\ &= \frac{1}{2}\left(1-\frac{C_0}{C}\right)\left(\frac{C_0}{\frac{C_0}{\overline{\overline{C}}}+1}\right)^2\frac{d^2\sigma_m}{dX_1^2} + \frac{1}{2}\left(1+\frac{C_0}{C}\right)\left(\frac{C_0}{\frac{C_0}{\overline{\overline{C}}}-1}\right)^2\frac{d^2\sigma_m}{dX_2^2}\end{aligned}$$

Note that in the above derivation we have used the conditions that $(d\sigma_m/dX) = 0$ when $(\partial\bar{\sigma}/\partial t = 0)$ [Eq. (4.69)], so that all the terms with $d\sigma_m/dX$ are omitted. When $M_2 \to M_1$ so that $M \to M_1$, we get the following relationship between the second order derivatives $\partial^2\bar{\sigma}/\partial t^2$ and $d^2\sigma_m/dt^2$ at the point M_1 on the unloading boundary $t = g(X)$:

$$\frac{\partial^2\bar{\sigma}}{\partial t^2} = \frac{1}{2}\left[\frac{1-\dfrac{C_0}{C}}{\left(\dfrac{1}{\overline{\overline{C}}}+\dfrac{1}{C}\right)^2} + \frac{1+\dfrac{C_0}{C}}{\left(\dfrac{1}{\overline{\overline{C}}}-\dfrac{1}{C}\right)^2}\right]\frac{d^2\sigma_m}{dX^2} \tag{4.76}$$

Similarly, in the plastic loading region, differentiate Eq. (4.72) with respect to the variable t, and let $M_2 \to M_1$. This way we get a relationship between the second order derivatives $\partial^2\sigma/\partial t^2$ and $d^2\sigma_m/dX^2$ at the point M_1 on $t = g(X)$, as shown in the following:

$$\frac{\partial^2\sigma}{\partial t^2} = \frac{\dfrac{d^2\sigma_m}{dX^2}}{\left(\dfrac{1}{\overline{C}} - \dfrac{1}{C}\right)^2} \tag{4.77}$$

Eliminate the quantity $d^2\sigma_m/dX^2$ from Eqs. (4.76) and (4.77), we get the following relationship at the point M_1 on $t = g(X)$, connecting the quantity $\partial^2\bar{\sigma}/\partial t^2$ and $\partial^2\sigma/\partial t^2$ to $\overline{C}$:

$$\frac{\dfrac{\partial^2\sigma}{\partial t^2}}{\dfrac{\partial^2\bar{\sigma}}{\partial t^2}} = \frac{2}{\left(\dfrac{1}{\overline{C}} - \dfrac{1}{C}\right)^2 \left\{ \dfrac{1 - \dfrac{C_0}{C}}{\left(\dfrac{1}{\overline{C}} + \dfrac{1}{C}\right)^2} + \dfrac{1 + \dfrac{C_0}{C}}{\left(\dfrac{1}{\overline{C}} - \dfrac{1}{C}\right)^2} \right\}} \tag{4.78a}$$

$$= \frac{\left(C_0^2 - \overline{C}^2\right)^2}{\left(C - \overline{C}\right)^2 \left(C_0^2 + 2\dfrac{C_0^2}{C}\overline{C} + \overline{C}^2\right)} \cdot \frac{C^2}{C_0^2}$$

This is an algebraic equation of fourth degree with respect to the variable $\overline{C}$. Introducing the following nondimensional parameters:

$$\beta = \frac{\dfrac{\partial^2\sigma}{\partial t^2}}{\dfrac{\partial^2\bar{\sigma}}{\partial t^2}}, \qquad a = \frac{C}{C_0}, \qquad \bar{a} = \frac{\overline{C}}{C_0}$$

then Eq. (4.78a) can be rewritten in the nondimensional form as following:

$$\beta = \frac{2a}{\left(1 - \dfrac{\bar{a}}{a}\right)^2 \left[\dfrac{1+a}{(1-\bar{a})^2} - \dfrac{1-a}{(1+\bar{a})^2}\right]} \tag{4.78b}$$

Or the equation can be written as an algebraic equation of fourth degree with respect to variable $\bar{a}$, as:

$$(\beta - a^2)\bar{a}^4 + 2\beta\left(\frac{1}{a} - a\right)\bar{a}^3 + \{(\beta + 2')a^2 - 3\beta\}\bar{a}^2 + (\beta - 1)a^2 = 0 \tag{4.78c}$$

Note that for a second order weak-discontinuous unloading boundary, $\partial^2\sigma/\partial t^2$ and $\partial^2\bar{\sigma}/\partial t^2$ must have the same signs, therefore $\beta \geq 0$.

Under the following four specific situations, the problem can be further simplified and an explicit expression of $\overline{C}$ can be obtained:

(i) When $\beta = 0$, from Eq. (4.78a) we get: $\overline{C} = C_0$;

(ii) When $(1/\beta) = 0$, from Eq. (4.78a) we get: $\overline{C} = C$;

(iii) When $\beta = 1$, which means that $(\partial^2\sigma/\partial t^2) = (\partial^2\bar{\sigma}/\partial t^2)$, because $(1 - a^2) \neq 0$, the Eq. (4.78c) can be reduced to:

$$\bar{a}^4 + \frac{2\bar{a}^3}{a} - 3\bar{a}^2 = \bar{a}^2\left(\bar{a}^2 + \frac{2\bar{a}}{a} - 3\right) = 0$$

from which the positive real root of $\overline{C}$ can be solved, which represents the propagating velocity of the rightward unloading boundary shown in Fig. 4.14:

$$\overline{C} = \frac{C_0^2}{C}\left[\sqrt{1 + 3\left(\frac{C}{C_0}\right)^2} - 1\right] \tag{4.79}$$

This result coincides with that obtained by Бидерман (Biderman) (1952).

(iv) When $\beta = a^2$, which means that $(\partial^2\varepsilon/\partial t^2) = (\partial^2\bar{\varepsilon}/\partial t^2)$, Eq. (4.78c) can be reduced to an cubic algebraic equation with respect to $\bar{a}$:

$$\bar{a}^3 + \frac{a}{2}\bar{a}^2 - \frac{a}{2} = 0$$

of which the following real positive root can be solved:

$$\bar{a} = \alpha + \sqrt[3]{\alpha^3 + \frac{3}{2}\alpha\left(1 + \sqrt{1 + \frac{4}{3}\alpha^2}\right)} + \sqrt[3]{\alpha^3 + \frac{3}{2}\alpha\left(1 - \sqrt{1 + \frac{4}{3}\alpha^2}\right)} \tag{4.80}$$

where a parameter $\alpha = a/6$ is introduced for the convenience of expression.

Similarly, if $(\partial^2\sigma/\partial t^2) = (\partial^2\bar{\sigma}/\partial t^2) = 0$, Eq. (4.78) becomes indefinite, then we need to investigate the features of the third derivatives $\partial^3\sigma/\partial t^3$, $\partial^3\bar{\sigma}/\partial t^3$ etc. By this analogy, if the derivatives $(\partial^i\sigma/\partial t^i) = (\partial^i\bar{\sigma}/\partial t^i) = 0, i = 1, 2, 3, \ldots, (n-1)$, then in order to understand the features of the unloading boundary we must investigate the nonzero nth derivatives $\partial^n\sigma/\partial t^n$ and $\partial^n\bar{\sigma}/\partial t^n$. By the similar mathematical deductions, we get a relationship that connects the quantities $\partial^n\sigma/\partial t^n$, $\partial^n\bar{\sigma}/\partial t^n$ and $\overline{C}$:

$$\frac{\dfrac{\partial^n\sigma}{\partial t^n}}{\dfrac{\partial^n\bar{\sigma}}{\partial t^n}} = \frac{2}{\left(\dfrac{1}{\overline{C}} - \dfrac{1}{C}\right)^n\left[\dfrac{1 - \dfrac{C_0}{C}}{\left(\dfrac{1}{\overline{C}} + \dfrac{1}{C}\right)^n} + \dfrac{1 + \dfrac{C_0}{C}}{\left(\dfrac{1}{\overline{C}} - \dfrac{1}{C}\right)^n}\right]} \tag{4.81}$$

which is an algebraic equation of $2n$th degree with respect to the variable $\overline{C}$.

4.7.2 Local linearization method

The above results not only provide a further understanding about the propagating features of an unloading boundary moving in a semi-infinite bar, but also provide an idea to solve the problem approximately by a new method, the so-called **local linearization method**.

The basic ideal is as follows: apply the results obtained in the subsection 4.7.1 to the end of the bar ($X = 0$), the initial propagating velocity of the unloading boundary, namely $\overline{C}_i = f'(t_0)$, at the point M_0 (the beginning point of the unloading boundary at $t = t_0$ as shown in Fig. 4.14) can be determined from Eq. (4.81), since the boundary loading $\sigma(0, t)$ and its derivatives with respect to time t including the least order partial derivatives of σ and $\bar{\sigma}$ which are not simultaneously zero, are known. In other words, the initial slope of the curve $X = f(t)$ at the point M_0 on the X–t plane (Fig. 4.14) can be determined. If the length of the initial segment of the curve $X = f(t)$, M_0A, is taken so short as it can be approximated by a short straight line with slope $\overline{C}_i$, then the location of M_0A is now approximately known. On the other hand, the stress and the particle velocity along this line segment can be determined by the physical states known in the plastic loading zone. Therefore the line segment M_0A is completely solved. As mentioned previously, once a short initial part of the curve $X = f(t)$ is known, the rest of the curve can be determined by applying the conjugate relationship [Eq. (4.37)] piece by piece.

An example to solve the unloading boundary by means of the local linearization method is given in Fig. 4.16 [Шапиро (Shapiro), 1946]. In this problem, the semi-infinite bar is made of linearly hardening plastic material. The bar is at natural, rest state before a parabolic-shape impact load is applied at its end:

$$\sigma_0(t) = -8Y\frac{t}{T}\left(1 - \frac{t}{T}\right)$$

where Y is the yield limit, T the loading duration. At the time $t = T/2$ the boundary load reaches its maximum $(\sigma_0)_{\max} = -2Y$, and the unloading starts. Since at this time $(\partial\sigma/\partial t) = (\partial\bar{\sigma}/\partial t) = 0$, while $\left(\partial^2\sigma/\partial t^2\right) = \left(\partial^2\bar{\sigma}/\partial t^2\right) \neq 0$, the initial propagating velocity of the unloading boundary can be calculated from Eq. (4.79). Assuming $C_1/C_0 = 1/4$, we have:

$$\overline{C}(M_0) = C_0(\sqrt{19} - 4) = 0.359C_0 = 1.44C_1$$

Now we can draw an initial short straight line M_0A_1 with the slope $\overline{C}(M_0)$. The stress at the point A_1, $\sigma_m(A_1)$, can be determined from the boundary loading $\sigma_0(t)$ at the intersection of the time axis and the rightward plastic characteristic line passing through the point A_1 with slope C_1. Practically, it is very close to the maximum boundary load $(\sigma_0)_{\max}$.

Once the initial boundary segment M_0A_1, has been determined, we can use the conjugate relationship to determine the following segments A_1A_2, A_2A_3, ... until the plastic wave vanishes at the point *K*. Figure 4.16(b) shows the stress–time curves at three sections of the bar. From this figure it is clear that during the unloading process the plastic stress gradually attenuates until it completely vanishes. As can be seen too, the profiles of the stress impulse, including the quantities $\partial\sigma/\partial t$ and $\partial\bar{\sigma}/\partial t$ at the maximum stress point, varies

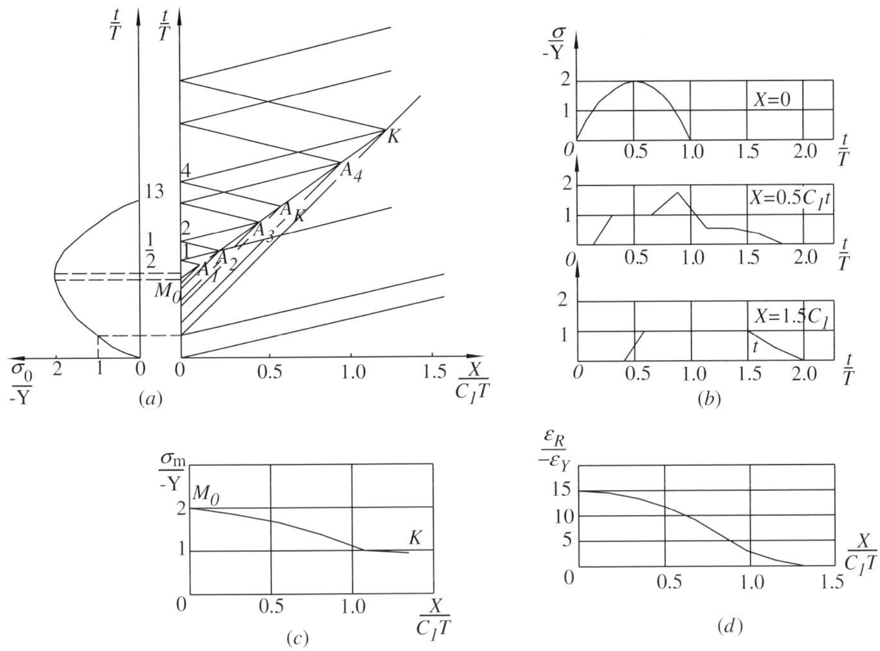

Fig. 4.16. A linearly-hardening plastic semi-infinite bar subjected to a parabolic-shape impact loading.

at different locations of the bar, particularly significantly different from those at the bar end. Figure 4.16(c) shows the maximum stress distribution along the unloading boundary, $\sigma_m(X)$; and Fig. 4.16(d) shows the residual strain distribution along the unloading boundary, $\varepsilon_R(X)$. The length of the plastic deformation region l is $1.32C_1T$, which is dependent on both the plastic wave speed C_1 and the loading duration T [Refer to Eq. (4.49)].

Note that on the σ–v plane, the mapping of the weak-discontinuous unloading boundary coincides with the mapping of the plastic simple waves, namely the line KM_0 shown in

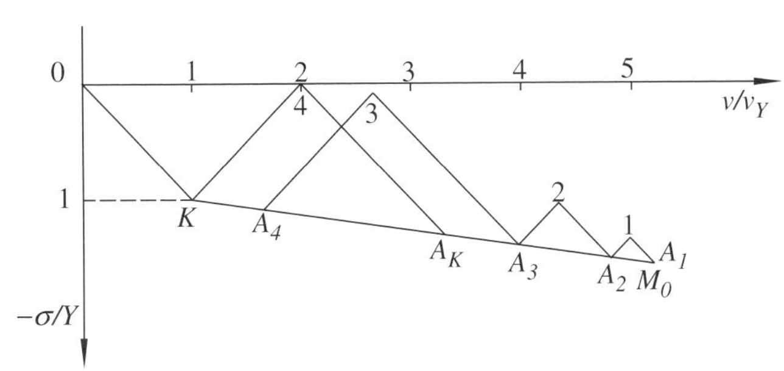

Fig. 4.17. The mapping of the weak-discontinuous unloading boundary coincides with the mapping of the plastic simple waves on σ–v plane.

Fig. 4.17. Each characteristic line in the unloading region on the X–t plane is corresponding to an unloading characteristic line on the σ–v plane. Together with the **diagram method**, the locations of points A_2, A_3, A_4, and A_K on the σ–v plane as shown in Fig. 4.17 can be easily determined. The locations of points 1, 2, 3, and 4 shown in the Fig. 4.17 are determined by the boundary stress levels at the corresponding points on the X–t plane shown in Fig. 4.16(a).

4.8 Head-On Unloading

So far our discussions about the problem of unloading waves are limited to the semi-infinite bar. The unloading disturbance and the plastic loading disturbance are propagating in the same direction (pursuing unloading). With regard to the interactions of two elastic–plastic waves propagating in the opposite directions, the situations when two head-on stress waves having the same signs have been discussed in the Section 4.1. In those cases, the interactions of two head-on waves result in a further plastic loading. In the present section, we will discuss the cases that the two head-on waves have different signs, and thus the interaction of these two waves will unload each other, so called a **head-on unloading** problem.

To better understand the differences between the head-on unloading and the pursuing unloading, it is useful to recall some known results in the case of linear elastic wave propagation. Consider an elastic bar with finite length. The bar is at natural, rest state before its left end is loaded by a sudden constant-velocity impact ($v_1 > 0$) and simultaneously its right end is loaded by another sudden constant-velocity impact ($v_2 > 0$), as shown in Fig. 4.18. Consequently, a compressive shock wave with the stress level σ_1 is propagating to the right direction in the bar, while a tensile shock wave with the stress level σ_2 is propagating to the left direction in the bar. Now let us consider two different unloading cases that have been discussed previously in Chapter 3. First, if the stress at the left end of the bar is unloaded to zero, then the particle velocity at the left end should also be zero. In the σ–v plane this process corresponds to the state-point 1 moving to the point 0. However, if the rightward compressive stress wave interacts with the leftward tensile stress wave and thus the stress σ_1 is unloaded to zero (under the situation that $\sigma_2 = -\sigma_1$), then the particle velocity after unloading should change from v_1 to $v_3 = 2v_1$. In σ–v plane this process corresponds to the state-point 1 jumps to the point 3. Therefore, in the above

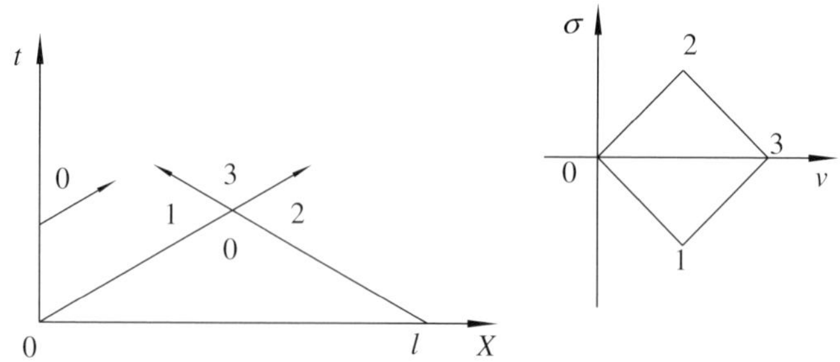

Fig. 4.18. Head-on unloading in a finite elastic bar.

two situations, although the stress σ_1 is unloaded to the same value (zero), the unloading particle velocity is actually different.

This result can be understood from another way. Assume the bar is in an initial state σ_1 and v_1, corresponding to the point 1 on the σ–v plane in Fig. 4.18. If the bar is unloaded to zero stress by a rightward unloading wave propagating from the left end, the unloading state corresponds to the point 0. However, if the bar is unloaded to zero stress by a leftward unloading wave propagating from the right end, the unloading state corresponds to the point 3. Note that the point 3 differs from the point 0 by a velocity difference $[-(2\sigma_1/\rho_0 C_0)]$. This difference originates from the fact that when the momentum conservation relationship [Eq. (2.35)] is applied to a wave, different signs must be taken respectively for the leftward wave and the rightward wave.

Now consider an elastic, linearly hardening plastic bar with a finite length and a stress-free, rest, initial state. When the bar is loaded by a sudden constant velocity v impact at its left end, a strong-discontinuous elastic wave and a strong-discontinuous plastic wave simultaneously propagate to the right direction in the bar. The state behind the plastic wave front corresponds to the point 2 on the σ–v plane in Fig. 4.19. Now if the bar is unloaded to zero stress by a rightward unloading wave propagating from the left end, by Eq. (2.35) the unloading state corresponds to the point L in Fig. 4.19 (referring to Fig. 4.5 too), the residual particle velocity v_L is [referring to Eq. (4.14)]:

$$v_L = v_2 + \frac{\sigma_2}{\rho_0 C_0} \tag{4.82}$$

However, if the bar is unloaded to zero stress by a leftward elastic unloading wave propagating from the right end, the unloading state corresponds to the point R. By Eq. (2.35) for a leftward wave, after unloading, the residual particle velocity v_R is:

$$v_R = v_2 - \frac{\sigma_2}{\rho_0 C_0} \tag{4.83}$$

Thus the latter unloading state differs from the former unloading state by a velocity difference $[-(2\sigma_2/\rho_0 C_0)]$.

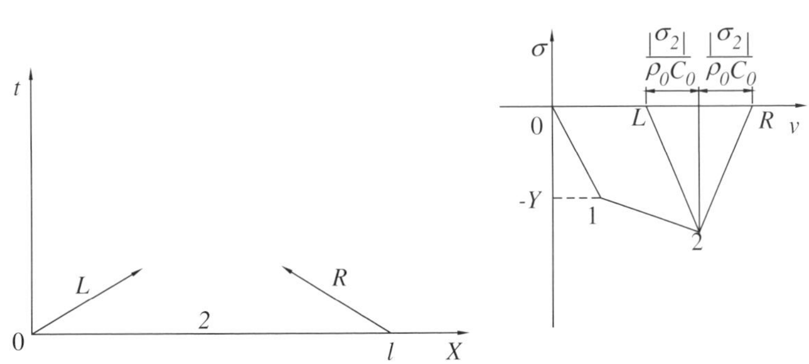

Fig. 4.19. Head-on unloading in a finite elastic-linearly hardening plastic bar.

With the above understanding, we now consider in detail the problem of head-on unloading when two elastic–plastic stress waves, having different stress signs and propagating in the opposite direction, interact. Assume a finite bar made of linearly hardening material, initially in a natural, rest state. Then at its left end a sudden constant velocity compressive impact (v_2) is loaded, and simultaneously a sudden constant-velocity tensile impact (v_4) is loaded at its right end. Consequently, two sets of strong-discontinuous elastic–plastic waves, having different stress sign and level, are propagating in the opposite direction within the bar. Before these two sets of waves meet, there exist four zones: the zones behind the rightward and the leftward elastic wave front respectively, i.e. the zone 1 and the zone 3; and the zones behind the rightward and the leftward plastic wave front respectively, i.e. the zone 2 and the zone 4, as shown in Fig. 4.20. The two elastic precursor waves first meet at the point a. As these two waves have the same absolute stress level (Y) but with different signs, after their interaction the stress is unloaded to zero: $\sigma_5 = 0$ but the particle velocity is: $v_5 = 2v_Y$, a state corresponding to the point 5 on the σ–v plane and the region 5 on the X–t plane. This stress-velocity state is similar to that after an elastic wave with stress level Y reflects from a stress-free boundary.

After the internal impact of the two elastic precursor waves, the leftward reflective wave ab, which is an unloading wave, meets the rightward plastic loading wave ob at the point b. Imagine that on the left side of the point b if the plastic loading region 2 is also unloaded to zero stress after the unloading wave ab has passed, then the unloading particle velocity will be v_R, as expressed by Eq. (4.83). If so, on the two sides of the point b, there will be a particle velocity difference ($v_R - v_5$), which must result in an internal impact at the point b. If $|v_R - v_5| \leq 2v_Y$, then the two internal reflective waves due to the internal impact are within the elastic range. The state behind the two reflective wave fronts is corresponding to the point 6 on the σ–v plane.

Similarly, after the internal impact of the elastic precursor waves, the rightward reflective wave ac, which is an unloading wave, meets the leftward plastic loading wave lc at the point c. Again an internal impact occurs and two internal reflective waves are induced. The state behind these two reflective wave fronts is corresponding to the point 8 on the σ–v plane, assuming that $|v_B - v_5| \leq 2v_Y$.

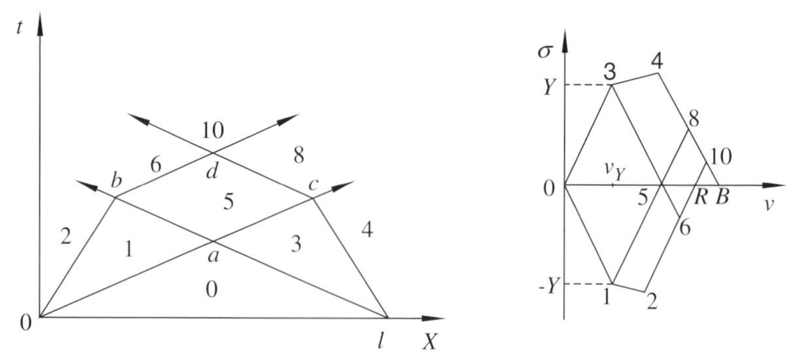

Fig. 4.20. Head-on unloading of two elastic-plastic stress waves with opposite stress signs in a finite elastic-linearly hardening plastic bar if $|v_R - v_5| > 2v_Y$.

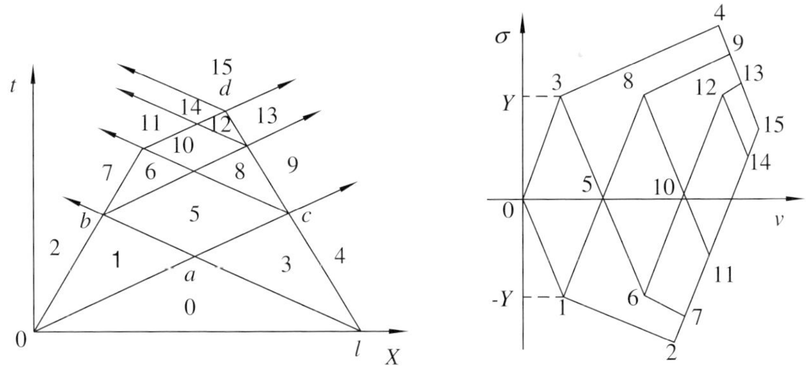

Fig. 4.21. Head-on unloading of two elastic-plastic stress waves with opposite stress signs in a finite elastic-linearly hardening plastic bar if $|v_R - v_5| > 2v_Y$.

Finally, the rightward reflective wave *bd* and the leftward reflective wave *cd* meet at the point *d* on *X–t* plane. Again an internal impact occurs and two reflective waves originate from this point. The state behind these two reflective wave fronts is corresponding to the point 10 on the σ–v plane. As the point 10 falls within the elastic range, both reflective waves emitting from the point *d* are elastic waves. Having developed to this point, the process of interaction between two different sign elastic–plastic waves propagating in opposite directions is completed.

However, if in the previous discussions, we have $|v_R - v_5| > 2v_Y$ or $|v_B - v_5| > 2v_Y$, then after the internal impact (at the points *b* or *c*) the induced internal reflective waves still contain plastic waves, though the intensities of these plastic waves are weakened. This situation is similar to what has been previously discussed on the pursuing unloading cases. Figure 4.21 shows one such example. Detailed explanations are omitted. The readers are suggested to analyze this situation.

Moreover, similar to the discussions about the pursuing unloading, the above results about the head-on unloading interaction between two strong-discontinuous, elastic–plastic waves, propagating with constant velocity, can be extended to the other situations, such as, that the head-on stress waves could be either shock waves (whose wave speed changes with wave intensity), or weak-discontinuous elastic–plastic waves. In the next section we will discuss the phenomena when two head-on shock waves unload each other in an increasingly hardening bar, through a detailed example. Here the head-on unloading process between two weak-discontinuous elastic–plastic waves is briefly discussed.

As previously discussed on the pursuing unloading problems, a weak-discontinuous elastic–plastic wave can be approximately considered as a series of incremental loading waves, each incremental wave is regarded as a strong-discontinuous wave with a small stress jump. Therefore the head-on unloading between two weak-discontinuous elastic–plastic waves can be treated as the head-on unloading between two series of incremental waves. Figure 4.22 shows such an example, which is corresponding to the problem shown in Fig. 4.20 but now the material is decreasingly hardening. On the *X–t* plane of Fig. 4.22,

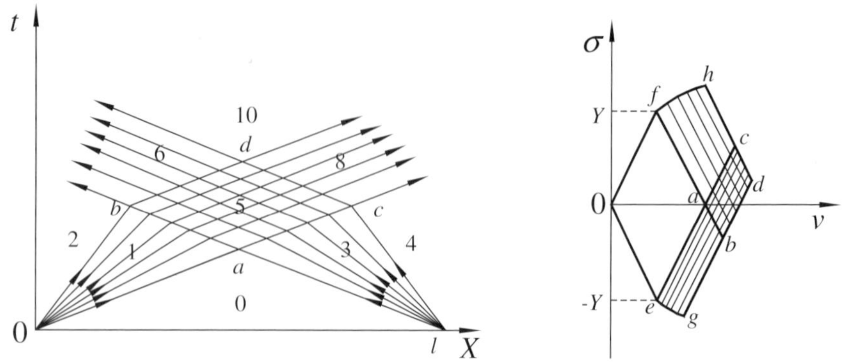

Fig. 4.22. Head-on unloading of two elastic-plastic stress waves with opposite stress signs in a finite elastic-increasingly hardening plastic bar.

the regions 0, 2, 4, and 10 are constant-value zones. Their mappings on the σ–t plane are the point o, g, h, and d, respectively. The regions 1, 3, 6, and 8 on the X–t plane are simple wave zones. Their mappings on the σ–t plane are the line-segments *eg*, *fh*, *bd*, and *cd*, respectively. The unloading zone *abcd* (the region 5) on the X–t plane, where the internal reflective waves propagate in both directions, is corresponding to the region *abcd* on the σ–t plane. It is noteworthy that stress and particle velocity at the point d, which indicates the terminating point of the above wave interactions, can be solved directly without knowing all the intermediate states. Viewing from the left side of the bar, the state d is arrived by following the loading line *oeg* (representing a rightward loading wave) and then the unloading line *gbd* (representing a leftward unloading wave) on the σ–t plane. By the momentum conservation relationship across the wave front [refer Eq. (2.63)], and noting that the signs are different for a leftward or a rightward wave, we have:

$$v_d = -\int_0^{\sigma_g} \frac{d\sigma}{\rho_0 C} + \int_{\sigma_g}^{\sigma_d} \frac{d\sigma}{\rho_0 C_0} = -\frac{1}{\rho_0 C_0}\int_0^{\sigma_g}\left(1+\frac{C_0}{C}\right)d\sigma + \frac{\sigma_d}{\rho_0 C_0}$$

Similarly, viewing from the right side of the bar, we have:

$$v_d = \int_0^{\sigma_h} \frac{d\sigma}{\rho_0 C} - \int_{\sigma_h}^{\sigma_d} \frac{d\sigma}{\rho_0 C_0} = \frac{1}{\rho_0 C_0}\int_0^{\sigma_h}\left(1+\frac{C_0}{C}\right)d\sigma - \frac{\sigma_d}{\rho_0 C_0}$$

Thus the quantities σ_d and v_d can be solved from the above two equations:

$$\left.\begin{aligned} v_d &= \frac{1}{2\rho_0 C_0}\left[\int_0^{\sigma_h}\left(1+\frac{C_0}{C}\right)d\sigma - \int_0^{\sigma_g}\left(1+\frac{C_0}{C}\right)d\sigma\right] \\ \sigma_d &= \frac{1}{2}\left[\int_0^{\sigma_h}\left(1+\frac{C_0}{C}\right)d\sigma + \int_0^{\sigma_g}\left(1+\frac{C_0}{C}\right)d\sigma\right] \end{aligned}\right\} \qquad (4.84)$$

Based on the above discussions for the head-on unloading problems, now we have accumulated experiences to treat different and even more complicated unloading problems, such as the reflection of an elastic–plastic wave on a free end, the reflection of an elastic–plastic wave on other surface that may cause unloading, the reflection and transmission of an elastic–plastic wave on an interface of different media; specifically the unloading reflection when an elastic–plastic wave propagates from a high wave-impedance media to a low wave-impedance media. The guiding principles to deal with this kind of elastic–plastic wave reflection problems are similar to those applied to analyze elastic wave reflections. Namely, on a boundary surface the prescribed boundary conditions must be satisfied; or on an internal surface between two media, the stress and particle velocity must be continuous across the interface. If an unloading disturbance appears and interacts with other waves, we need to distinguish whether the interaction is a pursuing unloading or a head-on unloading. Different unloading cases require different treatment. A further discussion will be given in the following section by analyzing in detail a practical problem, namely the elastic–plastic wave propagation within a finite length bar impacted on a rigid anvil.

4.9 High-Speed Impact of a Finite Bar onto a Rigid Target

The case that a flat-head projectile impacts onto a target is a practical problem of impact dynamics. We can view the projectile as a bar with a finite length, the target as a rigid anvil. Then the problem is a high-speed impact of a finite length bar on a rigid anvil. Researchers have used this case as a test method to determine the dynamic mechanical properties of the materials [Taylor, 1948; Whiffen, 1948; Ленский (Lenskii), 1951]. For such kind of impact problem a head-on unloading takes place, namely elastic–plastic waves will be reflected at the free end of the bar, and the reflected unloading waves will interact with the head-on elastic–plastic loading waves.

Suppose that the speed of the projectile is $-v^*$. When the projectile impacts on a static rigid target, the particle velocity at the impact end of the bar is suddenly reduced to zero. Therefore a set of elastic–plastic waves propagate toward the free end of the projectile. Now we view the problem in a coordinate system that moves with uniform velocity $-v^*$. In this system, the problem is transferred to another problem where a finite length bar is originally at a natural, rest state ($\sigma = v = 0$) and then is loaded by a sudden constant-value impact loading v^* at one end, while the other end of the bar is kept stress-free.

When the rightward elastic–plastic waves with intensity σ^* are reflected at the free end, the final particle velocity at the free end can be determined, by using the momentum conservation equation across the wave front, as:

$$v_{fs} = -\int_0^{\sigma^*} \frac{d\sigma}{\rho_0 C} + \int_{\sigma^*}^{0} \frac{d\sigma}{\rho_0 C_0} = \frac{1}{\rho_0} \int_{\sigma^*}^{0} \left(\frac{1}{C} + \frac{1}{C_0} \right) d\sigma = v^* - \frac{\sigma^*}{\rho_0 C_0} \tag{4.85}$$

When $C = C_0$, we get $v_{fs} = 2v^*$, which is the result of the elastic wave reflection. In case of plastic wave, we know that $(1/C_0) < (1/C)$, so $v_{fs} < 2v^*$. Therefore when the elastic–plastic waves are reflected at a free end, the end particle velocity is less than twice the particle velocity of incident wave. It is noteworthy that actually a plastic loading wave can never reach the free end. Because before it arrives at the free end, the forerunning elastic precursor wave has arrived at the free end and been reflected back as an unloading wave, which interacts with the after-running plastic wave and weakens it, until it vanishes completely. This phenomenon can be understood in another way. We know that the plastic waves always accompany with residual strains. Since the stress at the free end is always zero, plastic deformation is impossible to produce at the free end. It reveals that no plastic wave will ever arrive at the free end. In a finite length bar, this kind of head-on unloading, where the elastic unloading waves are successively reflected from the free end and repeatedly interact with the head-on plastic loading waves, are frequently seen and needed to be well understood.

4.9.1 Linear hardening bar

Let us first consider the case when the finite length bar is made of linear hardening bar [Ленский (Lenskii), 1949]. Under this situation both the elastic wave and plastic wave are strong-discontinuous waves. As shown in Fig. 4.23, before the elastic precursor wave arrives at the free end ($X = l$), the situation is same as that in a semi-infinite bar, so the states of the zones 0, 1, and 2 on the X–t plane are known:

$$v_0 = \sigma_0 = \varepsilon_0 = 0$$

$$v_1 = v_Y = \frac{Y}{\rho_0 C_0},\ \sigma_1 = -Y,\ \varepsilon_1 = -\varepsilon_Y = -\frac{Y}{E}$$

$$v_2 = v^*,\ \sigma_2 = \rho_0 C_1\left(\frac{Y}{\rho_0 C_0} - v^*\right) - Y,\ \varepsilon_2 = \frac{C_0\varepsilon_Y - v^*}{C_1} - \varepsilon_Y$$

where Y is the yield stress, $\varepsilon_Y (=Y/E)$ is the yield strain.

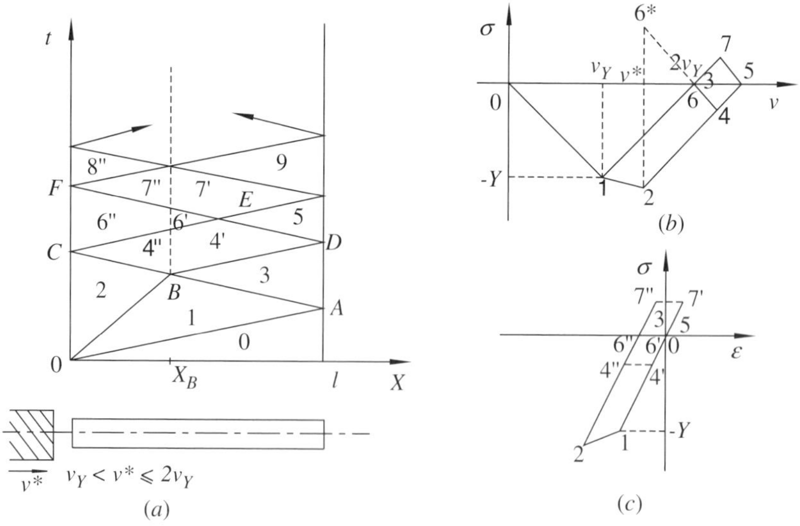

Fig. 4.23. Impact of a finite linearly-hardening bar onto a rigid target when (4.88) is satisfied.

At the time $t = l/C_0$, the elastic precursor wave arrives at the free end and is reflected. The state of the unloading zone 3 is:

$$v_3 = 2v_Y = \frac{2Y}{\rho_0 C_0}, \quad \sigma_3 = \varepsilon_3 = 0$$

At the point B, the reflected unloading wave AB meets with the incident plastic wave OB and unloads it. Imagine that the stress on the left side of the point B is unloaded to zero, then the particle velocity would be v_5 according to Eq. (4.83). However, this velocity v_5 differs from the particle velocity v_3 on the left side of the point B. Such velocity difference actually results in an internal impact, which produces two internal reflected waves BC and BD. The physical state behind the waves BC and BD actually correspond to the point 4 (not the presumed point 5) on the σ–v plane, since in the zone 4 the particle velocity and the stress should be continuous on both sides of the point B. Therefore, we have:

$$\begin{aligned} v_4 &= \frac{v_5 + v_3}{2} = \frac{1}{2}\left[v^* - \frac{C_1}{C_0}(v_Y - v^*) + v^Y + 2v^y\right] \\ &= \frac{\left(3 - \dfrac{C_1}{C_0}\right)(v_Y - v^*)}{2} + 2v^* \\ \sigma_4 &= -\frac{\rho_0 C_0 (v_5 - v_3)}{2} = -\frac{\rho_0 C_0}{2}\left[v^* - \frac{C_1}{C_0}(v_Y - v^*) + v_Y - 2v_Y\right] \\ &= \frac{\rho_0 C_0 \left(1 + \dfrac{C_1}{C_0}\right)(v_Y - v^*)}{2} \end{aligned} \tag{4.86}$$

Note that the strains on both sides of the point B are different. To the right side of the point B, the strain ε_4' is:

$$\varepsilon_{4'} = \frac{\sigma_4}{E} = \frac{\left(1 + \dfrac{C_1}{C_0}\right)(v_Y - v^*)}{2C_0}$$

while to the left side of the point B, the strain ε_4'' is:

$$\varepsilon_{4''} = \varepsilon_2 - \frac{\sigma_2 - \sigma_4}{E} = \frac{\left(2 + \dfrac{C_1}{C_0} - \dfrac{C_1^2}{C_0^2}\right)(v_Y - v^*)}{2C_1}$$

Therefore a stationary strong-discontinuity of strain exists at the section $X = X_B$. This section is the intersection of the point of lines OB and AB. Namely, $(X_B/C_1) = (2l - X_B)/(C_0)$. From this condition, the location of the stationary strong-discontinuity of strain X_B can be determined:

$$X_B = \frac{2C_1 l}{C_0 + C_1} \tag{4.87}$$

After the internal impact at the point B, the leftward internal reflective wave BC is an unloading wave. This wave releases the stress level from the state 2 to the state $4''$ on the σ–ε plot in Fig. 4.22(c). On the other side, the rightward internal reflective wave BD is a weakened loading wave. This wave raises the stress level from the unloading state 3 to the re-loading state $4'$. The concrete value of stress σ_4 is dependent on the initial impact velocity v^*.

If the initial impact velocity v^* is not much higher than the yield velocity v_Y, then the stress σ_4 is below the yield limit, namely $-\sigma_4 = |\sigma_4| \leq Y = \rho_0 C_0 v_Y$. From the expression of σ_4 [Eq. (4.86)], we can further get the inequality condition that v^* must correspondingly satisfy:

$$v_Y < v^* \leq \frac{3C_0 + C_1}{C_0 + C_1} v_Y \tag{4.88}$$

Under this condition, only one plastic zone and consequently one stationary strong-discontinuity of strain are produced within the bar. After the internal impact at the point B the rightward reflective wave BD is an elastic wave. Its reflection at the free end is same as that discussed in the elastic bar problem.

However, when the leftward internal reflective wave BC arrives at the impact end and is reflected back, the situation becomes complex. Depending on the magnitude of the initial impact velocity v^*, we should distinguish the following cases.

First, assume that the impact process still continues, namely the bar is not bounced away from the target. Therefore, after the unloading wave BC is reflected at the impact end (the zone 6), the particle velocity should be held as the prescribed one, i.e. $v_6 = v^*$, the stress σ_6 can be determined by the wave front momentum conservation equation along the rightward wave CE:

$$\sigma_6 = \sigma_4 - \rho_0 C_0 (v_6 - v_4) = \rho_0 C_0 (2v_Y - v^*) \tag{4.89}$$

From this equation, it is clear that if

$$v_Y < v^* \leq 2v_Y \tag{4.90}$$

then $\sigma_6 \geq 0$, showing a tensile stress state corresponding to the point 6^* in Fig. 4.23(b). Note that in practice an impact surface cannot carry any tensile loading, so actually the stress behind the reflected wave CE can only be unloaded to zero ($\sigma_6 = 0$). Now the impacted end of the bar becomes a free end. Corresponding to the zero stress state, the particle velocity is: $v_6 = 2v_Y > v^*$ (Note the inequality 4.90). This means that the bar bounces away from the target, and the assumption of the impact being continuous is untenable. The practical impact duration is $t_0 = (2l/C_0)$. Rather than the point 6^*, the physical state of the zone 6 actually corresponds to the point 6 in the σ–v plane, which coincides with the point 3.

Now both the waves CE and DE are elastic. They represent an elastic pulse propagating forward and reflected backward repeatedly in a finite length bar with both free ends.

In other words, the bar is in a free oscillation state. Note that the bar has a stationary strain-discontinuity at the section B. On the right side of B, the stress–strain state repeatedly changes between the state $4'$ to $7'$ in Fig. 4.23(c), while on the left side of B the stress–strain state repeatedly changes between the state $4''$ to $7''$ [see Fig. 4.23(c)].

We consider another region of the impact velocity. Assume:

$$2v_Y < v^* \leq \frac{3C_0 + C_1}{C_0 + C_1} v_Y \tag{4.91}$$

Now the stress in the zone 6, σ_6, calculated by Eq. (4.89) is a compressive stress ($\sigma_6 < 0$). It means that the impact process continues after the reflection at the point C (as illustrated in Fig. 4.24), though $|\sigma_4|$ is still less than Y. Through a discussion similar to the above one, it can be concluded that the impact process will finish at the point F and the impact duration is $t_F = 4l/(C_0 + C_1) = 2t_B$. There still exists one plastic zone deformed within the bar, of which the length is also determined by Eq. (4.87).

When $v^* = [(3C_0 + C_1)/(C_0 + C_1)]v_Y$, we have $\sigma_4 = -Y$. So if the initial impact velocity exceeds the value $[(3C_0+C_1)/(C_0+C_1)]v_Y$, then the stress level after the internal impact at the point B must be beyond the yield limit. In this case, a precursor elastic loading wave BD and a weakened plastic loading wave BB_1 propagate from the point B to the right end of the bar. Afterwards there are two possible cases. In the first case, the rightward elastic wave BD is reflected as a leftward unloading wave DB_1 at the right end (the free end), and then interacts with the incoming rightward plastic wave BB_1 in a head-on unloading way (Fig. 4.25). In the second case, the leftward elastic wave BC is reflected as a rightward unloading wave CB_2 at the left end (the impacted end), and then catches up with the forerunning plastic wave BB_2 in a pursuing unloading way (Fig. 4.26).

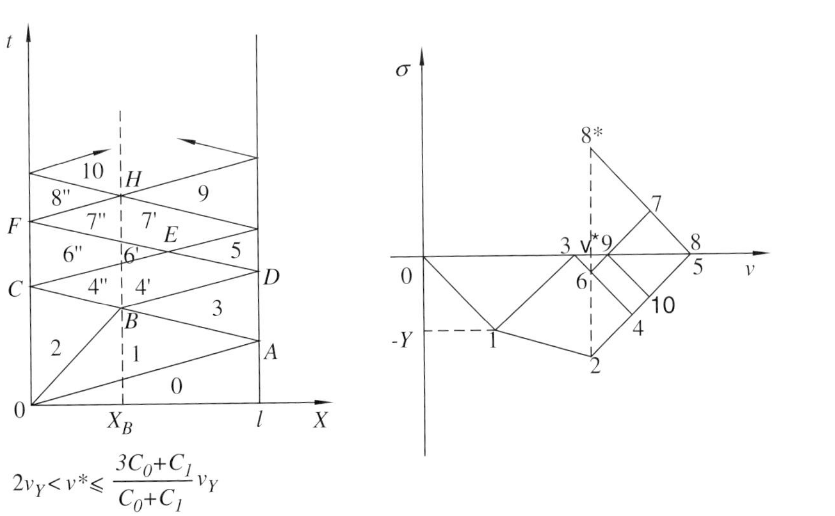

Fig. 4.24. Impact of a finite linearly-hardening bar onto a rigid target when (4.91) is satisfied.

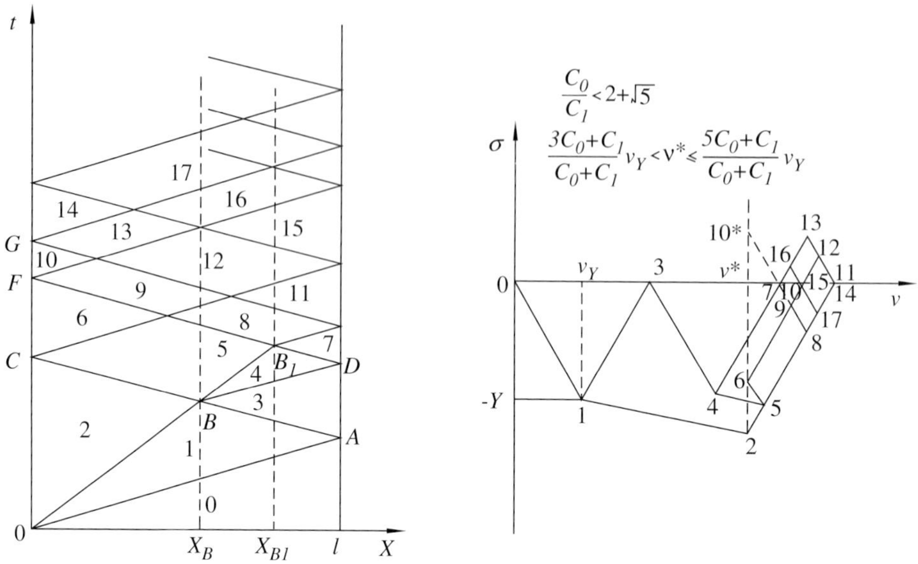

Fig. 4.25. Impact of a finite linearly-hardening bar onto a rigid target when (4.92a) is satisfied.

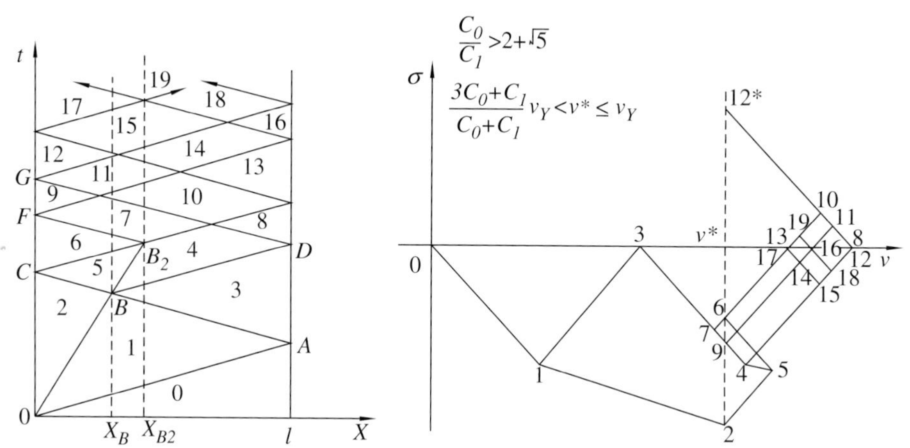

Fig. 4.26. Impact of a finite linearly-hardening bar onto a rigid target when (4.92b) is satisfied.

What actually happens, in practice, depends on the ratio of elastic wave speed to plastic wave speed, C_0/C_1. A critical situation is that the point B_1 in Fig. 4.25 coincides with the point B_2 in Fig. 4. 26. It is easy to prove that this critical situation corresponds to the condition of $C_0/C_1 = (2+\sqrt{5})$, and thus when $C_0/C_1 < (2+\sqrt{5})$, the first case occurs (Fig. 4.25), while when $C_0/C_1 > (2+\sqrt{5})$ the second case occurs (Fig. 4.26).

In the case of Fig. 4.25, it can be easily proved that when the initial impact velocity satisfies:

$$\frac{3C_0 + C_1}{C_0 + C_1} v_Y < v^* \leq \frac{5C_0 + C_1}{C_0 + C_1} v_Y \tag{4.92}$$

the plastic wave vanishes at the point B_1. The two internal reflective waves emitted from the point B_1 are both elastic waves. The second stationary strong-discontinuity of strain is located at

$$X_{B_1} = \frac{4C_0 C_1 l}{(C_0 + C_1)^2}$$

The impact process finishes at the point F. The impact duration is: $t_F = 4l/(C_0 + C_1)$.

In the case of Fig. 4.26, it can be easily proved that when the initial impact velocity v^* satisfies

$$\frac{3C_0 + C_1}{C_0 + C_1} v_Y < v^* \leq 3v_Y$$

the plastic wave vanishes at the point B_2. The two internal reflective waves emitted from the point B_2 are both elastic waves. No reflected plastic wave will be resulted at the point F (i.e. the zone 9 is an elastic zone). The second stationary strong-discontinuity of strain is located at

$$X_{B_2} = \frac{2C_1 l}{C_0 - C_1}$$

The impact process finishes at the point G. The impact duration is $t_G = 4l/(C_0 + C_1)$.

From the above discussions on the problems shown in Figs. 4.23–4.26, we see that in a finite-length bar loaded by a compressive impact, when two unloading waves interact in a head-on unloading way, a tensile stress will be induced. For example, the zone 7 in Figs. 4.23 and 4.24, the zone 12 and zone 13 in Fig. 4.25, the zone 10 in Fig. 4.26, etc., are all tensile stress zones. This phenomenon has been pointed out when discussing the interactions of elastic waves. Especially, the so-called **spalling** and other **reflective fracture** phenomena have been explained on this basis. Here it is worthwhile to point out that under a certain condition, the tensile stress induced by the interaction of two unloading waves can also result in a tensile yield, the so-called **yield in the reverse direction** or briefly **reverse yield**, comparative to the compressive yield occuring previously.

With regard to the tensile stress waves, in the previous analysis on the problems shown in Figs. 4.23–4.26, we have implicitly assumed that if the absolute value of a tensile stress is less than the initial yield stress Y or the hardened yield stress in compression, the tensile stress wave is elastic. In this context we have used the **isotropic hardening hypothesis**. This hypothesis assumes that the hardening effect due to compressive plastic deformation is kept when the material is under tensile loading. However, experimental results have practically shown that when a material has experienced plastic deformation and hardening in the positive direction (for example, the compressive direction), its yield stress tends to be lower under loading in the negative direction (the tensile direction). This is the so-called **Bauschinger effect**. The isotropic hardening hypothesis neglects the Bauschinger effect of the material.

In a material model where the Bauschinger effect is taken into account, it is generally assumed that the difference between the yield stress in the positive, tensile direction and the yield stress in the negative, compressive direction is always $2Y$, regardless of the development of plastic hardening. In the three-dimensional stress state this assumption is called a **kinematic hardening hypothesis**. Therefore, when the material has been hardened by compressive plastic stress waves, the tensile stress induced by a head-on unloading wave interaction may cause the material to yield in the negative direction (tensile direction) even this tensile stress level is not very high. Another possible situation is when tensile stress wave passes through a stationary strain discontinuity. Because the tensile yield stress across this discontinuity is different, when the incident wave is transmitted and reflected at the strain-discontinuous interface, a reverse yield may be induced. An example is illustrated in Fig. 4.27. When the leftward unloading elastic wave *BD* and the rightward elastic unloading wave *CD* meet at the point *D*, a reverse, tensile plastic yield (as shown by the point 6 on the σ–v plane) takes place due to the Bauschinger effect. Similarly, when the rightward tensile plastic wave *DE* passes through the stationary strain-discontinuity X_B, because the materials across the point X_B have different reverse (i.e. tensile) yield stress, the rightward transmitting wave is an elastic loading wave, while the leftward reflected wave is a strengthened tensile plastic wave. The zone 8 and the zone $9''$ on the X–t plane in Fig. 4.27 are all such kind of reverse (tensile) plastic zones.

Although the discussions so far are about strong-discontinuous elastic–plastic waves with constant propagating velocities, just as we have pointed out in the previous sections (Section 4.8 and before), these discussions in principle can be extended to the shock waves propagating in an increasingly hardening bar, or the weak-discontinuous waves propagating in a decreasingly hardening bar. We will briefly describe these treatments in the following sections.

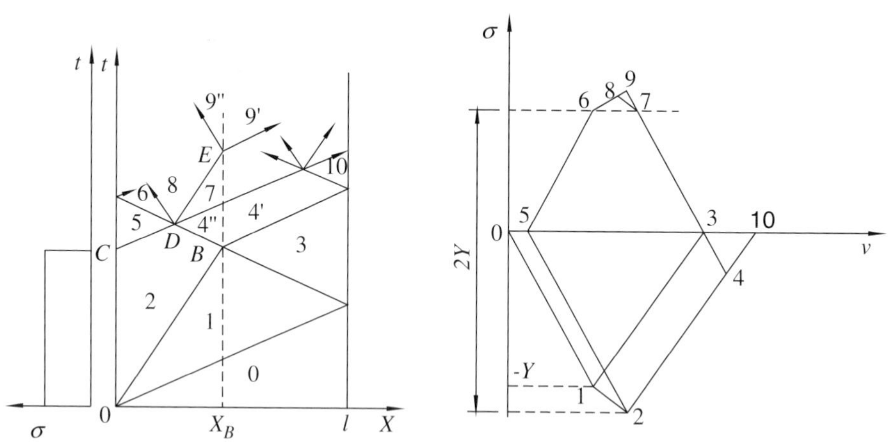

Fig. 4.27. An example showing a reverse, tensile plastic yield due to the Bauschinger effect.

4.9.2 Increasingly hardening bar

Now we consider an increasingly hardening bar impacted onto a rigid anvil with a constant velocity v^* (Lee and Tupper, 1954). When the problem is viewed from a coordinate system that moves with velocity $-v^*$, it is identical to the problem that a natural, rest bar is impacted at the left end by a sudden constant velocity v^*. The other end of the bar is kept free. Thus this problem is reduced to the head-on unloading of shock wave propagating in a finite length bar made of increasingly hardening material. The problem is a little more complex than the previous one because the speed of the plastic shock wave is now dependent on its intensity. Several key points to treat this problem are described below.

When the left end of the bar is impacted by a constant velocity v^*, a strong-discontinuous elastic precursor wave *oa* and a plastic shock wave *o*1 are propagating right with velocity C_0 and $\mathscr{D}_{JK}$, respectively, as shown in Fig. 4.28. The shock wave velocity $\mathscr{D}_{JK}$ is determined by the slope of the chord that connects the point J and K on the compressive stress–strain curve. Here the point J is the yield point, $\sigma_J = -Y$, while the point K represents the final state behind the shock wave, which is determined by the impact velocity v^*. Therefore, the formula to calculate $\mathscr{D}_{JK}$ is:

$$\mathscr{D}_{JK} = \sqrt{\frac{1}{\rho_0}\frac{\sigma_K + Y}{\varepsilon_K - \varepsilon_Y}}$$

while the σ_K should satisfy the momentum conservation equation across the shock wave front:

$$\sigma_K = -Y - \rho_0 \mathscr{D}_{JK}(v^* - v_Y)$$

Strictly speaking, the σ–ε relationship used to calculate the velocity of a shock wave should be the **shock adiabat** in the form of σ–ε, rather than the usual **isothermal** or **isentropic** stress–strain curve (refer to Section 7.7). However, when the shock wave intensity is not too high, the entropy-change during a shock process in solids is negligible. In this case, there are little differences between these three kinds of stress–strain relationships.

The rightward strong-discontinuous elastic loading wave *oa* is reflected at the free end ($X = L$) as a leftward strong-discontinuous elastic unloading wave. This unloading wave meets the rightward plastic shock wave *o*1 at the point 1. A head-on unloading process occurs. This unloading process can be analyzed by the approach stated in the Section 4.8. The results of this head-on unloading include: a stationary strong-discontinuity of strain is formed at X_1; a leftward internal reflective wave 1–A unloads the compressive stress in the zone K from σ_K to σ_N; while the rightward internal reflective wave, as its intensity is still higher than the yield limit, has a two-wave structure, namely, behind the rightward elastic precursor wave 1*b*, a weakened plastic shock wave *12* propagates in the zone M. Because the intensity of stress-jump across this shock wave has been lowered to $|\sigma_N + Y|$, the velocity of the shock wave is thus lowered to $\mathscr{D}_{MN}$, which can be determined by the slope of the chord MN' of the compressive stress–strain curve. (Note that on the σ–ε curve the point M coincides with the point J, and the stress level at the point N' is the same as

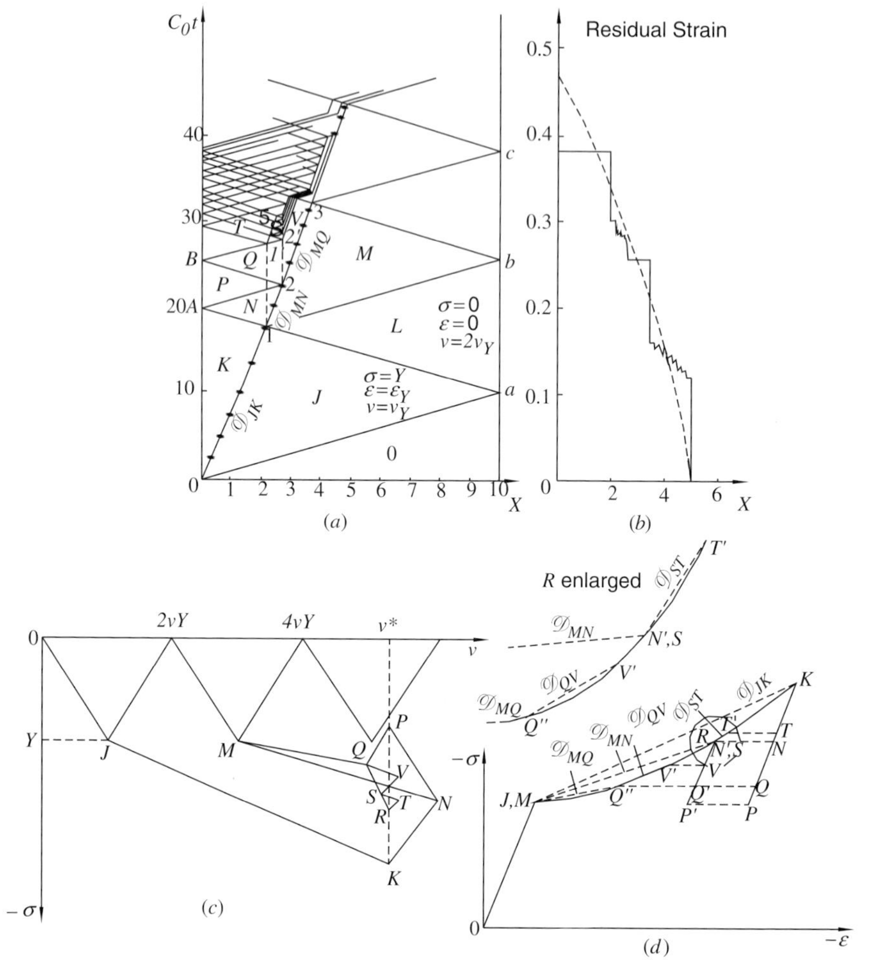

Fig. 4.28. Impact of a finite increasingly-hardening bar onto a rigid target with a constant velocity v^* [Reproduced from Lee & Tupper (1954) by permission of ASME].

that at the point N):

$$\mathscr{D}_{MN} = \sqrt{\frac{1}{\rho_0}\frac{\sigma_N + Y}{\varepsilon_{N'} - \varepsilon_Y}} \tag{4.93}$$

The quantities σ_N and $\mathscr{D}_{MN}$ are calculated by the so-called cut-and-trial method. In fact, regarding the leftward internal reflective wave $1A$, we have:

$$v_N - v_K = \frac{1}{\rho_0 C_0}(\sigma_N - \sigma_K)$$

On the other hand, regarding the rightward reflective plastic shock wave *12*, we have:

$$v_N - v_M = \frac{1}{\rho_0 \mathscr{D}_{MN}}(\sigma_N - \sigma_M)$$

Eliminate the quantity v_N from the above two equations, we obtain:

$$v_{K'} + \frac{1}{\rho_0 C_0}(\sigma_N - \sigma_K) = v_M - \frac{1}{\rho_0 \mathscr{D}_{MN}}(\sigma_N - \sigma_M) \tag{4.94}$$

where $v_K = v^*, \sigma_K = \sigma^*, v_M = 3v_Y$, and $\sigma_M = -Y$ are all known quantities. Equations (4.93) and (4.94) constitute a system of equations with regard to the unknown σ_N and $\mathscr{D}_{MN}$. These two equations can be solved by the cut-and-trial method.

The leftward internal reflective wave $1A$ is reflected at the impacted end. To satisfy the prescribed boundary velocity condition ($v = v^*$), the reflected wave $A2$ is an unloading wave, which unloads the stress σ_N in the zone N down to σ_P (in the zone P). This unloading wave catches up with the forerunning plastic shock wave *12* at the point 2. The corresponding wave interaction is a pursuing unloading. After this unloading, the rightward plastic shock wave *23* is weakened to $|\sigma_Q + Y|$. The leftward internal reflective loading wave $2B$ raises the unloading zone stress σ_P up to σ_Q. The values of σ_Q and the plastic shock wave velocity $\mathscr{D}_{MQ}$ can be determined by the similar cut-and-trial method mentioned above. To the left side of the stationary strain-discontinuity X_2, though σ_Q is higher than Y, it is lower than the maximum compressive stress ever reached ($|\sigma_K|$ and $|\sigma_N|$). Therefore the internal reflective wave $2B$ is still an elastic wave. This wave can passes through the stationary strain-discontinuity X_1 without any change.

As can be seen from the σ–v plot, $v_Q < v^*$, so that a reflective loading happens when the internal reflective elastic wave $2B$ is reflected at the impact-end, raising the compressive stress σ_Q in the unloading zone further to σ_R. However because $|\sigma_K| > |\sigma_R| > |\sigma_N|$, when the rightward elastic wave $B1'$ passes through the stationary strain-discontinuity X_1, a wave-transmission and -reflection occurs, just like the stress wave that propagates from a higher yield material to a lower yield material. To the right side of X_1 the material re-enters the plastic deformation zone, namely the transmitted waves are elastic–plastic waves with two-wave structure. The elastic precursor stress wave $1'2'$ raises the stress level from σ_Q to the yield stress $\sigma_N = \sigma_5$, while the plastic stress wave $1'5$ raises the stress σ_s in the zone S up to the stress σ_T in the zone T. The value of σ_T and the speed of shock wave $1'5$, $\mathscr{D}_{ST}$, can be determined by the cut-and-trial method again.

Similarly, when the rightward elastic wave $1'2'$ reaches the stationary strain-discontinuity X_2, because the right side of X_2 is in a plastic deformation state with stress level σ_Q and $|\sigma_S| > |\sigma_Q|$, a wave reflection and transmission will occur at point $2'$. Behind the rightward transmitted plastic shock wave the stress level is plastically compressed from σ_Q to σ_V, while behind the leftward reflected elastic unloading wave the stress level is lowered from σ_S to σ_V. The σ_V and $\mathscr{D}_{QV}$ are determined by the cut-and-trial method too.

By the analogy of this approach, we can determine the later wave reflections and transmissions at the stationary strain-discontinuity X_2, and a series of wave reflection and transmission at the stationary strain-discontinuity X_3, which is the third strain-discontinuity induced by the interaction between the reflective unloading wave *b*3 and the plastic shock wave *23*. These processes are repeated for the third reflective unloading wave reflected from the free end, *C*4, to interact successively with the plastic shock wave *34* and the following, until all plastic shock waves are unloaded, weakened, and finally vanish. At this stage the impact-end losses contact with the target. Afterwards only the remaining elastic waves propagate back-and-forth within the bar.

After the impact is ended, the residual plastic strain distribution is shown in Fig. 4.28(b). The plastic deformation notably concentrates at the impacted end of the bar. This reflects the phenomenon of plastic deformation localization induced by the effect of stress wave propagation, which is frequently seen in an impact process. Corresponding to the stationary strain-discontinuities at the points such as 1, 2, 3, the residual strain distribution displays relatively large jumps at these locations, respectively. Moreover, since an elastic incident wave may induce a transmitting plastic wave at these strain-discontinuities, some additional stationary strain-discontinuities may be formed such as at the point 5, but with relatively small jump of strain. Correspondingly, some relatively small jumps of residual plastic strain can be seen among the relatively large jumps of residual plastic strain, as shown in Fig. 4.28(b). Such additional residual plastic strain indicates that the **secondary plastic loading** appears in the elastic unloading zone. As has been discussed before, under a certain condition the material in an elastic unloading state can re-enter into a plastic deformation state due to the interaction between the elastic wave and the stationary strain-discontinuity. In the same plot, the dotted line represents the residual strain distribution that is calculated by the "rigid unloading analysis" method, which will be discussed in the following Chapter (see Section 5.3).

4.9.3 Decreasingly hardening bar

Finally, we will discuss an example of a finite length bar made of decreasingly hardening material and loaded by a sudden constant-velocity v^* at one end, as shown in Fig. 4.29.

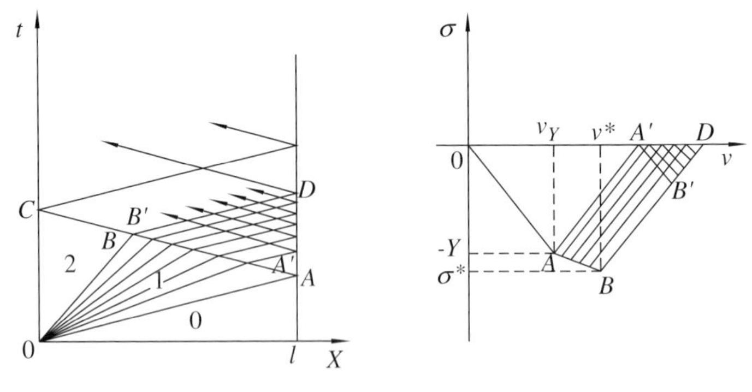

Fig. 4.29. A finite decreasingly-hardening bar subjected to a suddenly applied constant velocity v^*.

It is similar to the problem shown in Fig. 4.23, except that the plastic waves are now all weak-discontinuous waves.

Obviously, at the time $t = l/C_0$, the strong-discontinuous precursor elastic loading wave *OA* is reflected at the free end as a strong-discontinuous unloading wave *AB*. On the σ–v plot this reflection corresponds to the sudden jump from the point *A* to A'. Immediately after the reflection, the unloading wave *AB* interacts with a series of rightward weak-discontinuous centered plastic waves, namely, a series of internal impacts take place. As long as *AB* being strong discontinuous, according to the momentum conservation equation across this leftward strong-discontinuous wave front, we have:

$$\bar{\sigma} - \sigma = \rho_0 C_0(\bar{v} - v)$$

Since *AB* simultaneously is a leftward characteristic line in the unloading zone, we have:

$$\bar{\sigma} + \rho_0 C_0 \bar{v} = \bar{\sigma}(A') + \rho_0 C_0 \bar{v}(A') = 2\rho_0 C_0 v_Y$$

Here the point A' represents the states behind the strong-discontinuous unloading wave, which is corresponding to the line AA' on the σ–v plot in Fig. 4.29. Eliminating the quantity $\bar{\sigma}$ from the above two equations, and utilizing the simple wave relationship in the loading zone, we get:

$$\sigma - \bar{\sigma} = -Y + \frac{1}{2}\int_0^{\sigma}\left(1 - \frac{C_0}{C}\right)d\sigma \tag{4.95}$$

As can be seen, with the propagation of the reflected unloading wave from the free end towards the impact end, the head-on plastic waves are unloaded and simultaneously the intensity of the unloading wave itself $|\sigma - \bar{\sigma}|$ becomes smaller and smaller. On the σ–v plot, the line segment *AB* is the mapping on the plastic side of this unloading wave; and the line segment $A'B'$ is the mapping on the elastic unloading side of this unloading wave. Note that unlike the pursuing unloading case, the intensity of the plastic wave, which successively interacts with the unloading wave, is stronger in sequence. Suppose that until the point *B*, where the intensity of the centered plastic waves reaches the maximum (σ^*, v^*), the intensity of the unloading wave, namely the jump $|\sigma - \bar{\sigma}|$, is still larger than zero, or according to Eq. (4.95):

$$v^* + \frac{\sigma^*}{\rho_0 C_0} = \frac{1}{\rho_0}\int_0^{\sigma^*}\left(\frac{1}{C_0} - \frac{1}{C}\right)d\sigma < 2v_Y \tag{4.96}$$

then the strong-discontinuous unloading wave *AB* can pass through the whole plastic zone. This condition is the same as Eq. (4.16) in the pursuing unloading case, and is reduced to Eq. (4.88) when $C = C_1$.

If the inequality 4.96 is satisfied, then after the strong-discontinuous reflective unloading wave passes through the point *B*, its intensity (i.e. the stress-discontinuity across the unloading wave) is kept as a constant, which corresponds to the line segment *BC* on the X–t plot and corresponds to the jump from the point *B* to the point B' on the σ–v plot.

The state variables at the point B' can be calculated by Eq. (4.95) under the condition of $\sigma = \sigma^*$:

$$\bar{\sigma}(B') = \frac{1}{2}\left(\sigma^* - \rho_0 C_0 v^* + 2Y\right)$$

$$\bar{v}(B') = \frac{1}{2}\left(v^* - \frac{\sigma^*}{\rho_0 C_0} + 2v_Y\right)$$

Afterwards, the unloading wave *BC* is reflected at the impact-end denoted by the point C. The stress $\bar{\sigma}(C)$ and particle velocity $\bar{v}(C)$ after reflection is dependent on the magnitude of the initial impact velocity v^*. According to the momentum conservation equation across a rightward strong-discontinuous wave front, the following relationship between the quantities $\bar{\sigma}(C)$ and $\bar{v}(C)$ is obtained:

$$\bar{\sigma}(C) - \bar{\sigma}(B') = -\rho_0 C_0 \{\bar{v}(C) - \bar{v}(B')\}$$

If the impact process continues, $\bar{v}(C) = v^*$, then we have:

$$\bar{\sigma}(C) = 2Y - \rho_0 C_0 v^*$$

On the other hand, if the impact process continues, then $\bar{\sigma}(C)$ should be in a compressive state, thus we have,

$$v^* > 2v_Y \tag{4.97}$$

If the above inequalities are not satisfied, the impact process finishes at the point C. The impact-end becomes a free end, with $\bar{\sigma}(C) = 0$ and $\bar{v}(C) = 2v_Y$. This means that the bar is bounced away from the target by a velocity-difference, $2v_Y - v^*$. This situation is same as that discussed previously in Fig. 4.23 [refer to Eq. (4.90)].

If v^* simultaneously satisfies Eqs. (4.97) and (4.96), then although the strong-discontinuous unloading wave *AB* can pass through the plastic centered wave zone, the impact process does not finish at the point C. Under a certain circumstance the unloaded zone may be re-loaded plastically, forming a **secondary plastic zone**. For example, once the stress level of the internal reflective wave propagating along the rightward characteristic line *BD* is larger than the historical maximum stress $\sigma_m(X)$ at a certain section along the unloading boundary *AB*, a plastic re-loading will occur to form a secondary plastic zone. In the present situation, since the $\sigma_m(X)$ reaches its maximum at the point B and decreases with X towards the point A, the possibility of such secondary plastic re-loading is understandable.

If the impact velocity is so large that the inequality 4.96 is not satisfied, then the strong-discontinuous reflective unloading wave vanishes at a certain point K within the plastic loading zone. Namely the unloading wave is absorbed in the plastic zone. The stress level σ_K can be calculated by Eq. (4.95) under the condition of $\sigma_K - \bar{\sigma}_K = 0$:

$$\int_0^{\sigma_K} \left(1 - \frac{C_0}{C}\right) d\sigma = 2Y$$

After the unloading wave is absorbed, on the left side of the point K the rightward plastic waves are continuous to propagate; on the right side of the point K, it is also possible to form secondary plastic zones.

In order to better understand the secondary plastic zones, and to analyze the propagation of elastic–plastic boundaries under more general conditions, it is necessary to study further the general features of elastic–plastic boundary propagation, which will be discussed in the next section.

4.10 General Properties of the Loading–Unloading Boundary Propagation

4.10.1 Loading boundary and unloading boundary

In general, the boundary between the plastic loading zone and the elastic unloading zone varies with the propagation and interaction of stress waves. In the case of one-dimensional bar problem, the location of elastic–plastic boundary, changes with time. In the X–t plane if the locus of this boundary is expressed by a function $X = f(t)$, then the moving speed or propagating speed of this boundary, $\overline{C}$, is:

$$\overline{C} = \frac{dX}{dt} = f'(t) \tag{4.98}$$

Elastic–plastic boundaries can be categorized as two types. As shown in Fig. 4.30, when looking at the change of state across an elastic–plastic boundary with the increase of time, if the state changes from a plastic-loading state to an elastic unloading state, then this part of the boundary is called an **unloading boundary**, for which the propagating velocity is

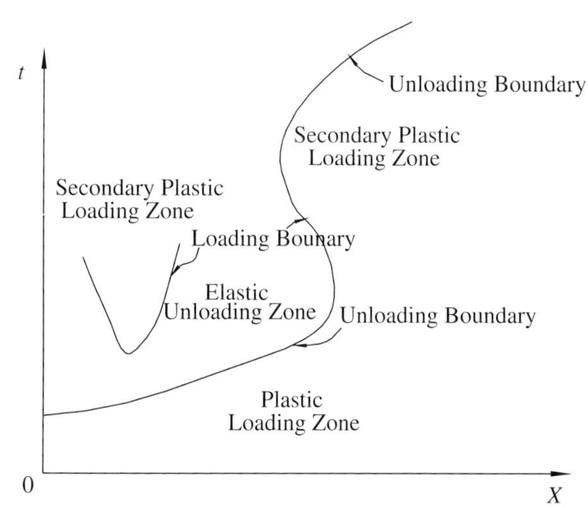

Fig. 4.30. Two categories of elastic-plastic boundaries: loading boundary and unloading boundary.

denoted by $\overline{C}_U$. On the contrary, if the state changes from an elastic state to a plastic-loading state, including the change from an elastic unloading state to a secondary-plastic loading state, then this part of the boundary is called a **loading boundary**, for which the propagating velocity is denoted by $\overline{C}_L$. For a continuous elastic–plastic boundary, once its slope, namely the moving speed $\overline{C}$, changes sign, it corresponds to a transition between these two types of boundaries.

It is worthwhile to emphasize again, unlike a characteristic line that represents the propagating locus of a mechanical disturbance (stress wave), an elastic–plastic boundary, in general, is not the propagating locus of a mechanical disturbance itself, but is the linking or moving locus of such critical points X, wherein due to the interaction of stress waves a transition occurs at time t either from the plastic loading state to the elastic unloading state, or from the elastic (either loading or unloading) state to the plastic loading state. The former is called an unloading boundary while the latter is called a loading boundary. Under certain specific conditions, as we have discussed before, an elastic–plastic boundary may coincide with a propagating locus of mechanical disturbance (namely a characteristic line). Except for those specific cases, in general, it is necessary to distinguish the boundary propagating velocity $\overline{C}$ from the elastic wave speed C_e and the plastic wave speed C_p.

With regard to the loading boundary, consider an arbitrary material particle on the boundary. Before the boundary passes, the particle is in an elastic **loading** state. After the boundary passes, the particle point is in a plastic **loading** state. Therefore, at the time when the boundary passes the particle, the slopes of the stress–time curves on both sides of the boundary have the same sign, as:

$$\frac{\partial \sigma^e}{\partial t} \Big/ \frac{\partial \sigma^p}{\partial t} \geq 0$$

Hereafter the superscripts (or subscripts) "e" and "p" stand for elastic and plastic, respectively.

However, in case that $(\partial \sigma^e/\partial t) = (\partial \sigma^p/\partial t) = 0$, we should consider the signs of the 2nd order derivatives. If, before and after the loading boundary passes the particle, the curvatures of the stress–time curves changes from nonpositive to nonnegative (tensile loading), or changes from nonnegative to nonpositive (compressive loading), then:

$$\frac{\partial^2 \sigma^e}{\partial t^2} \Big/ \frac{\partial^2 \sigma^p}{\partial t^2} \leq 0$$

With regard to the unloading boundary, consider an arbitrary material particle on the boundary. Now, when the boundary passes the particle, the state changes from a plastic **loading** process to an elastic **unloading** process, so we have:

$$\frac{\partial \sigma^e}{\partial t} \Big/ \frac{\partial \sigma^p}{\partial t} \leq 0$$

In case that $(\partial\sigma^e/\partial t) = (\partial\sigma^p/\partial t) = 0$, before and after the unloading boundary passes the particle, the curvatures of the stress–time curves have the same signs, as:

$$\frac{\partial^2\sigma^e}{\partial t^2} \bigg/ \frac{\partial^2\sigma^p}{\partial t^2} \geq 0$$

Summarizing the above discussions, on a loading boundary we have:

$$\frac{\partial\sigma^e}{\partial t} \bigg/ \frac{\partial\sigma^p}{\partial t} \geq 0 \quad \text{or} \quad \frac{\partial^2\sigma^e}{\partial t^2} \bigg/ \frac{\partial^2\sigma^p}{\partial t^2} \leq 0 \tag{4.99}$$

while on an unloading boundary we have:

$$\frac{\partial\sigma^e}{\partial t} \bigg/ \frac{\partial\sigma^p}{\partial t} \leq 0 \quad \text{or} \quad \frac{\partial^2\sigma^e}{\partial t^2} \bigg/ \frac{\partial^2\sigma^p}{\partial t^2} \geq 0 \tag{4.100}$$

4.10.2 Elastic–plastic boundary as a singular interface

In Section 4.7 we have discussed the propagation features of an unloading boundary by means of the characteristic line method, wherein the bar length is semi-infinite and consequently the plastic loading zone is a simple wave zone. Now we are going to introduce a new method – singular interface method – to discuss the problem in more general cases. This approach provides the basic relationships to determine the location of a loading or an unloading boundary. Once the location of an elastic–plastic boundary is determined, the wave problems that include unloading process and/or secondary plastic loading process can be solved easily.

When an elastic–plastic boundary passes through a material particle, the particle either is loaded from an elastic state into a plastic state, or is unloaded from a plastic state into an elastic state. In both cases, the mechanical properties of material display an essential change. Correspondingly, as we usually assume, a discontinuity of the slope of stress–strain curve, $d\sigma/d\varepsilon$ occurs, a kind of **singularity** with respect to the slope of the stress–strain curve. Consequently, a discontinuity of the wave speed C occurs, namely, we have

$$[C] = C_e - C_p \neq 0 \tag{4.101}$$

Or, by introducing a nondimensional parameter $\gamma = \overline{C}/C$, the discontinuity of wave speed can be re-expressed as:

$$[\gamma] = \gamma_e - \gamma_p \neq 0 \tag{4.101$'$}$$

Note that because of the mentioned singularity in the material constitutive relationship, the state variables σ and v or their derivatives are discontinuous across the elastic–plastic boundary. Therefore the elastic–plastic boundary is actually a kind of singular interface. This singular interface is different from the wave front. A wave front represents the propagation of a mechanical disturbance in media. It is the interface between disturbed media and undisturbed media, reflecting the singularities of mechanical variables. On the other hand, when an elastic–plastic boundary propagates in media, it carries the singularities of the material constitutive property, as well as the singularities of state variables. Note that during a propagating process of an elastic–plastic boundary, the singularities of the material constitutive property are invariable, while the singularities of state variables may vary. No matter what the situation, as will be proved, the state variable singularities and the changes of these singularities along the boundary, in the final analysis, are determined by the discontinuity of the stress–strain curve slope (i.e. the wave speed discontinuity). Thus, we shall call the wave speed discontinuity across an elastic–plastic boundary [expressed by Eq. (4.101)] the **basic singularity**. Accordingly, the hypothesis that the basic singularity exists is called the **basic assumption**. The basic assumption [Eq. (4.101)] is the basis for studying the propagation features of an elastic–plastic boundary (Wang *et al.*, 1983). Otherwise, if across a boundary $[C] = 0$, namely $d\sigma/d\varepsilon$ is continuous regardless of loading or unloading, then the problem is a simple nonlinear wave problem, which can be solved by usual approaches stated previously. In such cases, an elastic–plastic boundary coincides with the corresponding characteristic line, and there is no need to be treated specially.

Since the elastic–plastic boundary propagation is viewed as the propagation of a specific singularity interface, we can use the singularity interface method similar to that used in Section 2.7, in addition to the basic singularity expressed by Eq. (4.101), to deduce the kinematic compatibility equation and dynamic compatibility equation across an elastic–plastic boundary.

Thus, consider the discontinuity of an arbitrary physical quantity F across an elastic–plastic boundary

$$[F] = F^e - F^p$$

If we follow the propagation of the elastic–plastic boundary to observe the variance of the discontinuity of F, then similar to the "wave front derivative" in Eq. (2.51), now we have the "boundary derivative", namely the total derivative along the boundary:

$$\frac{d[F]}{dt} = \frac{\partial[F]}{\partial t} + \overline{C}\frac{\partial[F]}{\partial X} \tag{4.102}$$

This boundary derivative will be applied frequently to discuss the features of an elastic–plastic boundary.

4.10.3 *Strong-discontinuous elastic–plastic boundary*

First, we consider the case that the elastic–plastic boundary is a strong-discontinuity. Replacing the F in Eq. (4.102) with the particle displacement u and noting that the displacement across the boundary should be continuous, we get the **displacement continuity condition** (also called the **kinematic compatibility condition**) across the strong-discontinuous boundary, as:

$$[v] = -\overline{C}[\varepsilon] \tag{4.103}$$

Additionally, similar to Eq. (2.57), we have the **momentum conservation condition** (also called the **dynamic compatibility condition**) across the strong-discontinuous boundary, as:

$$[\sigma] = -\rho_0 \overline{C}[v] \tag{4.104}$$

From Eqs. (4.103) and (4.104), we obtain:

$$\overline{C} = \sqrt{\frac{1}{\rho_0}\frac{[\sigma]}{[\varepsilon]}} \tag{4.105}$$

The above equation does not contradict the Eq. (4.101), so a strong-discontinuous elastic–plastic boundary can exist. Under the specific condition that a jump of elastic unloading stress exists across the boundary, therefore:

$$\frac{[\sigma]}{[\varepsilon]} = E, \quad \overline{C} = C_e$$

which means that the propagating locus of an unloading boundary coincides with that of the strong-discontinuous unloading disturbance. This is the case that has been discussed in Section 4.9. Another specific case is that a jump of plastic loading stress exists across the boundary. In this case the propagating locus of a loading boundary coincides with that of the strong-discontinuous plastic loading disturbance (or plastic shock wave). In case of linear hardening material, we have:

$$\frac{[\sigma]}{[\varepsilon]} = E_1, \quad \overline{C} = C_1$$

This is the case that has been discussed in the problem shown in Fig. 4.12. If across the boundary, we have $[\sigma] = 0$ but $[\varepsilon] \neq 0$, then $\overline{C} = 0$. This represents a stationary strain-discontinuous interface.

4.10.4 First order weak-discontinuous boundary

Second, we consider the case that the elastic–plastic boundary is a weak-discontinuity. In this case the stress σ and the particle velocity v themselves are all continuous across the boundary but their derivatives are discontinuous. It should be emphasized again that, hereafter, when material is unloaded from a plastic state to an elastic state, or is loaded from an elastic loading/unloading state to a plastic state, we assume that the stress–strain curve is continuous, but the slope of stress–strain curve is discontinuous, including the case of initial yield. Thus, Eq. (4.101) must be satisfied on any elastic–plastic boundary. In fact, if the slope of stress–strain curve is continuous at the initial yield, then the elastic–plastic boundary coincides with the elastic characteristic line, and the problem can be easily solved.

Since σ and v are continuous across an elastic–plastic boundary, we can replace the F in Eq. (4.102) with σ and v, obtaining:

$$\begin{aligned} \frac{d[\sigma]}{dt} &= \left[\frac{d\sigma}{dt}\right] = \left[\frac{\partial\sigma}{\partial t}\right] + \overline{C}\left[\frac{\partial\sigma}{\partial X}\right] = 0 \\ \frac{d[v]}{dt} &= \left[\frac{dv}{dt}\right] = \left[\frac{\partial v}{\partial t}\right] + \overline{C}\left[\frac{\partial v}{\partial X}\right] = 0 \end{aligned} \tag{4.106}$$

Moreover, when the first order partial derivative governing equations are expressed in terms of the dependent variables σ and v, they have the same form as whatever an elastic zone or a plastic zone is concerned, or whatever a loading process or an unloading process is concerned [refer to Eqs. (2.13), (2.17), and (4.9)], namely we have:

$$\left.\begin{aligned} \frac{\partial v}{\partial X} &= \frac{1}{\rho_0 C^2}\frac{\partial\sigma}{\partial t} \\ \frac{\partial v}{\partial t} &= \frac{1}{\rho_0}\frac{\partial\sigma}{\partial X} \end{aligned}\right\} \quad (e,\ p) \tag{4.107}$$

Here (e, p) means that the equations are suitable to both the elastic zone and the plastic zone, although the wave speed should be chosen as C_e or C_p, respectively, depending on the zone concerned. Rewriting the Eq. (4.107) in the form of the discontinuous values (the jumps) across the elastic–plastic boundary and noting Eq. (4.101), we get:

$$\begin{aligned} \left[\frac{\partial v}{\partial X}\right] &= \frac{1}{\rho_0}\left[\frac{1}{C^2}\frac{\partial\sigma}{\partial t}\right] \\ \left[\frac{\partial v}{\partial t}\right] &= \frac{1}{\rho_0}\left[\frac{\partial\sigma}{\partial X}\right] \end{aligned} \tag{4.107$'$}$$

From Eq. (4.107), if we define:

$$\begin{aligned} K_1 &\equiv \frac{\partial\sigma}{\partial t} = \rho_0 C^2\frac{\partial v}{\partial X} \\ J_1 &\equiv \rho_0 C\frac{\partial v}{\partial t} = C\frac{\partial\sigma}{\partial X} \end{aligned} \tag{4.108}$$

substituting them into Eq. (4.106), then the Eq. (4.107$'$) is reduced to:

$$\begin{aligned} &[K_1 + \gamma J_1] = 0 \\ &\left[\frac{1}{C}(J_1 + \gamma K_1)\right] = 0 \end{aligned} \tag{4.109}$$

The above two equations are the basic equations that a weak-discontinuous elastic–plastic boundary must satisfy.

Eliminating the quantity J_1 from the above two equations, we get:

$$\left[\left(1-\gamma^2\right)K_1\right]=0 \tag{4.110}$$

which can be rewritten in the following expanded form:

$$\frac{\dfrac{\partial\sigma^p}{\partial t}}{\dfrac{\partial\sigma^e}{\partial t}}=\frac{1-\dfrac{\overline{C}^2}{C_e^2}}{1-\dfrac{\overline{C}^2}{C_p^2}} \tag{4.110$'$}$$

Note that because of the basic assumption that across an elastic–plastic boundary the wave speed is discontinuous, namely $C_e \neq C_p$, therefore from Eq. (4.110′) we have $(\partial\sigma^e/\partial t) \neq (\partial\sigma^p/\partial t)$, namely $[K_1] = [\partial\sigma/\partial t] \neq 0$. Consequently, from the first equation of Eq. (4.109) we also have: $[\gamma J_1] = \rho_0\overline{C}[\partial v/\partial t] \neq 0$. Finally from Eq. (4.106) we know $[\partial\sigma/\partial X] \neq 0$, $[\partial v/\partial X] \neq 0$. In other words, across the elastic–plastic boundary, all the first order derivatives of σ and v with respect to X and t must be discontinuous, except in the case that $(\partial\sigma^e/\partial t) = (\partial\sigma^p/\partial t) = 0$ (which will be discussed later). Thus we get an important Theorem of the elastic–plastic boundary.

Theorem 1: Unless on the whole elastic–plastic boundary $(\partial\sigma^e/\partial t) = (\partial\sigma^p/\partial t) = 0$, all the first order partial derivatives of σ and v are discontinuous, i.e. the elastic–plastic boundary is a first order weak-discontinuity of σ and v.

From Eq. (4.110), we can directly get the expression for the propagating velocity of this first order weak-discontinuity:

$$\overline{C}=\sqrt{\frac{\left[\dfrac{\partial\sigma}{\partial t}\right]}{\left[\dfrac{1}{C^2}\dfrac{\partial\sigma}{\partial t}\right]}} \tag{4.111}$$

which is the same as the one we deduced in Section 4.7 [Eq. (4.73)] for a semi-infinite bar by using characteristic line method. Here, the equation is deduced for general situations, applicable to both the unloading boundary and the loading boundary and no limitation of whether the plastic zone is a simple wave zone or not, etc. The equation expressed in the form of Eq. (4.111) was first obtained by Karman, Bohnenblust and Hyers (Karman *et al.*, 1942).

The first order weak-discontinuous boundary can be determined by the following approach, depending on the specific situation. Assume that in the X–t plot (Fig. 4.31), the solution before $t \leq t_1$ is known. Curve GF is a determined unloading boundary moving from the right lower direction toward the point F. We now try to determine the boundary point P at the next time-step $t + \Delta t$.

Starting from the point F, we draw five straight lines with slope of $\pm C_e$, $\pm C_p$, and 0 (the line FK), respectively. These lines divide the half plane $(t \geq t_1)$ into six zones: the zone I,

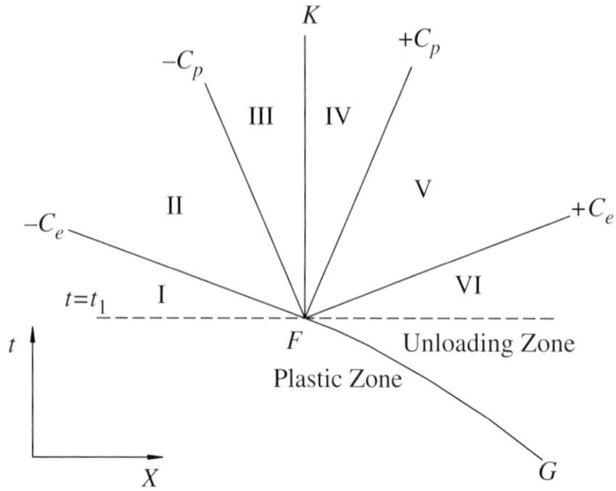

Fig. 4.31. Six possible zones exist for a weak-discontinuous boundary.

II, III, IV, V, and VI. If the next boundary segment FP lies to the left side of the line FK, meaning that $\overline{C}$ does not change sign, then FP is still an unloading boundary. On the contrary, if FP lies to the right side of the line FK, meaning that $\overline{C}$ changes sign, then FP is a loading boundary, forming the secondary plastic zone.

With regard to the unloading boundary, since $(\partial\sigma^e/\partial t)/(\partial\sigma^p/\partial t) \leq 0$ [Eq. (4.100)], then from Eq. (4.110′), $\overline{C}$ must satisfy the following inequality [comparing to Eq. (4.74)]:

$$C_e \geq |\overline{C}_U| \geq C_p \tag{4.112}$$

which means that the successive boundary segment FP must fall into the zone II. Note that if GF moves from the left lower direction toward the point F, then FP must fall into the zone V, which is the mirror symmetry of the zone II.

With regard to the loading boundary, including the secondary plastic loading boundary, since $(\partial\sigma^e/\partial t)/(\partial\sigma^p/\partial t) \geq 0$ [Eq. (4.99)], then from Eq. (4.110′), $\overline{C}$ must satisfy either of the following inequality:

$$\begin{cases} C_p \geq \left|\overline{C}_L\right| > 0 \\ \quad\text{or} \\ \left|\overline{C}_L\right| \geq C_e \end{cases} \tag{4.113}$$

which means that the successive boundary segment FP must fall into either zone IV, or zone VI, corresponding to the first inequality or the second inequality of (4.113), respectively. In case that GF moves from the left lower direction toward the point F, then FP must fall into the zone III or the zone I.

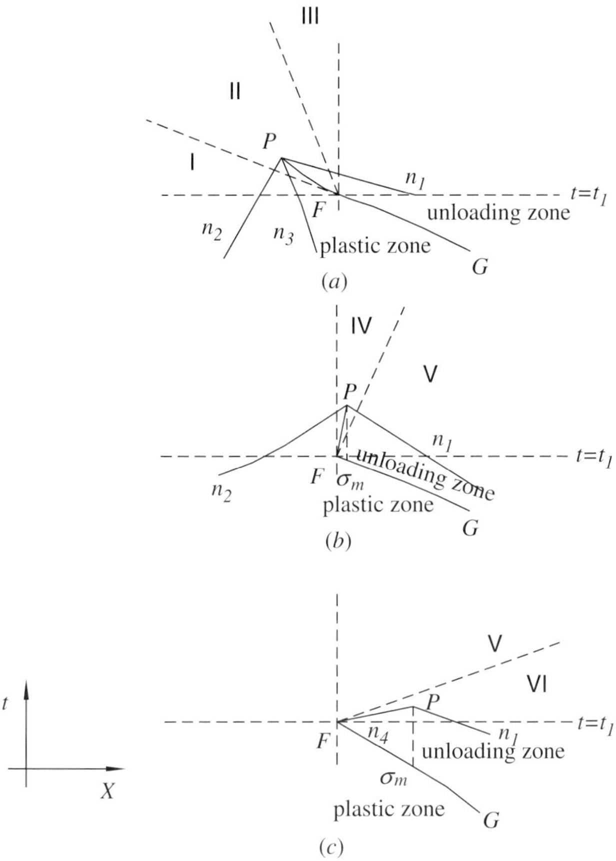

Fig. 4.32. Determination of the successive first order weak-discontinuity boundary in different zones.

What actually happened will be determined by the specific situation as solved before the time t_1. Whatever the zone the successive boundary segment *FP* falls into, all situations can be solved.

If *FP* is in **zone II**, then in the unloading zone a characteristic line n_1 can be drawn from the point *P* [see Fig. 4.32(a)]:

$$n_1: \quad dX + C_e dt = 0, \quad v^e + \frac{\sigma^e}{\rho_0 C_0} = \beta_1$$

and in the plastic zone two characteristic lines n_2 and n_3 can be drawn from the point *P*:

$$n_2: \quad dX - C_p dt = 0, \quad v^p - \phi(\sigma^p) = \alpha_2$$

$$n_3: \quad dX + C_p dt = 0, \quad v^p + \phi(\sigma^p) = \beta_3$$

The Riemann invariables $\beta_1, \alpha_2, \beta_3$ in the above expressions are all known. The above three equations, in addition to the stress continuous condition $\sigma^e = \sigma^p$ and the particle velocity continuous condition $v^e = v^p$, five equations totally are sufficient to solve the five unknown quantities: $\sigma^e, \sigma^p, v^e, v^p$, and $\overline{C}$. Therefore the point P is determined.

If FP is in **zone IV**, then from the point P a characteristic line n_1 in the elastic zone and a characteristic line n_2 in the plastic zone can be drawn [see Fig. 4.32(b)]. These give us two compatibility equations containing two known Riemann invariables. In addition, at the point P the stress must be continuous and satisfy the yield condition: $\sigma^e = \sigma^p = \sigma_m$ (two equations), and the particle velocity must be continuous: $v^e = v^p$. The five equations totally are sufficient to solve the five unknown quantities: σ^e, σ^p, v^e, v^p, and $\overline{C}$, or can be reduced to solving the following equation:

$$\rho_0 C_e \alpha_2 - \rho_0 C_e \beta_3 = \sigma_m + \rho_0 C_e \phi(\sigma_m)$$

which can be solved by a cut-and-trial method.

If FP is in **zone VI**, then from the point P two characteristic lines n_1 and n_4 in the unloading zone can be drawn [see Fig. 4.32(c)]. The expressions of n_1 have been given previously and the expression of the characteristic line n_4 is:

$$n_4: \quad dX - C_e dt = 0, \quad v^e - \frac{\sigma^e}{\rho_0 C_0} = \alpha_4$$

where the Riemann invariables α_4 is also known. In addition, at the point P we have conditions: $\sigma^e = \sigma^p = \sigma_m$, $v^e = v^p$. The five equations totally are sufficient to solve the five unknown quantities, or can be reduced to solving the following equation:

$$\rho_0 C_e(\beta_1 - \alpha_4) = 2\sigma_m$$

which can be solved by a cut-and-trial method.

As to the cases when GF moves from the left side, and thus FP falls into zone I, III, or V, the problem can be solved in the similar way.

The problem of determining the elastic–plastic boundary for a finite length bar made of decreasingly hardening material, under the boundary condition that one end is fixed and another end is loaded by a rectangular impact loading, has been discussed by Bohnenblust *et al.* (1942). For the problem of a decreasingly hardening bar impacted onto a rigid anvil, Lee analyzed the complex phenomenon wherein an elastic–plastic boundary propagates in the zone II, IV, and V (Lee, 1952). However, at that time they didn't adequately treat the situation when $\partial\sigma^e/\partial t$ and $\partial\sigma^p/\partial t$ are simultaneously zero.

4.10.5 Second order weak-discontinuous boundary

From the previous discussions about the first order weak-discontinuous boundary, we know that if across the boundary $(\partial\sigma^e/\partial t) = (\partial\sigma^p/\partial t) = 0$, then by Eq. (4.106) and Eq. (4.107′),

$$\left[\frac{\partial \sigma}{\partial t}\right] = \left[\frac{\partial \sigma}{\partial X}\right] = \left[\frac{\partial v}{\partial t}\right] = \left[\frac{\partial v}{\partial X}\right] = 0 \tag{4.114}$$

Further, from Eq. (4.107), which is a combination of the continuation equation and the constitutive relationship and is satisfied on both the elastic side and the plastic side across the boundary, therefore we have $(\partial v^e/\partial X) = (\partial v^p/\partial X) = 0$. This is another important property of the elastic–plastic boundary.

Theorem 2: If across the whole elastic–plastic boundary $(\partial \sigma^e/\partial t) = (\partial \sigma^p/\partial t) = 0$, then all the first order partial derivatives of σ and v are continuous everywhere along the boundary, and $(\partial v^e/\partial X)$ and $(\partial v^p/\partial X)$ must simultaneously be zero.

In this case, the boundary propagating velocity expressed by Eq. (4.111) becomes an indefinite $0/0$ form. To determine the propagation features of the boundary it is necessary to consider the higher order derivatives of σ and v.

Taking the boundary derivatives of the 1st order derivatives of σ and v, utilizing Theorem 2, we obtain:

$$\begin{aligned}
\frac{d}{dt}\left(\frac{\partial \sigma}{\partial t}\right) &= \frac{\partial^2 \sigma}{\partial t^2} + \overline{C}\frac{\partial^2 \sigma}{\partial t \partial X} = 0, \quad (e, p) \\
\left[\frac{d}{dt}\left(\frac{\partial \sigma}{\partial X}\right)\right] &= \left[\frac{\partial^2 \sigma}{\partial t \partial X}\right] + \overline{C}\left[\frac{\partial^2 \sigma}{\partial X^2}\right] = 0 \\
\frac{d}{dt}\left(\frac{\partial v}{\partial X}\right) &= \frac{\partial^2 v}{\partial t \partial X} + \overline{C}\frac{\partial^2 v}{\partial X^2} = 0, \quad (e, p) \\
\left[\frac{d}{dt}\left(\frac{\partial v}{\partial t}\right)\right] &= \left[\frac{\partial^2 v}{\partial t^2}\right] + \overline{C}\left[\frac{\partial^2 v}{\partial t \partial X}\right] = 0
\end{aligned} \tag{4.115}$$

where the (e, p) means that the equations are suitable to both the elastic zone and the plastic zone across the boundary.

Differentiate the first order partial differential governing equations Eq. (4.107), which are suitable to both the plastic zone and the elastic zone, with respect to variables X and t, we obtain:

$$\begin{aligned}
\frac{\partial^2 v}{\partial t \partial X} &= \frac{1}{\rho_0 C^2}\frac{\partial^2 \sigma}{\partial t^2} + \frac{1}{\rho_0}\frac{\partial \sigma}{\partial t}\frac{\partial}{\partial t}\left(\frac{1}{C^2}\right) \\
\frac{\partial^2 v}{\partial X^2} &= \frac{1}{\rho_0 C^2}\frac{\partial^2 \sigma}{\partial t \partial X} + \frac{1}{\rho_0}\frac{\partial \sigma}{\partial t}\frac{\partial}{\partial X}\left(\frac{1}{C^2}\right) \\
\frac{\partial^2 v}{\partial t^2} &= \frac{1}{\rho_0}\frac{\partial^2 \sigma}{\partial t \partial X}
\end{aligned}$$

$$\frac{\partial^2 v}{\partial t \partial X} = \frac{1}{\rho_0} \frac{\partial^2 \sigma}{\partial X^2}$$

The above four equations contain six second order derivatives of σ and v. Since we have assumed that across the boundary $(\partial\sigma/\partial t) = 0\ (e, p)$, only two second order derivatives are independent quantities, others can be separated into two groups and interrelated by the following equations, by defining two parameters K_2 and J_2 for convenience:

$$\begin{aligned} K_2 &\equiv \frac{\partial^2 \sigma}{\partial t^2} = \rho_0 C^2 \frac{\partial^2 v}{\partial t \partial X} = C^2 \frac{\partial^2 \sigma}{\partial X^2} \\ J_2 &\equiv \rho_0 C \frac{\partial^2 v}{\partial t^2} = C \frac{\partial^2 \sigma}{\partial t \partial X} = \rho_0 C^3 \frac{\partial^2 v}{\partial X^2} \end{aligned} \tag{4.116}$$

Substituting them into Eq. (4.115), we get:

$$\begin{aligned} & K_2 + \gamma J_2 = 0, \quad (e, p) \\ & \left[\frac{1}{C}(J_2 + \gamma K_2) \right] = 0 \end{aligned} \tag{4.117}$$

The first equation of Eq. (4.117) is suitable to both sides across the boundary. The total three equations of Eq. (4.117) are the basic equations that a higher-than-first order weak-discontinuous elastic–plastic boundary must satisfy.

Eliminate quantity J_2 in Eq. (4.117), we get:

$$\left[(1 - \gamma^2) K_2 \right] = 0 \tag{4.118}$$

By the basic assumption, we have $[\gamma] \neq 0$ [Eq. (4.101)]. From the above equation we know that $[K_2] = [\gamma^2 K_2] \neq 0$ (unless $[K_2] = 0$ and consequently $(\partial^2\sigma^e/\partial t^2) = (\partial^2\sigma^p/\partial t^2) = 0$). Moreover from Eq. (4.117) we also know that $[\gamma J_2] \neq 0$. Finally from Eq. (4.115) we know that all the secondary partial derivatives of σ and v are discontinuous across the boundary, except in the case that $(\partial^2\sigma^e/\partial t^2) = (\partial^2\sigma^p/\partial t^2) = 0$ (which will be discussed further). Thus we get another important property of the elastic-plastic boundary.

Theorem 3: If across the whole elastic–plastic boundary $(\partial\sigma^e/\partial t) = (\partial\sigma^p/\partial t) = 0$, then unless on the boundary $\partial^2\sigma^e/\partial t^2$ and $\partial^2\sigma^p/\partial t^2$ are continuous, all the second order partial derivatives of σ and v are discontinuous. In this case the elastic-plastic boundary is a second order weak-discontinuity of σ and v.

From Eq. (4.118), the propagating velocity of this second order weak-discontinuity can be determined by the following formula:

$$\overline{C} = \sqrt{\frac{\left[\dfrac{\partial^2 \sigma}{\partial t^2} \right]}{\left[\dfrac{1}{C^2} \dfrac{\partial^2 \sigma}{\partial t^2} \right]}} \tag{4.119}$$

This formula was obtained by Clifton and Ting (1968) in a special case of elastic–plastic wave propagation. Nevertheless, their general formulae for a boundary with higher-than-second order discontinuity is not likely adequate (refer to the theorem 5 stated below).

From the first equation of Eq. (4.117), the boundary propagation velocity $\overline{C}$ should also simultaneously satisfy the following two equations:

$$\overline{C} = -\frac{\dfrac{\partial^2 \sigma^e}{\partial t^2}}{\dfrac{\partial^2 \sigma^e}{\partial t \partial X}} = -\frac{\dfrac{\partial^2 \sigma^e}{\partial t^2}}{\rho_0 \dfrac{\partial^2 v^e}{\partial t^2}}$$

$$\overline{C} = -\frac{\dfrac{\partial^2 \sigma^p}{\partial t^2}}{\dfrac{\partial^2 \sigma^p}{\partial t \partial X}} = -\frac{\dfrac{\partial^2 \sigma^p}{\partial t^2}}{\rho_0 \dfrac{\partial^2 v^p}{\partial t^2}} \qquad (4.119')$$

The above equations do not contradict the Eq. (4.119), so it is possible that an elastic–plastic boundary satisfies both Eqs. (4.119) and (4.119′). Therefore a second order weak-discontinuous boundary can exist realistically.

In the X–t plane, the locations of a second order weak-discontinuous boundary are rather different from that of a first order weak-discontinuous boundary.

With regard to the unloading boundary, since the curvatures of σ–t curve have the same signs on the elastic side and the plastic side across the boundary, namely $(\partial^2\sigma^e/\partial t^2)/(\partial^2\sigma^p/\partial t^2) \geq 0$ [Eq. (4.100)], so from Eq. (4.118), we know that $\overline{C}$ must satisfy either one of the following inequalities:

$$\begin{cases} C_p \geq |\overline{C}_U| \geq 0 \\ \quad \text{or} \\ |\overline{C}_U| \geq C^e \end{cases}$$

which means that the successive unloading boundary should fall into zone VI or IV, or their mirror symmetric zones I or III (Fig. 4.33).

With regard to the loading boundary, because the curvatures of σ–t curve change signs across the boundary, namely $\left(\partial^2\sigma^e/\partial t^2\right) / \left(\partial^2\sigma^p/\partial t^2\right) \leq 0$ [Eq. (4.99)], from Eq. (4.118), we know that $\overline{C}$ must satisfy the following inequality:

$$C_e \geq |\overline{C}_L| > C_p$$

which means that the successive unloading boundary should fall into zone II or its mirror symmetric zone V (Fig. 4.33).

By comparing the above results with Eqs. (4.112) and (4.113), it is thus clear that the possible scopes of an unloading boundary velocity $\overline{C}_U$, or a loading boundary velocity $\overline{C}_L$,

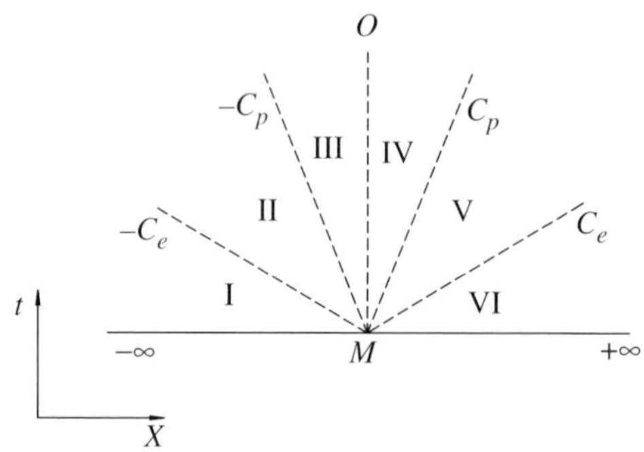

Fig. 4.33. Determination of the successive second order weak-discontinuity boundary in different zones.

in the case of a second order weak-discontinuous boundary are just the counterchange of that in the case of first order weak-discontinuous boundary. The combination of those for these two cases just cover the whole possible range on the X–t plot from $X = -\infty$ to $X = \infty$, as shown in Fig. 4.33, and correspondingly in Table 4.2.

4.10.6 Discussions on higher-than-second order weak-discontinuous boundaries

In the previous discussions, we have deduced the basic equations for the higher-than-first order weak-discontinuous boundary, Eq. (4.117), as:

$$K_2 + \gamma J_2 = 0, \ (e, p)$$

$$\left[\frac{1}{C}(J_2 + \gamma K_2)\right] = 0$$

If $K_2^e(\equiv \partial^2\sigma^e/\partial t^2)$ and $K_2^p(\equiv \partial^2\sigma^p/\partial t^2)$ are continuous everywhere along the boundary, then from Eq. (4.118) we must have $(\partial^2\sigma^e/\partial t^2) = (\partial^2\sigma^p/\partial t^2) = 0$. Furthermore from the first equation of Eq. (4.117) we get $(\gamma J_2)^e = (\partial^2 v^e/\partial t^2) = 0$ and $(\gamma J_2)^p = (\partial^2 v^p/\partial t^2) = 0$. Therefore the second equation of Eq. (4.117) is satisfied automatically.

Table 4.2. The possible scopes of both the first and the second order weak-discontinuous loading-unloading boundaries.

Elastic-plastic boundary	Loading boundary	Unloading boundary
1st order weak discontinuity	$\lvert\overline{C}_L\rvert \geq C_e$, (zone I & VI) $\lvert\overline{C}_L\rvert \leq C_p$, (zone III & IV)	$C_p \leq \lvert\overline{C}_U\rvert \leq C_e$, (zone II & V)
2nd order weak discontinuity	$C_p \leq \lvert\overline{C}_L\rvert \leq C_e$, (zone II & V)	$\lvert\overline{C}_U\rvert \geq C_e$, (zone I & VI) $\lvert\overline{C}_U\rvert \leq C_p$, (zone III & IV)

From the definitions of K_2 and J_2 [Eq. (4.116)] we can deduce that on both the elastic side and the plastic side across the boundary the following relations exist:

$$\frac{\partial^2 \sigma}{\partial t^2} = \frac{\partial^2 \sigma}{\partial t \partial X} = \frac{\partial^2 \sigma}{\partial X^2} = \frac{\partial^2 v}{\partial t^2} = \frac{\partial^2 v}{\partial t \partial X} = \frac{\partial^2 v}{\partial X^2} = 0, \quad (e, p)$$

Thus we obtain the fourth theorem of the elastic–plastic boundary as follows.

Theorem 4: If everywhere along the whole elastic–plastic boundary $(\partial \sigma^e/\partial t) = (\partial \sigma^p/\partial t) = 0$ and simultaneously $(\partial^2 \sigma^e/\partial t^2) = (\partial^2 \sigma^p/\partial t^2)$, then all the second order partial derivatives of σ and v are continuous and are all equal to zero.

In this case, the boundary propagating velocity expressed by either formula expressed in Eq. (4.119) becomes an indefinite 0/0 form. It seems that we need to consider the third order derivatives of σ and v in order to determine the propagating velocity of the boundary. However, we can immediately prove that because of the basic singularity expressed by Eq. (4.101), the higher-than-second order weak-discontinuous elastic–plastic boundary is actually impossible to exist, unless this boundary coincides with the characteristic line. The proof is stated below.

Suppose that this kind of weak-discontinuity exists. According to theorem 4, along the whole boundary, it must have

$$\begin{aligned} \frac{d}{dt}\left(\frac{\partial^2 \sigma}{\partial t^2}\right) &= \frac{d}{dt}\left(\frac{\partial^2 \sigma}{\partial t \partial X}\right) = \frac{d}{dt}\left(\frac{\partial^2 \sigma}{\partial X^2}\right) = \frac{d}{dt}\left(\frac{\partial^2 v}{\partial t^2}\right) \\ &= \frac{d}{dt}\left(\frac{\partial^2 v}{\partial t \partial X}\right) = \frac{d}{dt}\left(\frac{\partial^2 v}{\partial X^2}\right) = 0 \end{aligned} \tag{4.120}$$

On the other hand, in both the plastic zone and the elastic zone across the boundary, differentiating the governing equations [Eq. (4.107)] twice, and utilizing the preconditions that $(\partial \sigma/\partial t) = 0$ and $(\partial^2 \sigma/\partial t^2) = 0$, it can be shown that among the eight 3rd order partial derivatives of σ and v, only two of them are independent, others can be separated into two groups and interrelated by the following equations, by defining two parameters K_3 and J_3 for convenience:

$$\begin{aligned} K_3 &\equiv \frac{\partial^3 \sigma}{\partial t^3} = \rho_0 C \frac{\partial^3 v}{\partial t^2 \partial X} = C^2 \frac{\partial^3 \sigma}{\partial t \partial X^2} = \rho_0 C^4 \frac{\partial^3 v}{\partial X^3} \\ J_3 &\equiv \rho_0 C \frac{\partial^3 v}{\partial t^3} = C \frac{\partial^3 \sigma}{\partial t^2 \partial X} = \rho_0 C^3 \frac{\partial^3 \sigma}{\partial t \partial X^2} = C^3 \frac{\partial^3 \sigma}{\partial X^3} \end{aligned} \tag{4.121}$$

Substitute Eq. (4.121) into Eq. (4.120), and recall that the total derivative following the boundary, namely the "boundary derivative", is defined as: $(d/dt) = (\partial/\partial t) + \overline{C}(\partial/\partial X)$ [see Eq. (4.102)], we finally get:

$$\begin{aligned} K_3 + \gamma J_3 &= 0 \quad (e, p) \\ J_3 + \gamma K_3 &= 0 \quad (e, p) \end{aligned} \tag{4.122}$$

The above equations are suitable to both the elastic side and the plastic side across the boundary. So Eq. (4.122) is equivalent to a system of homogeneous linear algebraic equations with respect to four unknown quantities K_3^e, K_3^p, J_3^e, and J_3^p. The sufficient and necessary conditions that a nontrivial solution exists are that the determinant of coefficients should be zero:

$$\begin{vmatrix} 1 & 0 & \gamma_e & 0 \\ 0 & 1 & 1 & \gamma_p \\ \gamma_e & 0 & 1 & 0 \\ 0 & \gamma_p & 0 & 1 \end{vmatrix} = \left(1-\gamma_e^2\right)\left(1-\gamma_p^2\right) = 0$$

It is thus known that there exist four possible solutions for the boundary velocity $\overline{C}$:

$$1-\gamma_e^2=0, \quad \text{namely} \quad \overline{C}=\pm C_e$$

$$1-\gamma_p^2=0, \quad \text{namely} \quad \overline{C}=\pm C_p$$

This means that a 3rd order weak-discontinuous elastic–plastic boundary, if exists, should either coincide with an elastic characteristic line, or coincide with a plastic characteristic line. However, in order to simultaneously satisfy Eq. (4.122) and the basic assumption that wave speed is discontinuous across the boundary [Eq. (4.101)], if the boundary coincides with an elastic characteristic line, then must have $K_3^p = J_3^p = 0$; or if coincides with a plastic characteristic line, then must have $K_3^e = J_3^e = 0$. On the contrary, if the boundary neither coincides with an elastic characteristic line (namely $|\gamma_e| \neq 1$), nor a plastic characteristic line (namely $|\gamma_p| \neq 1$), then Eq. (4.122) only have trivial solutions as:

$$K_3^e = J_3^e = K_3^p = J_3^p = 0$$

which means that all the 3rd partial derivatives of σ and v must simultaneously be zero. In other words, there exists no 3rd order weak-discontinuous elastic–plastic boundary that does not coincide with a characteristic line. By the analogy of this approach, it is deduced that any higher order derivatives of σ and v cannot be discontinuous across an elastic–plastic boundary propagating in any noncharacteristic direction. To conclude, we have the following most important feature of elastic–plastic boundary (Yu *et al.*, 1981, 1982, 1984).

Theorem 5: Across a weak-discontinuous elastic–plastic boundary characterized by the basic singularity [Eq. (4.101)], either the 1st order or the 2nd order derivatives of σ and v must be discontinuous, while any higher-than-second order discontinuity is impossible, except that the boundary coincides with a characteristic line.

It is worthy to point out that the above important theorem comes from the basic singularity, namely the slope of material stress–strain curve $(d\sigma/d\varepsilon)$ and thus the wave speed being discontinuous across the boundary. In fact, suppose there exists a higher order weak-discontinuous boundary, say, a 3rd order weak-discontinuity. Then the 2nd order derivatives across the boundary must be continuous, namely Eq. (4.122) must be satisfied.

Rewriting Eq. (4.122) in the form of jumps across the boundary, as:

$$\left[\frac{K_3}{J_3} + \gamma\right] = 0$$

$$\left[\frac{J_3}{K_3} + \gamma\right] = 0$$

from which we obtain:

$$[\gamma] = -\left[\frac{K_3}{J_3}\right] = -\left[\frac{J_3}{K_3}\right] \tag{4.122$'$}$$

On the other hand, note that

$$\left[\frac{J_3}{K_3}\right] = \left[\frac{1}{\dfrac{K_3}{J_3}}\right] = \frac{\left[\dfrac{K_3}{J_3}\right]}{\left(\dfrac{K_3}{J_3}\right)^e \left(\dfrac{K_3}{J_3}\right)^p}$$

then if Eq. (4.112′) is held then the following equation must be satisfied:

$$[\gamma] = -\left[\frac{K_3}{J_3}\right] = 0$$

which is contradictory to the basic assumption [Eq. (4.101)]. In other words under the basic assumption, the expressions of Eq. (4.122) cannot be held simultaneously. Thus on the elastic–plastic boundary the 2nd order derivatives of σ and v are not continuous everywhere, namely the 3rd order weak-discontinuous boundary does not exist. Since the existence of an nth order weak-discontinuous boundary is based on the prerequisite that all the lower-than-n order partial derivatives of σ and v are continuous, any higher-than-second weak-discontinuous boundary is impossible, unless it coincides with a characteristic line.

4.10.7 Higher order isolated points on elastic–plastic boundary

Earlier we have proved that an elastic–plastic boundary can only be a first order or a second order discontinuity with respect to σ and v. This property does not exclude the possibility that at a certain isolated point of the boundary, say the point M, higher order derivatives of σ and v are discontinuous. Namely, there may exist the isolated nth order ($n \geq 2$) discontinuous points. Because the existence of an nth order ($n \geq 2$) discontinuous point only requires that the σ and v and their derivatives up to $(n-1)$th order are continuous across the boundary at **this point**, rather than everywhere of the whole boundary. For example, on a 1st order weak-discontinuous boundary there may exist 2nd or higher order isolated discontinuous points, or on a 2nd order weak-discontinuous boundary there may exist 3rd or higher order isolated discontinuous points, etc.

For a 1st order weak-discontinuous boundary, Eq. (4.109) must be satisfied at any point of the boundary, namely:

$$[K_1] = -\overline{C}\left[\frac{J_1}{C}\right]$$

$$\left[\frac{J_1}{C}\right] = -\overline{C}\left[\frac{K_1}{C^2}\right]$$

Surely they should be satisfied at an arbitrary point M on the boundary. If at and only **at this point** M: $K_1^e = K_1^p$, or written as $[K_1]_M = 0$, then at and only at the point M, we have:

$$[K_1]_M = \left[\frac{J_1}{C}\right]_M = \left[\frac{K_1}{C^2}\right]_M = 0$$

Therefore at the point M all the first order derivatives of σ and v are continuous. Because of the basic singularity, the wave speed at the point M is discontinuous, $[C] \neq 0$. This means that $K_1^e(M) = K_1^p(M) = 0$ too. If the second order derivatives of σ and v are discontinuous at the point M, then M is an isolated 2nd order weak-discontinuity point.

We now use the same approach to consider a 2nd order weak-discontinuous boundary. We know that on the boundary all the first order derivatives of σ and v are continuous everywhere, and $(\partial\sigma^e/\partial t) = (\partial\sigma^p/\partial t) = (\partial v^e/\partial X) = (\partial v^p/\partial X) = 0$. Also everywhere on the boundary, Eq. (4.117) must be satisfied, which can be rewritten as:

$$K_2 = -\overline{C}\frac{J_2}{C}, \quad (e, p)$$

$$\left[\frac{J_2}{C}\right] = -\overline{C}\left[\frac{K_2}{C^2}\right]$$

Certainly they should be also satisfied at an arbitrary point M on the boundary. If at and only at the point M: $K_2^e = K_2^p$, namely $[K_2]_M = 0$ (Noting that this is not the case at other points), then at and only at the point M, we have:

$$K_2^e = K_2^p, \quad \left(\frac{J_2}{C}\right)^e = \left(\frac{J_2}{C}\right)^p, \quad [K_2]_M = \left[\frac{K_2}{C^2}\right]_M = 0$$

Moreover, because of the basic singularity, the wave speed at the point M is discontinuous, $[C] \neq 0$. This means that:

$$K_2^e(M) = K_2^p(M) = J_2^e(M) = J_2^p(M) = 0$$

Therefore all the second order derivatives of σ and v are continuous at the point M. If the third order derivatives of σ and v are discontinuous at the point M, then M is an isolated 3rd order weak-discontinuity point. It is easy to show: at an arbitrary point M on a 1st order weak-discontinuous boundary, if both conditions $K_1^e(M) = K_1^p(M)$ and

$K_2^e(M) = K_2^p(M)$ are satisfied, but $K_3^e(M)$ and $K_3^p(M)$ are not simultaneously zero, then M is a 3rd order isolated point at which all the second order derivatives of σ and v are zero.

By the analogy of such argument, the nth isolated weak-discontinuous point can exist. Moreover, because of the basic singularity, we have the following important theorem for the nth isolated weak-discontinuous point:

Theorem 6: On an elastic–plastic boundary, the existence of an isolated nth ($n \geq 2$) weak-discontinuous point is possible. At this point and only at this point, all the partial derivatives of σ and v up to the $(n-1)$th order continuous and all these partial derivatives are zero except $\partial\sigma/\partial X$ and $\partial v/\partial t$.

By differentiating the governing Eq. (4.107) many times, we know that for an nth order weak-discontinuous point M, among the $(n+1)$ partial derivatives of nth order in either the elastic zone or the plastic zone, only two of them are independent:

$$\left(C^i \frac{\partial^n \sigma}{\partial X^i \partial t^{n-i}}\right)_M = \begin{cases} \left(\dfrac{\partial^n \sigma}{\partial t^n}\right)_M \equiv K_n, & (i - \textit{even number}) \\ \rho_0 \left(C \dfrac{\partial^n v}{\partial t^n}\right)_M \equiv J_n, & (i - \textit{odd number}) \end{cases}$$

$$\left(C^i \frac{\partial^n v}{\partial X^i \partial t^{n-i}}\right)_M = \begin{cases} \left(\dfrac{\partial^n v}{\partial t^n}\right)_M \equiv \dfrac{J_n}{\rho_0 C}, & (i - \textit{even number}) \\ \dfrac{1}{\rho_0}\left(\dfrac{1}{C} \dfrac{\partial^n \sigma}{\partial t^n}\right)_M \equiv \dfrac{K_n}{\rho_0 C}, & (i - \textit{odd number}) \end{cases}$$

$$(i = 0,\ 1,\ 2, \ldots, n) \tag{4.123}$$

With regard to the successive boundary following the nth order isolated point M, according to theorem 5, the boundary only can be either a 1st order weak-discontinuous boundary, or a 2nd order weak-discontinuous boundary.

First, we discuss the situation that the successive boundary is a 1st order weak-discontinuous boundary. In such a case, since everywhere on the boundary must have:

$$[\sigma] = 0, \quad [v] = 0$$

the nth order boundary derivatives (d^n/dt^n) of $[\sigma]$ and $[v]$ obviously should also be equal to zero everywhere, including at the point M, namely:

$$\left(\frac{d^n[\sigma]}{dt^n}\right)_M = 0, \quad \left(\frac{d^n[v]}{dt^n}\right)_M = 0$$

Note that by the definition of boundary derivative, $(d^n/dt^n) = [(\partial/\partial t) + \overline{C}(\partial/\partial X)]^n$. Using the binomial expansion, and taking into consideration the theorem 6, the above equations can be written as:

$$\sum_{i=0}^{n} C_n^i \left(\frac{\partial^n \sigma}{\partial X^i \partial t^{n-i}} \right)_M \overline{C}_M^i = 0$$
$$\sum_{i=0}^{n} C_n^i \left(\frac{\partial^n v}{\partial X^i \partial t^{n-i}} \right)_M \overline{C}_M^i = 0 \tag{4.124}$$

where $C_n^i = \frac{n!}{i!(n-i)!}$ is the binomial coefficient. Substitute Eq. (4.123) into Eq. (4.124), at the point M we have:

$$\left[C_n^0 K_n + C_n^1 \gamma J_n + C_n^2 \gamma^2 K_n + + C_n^{31} \gamma^3 J_n + \cdots + \text{the } n\text{th term} \right] = 0$$
$$\left[\frac{1}{C} \left(C_n^0 J_n + C_n^1 \gamma K_n + C_n^2 \gamma^2 J_n + + C_n^{31} \gamma^3 K_n + \cdots + \text{the } n\text{th term} \right) \right] = 0$$

Introducing the coefficient ϕ_n for the even terms and the coefficient ψ_n for the odd terms:

$$\phi_n = C_n^0 + C_n^2 \gamma^2 + C_n^4 \gamma^4 + \cdots = \frac{1}{2} \left\{ (1+\gamma)^n + (1-\gamma)^n \right\}$$
$$\psi_n = C_n^1 \gamma + C_n^3 \gamma^3 + C_n^5 \gamma^5 + \cdots = \frac{1}{2} \left\{ (1+\gamma)^n - (1-\gamma)^n \right\} \tag{4.125}$$

we obtain the basic equations for the nth order isolated point M following which the elastic–plastic boundary is a 1st order weak-discontinuity:

$$[\phi_n K_n + \psi_n J_n] = 0$$
$$\left[\frac{1}{C} (\phi_n J_n + \psi_n K_n) \right] = 0 \tag{4.126}$$

Note that between coefficients ϕ and ψ the following recursion relationships exist:

$$\phi_n = \phi_{n-1} + \gamma \psi_{n-1}$$
$$\psi_n = \psi_{n-1} + \gamma \phi_{n-1} \tag{4.127}$$

Substituting Eq. (4.127) into (4.126) and after some operations, we get:

$$\left[\left(1 - \gamma^2 \right) (\phi_{n-1} K_n + \psi_{n-1} J_n) \right] = 0 \tag{4.128}$$

or

$$\frac{\phi_{n-1}^p K_n^p + \psi_{n-1}^p J_n^p}{\phi_{n-1}^e K_n^e + \psi_{n-1}^e J_n^e} = \frac{1 - \dfrac{\overline{C}_M^2}{C_e^2}}{1 - \dfrac{\overline{C}_M^2}{C_p^2}} \tag{4.128'}$$

It is worthy to point out that Eqs. (4.126) and (4.128′) actually can be deduced from the basic equations for a 1st order weak-discontinuous boundary [Eq. (4.109)]:

$$[K_1 + \gamma J_1] = 0$$

$$\left[\frac{1}{C}(J_1 + \gamma K_1)\right] = 0$$

and the Eq. (4.110), $[(1-\gamma^2)K_1] = 0$, rewritten in the form:

$$\frac{K_1^p}{K_1^e} = \frac{1-(\gamma^e)^2}{1-(\gamma^p)^2} = \frac{1-\dfrac{\overline{C}^2}{C_e^2}}{1-\dfrac{\overline{C}^2}{C_p^2}}$$

by directly applying to the point M. The deduction procedures are shown below.

First we expand K_1 into a Taylor series of the partial derivatives of σ up to the nth order at the point M along the elastic–plastic boundary. Because the point M is an isolated nth order weak discontinuity, all partial derivatives lower than the nth order are zero (Theorem 6), therefore we finally get:

$$\begin{aligned}
K_1 = \frac{\partial\sigma}{\partial t} &= \frac{1}{(n-1)!}\frac{d^{n-1}}{dt^n}\left(\frac{\partial\sigma}{\partial t}\right)_M (t-t_M)^{n-1} \\
&= \frac{(t-t_M)^{n-1}}{(n-1)!}\sum_{i=0}^{n-1} C_{n-1}^i \left(\frac{\partial^n\sigma}{\partial X^i \partial t^{n-i}}\right)_M \overline{C}_M^i \\
&= \frac{(t-t_M)^{n-1}}{(n-1)!}\left(C_{n-1}^0 K_n + C_{n-1}^1 \gamma J_n + C_{n-1}^2 \gamma^n K_n + \cdots\right)_M \\
&= \frac{(t-t_M)^{n-1}}{(n-1)!}(\phi_{n-1}K_n + \psi_{n-1}J_n)
\end{aligned}$$

Similarly, for J_1 we have:

$$J_1 = \rho C\left(\frac{\partial v}{\partial t}\right)_M = \frac{(t-t_M)^{n-1}}{(n-1)!}(\phi_{n-1}J_n + \psi_{n-1}K_n)$$

Substituting these expressions into Eqs. (4.109) and (4.110), taking into consideration the theorem 6 and the recursion relationship 4.127, we can deduce the Eqs. (4.126) and (4.128′). Note that the deduction of Eq. (4.128′) is actually the same as to calculate the indeterminate form $(K_1^p/K_1^e) = (0/0)$ by the L'hopital's rule.

Among the three equations of Eqs. (4.126) and (4.128′), only two, and any two, are independent. These two equations contain five quantities K_n^e, K_n^p, J_n^e, J_n^p, and $\overline{C}_M$. Among the four partial derivatives, if we have known three, then from these two equations we can calculate the fourth derivative, and $\overline{C}_M$. Or eliminating the fourth derivative from these two equations, the equation for solving $\overline{C}_M$ can be obtained, which is a $(2n)$th algebric equation. For example, if the quantities K_n^e, K_n^p, and J_n^p are known, eliminating the quantity J_n^e, then the equation for $\overline{C}_M$ is:

$$\begin{aligned}
&\left\{\frac{C_e+C_p}{2}(1-\gamma_e)^n-\frac{C_e-C_p}{2}(1-\gamma_e)^n\right\}(1+\gamma_p)^n\left(K_n^p+J_n^p\right)\\
&+\left\{\frac{C_e+C_p}{2}(1+\gamma_e)^n-\frac{C_e-C_p}{2}(1-\gamma_e)^n\right\}(1-\gamma_p)^n\left(K_n^p-J_n^p\right)\\
&-2C_p\left(1-\gamma_e^2\right)^n K_n^e=0
\end{aligned}\tag{4.129}$$

Alternatively, if the quantities K_n^e, K_n^p, and J_n^e are known, we eliminate the quantity J_n^p, and then the equation for $\overline{C}_M$ is:

$$\begin{aligned}
&\left\{\frac{C_e+C_p}{2}(1-\gamma_p)^n-\frac{C_e-C_p}{2}(1+\gamma_p)^n\right\}(1+\gamma_e)^n\left(K_n^p+J_n^p\right)\\
&+\left\{\frac{C_e+C_p}{2}(1+\gamma_p)^n+\frac{C_e-C_p}{2}(1-\gamma_p)^n\right\}(1-\gamma_e)^n\left(K_n^e-J_n^e\right)\\
&-2C_e\left(1-\gamma_p^2\right)^n K_n^p=0
\end{aligned}\tag{4.130}$$

When the boundary conditions at the end of bar (K_n^e and K_n^p) are known, Eq. (4.129) can be applied to determine the initial propagating velocity of an unloading boundary, across which a known plastic zone (i.e. J_n^p is known) is unloaded to an elastic zone. Similarly, Eq. (4.130) can be applied to determine the initial propagating velocity of a loading boundary, across which a known elastic zone (i.e. J_n^e is known) is loaded to a plastic zone.

If the point M is a second order weak-discontinuity ($n=2$) and the successive boundary is a 1st order weak-discontinuous, then Eq. (4.126) is reduced to:

$$\begin{aligned}
&\left[\left(1+\gamma^2\right)K_2+2\gamma J_2\right]=0\\
&\left[\frac{1}{C}\left(\left(1+\gamma^2\right)J_2+2\gamma K_2\right)\right]=0
\end{aligned}\tag{4.131}$$

while Eq. (4.128) is reduced to:

$$\left[\left(1-\gamma^2\right)\left(K_2+\gamma J_2\right)\right]=0\tag{4.132}$$

Of the above three equations, two of them are independent. In discussions on the problem where the end of the bar is a 2nd order weak-discontinuity and the successive elastic–plastic boundary is a 1st order weak-discontinuous, Ting first deduced the first and the third equations although in a different form (Ting, 1971).

Therefore, to a 1st order weak-discontinuous boundary, no matter which order of discontinuity the isolated point M is, there always exists two independent equations which connect the boundary propagating velocity $\overline{C}_M$ to the four quantities, the $\partial^n\sigma/\partial t^n$ and $\partial^n v/\partial t^n$ at both the elastic side and the plastic side across the boundary. In general situation, once three of the four partial derivatives are known, then $\overline{C}_M$ can be calculated. In the case of simple wave propagation, the problem is reduced to that, once K_n^e and K_n^p are known then $\overline{C}_M$ can be determined, just as stated in Section 4.7.

Similarly, we discuss the situation that the isolated point M is an nth order discontinuity and the successive boundary is a 2nd order weak-discontinuous. According to the theorem 2, everywhere on the boundary we have:

$$\frac{\partial\sigma}{\partial t} = \frac{\partial v}{\partial X} = 0$$
$$\left[\frac{\partial\sigma}{\partial X}\right] = \left[\frac{\partial v}{\partial t}\right]$$

Obviously the $(n-1)$th order boundary derivatives (d^{n-1}/dt^{n-1}) of them should also be equal to zero everywhere, including at the point M, namely:

$$\begin{aligned} \frac{d^{n-1}}{dt^{n-1}}\left(\frac{\partial\sigma}{\partial t}\right)_M &= \frac{d^{n-1}}{dt^{n-1}}\left(\frac{\partial v}{\partial X}\right)_M = 0 \\ \left[\frac{d^{n-1}}{dt^{n-1}}\left(\frac{\partial\sigma}{\partial X}\right)\right]_M &= \left[\frac{d^{n-1}}{dt^{n-1}}\left(\frac{\partial v}{\partial t}\right)\right]_M = 0 \end{aligned} \tag{4.133}$$

After expanding the above equations to binomial expansions, taking into consideration the theorem 6 about the isolated point M, and then substituting Eqs. (4.123) and (4.124) in the expansions, we get the following basic equations for the isolated nth order weak-discontinuous point M on a 2nd order weak-discontinuous successive boundary:

$$\begin{aligned} &\phi_{n-1}K_n + \psi_{n-1}J_n = 0, \quad (e, p) \\ &\left[\frac{1}{C}\left(\phi_{n-1}J_n + \psi_{n-1}K_n\right)\right] = 0 \end{aligned} \tag{4.134}$$

Substitute the recursive relationship for ϕ and ψ [Eq. (4.127)] into Eq. (4.134), and after some operations, we have:

$$\left[\left(1-\gamma^2\right)\left(\phi_{n-2}K_n + \psi_{n-2}J_n\right)\right] = 0 \tag{4.135}$$

which may be rewritten as:

$$\frac{\phi_{n-2}^{p} K_{n}^{p} + \psi_{n-2}^{p} J_{n}^{p}}{\phi_{n-2}^{e} K_{n}^{e} + \psi_{n-2}^{e} J_{n}^{e}} = \frac{1 - \dfrac{\overline{C}_{M}^{2}}{C_{e}^{2}}}{1 - \dfrac{\overline{C}_{M}^{2}}{C_{p}^{2}}} \tag{4.135$'$}$$

Obviously, among the four equations expressed in Eqs. (4.134) and (4.135), only three of them are independent.

4.10.8 Supplementary conditions on loading boundary

The above discussions are applicable to both a loading boundary and an unloading boundary. However, in the case of loading boundary, the yield condition should be added. Note that when the material is loaded from an elastic state (or unloading state) to a plastic state, since the slope of stress–strain curve $d\sigma/d\varepsilon$ is discontinuous [Eq. (4.101)], the wave speed under the yield stress is discontinuous too, namely is two-valued. Thus the yield criterion such as Mises criterion or Tresca criterion usually used in the static plasticity is only the necessary condition but not the sufficient condition for yielding. An example is shown in Fig. 4.34 wherein a set of elastic-plastic simple waves is propagating in a semi-infinite bar. When the boundary stress is just equal to the initial yield stress Y_0, as it has been discussed in Chapter 2 about the two-wave structure, a precursor elastic wave with stress level $\sigma = Y_0$ propagates by a faster velocity of C_e, and simultaneously a plastic wave with the same stress level $\sigma = Y_0$ propagates by a slower velocity $C_p(Y_0)$. Between these two wave fronts, is formed a constant-value zone with $\sigma = Y_0$. Note that this constant-value zone is essentially an elastic zone, since the plastic disturbance, which is propagated by the successive plastic wave, has not arrived yet. Therefore, much different from the static

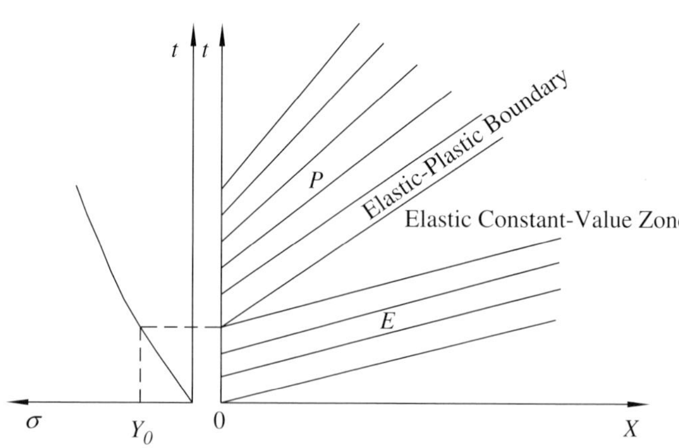

Fig. 4.34. Elastic-plastic boundary in case of a set of elastic-plastic simple waves propagating in a semi-infinitive bar.

plasticity, in the dynamic plasticity related to elastic–plastic wave propagation we could not say that any zone that satisfies the usual yield condition of $\sigma = Y_0$ is a plastic zone. Strictly speaking, when the slope of stress–strain curve $d\sigma/d\varepsilon$ is discontinuous at the yield point (4.101), the yield condition on a loading elastic–plastic boundary $t = g(X)$ should be expressed as:

$$\sigma(X, g(X)) = \sigma_m(X), \quad \text{and} \begin{cases} \dfrac{\partial|\sigma^p|}{\partial t} > 0 & \text{(For a first order boundary)} \\ \dfrac{\partial^2|\sigma^p|}{\partial t^2} > 0 & \text{(For a second order boundary)} \end{cases} \tag{4.136}$$

where $\sigma_m(X)$ is the initial yield limit, $\sigma_m(X) = Y_0$ during the initially yielding, otherwise $\sigma_m(X)$ is the maximum stress ever experienced at the point X.

Another feature can be seen from Fig. 4.34. Because the stress wave velocity has two values at yield stress, the boundary derivative of stress $\partial\sigma/\partial t$ just before the yielding (noted by $(\partial\sigma^b/\partial t)$) and the boundary derivative of stress $\partial\sigma/\partial t$ just after the yielding (noted by $(\partial\sigma^a/\partial t)$) propagate separately along different characteristic lines by different propagating velocities. These values do not represent respectively the values of $\partial\sigma^e/\partial t$ and $\partial\sigma^p/\partial t$ across an elastic–plastic loading boundary at this point. Namely, we cannot substitute the mentioned $\partial\sigma^b/\partial t$ and $\partial\sigma^a/\partial t$ values directly into Eq. (4.111) to calculate the propagating velocity of an elastic–plastic loading boundary, $\overline{C}_L$. In fact, in a general case, according to the inequality 4.113, the varied range of $\overline{C}_L$ has two possibilities: (1) $C_p \geq |\overline{C}_L| \geq 0$ and (2) $|\overline{C}_L| \geq C_e$.

In the first case ($|\overline{C}_L| \leq C_p$), as shown in Fig. 4.35(a), the $\partial\sigma^b/\partial t$ actually propagates along the elastic characteristic lines, while the $\partial\sigma^e/\partial t$ on the elastic side across the loading boundary corresponds to the $\partial\sigma^m/\partial t$ in the zone m, which needs to be determined.

In the second case ($|\overline{C}_L| \geq C_e$), as shown in Fig. 4.35(b), the $\partial\sigma^a/\partial t$ actually propagates along the plastic characteristic lines, while the $\partial\sigma^p/\partial t$ on the plastic side across the loading boundary corresponds to the $\partial\sigma^m/\partial t$ in the zone m, which needs to be determined.

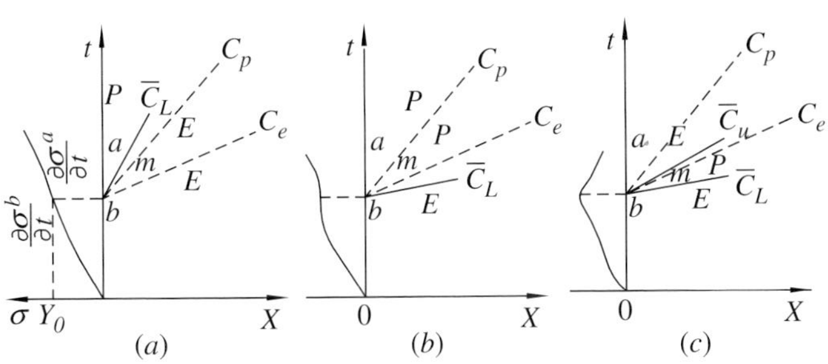

Fig. 4.35. Multiple-valued propagation speeds of a first-order weak-discontinuous elastic-plastic boundary.

Therefore, physically or mathematically speaking, to determine $\overline{C}_L$ only Eq. (4.109) is not sufficient and other conditions (namely the yield conditions) must be supplemented.

Thus, for a first order weak-discontinuous loading boundary, the supplemental yield conditions expressed by Eq. (4.136), or equivalently the corresponding differential form expressed by Eq. (4.137), should be satisfied too:

$$\frac{d\sigma_m(X)}{dX} = \frac{\partial \sigma}{\partial X} + \frac{1}{\overline{C}_L}\frac{\partial \sigma}{\partial t} = \frac{1}{\overline{C}_L}(\phi_1 K_1 + \psi_1 J_1) \tag{4.137}$$

or

$$\overline{C}_L = \frac{\dfrac{\partial \sigma}{\partial t}}{\dfrac{d\sigma_m}{dX} - \dfrac{\partial \sigma}{\partial X}} \tag{4.137$'$}$$

Moreover we have the following compatibility equation along the elastic or plastic characteristic line, across which are the zone m and the zone i:

$$\frac{\partial \sigma^m}{\partial t} - \frac{\partial \sigma^i}{\partial t} = -C_i\left(\frac{\partial \sigma^m}{\partial X} - \frac{\partial \sigma^i}{\partial X}\right) \tag{4.138}$$

where, if it is the case shown in Fig. 4.35(a), the superscript i should be substituted by superscript a, accordingly C_i takes the value C_e; while if it is the case shown in Fig. 4.35(b), the superscript i should be substituted by superscript b, and C_i takes the value C_p. Thus, Eqs. (4.137) and (4.138), in addition to the two equations in Eq. (4.109), totally four equations are enough to solve the four unknown quantities: $\overline{C}_L$, $\partial\sigma^m/\partial t$, $\partial\sigma^m/\partial X$, and $\partial\sigma^a/\partial X$. The problem is solvable.

Now consider the case for a second order weak-discontinuous loading boundary (Fig. 4.36). Similar to the previous case, the yield condition 4.136 in the second order differential form is:

$$\frac{d^2\sigma_m}{dX^2} = \frac{1}{\overline{C}_L^2}(\phi_2 K_2 + \psi_2 J_2) = \left(\frac{1}{C^2} + \frac{1}{\overline{C}_L^2}\right)\frac{\partial^2 \sigma}{\partial t^2} + \frac{2}{\overline{C}_L}\frac{\partial^2 \sigma}{\partial X \partial t} \tag{4.139}$$

Or, considering the Eq. (4.119$'$) the above equation can be rewritten as:

$$\frac{d^2\sigma_m}{dX^2} = \left(\frac{1}{C^2} - \frac{1}{\overline{C}_L^2}\right)\frac{\partial^2 \sigma}{\partial t^2} \tag{4.139$'$}$$

Moreover, similar to Eq. (4.138), we have the following compatibility equation along the elastic characteristic line, across which are the zone b and the zone r in Fig. 4.36:

$$\frac{\partial \sigma^r}{\partial t} - \frac{\partial \sigma^b}{\partial t} = -C_e\left(\frac{\partial \sigma^r}{\partial X} - \frac{\partial \sigma^b}{\partial X}\right) \tag{4.140}$$

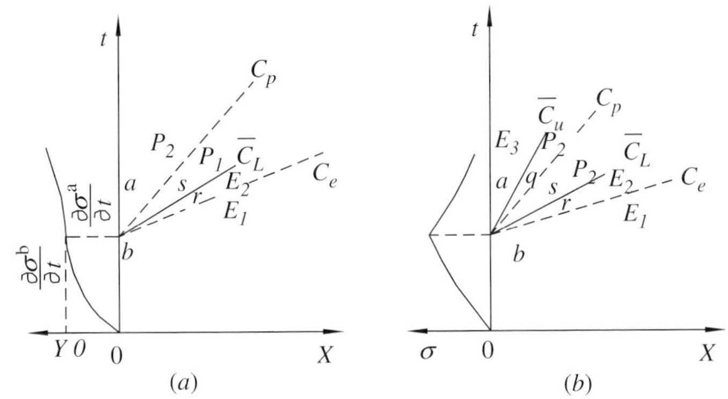

Fig. 4.36. Multiple-valued propagation speeds of a second-order weak-discontinuous elastic-plastic boundary.

and, the following compatibility equation along the plastic characteristic line, across which are the zone s and the zone a in Fig. 4.36:

$$\frac{\partial \sigma^a}{\partial t} - \frac{\partial \sigma^s}{\partial t} = -C_p\left(\frac{\partial \sigma^a}{\partial X} - \frac{\partial \sigma^s}{\partial X}\right) \tag{4.141}$$

Thus, Eqs. (4.139)–(4.141), in addition to the three basic equations in Eq. (4.117), totally six equations are enough to solve the six unknown quantities: $\overline{C}_L$, $\partial\sigma^r/\partial t$, $\partial\sigma^r/\partial X$, $\partial\sigma^s/\partial t$, $\partial\sigma^s/\partial X$, and $\partial\sigma^a/\partial X$. The problem is solvable.

If the point M being considered is a nth ($n \geq 2$) order weak-discontinuous point, then by differentiating Eq. (4.136) n times, we deduce that at the point M the following equation should be satisfied:

$$\frac{d^n \sigma_m}{dX^n} = \frac{1}{\overline{C}_L^n}\left(\phi_n K_n + \psi_n J_n\right) \tag{4.142}$$

regardless of whether the successive boundary is a 1st order discontinuity or a 2nd order discontinuity. Note that during the deduction of Eq. (4.142) we have already assumed:

$$\left(\frac{d^i \sigma_m}{dX^i}\right)_M = 0, \quad i = 2, 3, \ldots, (n-1)$$

otherwise we will get: $\overline{C}_M = 0$.

Based on the above discussions, it is worthwhile to point out another important theorem associated with the so-called **branching point** of boundary. For the 1st order weak-discontinuous loading boundary in the case of $\overline{C}_L \geq C_e$, since the $d\sigma_m/dX$ and the $\partial\sigma^b/\partial t$, $\partial\sigma^b/\partial X$ in the elastic zone are all known, the propagating velocity of this loading

Table 4.3. Basic relationships for determining propagation speeds of elastic-plastic boundaries in different cases.

The point / Successive boundary	1st order weak-discontinuous point	2nd order weak-discontinuous point	nth order weak-discontinuous point
1st order weak-discontinuous boundary	$[K_1 + \gamma J_1] = 0$ $\left[\frac{1}{C}(J_1 + \gamma K_1)\right] = 0$	$\left[(1+\gamma^2)K_2 + 2\gamma J_2\right] = 0$ $\left[\frac{1}{C}\left((1+\gamma^2)J_2 + 2\gamma K_2\right)\right] = 0$	$[\phi_n K_n + \psi_n J_n] = 0$ $\left[\frac{1}{C}(\phi_n J_n + \psi_n K_n)\right] = 0$
	$\left[(1-\gamma^2)K_1\right] = 0$ Karman *et al.* (1942)	$\left[(1-\gamma^2)(K_2 + \gamma J_2)\right] = 0$ Ting (1971)	$\left[(1-\gamma^2)(\phi_{n-1} K_n + \psi_{n-1} J_n)\right] = 0$
2nd order weak-discontinuous boundary		$K_2 + \gamma J_2 = 0$ $\left[\frac{1}{C}(J_2 + \gamma K_2)\right] = 0$	$\phi_{n-1} K_n + \psi_{n-1} J_n = 0$ $\left[\frac{1}{C}(\phi_{n-1} J_n + \psi_{n-1} K_n)\right] = 0$
		$\left[(1-\gamma^2)K_2\right] = 0$ Clifton and Ting (1968)	$\left[(1-\gamma^2)(\phi_{n-2} K_n + \psi_{n-2} J_n)\right] = 0$
Supplementary conditions on loading boundary	$\frac{d\sigma_m}{dX} = \frac{1}{\bar{C}}(K_1 + \gamma J_1)$	$\frac{d^2\sigma_m}{dX^2} = \frac{1}{\bar{C}^2}\left\{(1+\gamma^2)K_2 + 2\gamma J_2\right\}$	$\frac{d^n\sigma_m}{dX^n} = \frac{1}{\bar{C}^n}(\phi_n K_n + \psi_n J_n)$

boundary $\overline{C}_L$ can be determined immediately from Eq. (4.137). It means that the determination of $\overline{C}_L$ is independent of $\partial\sigma^a/\partial t$, even though $(\partial\sigma^a/\partial t) < 0$, namely, the zone a is in an unloading situation. If so, then another unloading boundary can be formed in the plastic zone behind this loading boundary, as shown in Fig. 4.35(c). The same situation can appear in the case of a 2nd order weak-discontinuous loading boundary, as shown in Fig. 4.36(b). Thus there may be more than one elastic–plastic boundaries propagating through the same point M considered. In other words, the boundary propagating velocity $\overline{C}$ has multi-values at the point M considered. Such a point is called the **branching point** of boundary, which is a singular point with respect to the elastic–plastic boundary itself (Yu *et al.*, 1984). Such phenomena, as can be understood from the previous discussion, are closely related to or originated from the basic singularity of an elastic–plastic boundary $[C] \neq 0$.

Whether or not a branching point appears on an elastic–plastic boundary, the approaches stated in this section are sufficient to determine the propagating velocity of an elastic–plastic boundary under all the possible situations mentioned (Yu *et al.*, 1984). The corresponding basic relationships are summarized in Table 4.3. The readers, who are interested in a further study on the discontinuities across elastic/plastic boundary in elastic–plastic wave propagation, are suggested to read the review paper written by Ting (Ting, 1990).

CHAPTER 5

Rigid Unloading Approximation

All the elastic–plastic wave propagation containing unloading processes discussed in Chapter 4 are based on the assumption of **elastic unloading**. In such cases, unloading disturbances propagate with elastic wave velocity C_0. If the elastic modulus E is much larger than the slope of plastic stress–strain curve $d\sigma/d\varepsilon$; or in other words, if the elastic wave velocity is much larger than the plastic wave velocity C (i.e. $C_0 \gg C$), then relative to the plastic wave velocity C, the C_0 can be approximately regarded as infinite (i.e. $C_0/C \approx \infty$). When unloading problems are dealt with, this is corresponding to that of the elastic unloading stress–strain curve which is approximately regarded as a straight line parallel to the σ axis (Fig. 5.1), or corresponding to that of the elastic–plastic body once unloading is regarded as a rigid body but the inertia effect of rigid part should be still taken into account. By such a "rigid unloading" assumption, the elastic–plastic wave propagation problems containing unloading processes usually can be simplified a lot.

5.1 Rigid Unloading in a Semi-Infinite Bar

5.1.1 Rigid unloading in linear hardening plastic bars

At the outset, let us discuss a very simple example. Consider a semi-infinite bar with a stress–strain relationship of the so-called **linear hardening plastic loading–rigid unloading**, as shown in Fig. 5.1(a), which can be regarded as a simplified model of the linear hardening plastic loading–elastic unloading stress–strain relationship of soil shown in Fig. 4.13. A constant-velocity impact loading v^* is suddenly applied on the bar end ($X = 0$), and at time t_1 the bar end is suddenly unloaded to make it stress-free ($\sigma = 0$), as shown in Fig. 5.2.

Obviously, before unloading ($t < t_1$) the situation is the same as what we have discussed earlier on plastic loading waves, namely a strong discontinuous wave is propagating with a constant wave velocity $C_1 = \sqrt{E_1/\rho_0}$. In case of zero initial condition, the states in

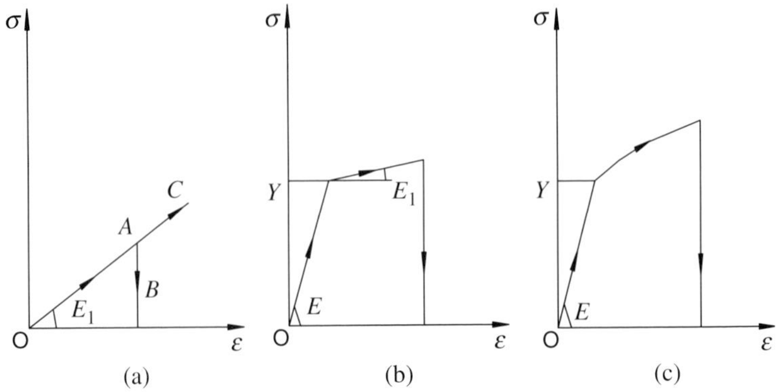

Fig. 5.1. Some stress–strain relationships under rigid–unloading assumption: (a) linear loading–rigid unloading; (b) elastic–linear hardening plastic loading–rigid unloading; (c) elastic-decreasing hardening plastic loading–rigid unloading.

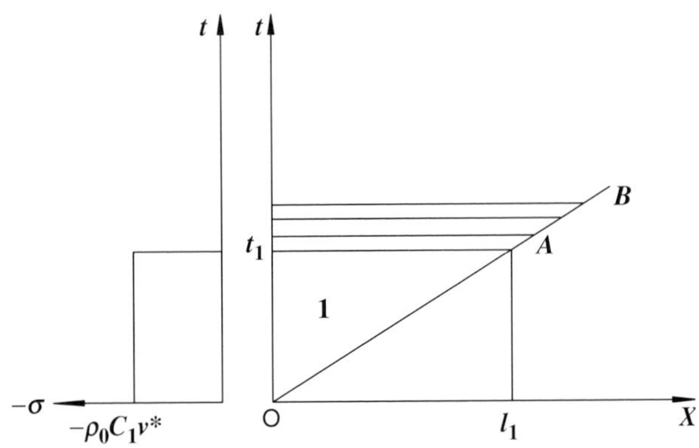

Fig. 5.2. In a linear loading–rigid unloading material [Fig. 5.1(a)], a strong discontinuous wave propagates with a constant wave velocity when a boundary loading is suddenly applied, and unloading disturbances propagate with an infinite wave velocity when a boundary unloading is suddenly applied.

region *1* are:

$$v = v^*, \ \sigma_1 = -\rho_0 C_1 v_1 = -\rho_0 C_1 v^*, \ \varepsilon_1 = \frac{\sigma_1}{E_1} \tag{5.1}$$

At $t = t_1$, the plastic loading wave front arrives at location $l_1 = C_1 t_1$, while an unloading suddenly takes place at the bar end. According to the rigid unloading assumption, the unloading disturbance propagates with an infinite wave velocity, so that all particles of the segment behind the plastic loading wave front ($X \leq l_1$) are instantaneously unloaded and move as a rigid body, while the amplitude of the plastic loading wave begins to attenuate due to the interaction with the unloading disturbances. Similar to the elastic unloading process discussed earlier, such an attenuation process results from the continuous interaction between the plastic loading disturbances, which propagate with a velocity of C_1, and the unloading disturbances, which propagate with infinite velocity. Note that as the propagation of the continuously attenuating plastic wave goes on, the length of the unloaded rigid segment l continuously increases ($l = C_1 t$). Obviously, at $t = t_1$ the unloading boundary which propagates with infinite velocity, is represented by the line $t_1 A$ in Fig. 5.2 ($\overline{C} = \infty$), while at $t > t_1$ the unloading boundary coincides with the propagating locus AB of plastic loading wave and so the boundary propagating velocity is equal to the plastic wave velocity ($\overline{C} = C_1$).

For the strong discontinuous plastic wave, according to the dynamic compatibility condition across a strong discontinuity, we have the following equation to relate its stress amplitude σ_m with its particle velocity amplitude v_m:

$$\sigma_m = -\rho_0 C_1 v_m \tag{5.2}$$

the subscript m used here is to denote that these quantities are the maximum that have been attained in the history of plastic deformation for each particle. In general they are also the quantities along the unloading boundary.

On the other hand, for the unloaded rigid segment, according to the Newton's second law, we have the following motion equation:

$$\sigma_m = \rho_0 l \frac{dv_m}{dt} \tag{5.3}$$

From Eqs. (5.1) and (5.2) and noticing $l = C_1 t$, it is clear that the particle velocity $v_m(t)$ for the rigid segment should satisfy the following ordinal differential equation:

$$\frac{dv_m}{dt} + \frac{v_m}{t} = 0 \tag{5.4}$$

After determining the integration constant by the condition $v_m(t_1) = v_1$, the solution of this ordinary differential equation (ODE) can be obtained as:

$$v_m t = v_1 t_1 \tag{5.5a}$$

or noticing $l/t = l_1/t_1 = C_1$, it can be rewritten as

$$v_m l = v_1 l_1 \tag{5.5b}$$

By substituting Eqs. (5.1) and (5.2) into Eq. (5.5b), we obtain the following hyperbolic functions which show how σ_m and ε_m attenuate with the unloading boundary propagating location l, or with time t:

$$\sigma_m l = \sigma_1 l_1, \quad \sigma_m t = \sigma_1 t_1 \tag{5.6}$$

$$\varepsilon_m l = \varepsilon_1 l_1, \quad \varepsilon_m t = \varepsilon_1 t_1 \tag{5.7}$$

Recall that in the case of elastic unloading, a similar hyperbolic relationship between σ_m and the locations of stationary strain-discontinuous interfaces l' has been obtained (see Section 4.4.1 and Fig. 4.7). By comparing these two cases, it is easy to find that the discrete stationary strain-discontinuous interfaces in the case of elastic unloading now become continuously distributed in the present case of rigid unloading.

5.1.2 Rigid unloading in linear elastic–linear hardening plastic bars

Consider now a semi-infinite bar with a stress–strain relationship of the so-called **linear elastic–linear hardening plastic loading–rigid unloading**, as shown in Fig. 5.1(b). In such a case, the elastic precursor wave ahead of the plastic loading wave should be taken into consideration. If the boundary condition at bar end is not sudden unloading but gradual unloading as shown in Fig. 5.3, then the loads applied on the rigid segment are not only the σ_m acting from the unloading boundary ($X = l$), as given on the left-hand side of Eq. (5.3), but also the $\sigma_0(t)$ acting at the bar end ($X = 0$). Nevertheless, the problem can be solved by similar procedures mentioned above.

Actually, in such a case, Eqs. (5.2) and (5.3) become respectively to:

$$v_m = v_Y - \frac{1}{\rho_0 C_1}(\sigma_m - Y) \tag{5.8}$$

$$\sigma_m - \sigma_0 = \rho_0 l \frac{dv_m}{dt} \tag{5.9}$$

Substituting Eq. (5.8) into Eq. (5.9), we have

$$\sigma_m - \sigma_0(t) = -\frac{1}{C_1}\frac{d\sigma_m}{dt} \tag{5.10}$$

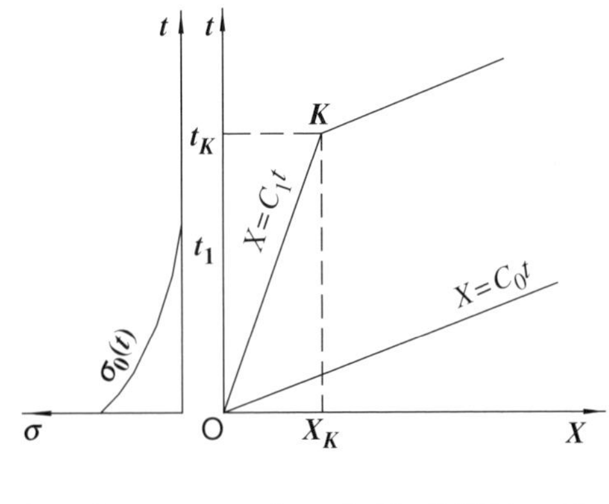

Fig. 5.3. A semi-infinite bar with σ–ε curve shown in Fig. 5.1(b) is subjected to a sudden boundary loading followed by gradual unloading.

Noticing $l = C_1 t$, the solution of this equation can be obtained as:

$$\sigma_m = \frac{1}{t}\int_0^t \sigma_0(t)dt \tag{5.11}$$

Since $\sigma_0(t)$ is a known monotonously decreasing function presenting gradual unloading, Eq. (5.11) shows how σ_m attenuates with time. If at $t \geq t_1$, $\sigma_0(t) = 0$, then by Eq. (5.11) σ_m at $t = t_1$ is

$$\sigma_1 = \frac{1}{t_1}\int_0^{t_1} \sigma_0(t)dt \tag{5.12}$$

Therefore, from time $t \geq t_1$, the attenuation of σ_m follows the same rule of Eq. (5.6).

The time t_K in which the plastic wave begins to disappear due to the interaction with the unloading disturbances, and the corresponding location $X_K = C_1 t_K$ can be determined according to the condition:

$$\sigma_m(t_K) = Y$$

After that time, the unloading of the bar segment which has undergone elastic deformation can be dealt with as we have discussed earlier on elastic unloading.

Once the attenuation of the maximum stress $\sigma_m(t)$ is obtained, from the dynamic compatibility condition across a strong discontinuity [Eq. (5.8)] and the linear hardening plastic stress–strain relationship $\varepsilon_m = \varepsilon_Y + ((\sigma_m - Y)/E_1)$, the corresponding attenuation of the maximum particle velocity $v_m(t)$ and the maximum strain $\varepsilon_m(t)$ can be obtained respectively as:

$$v_m(t) = \frac{1}{\rho_0 C_1}\left[\left(1 - \frac{1}{\mu_1}\right)Y - \frac{1}{t}\int_0^t \sigma_0(t)dt\right] \tag{5.13a}$$

$$\varepsilon_m(t) = \frac{1}{E_1}\left[\left(\frac{1}{\mu_1^2} - 1\right)Y - \frac{1}{t}\int_0^t \sigma_0(t)dt\right] \tag{5.13b}$$

where

$$\mu_1 = \frac{C_0}{C_1} = \sqrt{\frac{E}{E_1}}$$

Especially, if $\sigma_0(t)$ linearly decreases with time, as expressed by

$$\sigma_0(t) = -p_m\left(1 - \frac{t}{T}\right)$$

where T is the time when $\sigma_0(t) = 0$, then from Eq. (5.11) it can be shown that at $t \leq T$, σ_m is a linear function of t or X:

$$\sigma_m(t) = -p_m\left(1 - \frac{t}{2T}\right) = -p_m\left(1 - \frac{\mu_1}{2}\frac{X}{C_0 T}\right) \tag{5.14}$$

Recall that in Chapter 4, the following equation has been obtained for the case of elastic unloading [Eq. (4.50)]

$$\sigma_m(t) = -p_m\left(1 - \frac{\mu^2 - 1}{2\mu_1}\cdot\frac{X}{C_0T}\right) = -p_m\left(1 - \frac{\mu_1 - \dfrac{1}{\mu_1}}{2}\frac{X}{C_0T}\right)$$

Comparing this equation with Eq. (5.14), it can be seen that the difference between these two equations decreases with increasing μ_1. If we introduce $B_R = \mu_1/2$ and $B_E = (\mu_1^2 - 1)/(2\mu_1)$ to denote the linear coefficient of the term (X/C_0T) in these two equations [see Eq. (4.53)], respectively, then the relative difference of these two coefficients is:

$$\delta_B = \frac{B_R - B_E}{B_R} = \frac{1}{\mu_1^2}$$

For metallic materials, the elastic wave velocity is approximately 10 times the plastic wave velocity, namely $\mu_1 \approx 10$, so that $\delta_B \approx 1\%$. It means that the error induced by using rigid unloading simplification where the elastic wave velocity is approximately regarded as infinite, is small enough, if the elastic wave velocity C_0 is much larger than the plastic wave velocity C_1. Then the complicated elastic unloading problem, which is attributed to solving simultaneously two different groups of partial differential equations respectively in the plastic region and the elastic region, can be simplified to solve the problem which only contains ordinal differential equation.

The example discussed above is a problem where the rigid unloading boundary coincides with the propagating locus of strong discontinuous plastic loading wave propagating with constant velocity. If the plastic loading wave is a continuous wave, then the unloading boundary is a weak discontinuous boundary and no longer coincides with the propagating loci of plastic waves, so that the problem will be more complicated. Nevertheless, the problem can still be solved by a similar treatment as shown below.

5.1.3 Rigid unloading in linear elastic–decreasing hardening plastic bars

Consider now a semi-infinite bar with a stress–strain relationship of the so-called **linear elastic–decreasing hardening plastic loading–rigid unloading**, as shown in Fig. 5.1(c). Assume a gradual loading denoted by $\sigma_{0l}(t)$ and then a gradual unloading denoted by $\sigma_{0u}(t)$, without any strong discontinuity, are applied at the bar end, which are expressed respectively as (Fig. 5.4).

$$\begin{cases} \sigma(0,t) = \sigma_{0l}(t), & \dfrac{d|\sigma_{0u}|}{dt} \geq 0, \quad 0 \leq t \leq t_0 \\ \sigma(0,t) = \sigma_{0u}(t), & \dfrac{d|\sigma_{0u}|}{dt} \leq 0, \quad t \geq t_0 \end{cases}$$

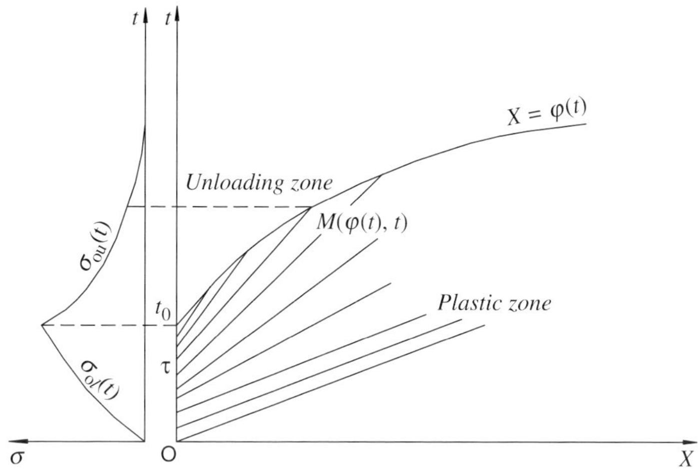

Fig. 5.4. A semi–infinite bar with σ–ε curve shown in Fig. 5.1(c) is subjected to a gradual boundary loading up to a maximum $\sigma_{\max}$, followed by a gradual unloading.

In such a case, the unloading boundary is a weak discontinuous one $X = \varphi(t)$, as shown in Fig. 5.4. Correspondingly, the previous compatibility condition across a strong discontinuous interface, Eq. (5.8), now should be replaced by the following differential relation for the rightward plastic simple waves:

$$dv_m = -\frac{d\sigma_m}{\rho_0 C} \tag{5.15}$$

Noticing that the length of rigid segment l in the motion equation of rigid segment [Eq. (5.9)] now should be replaced by $\varphi(t)$, then Eq. (5.10) should be rewritten as:

$$\frac{\varphi(t)}{C}\frac{d\sigma_m}{dt} + \sigma_m - \sigma_{0u} = 0, \; t \geq t_0 \tag{5.16}$$

For plastic simple waves, there is one more relation:

$$\varphi(t) = C(\sigma_m) \cdot [t - \tau_l(\sigma_m)] \tag{5.17}$$

where $\tau_l(\sigma)$ is the inverse function of $\sigma_{0l}(t)$; or we have

$$\sigma_m(t) = \sigma_{0l}\left(t - \frac{\varphi(t)}{C(\sigma_m)}\right) \tag{5.18}$$

Substituting Eq. (5.7) into Eq. (5.16), eliminating $\varphi(t)$, $\sigma_m(t)$:

$$\frac{d\sigma_m}{dt} + \frac{\sigma_m - \sigma_{0u}}{t - \tau_l(\sigma_m)} = 0 \tag{5.19}$$

Once $\sigma_m(t)$ is solved, substitute it into Eq. (5.17), the unloading boundary $X = \varphi(t)$ can be determined. Alternatively, substituting Eq. (5.18) into Eq. (5.16), eliminating $\sigma_m(t)$, then the problem is summed up to solve a first-order ordinal differential equation with respect to $\varphi(t)$. Once $\varphi(t)$ is solved, substitute it into Eq. (5.18), $\sigma_m(t)$ can be determined.

For example, if $\sigma_{0l}(t)$ is a linear increasing function of time (Nowaski, 1978):

$$\sigma_{0l}(\tau) = \sigma_{\max}\frac{\tau}{t_0},$$

then Eqs. (5.18) and (5.19) are respectively:

$$\sigma_m(t) = \frac{\sigma_{\max}}{t_0}\left[t - \frac{\varphi(t)}{C}\right], \tag{5.20}$$

$$\frac{d\sigma_m}{dt} + \frac{\sigma_m - \sigma_{0u}}{t - \dfrac{\sigma_m - \sigma_{0u}}{\sigma_{\max}}} = 0 \tag{5.21}$$

Furthermore, assuming $C = C_1$ (linear hardening plastic material), Eq. (5.20) reduces to

$$\begin{aligned} \sigma_m(t) &= \frac{\sigma_{\max}}{t_0}\cdot\left[t - \frac{\varphi(t)}{C_1}\right] \\ \frac{d\sigma_m}{dt} &= \frac{\sigma_{\max}}{t_0}\cdot\left(1 - \frac{\varphi'}{C_1}\right) \end{aligned} \tag{5.22}$$

Substituting the above equation into Eq. (5.21), we have

$$\varphi\varphi' = C_1^2\cdot\left(t - \frac{t_0\sigma_{0u}}{\sigma_{\max}}\right) \tag{5.23}$$

Thus the unloading boundary $X = \varphi(t)$ can be solved

$$\varphi(t) = C_1\cdot\left[t^2 - t_0^2 - \frac{2t_0}{\sigma_{\max}}\int_{t_0}^{t}\sigma_{0u}(t)dt\right]^{1/2} \tag{5.24}$$

where the condition $\varphi(t_0) = 0$ has been used to determine the integrate constant.

If $\sigma_{0u}(t)$ is also a linear increasing function of time, as shown in Fig. 5.5:

$$\sigma_{0u}(t) = \sigma_{\max}\cdot\frac{t_1 - t}{t_1 - t_0} \tag{5.25}$$

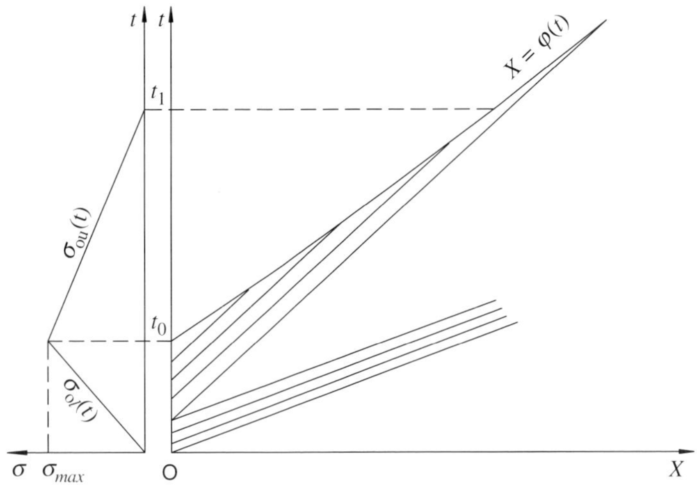

Fig. 5.5. A semi–infinite bar with σ–ε shown in Fig. 5.1(a) is subjected to a linear increasing boundary loading up to a maximum $\sigma_{\max}$ followed by a linear decreasing boundary unloading.

then from Eq. (5.24), we have

$$\varphi(t) = \begin{cases} C_1 \cdot \sqrt{\dfrac{t_1}{t_1 - t_0}}(t - t_0), & t_0 \le t \le t_1 \\ C_1 \cdot \sqrt{t^2 - t_0 t_1}, & t \ge t_1 \end{cases} \tag{5.26}$$

Thus it can be seen, at $t_0 \leqq t \leq t_1$, the unloading boundary is a straight line propagating with a constant velocity $\overline{C} = \varphi'(t) = C_1 \left[t_1/(t_1 - t_0)\right]^{1/2} > C_1$, while at $t \ge t_1$, the unloading boundary is a hyperbolic curve propagating with a velocity $\overline{C}$ which gradually decreases and approaches C_1.

Substituting Eq. (5.26) into Eq. (5.22), the stress along the unloading boundary $\sigma_m(t)$ can be obtained, and furthermore from Eq. (5.15) the corresponding particle velocity $v_m(t)$ can be obtained, respectively as:

for $t_0 \le t \le t_1$

$$\begin{cases} \sigma_m(t) = \dfrac{\sigma_{\max}}{t_0}\left[t - \sqrt{\dfrac{t_1}{t_1 - t_0}}(t - t_0)\right], \\ v_m(t) = -\dfrac{Y}{\rho_0 C_0}(1 - \mu_1) - \dfrac{\sigma_{\max}}{\rho_0 C_1 t_0}\left[t - \sqrt{\dfrac{t_1}{t_1 - t_0}}(t - t_0)\right], \end{cases} \tag{5.27}$$

for $t \geq t_1$

$$\begin{cases} \sigma_m(t) = \dfrac{\sigma_{\max}}{t_0}\left(t - \sqrt{t^2 - t_0 t_1}\right) \\ v_m(t) = -\dfrac{Y}{\rho_0 C_0}(1-\mu_1) - \dfrac{\sigma_{\max}}{\rho_0 C_1 t_0}\left(t - \sqrt{t^2 - t_0 t_1}\right) \end{cases} \tag{5.28}$$

where $\mu_1 = C_0/C_1$. It should be noted that the particle velocities of the unloading segment (as a rigid segment) of bar are uniform along the bar at any time, namely in the unloading region we have

$$\bar{v}(X,t) = v_m(t),$$

while the stresses are non-uniform along the bar in the unloading region due to the inertia effect of rigid unloading segment, namely in the unloading region, similar to Eq. (5.9), we have

$$\overline{\sigma}(X,t) = \rho_0 X \frac{d\bar{v}}{dt} + \sigma_{0u} = \rho_0 X \frac{dv_m}{dt} + \sigma_{0u}.$$

Substituting Eqs. (5.25), (5.27), and (5.28) into the above equation, we obtain

$$\begin{cases} \overline{\sigma}(X,t) = -\dfrac{\sigma_{\max}}{C_1 t_0}\left(1 - \sqrt{\dfrac{t_1}{t_1 - t_0}}\right) X + \sigma_{\max} \cdot \dfrac{t_1 - t}{t_1 - t_0}, & t_0 \leq t \leq t_1 \\ \overline{\sigma}(X,t) = -\dfrac{\sigma_{\max}}{C_1 t_0}\left(1 - \dfrac{t}{\sqrt{t^2 - t_1 t_0}}\right) X, & t \geq t_1 \end{cases} \tag{5.29}$$

Moreover, since the strains in the unloading region are entirely determined by the maximum stresses σ_m having reached before unloading

$$\varepsilon_m = \varepsilon_Y + \frac{\sigma_m - Y}{E_1} = \frac{Y}{E_0}\left(1 - \frac{E_0}{E_1}\right) + \frac{\sigma_m}{E_1}$$

thus the strain distribution in the unloading region can be determined too as

$$\begin{cases} \varepsilon_m(X) = \dfrac{Y}{\rho_0 C_0^2}(1-\mu_1^2) + \dfrac{\sigma_m}{\rho_0 C_1^2 t_0} \cdot \left[t_0 + \dfrac{X}{C_1}\left(\dfrac{1}{\sqrt{\dfrac{t_1}{t_1 - t_0}}} \right) \right], & t_0 \leq t \leq t_1 \\ \varepsilon_m(X) = \dfrac{Y}{\rho_0 C_0^2}(1-\mu_1^2) + \dfrac{\sigma_m}{\rho_0 C_1^2 t_0} \left[\sqrt{\dfrac{X^2}{C_1^2} + t_0 t_1} - \dfrac{X}{C_1} \right], & t \geq t_1 \end{cases} \tag{5.30}$$

With regard to the **general propagating properties** of the **rigid unloading boundary** in semi-infinite bars, all can be directly deduced from the corresponding results for the elastic unloading boundary discussed earlier in Chapter 4 by letting $C_0 \to \infty$. For example, for the one-order weak discontinuous rigid unloading boundary, let $C_0 \to \infty$ in Eq. (4.73), we obtain the following relation

$$\frac{\partial\sigma/\partial t}{\partial\overline{\sigma}/\partial t} = \frac{1}{1 - \dfrac{\overline{C}^2}{C^2}} \tag{5.31a}$$

or

$$\overline{C} = C\sqrt{1 - \frac{(\partial\overline{\sigma}/\partial t)}{(\partial\sigma/\partial t)}} \tag{5.31b}$$

For the unloading boundary, there must be $\partial\sigma/\partial t \geq 0$ and $\partial\overline{\sigma}/\partial t \leq 0$, therefore

$$\overline{C} \geq C.$$

Obviously, if $\partial\sigma/\partial t = \infty$, $\partial\overline{\sigma}/\partial t \neq \infty$, or if $\partial\overline{\sigma}/\partial t = 0$, $\partial\sigma/\partial t \neq 0$, then we have $\overline{C} = C$; while if $\partial\sigma/\partial t = 0$, $\partial\overline{\sigma}/\partial t \neq 0$, or if $\partial\overline{\sigma}/\partial t = -\infty$, $\partial\sigma/\partial t \neq \infty$, then we have $\overline{C} = \infty$.

5.2 Rigid Unloading in Finite Bars

In this section, we discuss an example for stress wave propagation in a finite bar characterized by rigid unloading (Nowaski, 1978). The bar consists of two segments: one ($0 \leq X \leq l$) is made of A material with linear hardening plastic loading–rigid unloading behavior, as shown in Fig. 5.1(a), denoting its density and plastic wave velocity as ρ_A and C_A, respectively; while another ($X \geq l$) is made of B material with linear elastic behavior, denoting its density and elastic wave velocity as ρ_B and C_B, respectively. In addition, a rigid mass exists between these two segments, as shown in Fig. 5.6. Originally, the bar is at rest and stress-free. At time $t = 0$, the bar end ($X = 0$) is suddenly loaded and then gradually unloaded.

The whole problem can be separated into several regions as shown in Fig. 5.6, and the solution in each region is discussed as follows.

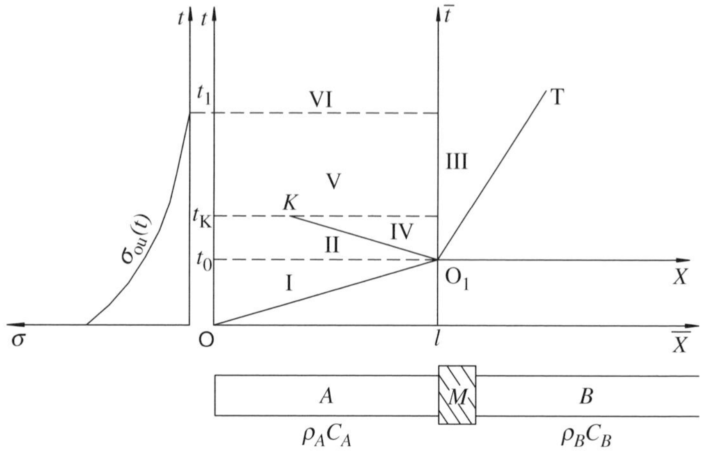

Fig. 5.6. Stress wave propagations in a finite bar characterized by rigid unloading.

Region I: Before the rightward plastic wave OO_1 arrives at the interface $X = l$, the solution is the same as that in a semi-infinite bar discussed earlier, namely:

$$\begin{cases} \sigma_{m1}(t) = \frac{1}{t}\int_0^t \sigma_0(t)dt, \ \sigma_{m1}(t) = \dfrac{C_A}{X}\int_0^{X/C_A} \sigma_0(t)dt, \\ v_{m1}(t) = -\dfrac{1}{\rho_A C_A t}\int_0^t \sigma_0(t)dt, \\ \overline{\sigma_1}(X,t) = \left(1 - \dfrac{X}{C_A t}\right)\sigma_0(t) + \dfrac{X}{C_A t^2}\int_0^t \sigma_0(t)dt, \\ \varepsilon_{m1}(X) = \dfrac{1}{\rho_A C_A X}\int_0^{X/C_A} \sigma_0(t)dt \end{cases} \quad (5.32)$$

The unloading boundary coincides with the propagating locus of the strong discontinuous plastic loading wave OO_1.

At $t = t_0 = l/C_A$, the strong discontinuous plastic wave arrives at the interface, and the wave reflection and transmission take place. Because a rigid mass exists at the interface, the interface acceleration is finite, and moreover the B material is linearly elastic, so the reflected wave O_1K is a strong discontinuous plastic wave, while the transmitted wave O_1T is a weak discontinuous elastic wave. The propagating locus of the reflected strong discontinuous plastic wave is expressed as

$$X = 2l - C_A t \quad (5.33)$$

The region in front of it is the region II that has experienced rigid unloading and is now plastically reloaded, while the region behind it is the region IV, a new rigid unloading region caused by the movement of rigid mass *M*.

Region II: In this region, the stress $\sigma_2(X,t)$ should satisfy the following motion equation:

$$\sigma_2(X,t) = \rho_A X \frac{dv_2}{dt} + \sigma_0(t). \tag{5.34}$$

From this equation, along the wave front O_1K, we have

$$\sigma_{m2}(t) = \sigma_2\left[\varphi(t), t\right] = \rho_A(2l - C_A t)\frac{dv_{m2}}{dt} + \sigma_0(t). \tag{5.35}$$

Note that when a rigid unloading state is to be reloaded into a plastic state, the stress should reach the maximum which is reached before unloading [point A on Fig. 5.1(a)], namely

$$\sigma_{m2}(X) = \sigma_{m1}(X).$$

Then from Eqs. (5.32) and (5.33), $\sigma_{m2}(X)$ can be expressed as

$$\sigma_{m2}(t) = \frac{C_A}{2l - C_A t}\int_0^{((2l/C_A)-t)} \sigma_0(t)dt.$$

Substituting it into Eq. (5.35) and considering

$$v_2(t_0) = v_{m1}(t_0) = -\frac{1}{\rho_A C_A t_0}\int_0^{t_0} \sigma_0(t)dt$$

we have $v_{m2}(t)$, namely $v_2(t)$:

$$v_2(t) = -\frac{1}{\rho_A}\int_{t_0}^{t} \frac{\sigma_0(\tau)}{2l - C_A\tau} + \frac{1}{\rho_A}\int_{t_0}^{t}\left\{\frac{C_A}{(2l - C_A\eta)^2}\int_0^{((2l/C_A)-\eta)} \sigma_0(\tau)d\tau\right\}d\eta$$
$$- \frac{1}{\rho_A C_A t_0}\int_0^{t_0}\sigma_0(\tau)d\tau \tag{5.36}$$

Thus, substituting Eq. (5.36) into Eq. (5.34), the stress distribution in region II, $\sigma_2(X,t)$, can be obtained as:

$$\sigma_2(X,t) = \left(1 - \frac{X}{2l - C_A t}\right)\sigma_0(t) + \frac{C_A X}{(2l - C_A t)^2}\int_0^{((2l/C_A)-t)} \sigma_0(\tau)d\tau. \tag{5.37}$$

The strain distribution in region II is the same as in region I.

$$\varepsilon_2(X) = \varepsilon_1(X)$$

Regions III and IV: The solutions in regions III and IV are coupled since the rigid mass M should simultaneously satisfy the following two equations, representing the continuity condition and momentum conservation condition, respectively:

$$v_4(\bar{t}) = v_3(l, \bar{t}) \tag{5.38}$$

$$M\frac{dv_4}{d\bar{t}} = \sigma_3(l, \bar{t}) - \sigma_4(l, \bar{t}) \tag{5.39}$$

here the notation $\bar{t} = t - (l/C_A)$ has been introduced. The region III is an elastic simple wave region, hence we have

$$\sigma_3(l, \bar{t}) = -\rho_B C_B v_3(l, \bar{t}) \tag{5.40}$$

On the other hand, the region IV is a rigid unloading region, so the following equation should be satisfied.

$$\sigma_4(l, \bar{t}) - \sigma_{m4}(\bar{t}) = \rho_B C_B \bar{t}\frac{dv_4}{d\bar{t}} \tag{5.41}$$

and according to the conservation condition across the strong discontinuous plastic wave front O_1K, $\sigma_{m4}(\bar{t})$, and $v_4(\bar{t})$ should furthermore satisfy

$$\sigma_{m4}(\bar{t}) = \sigma_{m2}(\bar{t}) + \rho_A C_A\{v_4(\bar{t}) - v_2(\bar{t})\} \tag{5.42}$$

where σ_{m2} and v_2 have been obtained from the solutions in region II. From the five equations given in Eqs. (5.38)–(5.42), eliminating $\sigma_3(l, \bar{t})$, $\sigma_4(l, \bar{t})$, $\sigma_{m4}(l, \bar{t})$, and $v_3(l, \bar{t})$, the following linear ordinal differential equation with respect to $v_4(\bar{t})$ can be obtained

$$\frac{dv_4(\bar{t})}{d\bar{t}} + f_1(\bar{t})v_4(\bar{t}) = f_2(\bar{t}) \tag{5.43}$$

where

$$f_1(\bar{t}) = \frac{\rho_A C_A + \rho_B C_B}{\rho_A C_A \bar{t} + M}$$

$$f_2(\bar{t}) = \frac{\rho_A C_A v_2(\bar{t}) + \sigma_{m2}(\bar{t})}{\rho_A C_A \bar{t} + M}$$

The solution of Eq. (5.43) is

$$v_4(\bar{t}) = \int_0^{\bar{t}} f_2(\eta)\exp\left[-\int_\eta^{\bar{t}} f_1(\tau)d\tau\right]d\eta \tag{5.44}$$

here the condition $v_4(0) = 0$ has been used to determine the integrate constant. Once $v_4(\bar{t})$ is determined, other variables in region IV can be determined too from Eqs. (5.41) and (5.42). Consequently, $v_3(l, \bar{t})$ and $\sigma_3(l, \bar{t})$ can be determined from Eqs. (5.38) and (5.40), and thereby the region III (elastic simple wave region) is solved.

Because of the unloading effect, the strong discontinuous amplitude of the leftward reflected plastic wave O_1K continuously decreases in its propagation. If at $t = t_K < t_1$, the following condition is satisfied

$$v_4(t_k) = v_2(t_k),\ \sigma_{m4}(t_k) = \sigma_{m2}(t_k) \tag{5.45}$$

It means that the strong discontinuous plastic wave disappears. Henceforth, the whole bar segment A is in an unloading process. Since $v_4(t)$ and $v_2(t)$ have been obtained, t_K can be determined from Eq. (5.45).

Region V: The region V still should be simultaneously solved with region III. Similar to Eqs. (5.38)–(5.41), we have

$$\left.\begin{aligned} &v_5(\bar{t}) = v_3(l,\bar{t}),\\ &M\frac{dv_5}{dt} = \sigma_3(l,\bar{t}) - \sigma_5(l,\bar{t}),\\ &\sigma_3(l,\bar{t}) = -\rho_B C_R v_3(i,\bar{t}),\\ &\sigma_5(l,\bar{t}) - \sigma_0(\bar{t}) = \rho_A l\frac{dv_5}{d\bar{t}} \end{aligned}\right\} \tag{5.46}$$

Eliminating $\sigma_3(l,\bar{t}), \sigma_5(l,\bar{t})$, and $v_3(l,\bar{t})$ from the above four equations, we obtain the following linear ordinal differential equation with respect to $v_5(\bar{t})$:

$$\frac{dv_5(\bar{t})}{d\bar{t}} + f_3 v_5(\bar{t}) = f_4\sigma_0(\bar{t}) \tag{5.47}$$

$$f_3 = \frac{\rho_B C_B}{\rho_A l + M}$$

$$f_4 = -\frac{1}{\rho_A l + M}$$

of which the solution is

$$v_5(\bar{t}) = \exp[-f_3\cdot(\bar{t}-\bar{t}_K)]\left\{f_4\cdot\int_{\bar{t}_K}^{\bar{t}}\sigma_0(\eta)\cdot\exp[f_3\cdot(\eta-\bar{t}_K)]d\eta + v_2(\bar{t}_K)\right\}$$

where the condition $v_5(\bar{t}_K) = v_2(\bar{t}_K)$ has been used to determine the integrate constant. Once $v_5(\bar{t}_K)$ is determined, similar to the situation discussed for region IV, both region V and region III between time t_K and t_1 (i.e. $t_K \le t \le t_1$) are entirely solved.

Region VI: At $t \ge t_1$, except $\sigma_0(t) \equiv 0$, all others are the same as discussed on region V. Therefore, let all $\sigma_0(\bar{t})$ in the associated equations in discussions on region V vanish ($\sigma_0(\bar{t}) = 0$), the solution for region VI can be obtained.

In the current example, if the rigid mass $M \equiv 0$, then the problem is reduced to solve the stress wave propagation in the laminated media composed by a linear hardening plastic loading–rigid unloading material (A) and a linear elastic material (B).

Alternatively, if the rigid mass $M \to \infty$, then the problem is reduced to solve the stress wave propagation in a finite bar made of a linear hardening plastic loading–rigid unloading material and with a fixed end boundary condition. In such a case, we have $f_1 = f_2 = f_3 = 0$, $v_4(X,t) = 0$, and

$$\sigma_4(X,t) = \sigma_{m4}(t) = \frac{C_A}{2l - C_A t}\int_0^{((2l/C_A)-t)} \sigma_0(\tau)d\tau + \int_{t_0}^{t} \frac{C_A}{2l - C_A\tau}\sigma_0(\tau)d\tau$$

$$- \int_{t_0}^{t}\left[\frac{C_A^2}{(2l - C_A\eta)^2}\int_0^{((2l/C_A)-\eta)} \sigma_0(\tau)d\tau\right]d\eta + \frac{1}{t_0}\int_0^{t_0} \sigma_0(t)dt$$

5.3 Rigid Unloading Analyses for Shock Wave Propagation

In the previous two sections, the strong discontinuous plastic loading waves concerned are those formed in linear hardening plastic materials under the boundary condition of suddenly applied loading. In such cases, the strong discontinuous wave velocity is constant, so the problem is relatively easier to deal with. However, if the shock waves form in a gradually increasing hardening plastic material, wherein the shock wave velocity varies with shock intensity, the problem is then relatively complicated. In the present section, we discuss the rigid unloading in such a case.

In the following, we first discuss an example of semi-infinite bar (Kaliski *et al.*, 1967), and then discuss an example of finite bar (Lee and Tupper, 1954).

5.3.1 Rigid unloading of shock wave propagating in semi-infinite bars

Consider a semi-infinite bar with a stress–strain relationship as shown in Fig. 5.7(a), namely, for the loading:

$$\begin{cases} \sigma = E_1\varepsilon, & \sigma \le \sigma_A \\ \varepsilon = \varepsilon_A, & \sigma \ge \sigma_A \end{cases}$$

therefore if $\sigma \le \sigma_A$, then the plastic velocity is constant $C_1 = (E_1/\rho_0)^{1/2}$, and if $\sigma > \sigma_A$, then a shock wave will form and its velocity D is:

$$\mathrm{D} = \sqrt{\frac{\sigma}{\rho_0\varepsilon_A}}$$

while for unloading, it shows a rigid unloading behavior. Such a stress–strain relationship can be regarded as a simplified model of the actual stress–strain relationship for soils as shown in Fig. 5.7(b).

Assume that the semi-infinite bar is originally at rest and stress-free, at time $t = 0$ it is suddenly loaded by a boundary stress larger than σ_A, and then gradually unloaded. Thus,

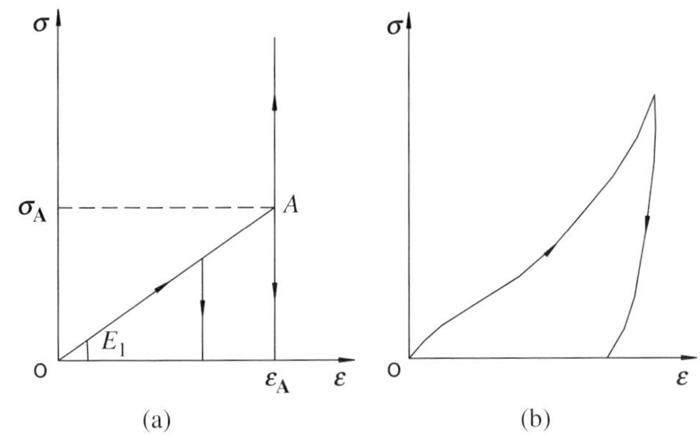

Fig. 5.7. Two stress–strain models that may result in shock waves: (a) linear hardening plastic loading–rigid unloading; (b) stress–strain relation for soils. Note that (a) can be regarded as a simplified model of (b).

from the bar end, a gradually attenuating plastic shock wave propagates rightwards. Its propagating locus $X = \varphi(t)$ is also the rigid unloading boundary, as plotted in Fig. 5.8.

According to the kinematic compatibility condition and the dynamic compatibility condition across a shock wave front, the plastic stress value $\sigma_m(t)$, strain ε_A, particle velocity $v_m(t)$, and the shock wave velocity $\varphi'(t)$ should satisfy the following relations:

$$v_m(t) = -\varphi'(t)\varepsilon_A \tag{5.48}$$

$$\sigma_m(t) = -\rho_0\varphi'(t)v_m(t) \tag{5.49}$$

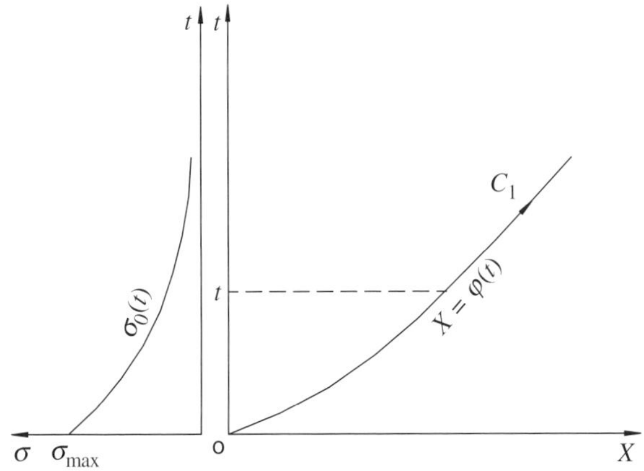

Fig. 5.8. Gradual attenuation of a plastic shock wave propagating in a semi–infinite bar due to rigid unloading.

On the other hand, the rigid unloading segment of bar should comply with

$$\rho_0 X \frac{d\bar{v}}{dt} = \overline{\sigma}(X,t) - \sigma_0(t). \tag{5.50}$$

Applying it along the unloading boundary $X = \varphi(t)$, noticing $\bar{v}(t) = v_m(t)$, we have

$$\rho_0 \varphi \frac{dv_m}{dt} = \sigma_m(X,t) - \sigma_0(t).$$

Substituting Eqs. (5.48) and (5.49) into the above equation and after rearranging, we obtain

$$\varphi\varphi'' + (\varphi')^2 = \frac{\sigma_0(t)}{\rho_0 \varepsilon_A} = f(t) \tag{5.51}$$

or namely

$$\left(\varphi^2\right)'' = 2f(t).$$

Integrating it twice and using the initial conditions $\varphi(0) = 0$, $(\varphi^2)'\,|_{t=0} = 2\varphi\varphi'\big|_{t=0} = 0$ to determine the integrate constants, the unloading boundary can be solved as:

$$\varphi(t) = \left[2\int_0^t d\tau \int_0^\tau f(\xi)d\xi\right]^{1/2} \tag{5.52}$$

Correspondingly, the shock wave velocity, or namely the propagating velocity of unloading boundary $\varphi'(t)$ is

$$\varphi'(t) = \frac{1}{\varphi}\int_0^t f(\xi)d\xi = \frac{\int_0^t f(\xi)d\xi}{\left[2\int_0^t d\tau \int_0^\tau f(\xi)d\xi\right]^{1/2}}$$

Substituting it into Eqs. (5.48)–(5.50), the particle velocity $v_m(t)$ and plastic stress $\sigma_m(t)$ along the unloading boundary, as well as the stress $\overline{\sigma}(X,t)$ of the rigid unloading segment, which is varied with X and t, can be obtained, respectively.

$$v_m(t) = -\varepsilon_A \frac{\int_0^t f(\xi)d\xi}{\left[2\int_0^t d\tau \int_0^\tau f(\xi)d\xi\right]^{1/2}}$$

$$\sigma_m(t) = \rho_0 \varepsilon_A \frac{\int_0^t f(\xi)d\xi}{\left[2\int_0^t d\tau \int_0^\tau f(\xi)d\xi\right]^{1/2}}$$

$$\overline{\sigma}(X,t) = \sigma_0(t) - \rho_0 \varepsilon_A X \frac{f(t)\left[2\int_0^t d\tau \int_0^\tau f(\xi)d\xi\right]^{1/2} - \left[\int_0^t f(\xi)d\xi\right]^2 \left[2\int_0^t d\tau \int_0^\tau f(\xi)d\xi\right]^{-1/2}}{2\int_0^t d\tau \int_0^\tau f(\xi)d\xi}$$

Obviously, the rightward propagating shock wave continuously attenuates due to the rigid unloading effect. If at time $t = t^*$,

$$\sigma_m(t^*) = \sigma_A$$

then after that the plastic wave velocity is constant C_1, the problem can be dealt with as what we discussed earlier for the linear hardening materials.

5.3.2 Rigid unloading of shock wave propagating in finite bars

Finally, consider an example of **rigid unloading of shock wave propagating in a finite bar** (Lee and Tupper, 1954). With regard to the example discussed earlier in Section 4.9, namely, for a finite, increasingly hardening plastic bar impacts on a rigid target as shown in Fig. 4.28, if the elastic wave velocity, either at loading or at unloading, is relatively much larger than that of the plastic wave velocity and consequently can be regarded as infinite, and moreover the elastic deformation is relatively much smaller than the plastic deformation and consequently can be neglected, then the elastic-plastic stress–strain relationship shown in Fig. 4.28(d) can be replaced by the rigid-plastic stress–strain relationship shown in Fig. 5.9(a). This is equivalent to such a problem of stress wave analysis: all the bar segments where the plastic shock wave front has not arrived yet, can be regarded as a rigid body without any deformation, while all the bar segments where the plastic shock wave front has passed also can be regarded as a rigid body without any elastic recovery deformation.

If the finite bar impacts on the rigid target with a constant velocity v^* [Fig. 5.9(b)], then a rigid plastic shock wave propagates leftward with a shock velocity D from the bar end $X = l$, then the stress jump across the wave front is $(\sigma - Y)$, the strain jump is $(\varepsilon - 0)$, and the particle velocity jump is $(0 - v)$. According to the jump conditions across the leftward

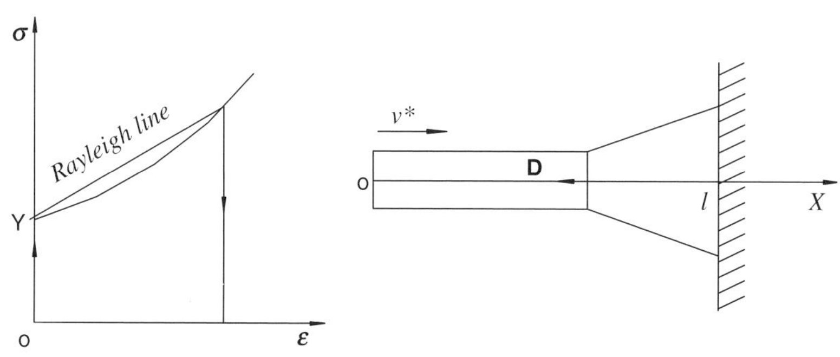

Fig. 5.9. Rigid unloading of a plastic shock wave propagation in a finite bar: (a) the loading and unloading stress–strain curve; (b) a finite bar impacts onto a rigid target with a constant velocity v^*.

shock wave front, we have

$$v = -\mathrm{D}\varepsilon \tag{5.53}$$

$$\sigma - Y = -\rho_0 \mathrm{D} v \tag{5.54}$$

Eliminating the v from these equations, the plastic shock wave velocity D is obtained:

$$\mathrm{D}^2 = \frac{\sigma - Y}{\rho_0 \varepsilon} \tag{5.55}$$

while eliminating D from Eqs. (5.53) and (5.54), we have

$$\rho_0 v^2 = (\sigma - Y)\varepsilon \tag{5.56}$$

which gives the relation between v and σ, when the $\sigma \sim \varepsilon$ relationship is given. For example, from the initial impact velocity v^*, the corresponding initial shock amplitude σ^* can be determined.

The bar segment where the plastic shock wave has not yet arrived, should satisfy the following motion equation for rigid body:

$$\rho_0 X \frac{dv(t)}{dt} = Y \tag{5.57}$$

where X is the length of the bar segment in front of shock wave. Noticing that the propagating locus of the shock wave front $X = \varphi(t)$ or $t = \psi(X)$ is also the locus of the rigid boundary, Eq. (5.57) can be rewritten as:

$$-\rho_0 X \mathrm{D} \frac{dv[\psi(X)]}{dX} = Y$$

Eliminating D from Eq. (5.53), we have

$$\rho_0 X v \frac{dv}{dX} = Y\varepsilon$$

$$\rho_0 X \frac{d(v^2)}{dX} = 2Y\varepsilon$$

Moreover, eliminating v^2 from Eq. (5.55) gives

$$\frac{d[(\sigma - Y)\varepsilon]}{Y\varepsilon} = 2\frac{dX}{X}$$

Integrating it, noting $\sigma = \sigma^*$ at $X = l$, and introducing

$$F(\sigma) = \int_0^{(\sigma - Y)\varepsilon} \frac{d[(\sigma - Y)\varepsilon]}{Y\varepsilon}$$

we obtain

$$\ln\left(\frac{X}{l}\right)^2 = F(\sigma) - F(\sigma^*) \tag{5.58}$$

After some mathematical calculations, and noticing Eq. (5.55), $F(\sigma)$ can be expressed in the following form:

$$F(\sigma) = \int_0^{(\sigma-Y)\varepsilon} \frac{d\left[(\sigma - Y)\,\varepsilon\right]}{Y\varepsilon} = \int_Y^{\sigma} \frac{d\sigma}{Y} + \int_0^{\varepsilon(\sigma)} \frac{\sigma - Y}{Y\varepsilon} d\varepsilon$$

$$= \frac{1}{Y}\left[\sigma + \rho_0 \int_0^{\varepsilon(\sigma)} \mathrm{D}^2 d\varepsilon\right] - 1 = \frac{1}{Y}\left[\sigma + \int_Y^{\sigma} \frac{\mathrm{D}^2}{C^2} d\sigma\right] - 1$$

Since for a given $\sigma \sim \varepsilon$ curve, we knew $\mathrm{D}^2 = \frac{1}{\rho_0}\frac{\sigma - Y}{\varepsilon}$ and $C^2 = \frac{1}{\rho_0}\frac{d\sigma}{d\varepsilon}$, this is also a known function. Thus Eq. (5.58) determines the stress distribution $\sigma_m(X)$ and the corresponding strain distribution $\varepsilon_m(X)$ along the rigid plastic boundary. In addition, it can be seen from Eq. (5.58) that the distributions of maximum plastic stress and strain are merely a function of X/l.

Let $\sigma = Y$ in Eq. (5.58), the location where the plastic wave vanishes, and consequently the total length of the plastic deformation distributing in bar is determined. Obviously, the plastic shock wave must vanish before it arrives at the stress-free end. This is understandable: in the process the plastic shock wave propagates towards the stress-free end, the shock wave continuously interacts with the rigid unloading waves reflected from the stress-free end with a wave velocity of ∞, till the shock wave vanishes.

The results obtained by both the elastic–plastic wave theoretical analysis and the rigid unloading analysis for the same problem are compared as shown in Fig. 4.28 (Lee and Tupper, 1954). In Fig. 4.28(a), the propagating locus of shock wave front given by the rigid unloading analysis is marked by solid dots, and the locus given by the elastic–plastic wave analysis is shown by the solid line. As can be seen, the former basically coincides with the latter. The residual strain distribution is shown in Fig. 4.28(b), wherein the result given by the rigid unloading analysis is plotted by dashed line while the result given by the elastic–plastic wave theory is plotted by the solid line. It is shown that the rigid unloading analysis **averages** the original discontinuous distribution of residual strain obtained in the elastic–plastic wave analysis so that its distribution is continuous. Obviously, with increasing impact velocity, the plastic deformation of bar must increase and the number of unloading wave reflected from the stress-free end must increase, then the approximate result obtained by the rigid unloading analysis is closer to the elastic–plastic wave solution.

CHAPTER 6

One-Dimensional Visco-Elastic Waves and Elastic-Visco-Plastic Waves

The results of elastic–plastic wave propagation discussed so far are all based on the basic assumption that **stress is only dependent on strain and not on strain rate**. This kind of stress wave theory is called **rate-independent theory**. However, the mechanical behavior of real material is, more or less, rate dependent or time dependent. For example, the strain increase with increasing time under constant stress (the phenomenon called **creep**), the stress decrease with increasing time under constant strain (the phenomenon called **stress relaxation**), the **hysteresis loop** phenomenon in a loading–unloading or stress cyclic process, the **absorption** and **dispersion** phenomenons in stress wave propagation, and the **strengthening** and **embrittlement** of materials under high strain rates, etc., all show that the constitutive response of real material is essentially time dependent or rate dependent. In such a situation, to describe the rate-dependent constitutive relationship of material, a curve on the (σ, ε) plane, $f(\sigma, \varepsilon) = 0$, is no longer appropriate; at least a curved surface in the $(\sigma, \varepsilon, \dot{\varepsilon})$ space, $f(\sigma, \varepsilon, \dot{\varepsilon}) = 0$, should be used. The stress wave theory based on such rate-dependent constitutive relationship is called **rate-dependent theory** or **strain-rate-dependent theory**.

The mechanical response of rate-dependent materials in general can be separated into two parts: the time-independent **instantaneous response** and the time-dependent **non-instantaneous response**. Macroscopically, all kinds of noninstantaneous responses may be attributed to or equivalent to some kinds of viscous responses induced by the so-called internal dissipative force. Thus, based on the elastic response to take such viscous effects into account, the **visco-elastic wave theory** has been correspondingly developed, and has been extensively applied to study polymer materials. In wave propagation, the viscous effect mainly results in the dispersion phenomenon (i.e. the wave velocity depends on the wave frequency) and the absorption phenomenon (i.e. the wave amplitude attenuates with the propagating distance). Moreover, based on the elastic–plastic response to take the viscous effects into account for both the elastic part and the plastic part, the corresponding **visco-elastic-plastic wave theory** has been developed. However, if the viscous effect is taken into account only for the plastic part, then correspondingly the **elastic-visco-plastic**

wave theory has been developed, which has become one of the most important rate-dependent plastic wave theories since the Соколовский–Malvern model was proposed around the 1950s.

In the following, we will mainly discuss the one-dimensional longitudinal visco-elastic waves and elastic-visco-plastic waves.

6.1 Linear Visco-Elastic Constitutive Relationship

The development of visco-elastic theory is closely related to the studies on the mechanical behavior of high-molecular polymers. It is well known that even in a usual range of temperature or time change, polymers display a markedly viscous behavior, showing almost all the behavior patterns of the typical viscous fluid to the typical elastic solid.

The simplest constitutive relationship for describing the elastic solid is the linear elastic stress–strain law (Hooke's law), while the simplest constitutive relationship for describing the viscous fluid is the linear viscous law between the stress and the velocity gradient (Newton's law), of which the one-dimensional forms are expressed, respectively, as

$$\sigma = E\varepsilon \tag{6.1}$$

$$\sigma = \eta \frac{\partial v}{\partial x} \tag{6.2}$$

where η is the viscous coefficient, and others have been defined previously. Under the condition of **small deformation**, the difference between Lagrange variables and Euler variables can be neglected; then, the velocity gradient and the strain rate are the same:

$$\frac{\partial v}{\partial x} \approx \frac{\partial v}{\partial X} = \frac{\partial^2 u}{\partial X \partial t} = \frac{\partial \varepsilon}{\partial t} = \dot{\varepsilon}$$

so Eq. (6.2) can be rewritten as:

$$\sigma = \eta \dot{\varepsilon} \tag{6.3}$$

It means that the relationship between the stress and strain rate is linear.

In the scope of linear constitutive relationship, a medium having both elastic and viscous character can be imaged as a linear combination of linear elastic solids and linear viscous fluids, called **linear visco-elastic body**, of which the constitutive relationship is a kind of linear combination of Eqs. (6.1) and (6.3). If the linear elastic body is represented by a spring, as shown in Fig. 6.1(a), and the linear viscous fluid by a dashpot, as shown in Fig. 6.1(b), then the simplest form of linear combination is either a body connected to a spring with a dashpot in series, called **Maxwell body**, as shown in Fig. 6.2, or a body connected to a spring with a dashpot in parallel, called **Kelvin–Voigt body**, as shown in Fig. 6.3.

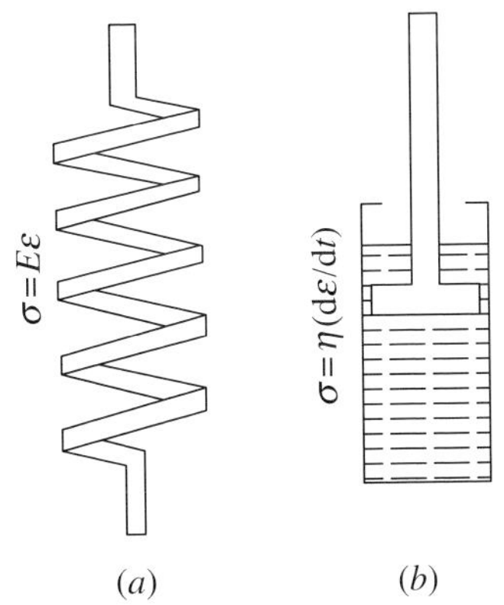

Fig. 6.1. (a) Spring model and (b) Dash-pot model.

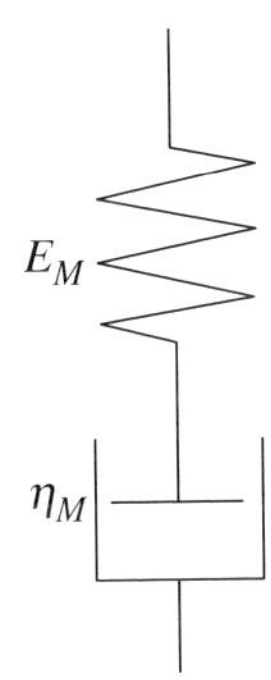

Fig. 6.2. Maxwell model.

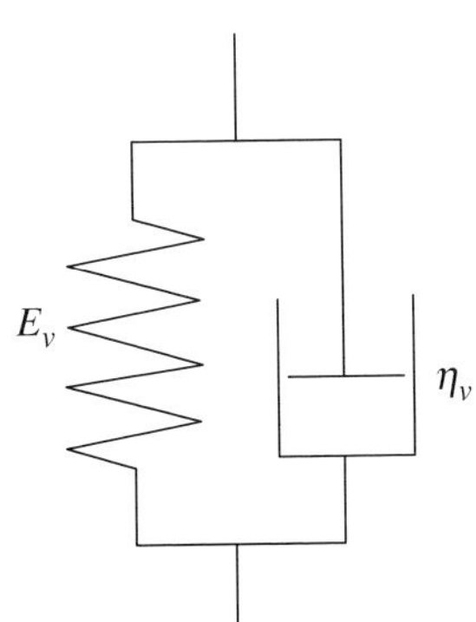

Fig. 6.3. Kelvin–Voigt model.

6.1.1 Maxwell model

For the Maxwell model (Fig. 6.2), the stress carried by the spring equals that carried by the dashpot:

$$\sigma = \sigma_E = \sigma_\eta$$

while the total strain is the sum of that carried by the spring and the dashpot, respectively:

$$\varepsilon = \varepsilon_E + \varepsilon_\eta$$

Consequently, the total strain rate is the sum of both parts too:

$$\dot{\varepsilon} = \dot{\varepsilon}_E + \dot{\varepsilon}_\eta$$

Substituting Eqs. (6.1) and (6.3) into the above equation, we obtain the following constitutive equation for Maxwell model:

$$\dot{\varepsilon} = \frac{\dot{\sigma}}{E_M} + \frac{\sigma}{\eta_M} \tag{6.4}$$

Thus, under the constant strain condition ($\dot{\varepsilon} = 0$), stress will attenuate with time from the initial stress σ_0 according to the following exponential function:

$$\sigma(t) = \sigma_0 \exp\left(-\frac{E_M t}{\eta_M}\right) = \sigma_0 \exp\left(-\frac{t}{\theta_M}\right) \tag{6.5}$$

which approximately describes the stress relaxation behavior of materials, as shown in Fig. 6.4, where $\theta_M = \eta_M / E_M$ represents the time required to let the stress relax from σ_0 to its 1/e (36.79%), called the **relaxation time**. It is an important material parameter used to characterize visco-elastic materials.

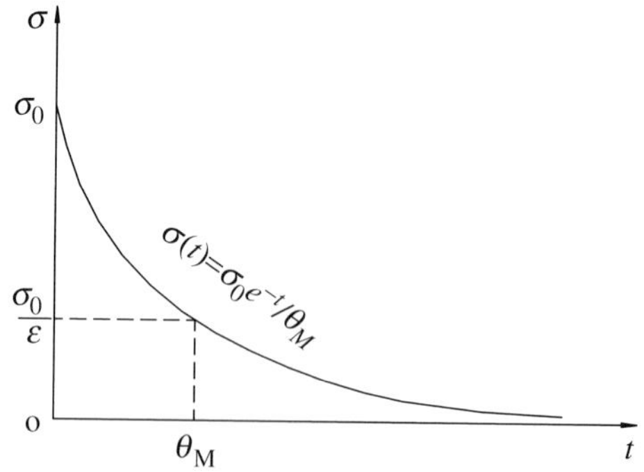

Fig. 6.4. Stress relaxation under constant strain ($\dot{\varepsilon} = 0$).

A **relaxation modulus** is defined by

$$E_r(t) = \frac{\sigma(t)}{\varepsilon} \tag{6.6}$$

Equation (6.5) can be rewritten as:

$$E_r(t) = \frac{\sigma_0}{\varepsilon}\exp\left(-\frac{t}{\theta_M}\right) = E_M\exp\left(-\frac{t}{\theta_M}\right) \tag{6.7}$$

Therefore, the relaxation character is independent of the value of applied constant strain.

Of course, the Maxwell model is a simplified ideal model. In fact, to describe the practical stress relaxation behavior of most true materials as close to the experimental results as possible, a simplified model containing only one exponential attenuation term is insufficient, and usually a sum of a series of exponential terms with different relaxation times should be used. In other words, not a single relaxation time but a series of relaxation times θ_i $(i = 1, 2, 3, \ldots n)$ should be used to describe the true visco-elastic behavior of materials. Therefore, under the condition of constant strain, the relation of how stress relaxes with time should be expressed as:

$$\sigma(t) = \varepsilon\sum_{i=1}^{n} E_i\exp\left(-\frac{t}{\theta_i}\right) \tag{6.8a}$$

or, expressed by relaxation modulus, we have

$$E_r(t) = \sum_{i=1}^{n} E_i\exp\left(-\frac{t}{\theta_i}\right) \tag{6.8b}$$

This is equivalent to a visco-elastic body that is combined by a series of Maxwell elements connected in parallel, each with different E_i and θ_i (or η_i), called **generalized Maxwell body**, as shown in Fig. 6.5. The E_i in the equation can be regarded as the **contribution** of the ith Maxwell element with relaxation time θ_i to the total stress relaxation, the **weight**. When $n \to \infty$, the series of discrete relaxation time approaches continuous distribution, and the E_i is replaced by the weight function $E(\theta)\,d\theta$. Thus, the above equation in the sum form transfers into the following equation in the integral form:

$$E_r(t) = \int_0^{\infty} E(\theta)\exp\left(-\frac{t}{\theta}\right)d\theta \tag{6.9}$$

$E(\theta)$ is called **relaxation-time spectrum**, or **relaxation spectrum** for short. Sometimes, the distribution of relaxation may be very wide, and it will be more convenient to use $\ln(\theta)$ as an independent variable to investigate the relaxation spectrum. For this reason, introduce

$$H(\ln\theta)\,d\ln\theta = E(\theta)\,d\theta$$

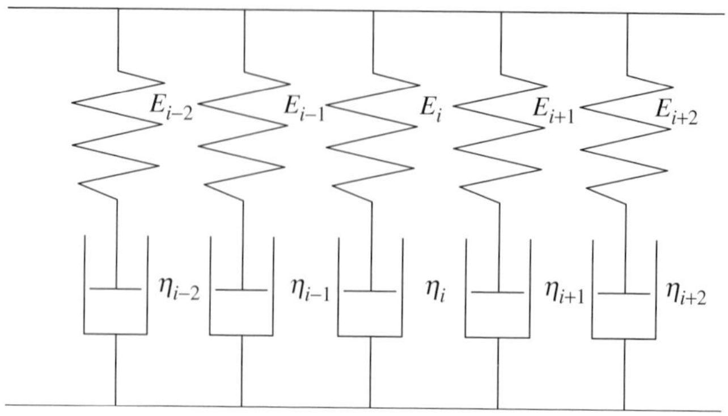

Fig. 6.5. Generalized Maxwell model.

Equation (6.9) is then rewritten as:

$$E_r(t) = \int_{-\infty}^{\infty} H(\ln\theta)\exp\left(-\frac{t}{\theta}\right)d\ln\theta = \int_{0}^{\infty} H(\theta)\frac{1}{\theta}\exp\left(-\frac{t}{\theta}\right)d\theta \tag{6.10}$$

where $H(\ln\theta)$ is the relaxation spectrum with respect to $\ln\theta$.

6.1.2 Kelvin–Voigt model

For the Kelvin–Voigt model (Fig. 6.3), strain experienced by the spring equals strain experienced by the dashpot:

$$\varepsilon = \varepsilon_E = \varepsilon_\eta$$

while the total stress is the sum of both parts:

$$\sigma = \sigma_E + \sigma_\eta$$

Substituting Eqs. (6.1) and (6.2) into the above equation, we obtain the following constitutive relationship for Voigt model:

$$\sigma = E_v\varepsilon + \eta_v\dot{\varepsilon} = \left(E_v + \eta_v\frac{\partial}{\partial t}\right)\varepsilon \tag{6.11}$$

which can be formally obtained from Hooke's law by using the operator $\left(E_v + \eta_v\frac{\partial}{\partial t}\right)$ instead of the elastic modulus E.

Thus, under the condition of constant stress σ, assuming a zero initial strain, the strain increases with time according to the following rule:

$$\left.\begin{aligned} \varepsilon(t) &= \frac{\sigma}{E_v}\left[1-\exp\left(-\frac{t}{\tau_v}\right)\right] \\ \tau_v &= \frac{\eta_v}{E_v}, \end{aligned}\right\} \tag{6.12}$$

which approximately describes the creep behavior of materials, as shown in Fig. 6.6, where $\tau_v = \eta_v/E_v$. If the stress is suddenly unloaded from σ to zero at $\varepsilon = \varepsilon_0$, Eq. (6.11) shows that the strain displays a delayed recovery, according to the following rule:

$$\varepsilon(t) = \varepsilon_0 \exp\left(-\frac{t}{\tau_v}\right) \tag{6.13}$$

at $t = \tau_v$, the strain is recovered to 1/e (36.79%) of its initial value, where τ_v is called **retardation time**. It is another important material parameter used to characterize visco-elastic materials.

Similar to the definition of relaxation modulus, **creep compliance** is defined by:

$$J_c(t) = \frac{\varepsilon(t)}{\sigma} \tag{6.14}$$

then Eq. (6.12) is reduced to:

$$J_c(t) = \frac{1}{E_v}\left[1-\exp\left(-\frac{t}{\tau_v}\right)\right] = J_v\left[1-\exp\left(-\frac{t}{\tau_v}\right)\right] \tag{6.15}$$

where $J_v = 1/E_v$ is the compliance of spring element in Voigt model.

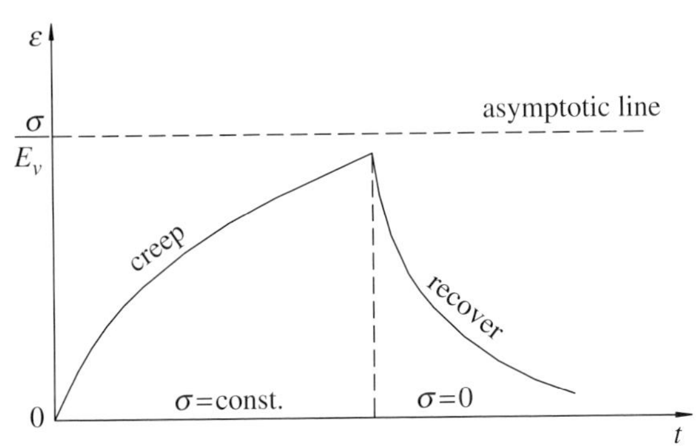

Fig. 6.6. Creep under constant stress and a delayed recover of strain when stress is suddenly unloaded from σ to zero.

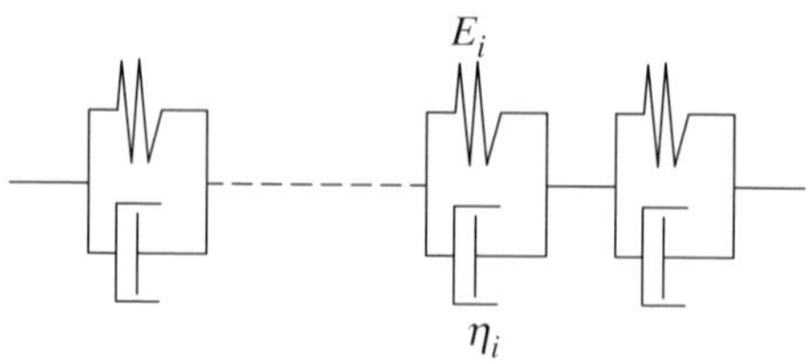

Fig. 6.7. Generalized Kelvin–Voigt model.

Similar to what we have discussed on Maxwell body, a Kelvin–Voigt body with only a single retardation time is insufficient to precisely describe the creep behavior of true materials, so we introduce the **generalized Kelvin–Voigt body** combined by a series of Kelvin–Voigt elements connected in series, each with different retardation time, as shown in Fig. 6.7. Then, similar to Eqs. (6.9) and (6.10), we have

$$J_c(t) = \int_0^{\infty} J(\tau)\left[1 - \exp\left(-\frac{t}{\tau}\right)\right] d\tau = \int_{-\infty}^{\infty} L(\ln\tau)\left[1 - \exp\left(-\frac{t}{\tau}\right)\right] d\ln\tau \quad (6.16)$$

$J(t)$ and $L(\ln\tau)$ are the **retardation time spectrum** as functions of τ and $\ln\tau$, respectively.

6.1.3 Standard linear solid model

This points out that, although Maxwell body can approximately describe stress relaxation, the strain rate is constant under the constant stress condition, according to Eq. (6.4). Therefore, this, in fact, just reflects Newton viscous flow and is insufficient to describe the more complicated true creep phenomenon. Moreover, when it is used to describe stress relaxation, according to Eq. (6.5), after a sufficiently long time period, the stress will relax to zero, while in practice the relaxation stress of true materials usually approaches a finite value under the constant strain condition.

On the other hand, although the Kelvin–Voigt body can approximately describe the creep phenomenon, the stress, according to Eq. (6.11), is constant under constant strain condition and is unable to reflect the stress relaxation behavior.

In order to more satisfactorily describe stress relaxation as well as the creep, more complicated models, combined with springs and dashpots, are obviously required. Among these, the three-element model is attractive and not too complicated, which may be either combined with a Maxwell body connected to a spring E_a in parallel [Fig. 6.8(a)], or combined with a Kelvin–Voigt body connected to a spring E'_a in series [Fig. 6.8(b)]. The constitutive equations for them are, respectively, expressed as:

$$E_M\sigma + \eta_M\dot{\sigma} = E_M E_a\varepsilon + (E_M + E_a)\eta_M\dot{\varepsilon} \quad (6.17)$$

$$\left(E'_a + E_a\right)\sigma + \eta_v\dot{\sigma} = E'_a E_a\varepsilon + E'_a\eta_v\dot{\varepsilon} \quad (6.18)$$

Obviously, if $E'_a = E_a + E_M$, $E_v = E_a(1 + (E_a/E_M))$, and $\eta_v E_M E_a = \eta_M E_v E'_a$, then these two equations, or rather these two models, are equivalent to each other. In fact,

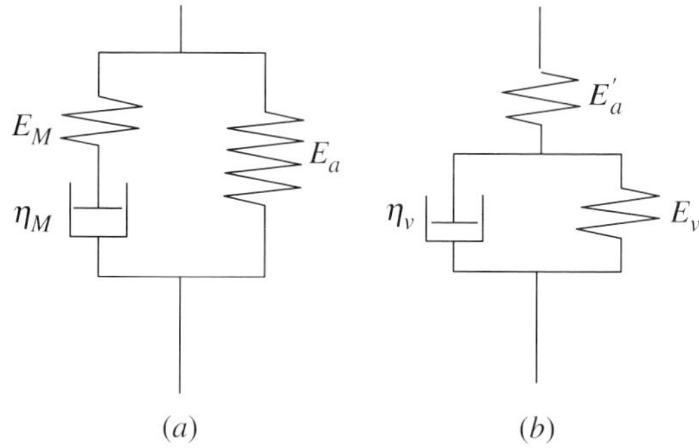

Fig. 6.8. Standard linear solid model.

both of them can be regarded as a linear combination of Eqs. (6.4) and (6.11):

$$a_0\sigma + a_1\dot{\sigma} = b_0\varepsilon + b_1\dot{\varepsilon}$$

This model is sometimes called a **standard linear solid model**. Of course, in order to more precisely describe the general behavior of true visco-elastic bodies, usually a more general linear visco-elastic model is required, of which the constitutive equation can be written as:

$$\sum_{i=0}^{n} a_i \frac{\partial^i \sigma}{\partial t^i} = \sum_{i=0}^{n} b_i \frac{\partial^i \varepsilon}{\partial t^i}$$

where a_i and b_i $(i = 1, 2, 3, \ldots n)$ are all material parameters.

6.2 Stress Waves Propagating in Linear Visco-Elastic Bars

Let us now discuss the stress waves propagating in linear visco-elastic bodies, and emphatically investigate how the viscous terms in constitutive equations influence stress wave propagation. For convenience sake, in the following, we will mainly discuss the longitudinal waves propagating in bars made of Maxwell body, Kelvin–Voigt body, and the standard linear solid body.

6.2.1 Longitudinal visco-elastic waves propagating in Kelvin–Voigt bars

The control equations for the longitudinal visco-elastic waves propagating in Kelvin–Voigt bars consist of Eqs. (2.12), (2.13), and (6.11), namely:

$$\left.\begin{aligned} \frac{\partial v}{\partial X} &= \frac{\partial \varepsilon}{\partial t} \\ \rho_0 \frac{\partial v}{\partial t} &= \frac{\partial \sigma}{\partial X} \\ \sigma &= E_v \varepsilon + \eta_v \frac{\partial \varepsilon}{\partial t} \end{aligned}\right\} \tag{6.19}$$

If the displacement u is taken as the unknown variable, then equivalently we have

$$\rho_0 \frac{\partial^2 u}{\partial t^2} = E_v \frac{\partial^2 u}{\partial X^2} + \eta_v \frac{\partial^3 u}{\partial X^2 \partial t} \tag{6.20a}$$

which, by introducing $C_v^2 = E_v/\rho_0$, can be rewritten as:

$$\frac{\partial^2 u}{\partial t^2} - C_v^2 \frac{\partial^2 u}{\partial X^2} - C_v^2 \tau_v \frac{\partial^3 u}{\partial X^2 \partial t} = 0 \tag{6.20b}$$

This is a third-order partial differential equation with respect to u. When the retardation time τ_v is so small that the third viscous term could be neglected, the Eq. (6.20b) is reduced to the second-order partial differential equation.

Because of the existence of the viscous term, Eq. (6.20b) is no longer satisfied by the elastic solution $u = F(X \pm C_0 t)$. Let us try the following harmonic wave solution:

$$u(X, t) = A \cdot \exp[\mathrm{i}(\omega t - k_1 X)]$$

where A is the wave amplitude, $\omega(=2\pi f)$ the radial frequency (f the frequency), and k_1 the wave number. If the above solution could satisfy Eq. (6.20), then we must have

$$\rho_0 \omega^2 = E_v k_1^2 + \mathrm{i}\eta_v k_1^2 \omega \tag{6.21}$$

This is an expression in complex form. Since it has been assumed that the radial frequency ω is a real number, then k_1 must be a complex number, say

$$k_1 = k + \mathrm{i}\alpha$$

Substituting it into Eq. (6.21), let the real part and the imaginary part on the left side of the equation equal those on the right side, respectively. We have

$$\left.\begin{aligned} \rho_0 \omega^2 &= E_v (k^2 - \alpha^2) - 2\eta_v \omega \alpha k \\ 2E_v \alpha k &= -\eta_v \omega (k^2 - \alpha^2) \end{aligned}\right\} \tag{6.22}$$

Thus, the solutions are obtained as follows:

$$k^2 = \frac{\rho_0 E_v \omega^2}{2(E_v^2 + \eta_v^2 \omega^2)} \left[\sqrt{1 + \frac{\eta_v^2 \omega^2}{E_v^2}} + 1\right] = \frac{\omega^2}{2C_v^2(1 + \omega^2 \tau_v^2)} \left[\sqrt{1 + \omega^2 \tau_v^2} + 1\right] \tag{6.23a}$$

$$\alpha^2 = \frac{\rho_0 E_v \omega^2}{2(E_v^2 + \eta_v^2 \omega^2)} \left[\sqrt{1 + \frac{\eta_v^2 \omega^2}{E_v^2}} - 1\right] = \frac{\omega^2}{2C_v^2(1 + \omega^2 \tau_v^2)} \left[\sqrt{1 + \omega^2 \tau_v^2} - 1\right] \tag{6.23b}$$

where α could be positive or negative. However, the positive root implies that the wave amplitude increases with X, with no physical meaning (violating the energy conservation);

therefore, α must be negative and for convenience could be rewritten as $(-\alpha)$. Thus, the solution of Eq. (6.20) is

$$u(X,t) = A\exp(-\alpha X)\exp[\mathrm{i}(\omega t - kX)] \tag{6.24}$$

It is thus clear that the α describes the wave amplitude attenuation exponential with increasing propagating distance. This phenomenon is called **absorption**, and α is called the **attenuation factor**. Equation (6.23b) shows that α varies with ω. For the low-frequency waves, of which the radial frequency $\omega \ll 1/\tau_v$ (namely, the period $T = 2\pi/\omega$ is much larger than the retardation time τ_v), α is proportional to ω^2.

Equation (6.23a) shows that the phase velocity of disturbance propagation, $C(=\omega/k)$, depends on the radial frequency ω (or the period $T = 1/f = \omega/(2\pi)$):

$$C^2 = \frac{2C_v^2(1+\omega^2\tau_v^2)}{1+\sqrt{1+\omega^2\tau_v^2}} = \frac{2C_v^2\left[1+\left(\frac{2\pi\tau_v}{T}\right)^2\right]}{1+\sqrt{1+\left(\frac{2\pi\tau_v}{T}\right)^2}} \tag{6.25}$$

Thus, the profile of wave discussed must distort in the wave propagation process. This phenomenon is called **dispersion** (**chromatic dispersion**). For the low-frequency waves satisfying $\omega \ll 1/\tau_v$, the phase velocity equals the longitudinal wave velocity in elastic bars, $C = C_0$, while the phase velocity increases with increasing frequency.

6.2.2 Longitudinal visco-elastic waves propagating in Maxwell bars

The longitudinal waves in Maxwell bars can be discussed in a similar way, for which the governing equations consist of Eqs. (2.12), (3.13), and (6.4), namely:

$$\left.\begin{aligned} \frac{\partial v}{\partial X} &= \frac{\partial \varepsilon}{\partial t} \\ \rho_0\frac{\partial v}{\partial t} &= \frac{\partial \sigma}{\partial X} \\ \frac{\partial \varepsilon}{\partial t} &= \frac{1}{E_M}\frac{\partial \sigma}{\partial t} + \frac{\sigma}{\eta_M} \end{aligned}\right\} \tag{6.26}$$

or, in terms of the displacement u as unknown variable, equivalently we have

$$\rho_0\frac{\partial^3 u}{\partial t^3} - E_M\frac{\partial^3 u}{\partial X^2 \partial t} + \frac{\rho_0 E_M}{\eta_M}\frac{\partial^2 u}{\partial t^2} = 0 \tag{6.27a}$$

Introducing $C_M^2 = E_M/\rho_0$, the above equation also can be rewritten as:

$$\frac{\partial^2 u}{\partial t^2} - C_M^2\frac{\partial^3 u}{\partial X^2 \partial t} + \frac{1}{\theta_M}\frac{\partial^2 u}{\partial t^2} = 0 \tag{6.27b}$$

It is easy to demonstrate that the above equation also is satisfied by the harmonic solution [Eq. (6.24)], although the wave number k and the attenuation factor α should be determined by the following equations, respectively:

$$k^2 = \frac{\omega^2}{2C_M^2}\left[\sqrt{1+\frac{1}{\omega^2\theta_M^2}}+1\right] \tag{6.28a}$$

$$\alpha^2 = \frac{\omega^2}{2C_M^2}\left[\sqrt{1+\frac{1}{\omega^2\theta_M^2}}-1\right] \tag{6.28b}$$

Thus, similar to what we have discussed on Kelvin–Voigt bars, the waves in Maxwell bars also display the dispersion and absorption phenomena. However, for the high-frequency waves, when $\omega \gg 1/\theta_M$, the phase velocity equals the longitudinal wave velocity in elastic bars, and, in such cases, the attenuation factor α is independent of the frequency:

$$\alpha = \frac{1}{2C_M\theta_M} = \frac{\rho_0 C_M}{2\eta_M} \tag{6.29}$$

6.2.3 Longitudinal visco-elastic waves propagating in standard linear solid bars

For the standard linear solid bar shown in Fig. 6.8(a), similarly, from Eqs. (2.12), (2.13), and (6.17), the governing equation in terms of u as an unknown variable can be deduced as:

$$\rho_0\frac{\partial^3 u}{\partial t^3}+\frac{\rho_0}{\theta_M}\frac{\partial^2 u}{\partial t^2} = (E_a+E_M)\frac{\partial^3 u}{\partial t\partial x^2}+\frac{E_a}{\theta_M}\frac{\partial^2 u}{\partial x^2} \tag{6.30}$$

where $\theta_M = \eta_M/E_M$. Moreover, it is easy to demonstrate that the above equation is also satisfied by the harmonic solution [Eq. (6.24)], although the wave number k and attenuation factor α should be determined by the following equations, respectively

$$k^2 = \frac{\rho_0\omega^2}{2E_a}\left\{\left[\frac{1+\omega^2\theta_M^2}{1+\left(1+\dfrac{E_M}{E_a}\right)^2\omega^2\theta_M^2}\right]^{1/2}+\frac{1+\left(1+\dfrac{E_M}{E_a}\right)\omega^2\theta_M^2}{1+\left(1+\dfrac{E_M}{E_a}\right)^2\omega^2\theta_M^2}\right\} \tag{6.31a}$$

$$\alpha^2 = \frac{\rho_0\omega^2}{2E_a}\left\{\left[\frac{1+\omega^2\theta_M^2}{1+\left(1+\dfrac{E_M}{E_a}\right)^2\omega^2\theta_M^2}\right]^{1/2}-\frac{1+\left(1+\dfrac{E_M}{E_a}\right)\omega^2\theta_M^2}{1+\left(1+\dfrac{E_M}{E_a}\right)^2\omega^2\theta_M^2}\right\} \tag{6.31b}$$

On the other hand, for the standard linear solid bar shown in Fig. 6.8(b), similarly, from Eqs. (2.12), (2.13), and (6.18), the governing equation in terms of u as unknown variable can be deduced as:

$$\rho_0\eta_v\frac{\partial^3 u}{\partial t^3}+\rho_0\left(E_a'+E_v\right)\frac{\partial^2 u}{\partial t^2}-\eta_v E_a'\frac{\partial^3 u}{\partial X^2\partial t}-E_v E_a'\frac{\partial^2 u}{\partial X^2}=0 \tag{6.32}$$

Moreover, it is easy to demonstrate that the above equation is also satisfied by the harmonic solution [Eq. (6.24)], although the wave number k and attenuation factor α should be determined by the following equations, respectively:

$$k^2=\frac{\rho_0\omega^2}{2E_cE_a'}\left[\sqrt{\frac{E_a'^2+E_c^2\omega^2\tau_v^2}{1+\omega^2\tau_v^2}}+\frac{E_a'+E_c\omega^2\tau_v^2}{1+\omega^2\tau_v^2}\right] \tag{6.33a}$$

$$\alpha^2=\frac{\rho_0\omega^2}{2E_cE_a'}\left[\sqrt{\frac{E_a'^2+E_c^2\omega^2\tau_v^2}{1+\omega^2\tau_v^2}}-\frac{E_a'+E_\alpha\omega^2\tau_v^2}{1+\omega^2\tau_v^2}\right] \tag{6.33b}$$

where E_c is the elastic modulus of the springs E_v and E_a' connected in series:

$$\frac{1}{E_c}=\frac{1}{E_a'}+\frac{1}{E_v}$$

According to either Eq. (6.31a) or Eq. (3.33a), the phase velocity of disturbance propagation, $C(=\omega/\mathrm{k})$, is dependent on the frequency, which describes the dispersion (chromatic dispersion) phenomenon. On the other hand, according to Eqs. (6.31b) or (6.33b), the attenuation factor α is also dependent on the frequency ω, which describes the absorption phenomenon of visco-elastic waves in standard linear solid bars. In other words, the visco-elastic waves propagating in the standard linear solid bars must distort according to the rules expressed by Eqs. (6.31) and (6.33).

For the high-strain-rate loading such as in the cases of explosion and impact, more attention is paid to the high-frequency waves. In the following, especially such cases are analyzed in a little detail. For example, for the standard linear solid body shown in Fig. 6.8(a), in the case of high-frequency waves, because of $\omega^2 \gg 1/\theta_M^2$, Eq. (6.31a) is reduced to:

$$k^2\approx\frac{\rho_0\omega^2}{E_a+E_M}$$

Consequently, the corresponding phase velocity is expressed as:

$$C^2=\frac{\omega^2}{k^2}=\frac{E_a+E_M}{\rho_0}=C_V^2 \tag{6.34}$$

which shows that the phase velocity for high-frequency waves is independent of the frequency. In other words, it means that for the standard linear solid shown in Fig. 6.8(a), in the case of high frequency, the dashpot in Maxwell element does not take effect, and then the standard linear solid body is reduced to an elastic body that consists of a spring E_a

connected with a spring E_M in parallel. Therefore, the wave velocity is just equal to the elastic wave velocity C_V in such an elastic body, as expressed by Eq. (6.34).

To discuss the attenuation factor α in the case of high frequency, if we directly start from Eq. (6.31b) and eliminate the term $(\omega^2\theta_M^2)$ in both the numerator and the denominator in the right side of the equation, then an incorrect conclusion ($\alpha C_0/\omega = 0$ in the case of high frequency) will be reached[1]. In order to obtain a correct conclusion, Eq. (6.31b) is first rewritten in the following form (Wang *et al.*, 1994):

$$\alpha^2 = \frac{\rho_0\omega^2}{2E_a\left(1+\dfrac{E_M}{E_a}\right)}\left\{\left[\frac{1+\dfrac{1}{\omega^2\theta_M^2}}{1+\dfrac{1}{\left(1+\dfrac{E_M}{E_a}\right)^2\omega^2\theta_M^2}}\right]^{1/2} - \frac{1+\dfrac{1}{\left(1+\dfrac{E_M}{E_a}\right)\omega^2\theta_M^2}}{1+\dfrac{1}{\left(1+\dfrac{E_M}{E_a}\right)^2\omega^2\theta_M^2}}\right\} \tag{6.35}$$

For high-frequency waves, $1/\omega^2\theta_M^2 \ll 1$, the following terms in Eq. (6.35) can be expanded into a series by neglecting the small quantities of higher orders:

$$\left(1+\frac{1}{\omega^2\theta_M^2}\right)^{1/2} \approx 1+\frac{1}{2}\left(\frac{1}{\omega^2\theta_M^2}\right)$$

$$\left[1+\frac{1}{(E_M/E_a)^2\omega^2\theta_M^2}\right]^{1/2} \approx 1-\frac{1}{2}\left[\frac{1}{(1+(E_M/E_a))^2\,\omega^2\theta_M^2}\right]$$

$$\frac{1}{1+\dfrac{1}{(1+(E_M/E_a))^2\omega^2\theta_2^2}} \approx 1-\frac{1}{(1+(E_M/E_a))^2\omega^2\theta_M^2}$$

We then obtain the attenuation factor for the high-frequency waves in standard linear solid (denoted as α_{LS} in the following):

$$\alpha_{LS}^2 = \frac{\rho_0\omega^2}{2(E_a+E_M)}\left\{\frac{1}{2}\left(\frac{1}{\omega^2\theta_M^2}\right)-\frac{1}{(1+(E_M/E_a))\,\omega^2\theta_M^2}+\frac{1}{2}\left[\frac{1}{(1+(E_M/E_a))^2\,\omega^2\theta_M^2}\right]\right\}$$
$$= \frac{\rho_0}{4\theta_M^2(E_a+E_M)}\left[1-\frac{1}{(1+(E_M/E_a))}\right]^2 = \frac{\rho_0 E_M^2}{4\theta_M^2(E_a+E_M)^3} \tag{6.36a}$$

This means that the attenuation factor for high-frequency waves is neither dependent on frequency nor equal to zero. Convenient for comparing with the attenuation factor

[1]See Hillier (1949), or as cited by Kolsky (1953) and Graff (1975).

for the high-frequency wave in Maxwell body (denoted as α_M in the following), the α in Eq. (6.29), Eq. (6.36a) can be rewritten in the following several different forms by noticing $\theta_M = \eta_M/E_M$, $C_M^2 = E_M/\rho_0$, and $C_V^2 = (E_a + E_M)/\rho_0$ (Eq. 6.34):

$$\alpha_{LS} = \frac{E_M}{2\theta_M C_v(E_a + E_M)} = \frac{\rho_0 C_v}{2\eta_M\left(1+\dfrac{E_a}{E_M}\right)^2} = \frac{\rho_0 C_M}{2\eta_M\left(1+\dfrac{E_a}{E_M}\right)^{3/2}} = \frac{\alpha_M}{\left(1+\dfrac{E_a}{E_M}\right)^{3/2}} \tag{6.36b}$$

Since $E_a/E_M > 0$, it follows that $\alpha_{LS} < \alpha_M$. It means that the high-frequency wave attenuation in the standard linear solid body is slower than that in the corresponding Maxwell body.

It should distinguish the dispersion due to the transverse inertia in bars, as discussed in Section 2.8, from the dispersion due to the viscous effect, as discussed in the present section. The former results from the bar's geometrical factor and is sometimes called **geometrical dispersion**, while the latter results from the viscosity of material and is sometimes called **constitutive viscous dispersion**. Note that the frequency effect in these two kinds of dispersion is expressed in two opposite directions. For visco-elastic waves, high-frequency wave velocity is faster than low-frequency wave velocity, while for geometrical dispersion, the situation is contrary (see Eq. 2.68).

6.2.4 Solutions of linear visco-elastic longitudinal waves propagating in bars by the characteristics method

In the following, we discuss the **characteristics method** used to solve the **linear visco-elastic longitudinal waves** propagating in linear visco-elastic bars. It can be regarded as a development of the characteristics method that we have discussed for rate-independent elastic and elastic–plastic waves; and, on the other hand, it can be regarded as a starting step for the further studies on the rate-dependent elastic-visco-plastic waves in Section 6.6.

For convenience, we take the bar made of Maxwell body as a typical example.

Multiplying each equation of the governing equations [Eq. (6.26)], respectively, by the indeterminate coefficients L, M, and N, and then adding them, the characteristic equations and the corresponding characteristic compatibility relations are obtained (herein the subscript M for the Maxwell model is temporarily neglected):

$$(L+N)\frac{\partial \varepsilon}{\partial t} + \left(M\rho_0\frac{\partial}{\partial t} - L\frac{\partial}{\partial X}\right)v + \left(-\frac{N}{E}\frac{\partial}{\partial t} - M\frac{\partial}{\partial X}\right)\sigma - N\frac{\sigma}{\eta} = 0 \tag{6.37}$$

For the above equation to contain only the directional derivatives along the characteristics line, these indeterminate coefficients should satisfy the following conditions:

$$\frac{dX}{dt} = \frac{0}{L+N} = \frac{-L}{M\rho_0} = \frac{ME}{N} \tag{6.38}$$

of which the solutions are:

$$L + N = 0, \quad \rho_0 E M^2 = -LN \tag{6.39a}$$

$$L = M = 0, \quad N \neq 0 \tag{6.39b}$$

Substituting Eq. (6.39a) into Eqs. (6.38) and (6.37), we obtain two families of real characteristic line and the corresponding characteristic compatibility relationships:

$$dX = \pm\sqrt{\frac{E}{\rho_0}}dt = \pm C_0 dt \tag{6.40a}$$

$$dv = \pm\frac{1}{\rho_0 C_0}d\sigma \pm \frac{C_0}{\eta}\sigma\, dt = \pm\frac{1}{\rho_0 C_0}d\sigma \pm \frac{C_0}{\eta}dX \tag{6.40b}$$

which represent the loci of wave fronts propagating with the wave velocity C_0 determined by the instantaneous response of material.

Substituting Eq. (6.39b) into Eqs. (6.38) and (6.37), we obtain the third set of real characteristic line and the corresponding characteristic compatibility relationship:

$$dX = 0 \tag{6.41a}$$

$$d\varepsilon - \frac{d\sigma}{E} - \frac{\sigma\, dt}{\eta} = 0 \tag{6.41b}$$

which represent the loci of particles movement, and the characteristic compatibility relation [Eq. (6.41b)] is actually another expression of the material derivative of the constitutive relation of material [Eq. (6.4)] along the particle motion locus.

Different from the situation for elastic waves, note that the terms having dt or dX appear in the compatibility conditions, Eqs. (6.40b) and (6.41b). They are the terms that describe the rate (time) dependence of visco-elastic wave propagation and reflect the dispersion and dissipation characters of visco-elastic waves. Although the calculation of compatibility relations along characteristics becomes more complicated due to the existence of these terms, the corresponding three families of characteristics are all in the form of straight lines, which are very convenient for numerical calculation.

So, we use the characteristics method to solve the problem of visco-elastic wave propagation. Through an arbitrary point N_i on the X–t plane, there are three characteristic lines as shown in Fig. 6.9 and three corresponding characteristic relations. If we write those relations in the difference form instead of the differential form, the three unknown variables σ, v, and ε at the point N_i can be determined according to the known initial and boundary conditions (by using the algebra equations). The specific procedures are similar to what discussed previously. We will illustrate the details by an example, the visco-elastic wave propagation in a semi-infinite Maxwell bar, as follows.

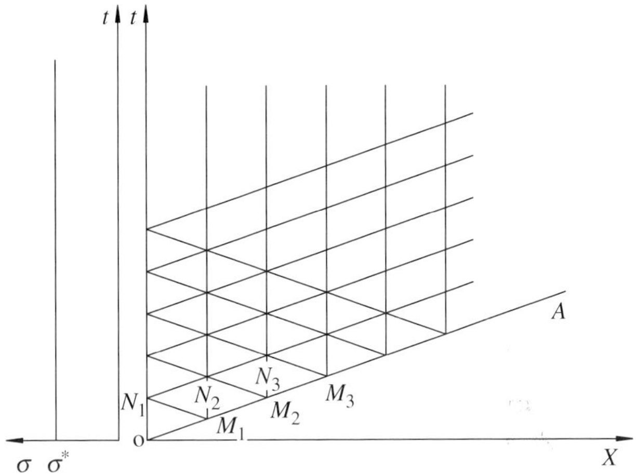

Fig. 6.9. Three families of characteristics for the linear visco-elastic longitudinal waves propagating in bars.

Consider a semi-infinite Maxwell bar, which is originally at rest and in a stress-free state ($v_0 = \sigma_0 = \varepsilon_0 = 0$), and suddenly suffers a constant stress σ^* at the bar end ($X = 0$). Then a strong discontinuous wave propagates with wave velocity $\mathscr{D}$ along the line *OA*, as shown in Fig. 6.9. According to the kinematic compatibility condition (Eq. 2.55) and the dynamic compatibility condition (Eq. 2.57) across a strong discontinuous interface deduced in Chapter 2, we have, respectively,

$$[v] = -\mathscr{D}[\varepsilon]$$

$$[\sigma] = -\rho_0\mathscr{D}[v]$$

Since both are independent of the material characters, they are also tenable for the strong discontinuous visco-elastic waves. Moreover, since $\dot{\varepsilon} \to \infty$ across a strong discontinuous interface, only the instantaneous response part of the constitutive relation takes effect. Thus, for Maxwell body, we have $[\sigma]/[\varepsilon] = E$, so the strong discontinuous wave velocity $\mathscr{D}$ in Maxwell bar is

$$\mathscr{D} = \sqrt{\frac{1}{\rho_0}\frac{[\sigma]}{[\varepsilon]}} = \sqrt{\frac{E}{\rho_0}} = C_0$$

which equals the elastic wave velocity in bars.

Thus, along the line *OA*, according to the dynamic compatibility condition and the kinematic compatibility condition, respectively, we have

$$v = -\frac{\sigma}{\rho_0 C_0} \tag{6.42a}$$

$$v = -C_0\varepsilon \tag{6.42b}$$

On the other hand, since the line *OA* is a characteristic line, according to the corresponding characteristic compatibility condition (Eq. 6.40b), we have

$$dv = \frac{d\sigma}{\rho_0 C_0} + \frac{\sigma}{\eta} dX$$

Substituting Eq. (6.42a) into the above equation, it is reduced to a first-order ordinary differential equation with respect to σ

$$\frac{d\sigma}{dX} + \frac{\rho_0 C_0}{2\eta}\sigma = 0$$

Determining the integration constant by using the boundary condition $\sigma|_{X=0} = \sigma^* = -\rho_0 C_0 v^*$, we obtain the solution along the line *OA*:

$$\sigma = \sigma^* \exp\left[-\frac{\rho_0 C_0}{2\eta} X\right] \tag{6.43a}$$

and from Eq. (6.42) we obtain

$$v = -\frac{\sigma^*}{\rho_0 C_0} \exp\left[-\frac{\rho_0 C_0}{2\eta} X\right] = v^* \exp\left[-\frac{\rho_0 C_0}{2\eta} X\right] \tag{6.43b}$$

$$\varepsilon = \frac{\sigma^*}{E} \exp\left[-\frac{\rho_0 C_0}{2\eta} X\right] \tag{6.43c}$$

It shows that the amplitude of a strong discontinuous wave propagating in Maxwell bar attenuates exponentially (an absorption phenomenon), and the corresponding attenuation factor α is

$$\alpha = \frac{\rho_0 C_0}{2\eta}$$

which coincides with the result for high-frequency waves expressed by Eq. (6.29).

Since the solutions along the line *OA* have been given by Eq. (6.43), the remaining solutions in the region *AOt* shown in Fig. 6.9 can be further determined. In fact, the σ along another characteristic line *Ot* are given by the boundary condition at the bar end, so the problem in the region *AOt* is attributed to solve a **characteristics boundary-value problem** for visco-elastic waves. However, it should be noticed that the region *AOt* is no longer a constant wave region as in the situation for elastic waves, but a region wherein a series of weak discontinuous visco-elastic waves propagate.

Let us now first discuss the solution at an arbitrary **boundary point** along *Ot*, e.g. the point N_1 in Fig. 6.9, where the strain $\varepsilon(N_1)$ and the particle velocity $v(N_1)$ are unknown, while the stress $\sigma(N_1)$ has been given by the boundary condition. From Eqs. (6.41a) and (6.40a), it is known that two characteristics ON_1 and M_1N_1 can always be found to link the unknown boundary point N_1 with two known points O and M_1. Then, we have two compatibility conditions Eqs. (6.41b) and (6.40b) along ON_1 and MN_1, which, in the form of finite differences, are

$$v(N_1) - v(M_1) = -\frac{1}{\rho_0 C_0}[\sigma(N_1) - \sigma(M_1)] + \frac{\sigma(M_1)}{\eta}[X(N_1) - X(M_1)] \tag{6.44a}$$

$$\varepsilon(N_1) - \varepsilon(0) - \frac{1}{E}[\sigma(N_1) - \sigma(0)] - \frac{\sigma(0)}{\eta}[t(N_1) - t(0)] = 0 \tag{6.44b}$$

by which the particle velocity $v(N_1)$ and strain $\varepsilon(N_1)$ at the point N_1 can be found.

Similarly, the unknown stress $\sigma(N_2)$, particle velocity $v(N_2)$, and strain $\varepsilon(N_2)$ at an arbitrary **interior point** N_2 can be solved by the following three characteristic compatibility conditions along the corresponding characteristic lines, N_1N_2, M_2N_2 and M_1N_2, respectively:

$$v(N_2) - v(N_1) = \frac{1}{\rho_0 C_0}[\sigma(N_2) - \sigma(N_1)] + \frac{\sigma(N_1)}{\eta}[X(N_2) - X(N_1)] \tag{6.45a}$$

$$v(N_2) - v(M_2) = -\frac{1}{\rho_0 C_0}[\sigma(N_2) - \sigma(M_2)] + \frac{\sigma(M_2)}{\eta}[X(N_2) - X(M_2)] \tag{6.45b}$$

$$\varepsilon(N_2) - \varepsilon(M_1) - \frac{1}{E}[\sigma(N_2) - \sigma(M_1)] - \frac{\sigma(M_1)}{\eta}[t(N_2) - t(M_1)] = 0 \tag{6.45c}$$

On the analogy of this, the solutions in the whole region AOt can be obtained, so that the problem of visco-elastic waves propagating in a semi-infinite bar is solvable.

6.2.5 Split Hopkinson visco-elastic bar

In the preceding subsection, it was shown that, on the prerequisite that both the visco-elastic constitutive equation of material and the initial boundary conditions are known, the problem of visco-elastic wave propagation is solvable by using the characteristics method. This kind of problem is the so-called "**direct problem**". In the current subsection, taking the "split Hopkinson visco-elastic bar" as an example, we will discuss how the boundary condition can be determined from the experimentally measured visco-elastic wave profiles when the visco-elastic constitutive equation of material and the initial condition are given. This kind of problem belongs to the category of the so-called "**first kind of inverse problem**". In Section 6.4, we will discuss the so-called "**second kind of inverse problem**", namely, how the visco-elastic constitutive equation of material can be determined from the experimentally measured visco-elastic wave profiles when the initial and boundary conditions are given.

The classical split Hopkinson pressure bar (SHPB) technique is based on the principle of one-dimensional linear elastic wave propagating in bars, as has been briefly discussed in Section 3.7. Traditionally, the pressure bars used are made of high-strength steel with a high elastic limit $\sigma_s > 1$ GPa and a high acoustic impedance $\rho_0 C_0$ around 40 MPa s/m (see Table 2.1), so that the specimen of test material that generally has lower strength and lower wave impedance will deform nonelastically, while the bars are still in the state of elastic deformation. However, if the wave impedance of specimen is as low as 0.1~1 MPa/m/s, much lower than that of bars, the signal of transmitted pulse will become too weak to be detected accurately. In such a case, it seems that a bar material with lower acoustic impedance, instead of high-strength steel, should be used. One preferable way is to use plastic (polymer) bars, such as a polymethyl methacrylate (PMMA) bar, but the **dispersion** and **dissipation** characters of visco-elastic wave propagation in polymer bars should be taken into account (Wang *et al.*, 1994).

Now, let us first analyze what the differences are between the split Hopkinson elastic bar and the split Hopkinson visco-elastic bar. In other words, what new problems will happen when the SHPB technique is generalized to use visco-elastic polymer bars.

As discussed in Section 3.7, the key of SHPB technique is how to determine the stress $\sigma(X_1, t)$ and the particle velocity $v(X_1, t)$ at the interface X_1 from the signals of incident wave $\sigma_i(X_{g1}, t)$ and the reflected wave signal $\sigma_r(X_{g1}, t)$ measured at X_{g1}, where the X_1 is the interface of the specimen with the input bar and the X_{g1} is the position of strain gauge G_1 on the input bar, as shown in Fig. 6.10; and also the stress $\sigma(X_2, t)$ and the particle velocity $v(X_2, t)$ at the interface X_2 of the specimen with the output bar from the transmitted wave signal $\sigma_t(X_{g2}, t)$ measured at X_{g2}, the position of strain gauge G_2 on the input bar. In fact, once $\sigma(X_1, t)$, $v(X_1, t)$, $\sigma(X_2, t)$, and $v(X_2, t)$ are obtained, the stress $\sigma_s(t)$, strain rate $\dot{\varepsilon}_s(t)$, and strain $\varepsilon_s(t)$ for the specimen can then be determined according to Eq. (3.16), respectively, as shown below again:

$$\sigma_s(t) = \frac{A}{2A_s}[\sigma(X_1, t) + \sigma(X_2, t)] = \frac{A}{2A_s}[\sigma_i(X_1, t) + \sigma_r(X_1, t) + \sigma_t(X_2, t)]$$

$$\dot{\varepsilon}_s(t) = \frac{v(X_2, t) - v(X_1, t)}{l_s} = \frac{v_t(X_2, t) - v_i(X_1, t) - v_r(X_1, t)}{l_s}$$

$$\varepsilon_s(t) = \int_0^t \dot{\varepsilon}_s(t)\, dt = \frac{1}{l_s}\int_0^t [v_t(X_2, t) - v_i(X_1, t) - v_r(X_1, t)]\, dt$$

Recall that in the case of elastic bars, according to the **linear propositional** relations between the stress, strain, and particle velocity for elastic waves, we have

$$\left.\begin{aligned}
\sigma_1 &= \sigma(X_1, t) = \sigma_i(X_1, t) + \sigma_r(X_1, t) = E[\varepsilon_i(X_1, t) + \varepsilon_r(X_1, t)] \\
\sigma_2 &= \sigma(X_2, t) = \sigma_t(X_2, t) = E\varepsilon_r(X_2, t) \\
v_1 &= v(X_1, t) = v_i(X_1, t) + v_r(X_1, t) = C_0[\varepsilon_i(X_1, t) - \varepsilon_r(X_1, t)] \\
v_2 &= v(X_2, t) = v_t(X_2, t) = C_0\varepsilon_t(X_2, t)
\end{aligned}\right\}$$

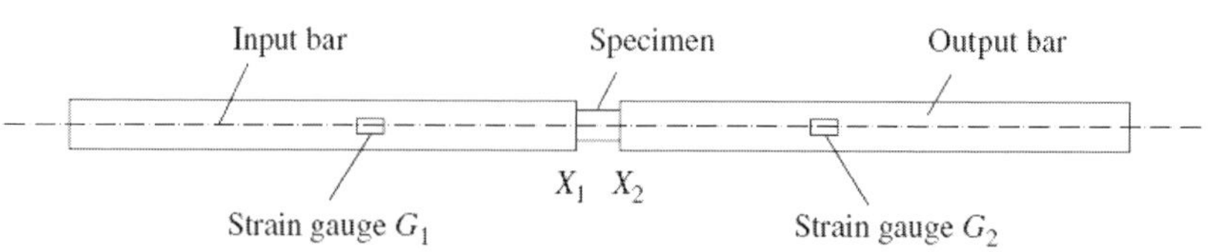

Fig. 6.10. Layout of input bar, specimen and output bar

then the dynamic stress $\sigma_s(t)$ and strain $\varepsilon_s(t)$ of the specimen can be determined only if the incident strain wave $\varepsilon_i(X_1, t)$ and the reflected strain wave $\varepsilon_r(X_1, t)$ at the interface X_1 as well as the transmitted strain wave $\varepsilon_t(X_2, t)$ at the interface X_2 are known. However, due to the **nondistortion character** of elastic wave propagating in bar, the incident strain wave $\varepsilon_i(X_1, t)$ and the reflected strain wave $\varepsilon_r(X_1, t)$ at the interface X_1 can be replaced by the incident wave signal $\varepsilon_i(X_{G1}, t)$, and the reflected wave signal $\varepsilon_r(X_{G1}, t)$ can be measured by the strain gauge G_1 at the position X_{G1} on the input bar. Also, the transmitted strain wave $\varepsilon_t(X_2, t)$ at the interface X_2 can be replaced by the transmitted wave signal $\varepsilon_t(X_{G2}, t)$ can be measured by the strain gauge G_2 at the position X_{G2} on the output bar. Thus, from the wave signals measured by the strain gauge G_1 at X_{G1} and the strain gauge G_2 at X_{G2}, respectively, the dynamic stress $\sigma_s(t)$ and strain $\varepsilon_s(t)$ for the specimen can be finally determined:

$$\sigma_s(t) = \frac{EA}{A_s}\varepsilon_t(X_{G2}, t) = \frac{EA}{A_s}[\varepsilon_i(X_{G1}, t) + \varepsilon_r(X_{G1}, t)]$$

$$\varepsilon_s(t) = -\frac{2C_0}{l_s}\int_0^t \varepsilon_r(X_{G1}, t)\,dt = \frac{2C_0}{l_s}\int_0^t [\varepsilon_i(X_{G1}, t) - \varepsilon_t(X_{G2}, t)]\,dt$$

However, in the case of visco-elastic bars, due to the **dispersion** and **attenuation** characters of visco-elastic waves, the linear propositional relations between the stress, strain, and particle velocity and, consequently, the nondistortion character no longer exist for visco-elastic waves. Thus, the keys of the current problem are attributed to: (1) how to determine the incident stress $\sigma_i(X_1, t)$ and incident particle velocity $v_i(X_1, t)$ at the interface X_1 from the incident strain wave signal $\varepsilon_i(X_{G1}, t)$ measured by the strain gauge G_1 at X_{G1}; (2) how to determine the reflected stress $\sigma_r(X_1, t)$ and reflected particle velocity $v_r(X_1, t)$ at the interface X_1 from the reflected strain wave signal $\varepsilon_r(X_{G1}, t)$ measured by the strain gauge G_1 at X_{G1}; (3) how to determine the transmitted stress $\sigma_t(X_1, t)$ and transmitted particle velocity $v_t(X_1, t)$ at the interface X_2 from the transmitted strain wave signal $\varepsilon_t(X_{G2}, t)$ measured by the strain gauge G_2 at X_{G2}; and then finally to determine the dynamic stress $\sigma_s(t)$ and strain $\varepsilon_s(t)$ for specimen according to Eq. (3.16). Among these, the first one is essentially attributed to solve the direct problem of visco-elastic wave propagation, similar to what we have discussed in Subsection 6.2.4; the other two are attributed to solve the so-called second kind of inverse problem of visco-elastic wave propagation, which will be discussed in detail in the following.

For convenience, let us discuss the SHPB bar made of PMMA as an example. Since deformation of the pressure bar itself is usually small, the problem can be simplified as a linear visco-elastic wave propagation analysis. In such a situation, the constitutive relationship of PMMA bar at high strain rates can be satisfactorily described by the following equation in the differential form (Wang *et al.*, 1994), which corresponds to the standard linear solid model, as shown in Fig. 6.8(a):

$$\frac{\partial \varepsilon}{\partial t} - \frac{1}{E_a + E_M}\frac{\partial \sigma}{\partial t} + \frac{E_a \varepsilon}{(E_a + E_M)\theta_M} - \frac{\sigma}{(E_a + E_M)\theta_M} = 0 \qquad (6.46)$$

which, together with the following movement equation and continuous equation,

$$\rho_0 \frac{\partial \upsilon}{\partial t} - \frac{\partial \sigma}{\partial X} = 0$$
$$\frac{\partial \varepsilon}{\partial t} - \frac{\partial \upsilon}{\partial X} = 0$$

constitute the governing equations of the current problem. Similar to what we discussed in Subsection 6.2.4 for the visco-elastic waves propagating in Maxwell bar by using the characteristics method, multiplying each of them by undetermined coefficients N, M, and L, respectively, and then adding together, we have

$$(L+N)\frac{\partial \varepsilon}{\partial t} + \left(M\rho_0 \frac{\partial}{\partial t} - L\frac{\partial}{\partial X}\right)v - \left(\frac{N}{E_a + E_M}\frac{\partial}{\partial t} + M\frac{\partial}{\partial X}\right)\sigma + \frac{N}{(E_a+E_M)\theta_M}(E_a\varepsilon - \sigma) = 0$$

In order to satisfy the requirement that the above equation consists of only the directional differential along characteristics $C(X,t)$, the undetermined coefficient, L, M, and N should satisfy the following relations:

$$\left.\frac{dX}{dt}\right|_c = \frac{0}{L+N} = -\frac{L}{M\rho_0} = \frac{M(E_a+E_M)}{N}$$

Obviously, two sets of solutions for L,M, and N exist. One is determined by the following equations

$$L + N = 0$$
$$\rho_0(E_a + E_M)M^2 = -LN \tag{6.47}$$

Then we obtain two families of characteristics:

$$\frac{dx}{dt} = \pm\sqrt{\frac{E_a + E_M}{\rho_0}} = C_v \tag{6.48a}$$

and, correspondingly, two sets of compatibility conditions along the characteristics

$$d\upsilon = \pm\frac{1}{\rho_0 C_v}d\sigma \pm \frac{\sigma - E_a\varepsilon}{\rho_0 C_v \theta_M}dt = \pm\frac{1}{\rho_0 C_v}d\sigma + \left[\frac{\sigma - E_a\varepsilon}{(E_a+E_M)\theta_M}\right]dX \tag{6.48b}$$

where the plus and minus signs are for rightward-propagating and leftward-propagating waves, respectively. Note that the propagation velocity along characteristics is the same as the phase velocity of high-frequency waves C_V [Eq. (6.34)].

Another set of solutions is determined by:

$$L = M = 0$$
$$N \neq 0 \tag{6.49}$$

Then, we obtain the third family of characteristics and the corresponding compatibility condition along it, respectively, as

$$dX = 0 \tag{6.50a}$$

$$d\varepsilon - \frac{d\sigma}{E_a + E_M} - \frac{\sigma - E_a\varepsilon}{(E_a + E_M)\theta_M}dt = 0 \tag{6.50b}$$

Obviously, Eq. (6.50a) coincides with the particle motion loci, and Eq. (6.50b) is, in fact, a special form of visco-elastic constitutive equation, Eq. (6.46), along the particle motion locus.

Thus, we use the characteristics method to solve the problem; similar to what we have discussed in Section 6.2.4 for the semi-infinite Maxwell bar, through an arbitrary point N_i on the X-t plane, there are three characteristic lines as shown in Fig. 6.9 and three corresponding characteristic relations. These relations are written in the difference form instead of the differential form, by which the three unknown variables σ, v, and ε at the point N_i can be determined according to the known initial and boundary conditions. The specific procedures are similar to those discussed previously for the Maxwell bar, only noting that the characteristics propagation velocity for the standard linear solid bar C_V should be used instead of the characteristics propagation velocity for the Maxwell bar C_0.

Now, let us consider a semi-infinite standard linear solid bar, which is originally at rest and in a stress-free state:

$$\sigma(X, 0) = \varepsilon(X, 0) = v(X, 0) = 0$$

and at the bar end ($X = 0$) suddenly suffered a constant stress σ^*. Then, as shown in Fig. 6.9, a strong discontinuous wave propagates along the line OA with wave velocity $\mathscr{D}$:

$$\mathscr{D} = \sqrt{\frac{1}{\rho_0}\frac{[\sigma]}{[\varepsilon]}} = \sqrt{\frac{E_a + E_M}{\rho_0}} \tag{6.51}$$

According to the kinematic [Eq. (2.55)] and dynamic [Eq. (2.57)] compatibility conditions across a strong discontinuous interface, as deduced in Chapter 2, and, moreover, considering that the line OA is also a characteristics so that the characteristic compatibility condition, Eq. (6.48b), should be satisfied, then, similar to Eq. (6.43), it could be easily shown that this strong discontinuous wave attenuates according to the following exponential:

$$\sigma = \sigma^* \exp\left[-\frac{\rho_0 C_V}{2\eta_M \left(1 + \frac{E_a}{E_M}\right)^2} X \right] = \sigma^* \exp(-\alpha_a X) \tag{6.52}$$

Completely similar results can be obtained for the particle velocity v and strain ε along the line *OA*. Once the solutions along the line *OA* are obtained, the remaining solutions in the region *AOt* in Fig. 6.9 are determinable, since it is attributed to solve a solvable **characteristics boundary-value problem** for visco-elastic waves. This mainly includes two kinds of basic operations: (1) to determine the solutions at boundary points, and (2) to determine the solutions at interior points.

For example, at an arbitrary boundary point N_1 in Fig. 6.9, the particle velocity $v(N_1)$ and strain $\varepsilon(N_1)$ can be determined from two characteristic compatibility conditions along the characteristics M_1N_1 and ON_1, respectively, namely, Eqs. (6.48) and (6.50). Writing in the finite difference form, we have

$$\begin{aligned} v(N_1) - v(M_1) &= -\frac{1}{\rho_0 C_V}[\sigma(N_1) - \sigma(M_1)] + \frac{E_a\varepsilon(M_1) - \sigma(M_1)}{\rho_0 C_V \theta_M}[t(N_1) - t(M_1)] \\ \varepsilon(N_1) - \varepsilon(0) - &\frac{1}{E_a + E_M}[\sigma(N_1) - \sigma(0)] + \frac{E_a\varepsilon(0) - \sigma(0)}{(E_a + E_M)\theta_M}[t(N_1) - t(0)] = 0 \end{aligned} \tag{6.53}$$

While at an arbitrary interior point N_2 in Fig. 6.9, the particle velocity $v(N_2)$, stress $\sigma(N_2)$, and strain $\varepsilon(N_2)$ can be determined from three characteristic compatibility conditions along the characteristics N_1N_2, M_1N_2, and M_2N_2, respectively. Writing in the finite difference form, we have

$$\begin{aligned} v(N_2) - v(N_1) &= \frac{1}{\rho_0 C_V}[\sigma(N_2) - \sigma(N_1)] - \frac{E_a\varepsilon(N_1) - \sigma(N_1)}{\rho_0 C_V \theta_M}[t(N_2) - t(N_1)] \\ v(N_2) - v(M_2) &= -\frac{1}{\rho_0 C_V}[\sigma(N_2) - \sigma(M_2)] + \frac{E_a\varepsilon(M_2) - \sigma(M_2)}{\rho_0 C_V \theta_M}[t(N_2) - t(M_2)] \\ \varepsilon(N_2) - \varepsilon(M_1) &= \frac{1}{E_a + E_M}[\sigma(N_2) - \sigma(M_1)] - \frac{E_a\varepsilon(M_1) - \sigma(M_1)}{(E_a + E_M)\theta_M}[t(N_2) - t(M_1)] \end{aligned} \tag{6.54}$$

Although the examples of Eqs. (6.53) and (6.54) that we deduced above are used to solve the so-called direct problem, i.e. to find the solution of visco-elastic wave propagation from the given initial and boundary conditions, the so-called inverse problem can be solved in a similar way without difficulty, i.e. to find the boundary condition from the known results of wave propagation and the given initial condition. For example, as shown in Fig. 6.9, if the stress, strain, and particle velocities at the points M_2 and N_3 have been known and if the stress, strain, and particle velocity at point M_1 have been given by the initial condition, then the stress, strain, and particle velocity at the point N_2 can be solved from the following three characteristic compatibility conditions along the characteristics N_3N_2, M_2N_2, and M_1N_2, respectively:

$$\begin{aligned}
v(N_2) - v(N_3) &= \frac{1}{\rho_0 C_V}[\sigma(N_2) - \sigma(N_3)] - \frac{E_a\varepsilon(N_3) - \sigma(N_3)}{\rho_0 C_V \theta_M}[t(N_3) - t(N_2)] \\
v(N_2) - v(M_2) &= -\frac{1}{\rho_0 C_V}[\sigma(N_2) - \sigma(M_2)] + \frac{E_a\varepsilon(M_2) - \sigma(M_2)}{\rho_0 C_V \theta_M}[t(N_2) - t(M_2)] \\
\varepsilon(N_2) - \varepsilon(M_1) &= \frac{1}{E_a + E_M}[\sigma(N_2) - \sigma(M_1)] - \frac{E_a\varepsilon(M_1) - \sigma(M_1)}{(E_a + E_M)\theta_M}[t(N_2) - t(M_1)]
\end{aligned} \tag{6.55}$$

From the above results, we can now discuss, in the case of the split Hopkinson visco-elastic bar, how to determine the dynamic stress–strain relationship of a specimen from the measured incident strain wave signals, the reflected strain wave signals, and the transmitted strain wave signals.

As mentioned previously, the specimens in SHPB tests are generally short enough so that the basic assumption of "uniform distribution of stresses along the specimen length" is satisfied, namely, $\sigma(X_2, t) = \sigma(X_1, t)$. Thus, any two measurements of the incident, reflected, and transmitted waves are sufficient to determine the dynamic stress–strain relationship of the specimen. In the case of visco-elastic bars, considering that the reflected wave is generally more difficult to separate from the dispersive tail-part of the incident visco-elastic wave, the incident signal $\varepsilon_i(X_{g1}, t)$ and the transmitted signal $\varepsilon_\tau(X_{g2}, t)$ are usually used in practice. Then, the dynamic stress–strain relationship of the specimen can be obtained by the following four successive subprocedures:

1. To determine the unknown incident stress $\sigma_i(X_1, t)$ and particle velocity $v_i(X_1, t)$ at the incident interface X_1 from the incident strain signal $\varepsilon_\tau(X_{g1}, t)$ measured by the strain gauge G_1 at X_{g1}.

   This subprocedure is attributed to solve a direct problem, i.e. to find the solution of visco-elastic wave propagation (in the positive X direction) for a given strain boundary condition and initial condition. Thus, the unknown stress and particle velocity at a boundary point (X_i, t_j), as shown in Fig. 6.11(a), can be determined by Eqs. (6.53). Or, in an explicit form, we have:

$$\begin{aligned}
\sigma(X_i, t_j) &= \sigma(X_i, t_{j-2}) + (E_a + E_M)[\varepsilon(X_i, t_j) - \varepsilon(X_i, t_{j-2})] \\
&\quad + [E_a\varepsilon(X_i, t_{j-2}) - \sigma(X_i, t_{j-2})]\frac{2\Delta t}{\theta_M} \\
v(X_i, t_j) &= v(X_{i+1}, t_{j-1}) + \frac{[\sigma(X_{i+1}, t_{j-1}) - \sigma(X_i, t_j)]}{\rho_0 C_V} \\
&\quad + \frac{[E_a\varepsilon(X_{i+1}, t_{j-1}) - \sigma(X_{i+1}, t_{j-1})]\Delta t}{\rho_0 C_V \theta_M}
\end{aligned} \tag{6.56}$$

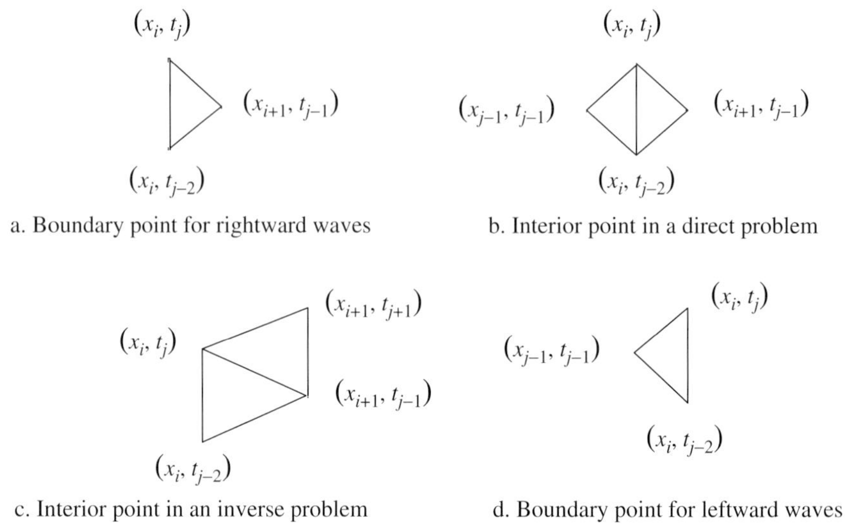

Fig. 6.11. The characteristics solutions for different cases in polymer SHPB.

where $\Delta t = t_j - t_{j-1}$ is the time difference, i.e. the time step length in numerical calculation. On the other hand, the unknown stress, strain, and particle velocity at an interior point (X_i, t_j), as shown in Fig. 6.11(b), can be determined by Eqs. (6.54). Or, in an explicit form, we have:

$$\begin{aligned}
\sigma(X_i,t_j)=\frac{1}{2}\Big\{&\sigma(X_{i+1},t_{j-1})+\sigma(X_{i-1},t_{j-1})+\rho_0C_V[v(X_{i-1},t_{j-1})\\
&-v(X_{i-1},t_{j-1})]+[E_a\varepsilon(X_{i+1},t_{j-1})-\sigma(X_{i+1},t_{j-1})\\
&+E_a\varepsilon(X_{i-1},t_{j-1})-\sigma(X_{i-1},t_{j-1})]\frac{\Delta t}{\theta_M}\Big\}
\end{aligned}$$

$$\begin{aligned}
v(X_i,t_j)=\frac{1}{2}\left\{\vphantom{\frac{a}{b}}\right.&\frac{\sigma(X_{i+1},t_{j-1})-\sigma(X_{i-1},t_{j-1})}{\rho_0C_V}+v(X_{i+1},t_{j-1})+v(X_{i-1},t_{j-1})\\
&\left.+\left(\frac{\begin{array}{c}[E_a\varepsilon(X_{i+1},t_{j-1})-\sigma(X_{i+1},t_{j-1})]\\ -[E_a\varepsilon(X_{i-1},t_{j-1})-\sigma(X_{i-1},t_{j-1})]\end{array}}{\rho_0C_V}\frac{\Delta t}{\theta_M}\right)\right\}
\end{aligned}$$

$$\begin{aligned}
\varepsilon(X_i,t_j)=\varepsilon(X_i,t_{j-2})&+\frac{\sigma(X_i,t_j)-\sigma(X_i,t_{j-2})}{E_a+E_M}\\
&-\frac{E_a\varepsilon(X_i,t_{j-2})-\sigma(X_i,t_{j-2})}{E_a+E_M}\frac{2\Delta t}{\theta_M}
\end{aligned}\tag{6.57}$$

2. To determine the unknown transmitted stress $\sigma_t(X_{g2}, t)$ and particle velocity $v_t(X_{g2}, t)$ at the strain gauge G_2 from the measured $\varepsilon_t(X_{g2}, t)$

   This subprocedure is also attributed to solve a direct problem for a given strain-boundary condition, similar to that shown in Fig. 6.11(b). Thus, Eqs. (6.56) and (6.57) can be used again in this subprocedure. Note that this subprocedure is a necessary preliminary step for the next subprocedure.

3. To determine the unknown transmitted stress $\sigma_t(X_2, t)$, strain $\varepsilon_t(X_2, t)$, and particle velocity $v_t(X_2, t)$ at the transmitted interface of specimen, X_2, from the known $\sigma_t(X_{g2}, t)$, $\varepsilon_t(X_{g2}, t)$, and $v_t(X_{g2}, t)$ at G_2.

   This subprocedure is attributed to solve an inverse problem, i.e. to find the unknown boundary condition from the given initial condition and the known results of visco-elastic wave propagation at a given point. Thus, Eq. (6.55) should be used to determine the stress, strain, and particle velocity at an interior point (X_i, t_j) as shown in Fig. 6.11(c). Or, in an explicit form, we have

$$
\begin{aligned}
\sigma(X_i,t_j) = \frac{1}{2}\Big\{ & \sigma(X_{i+1},t_{j-1})+\sigma(X_{i+1},t_{j+1})+\rho_0 C_V[v(X_{i+1},t_{j-1}) \\
& - v(X_{i+1},t_{j+1})]+[E_a\varepsilon(X_{i+1},t_{j-1})-\sigma(X_{i+1},t_{j-1}) \\
& + E_a\varepsilon(X_{i+1},t_{j+1})+\sigma(X_{i+1},t_{j+1})]\frac{\Delta t}{\theta_M}\Big\} \\
v(X_i,t_j) = \frac{1}{2}\Bigg\{ & \frac{\sigma(X_{i+1},t_{j-1})-\sigma(X_{i+1},t_{j+1})}{\rho_0 C_V}+v(X_{i+1},t_{j-1})+v(X_{i+1},t_{j+1}) \\
& + \left(\frac{\begin{array}{c}[E_a\varepsilon(X_{i+1},t_{j-1})-\sigma(X_{i+1},t_{j-1})] \\ +[E_a\varepsilon(X_{i+1},t_{j+1})-\sigma(X_{i+1},t_{j+1})]\end{array}}{\rho_0 C_V}\frac{\Delta t}{\theta_M}\right)\Bigg\} \\
\varepsilon(X_i,t_j) = \varepsilon(X_i, & t_{j-2})+\frac{\sigma(X_i,t_j)-\sigma(X_i,t_{j-2})}{E_a+E_M} \\
& -\frac{E_a\varepsilon(X_i,t_{j-2})-\sigma(X_i,t_{j-2})}{E_a+E_M}\frac{2\Delta t}{\theta_M}
\end{aligned}
\tag{6.58}
$$

Now, $\sigma_i(X_1, t)$, $\varepsilon_i(X_1, t)$, and $v_i(X_1, t)$ at the incident interface and $\sigma_t(X_2, t)$, $\varepsilon_t(X_2, t)$, and $v_t(X_2, t)$ at the transmitted interface of the specimen have all been determined. Then, by the assumption of "uniform distribution of stress along the specimen length" ($\sigma_i + \sigma_r = \sigma_t$), the reflected stress $\sigma_r(X_1, t)$ can be immediately determined:

$$
\sigma_r(X_1, t) = \sigma_t(X_2, t) - \sigma_i(X_1, t) \tag{6.59}
$$

The rest of quantities of reflected wave at the interface X_1 are determined by the fourth subprocedure, as follows.

4. To determine the reflected particle velocity $v_r(X_1, t)$ and strain $\varepsilon_r(X_1, t)$ at the incident interface of specimen, X_1, from the known $\sigma_r(X_1, t)$.

   This subprocedure is attributed to solve the direct problem of a visco-elastic wave propagating in negative X direction for a given stress-boundary condition. For the boundary points, the equations similar to Eq. (6.53) can be used, except that, due to the change of wave propagation direction, the corresponding change of sign in compatibility conditions should be taken into account. Thus, for an arbitrary boundary point (X_i, t_j), as shown in Fig. 6.11(d), we have

$$\begin{aligned}
\varepsilon(X_i,t_j) &= \varepsilon(X_i,t_{j-2}) + \frac{\sigma(X_i,t_j)-\sigma(X_i,t_{j-2})}{E_a+E_M} \\
&\quad - \frac{E_a\varepsilon(X_i,t_{j-2})-\sigma(X_i,t_{j-2})}{E_a+E_M}\frac{2\Delta t}{\theta_M} \\
v(X_i,t_j) &= v(X_{i-1},t_{j-1}) + \frac{[\sigma(X_{i-1},t_{j-1})-\sigma(X_i,t_j)]}{\rho_0 C_V} \\
&\quad + \frac{[E_a\varepsilon(X_{i-1},t_{j-1})-\sigma(X_{i-1},t_{j-1})]\Delta t}{\rho_0 C_V \theta_M}
\end{aligned} \tag{6.60}$$

   However, for the interior points as shown in Fig. 6.11(b), Eq. (6.54) can still be used, since they are derived from the compatibility conditions along both rightward and leftward characteristics and, consequently, are independent of the wave propagation direction.

Thus, we have obtained all the incident, reflected, and transmitted stresses, strains, and particle velocities at two interfaces of the specimen. Then, as discussed at the beginning of this subsection, by Eq. (3.16) in Chapter 3, the dynamic stress $\sigma_s(t)$, strain-rate $\dot{\varepsilon}_s(t)$, and strain $\varepsilon_s(t)$ of the specimen, and, consequently, the dynamic stress–strain relation of specimen at a high strain rate can finally be determined.

6.3 Nonlinear Visco-Elastic Constitutive Relationship

The linear visco-elastic constitutive relationship discussed in the previous section is no longer appropriate to describe the more complicated nonlinear behavior of visco-elastic materials such as polymers under **large deformation**. Thus, **non-linear visco-elastic consti-tutive relationships** are required and the corresponding **nonlinear visco-elastic wave propagation theories** are developed.

The experimental investigation by Zhu, Wang, and their coworkers in the recent 20 years for a variety of polymers, including polymethylmethacrylate (PMMA), polycarbonate (PC), polyamide (PA or nylon), acrylonitrile-butadiene-styrene (ABS), polybuthylene terephthalate (PBT), epoxy, and phenolics thermoset plastics, at strain rates from 10^{-4} to 10^3 s^{-1} showed that the mechanical behavior of polymers is highly sensitive to strain rates. The nonlinear visco-elastic behavior in uniaxial stress state for those polymers studied can be well described by the following Zhu-Wang-Tang (ZWT) nonlinear visco-elastic constitutive equation (Tang *et al.*, 1981; Wang and Yang, 1992), which in the differential form is:

$$\frac{\partial \sigma}{\partial t}=\frac{df_e(\varepsilon)}{d\varepsilon}\frac{\partial \varepsilon}{\partial t}+\frac{\partial \sigma_1}{\partial t}+\frac{\partial \sigma_2}{\partial t}=\left[\frac{df_e(\varepsilon)}{d\varepsilon}+E_1+E_2\right]\frac{\partial \varepsilon}{\partial t}-\frac{\sigma_1}{\theta_1}-\frac{\sigma_2}{\theta_2} \tag{6.61a}$$

and in the integral form is

$$\sigma=f_e(\varepsilon)+E_1\int_0^t \dot{\varepsilon}\exp\left(-\frac{t-\tau}{\theta_1}\right)d\tau+E_2\int_0^t \dot{\varepsilon}\exp\left(-\frac{t-\tau}{\theta_2}\right)d\tau \tag{6.61b}$$

where the first term

$$f_e(\varepsilon)=E_0\varepsilon+\alpha\varepsilon^2+\beta\varepsilon^3 \tag{6.61c}$$

describes the nonlinear elastic equilibrium response, and E_0, α, and β herein are the corresponding elastic constants; the next integral term in Eq. (6.61b) describes the viscoelastic response in low strain rates, and E_1, and θ_1 are the elastic constant and relaxation time, respectively, of the corresponding Maxwell element I; and the last integral term describes the viscoelastic response in high strain rates, and E_2 and θ_2 are the elastic constant and relaxation time, respectively, of the corresponding Maxwell element II. The corresponding rheological model is shown in Fig. 6.12, consisting of a nonlinear spring (non-Hookean) and two Maxwell elements in parallel connection.

For some typical engineering plastics such as epoxy, PMMA, and PC, the material parameters in ZWT equation obtained from the experiments are listed in Table 6.1.

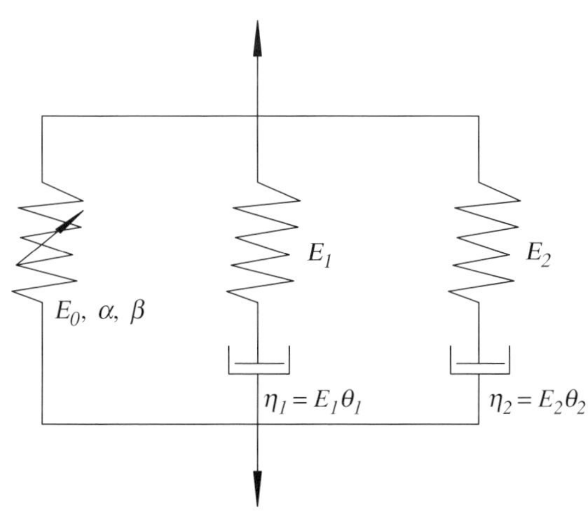

Fig. 6.12. The ZWT nonlinear visco-elastic constitutive model.

Table 6.1. Material parameters in ZWT equation for epoxy, PMMA, and PC.

	Epoxy	PMMA-1	PMMA-2	PMMA-3	PC
ρ_0, kg/m^3	1.20×10^3	1.19×10^3	1.19×10^3	1.19×10^3	1.20×10^3
E_0, GPa	1.96	2.04	2.19	2.95	2.20
α, GPa	4.12	4.17	4.55	10.9	23
β, GPa	−181	−233	−199	−96.4	−52
E_1, GPa	1.47	0.897	0.949	0.832	0.10
θ_1, s	157	15.3	13.8	7.33	470
E_2, GPa	3.43	3.07	3.98	5.24	0.73
θ_2, μs	8.57	95.4	67.4	40.5	140

Note that to avoid any unreal "elastic strain softening" that may be spuriously introduced by Eq. (6.61c), the nonlinear elastic equilibrium response $f_e(\varepsilon)$ can be equivalently described by the following exponential form (Zhou *et al.*, 1992):

$$f_e(\varepsilon) = \sigma_m \left[1 - \exp\left(-\sum_{i=1}^{n} \frac{(m\varepsilon)^i}{i}\right)\right] \tag{6.62}$$

where σ_m denotes the asymptotic maximum, m the ratio of E_0 and σ_m, and the positive integer n is a material parameter characterizing the initial linearity.

With regard to the relaxation time and in the ZWT equation, which describe the viscous character of the polymer, it is worthwhile to notice that each corresponds to an effective strain rate range or the so-called "**effective influence domain**". In fact (Chu *et al.*, 1985), for an arbitrary Maxwell element with elastic coefficient E_j and viscous coefficient η_j, according to Eq. (6.4), we have:

$$\dot{\varepsilon} = \frac{\dot{\sigma}_j}{E_j} + \frac{\sigma_j}{\eta_j}$$

At constant strain rate ($\dot{\varepsilon} = \text{const.}$, $\varepsilon = \dot{\varepsilon}t$), the stress relaxation response σ_j is given as

$$\sigma_j = E_j\theta_j\dot{\varepsilon}\left\{1 - \exp\left(-\frac{\varepsilon}{\theta_j\dot{\varepsilon}}\right)\right\} \tag{6.63}$$

Obviously, σ_j approaches its maximum, i.e. the **instantaneous response** $\sigma_I = E_j\varepsilon$, when $\dot{\varepsilon}$ approaches infinity, and its minimum, i.e. the **equilibrium response** $\sigma_E = 0$, when $\dot{\varepsilon}$ approaches zero. Introducing the nondimensional stress relaxation response defined as

$$\bar{\sigma}_j = \frac{\sigma_j}{\sigma_{\max} - \sigma_{\min}} = \frac{\sigma_j}{\sigma_I - \sigma_E} = \frac{\sigma_j}{E_j\varepsilon}$$

Eq. (6.63) can be rewritten as:

$$\bar{\sigma}_j = \frac{\theta_j\dot{\varepsilon}}{\varepsilon}\left\{1 - \exp\left(-\frac{\varepsilon}{\theta_j\dot{\varepsilon}}\right)\right\} = \frac{\theta_j}{t}\left\{1 - \exp\left(-\frac{t}{\theta_j}\right)\right\}$$

If we regard $\bar{\sigma}_j = 0.995$ as the beginning of a relaxation process, and $\bar{\sigma}_j = 0.005$ as the end of this relaxation process, then the **effective influence domain** (**EID**) of Maxwell element with relaxation time θ_j can be easily determined from the above equation. When it is in terms of time t, we have

$$10^{-2} \le \frac{t}{\theta_j} \le 10^{2.3} \tag{6.64a}$$

and when it is in terms of strain rate, assuming $\varepsilon = 1$ we have

$$10^{2} \ge \frac{\dot{\varepsilon}}{\left(\dfrac{\varepsilon}{\theta_j}\right)} \ge 10^{-2.3} \tag{6.64b}$$

It means that the EID for any relaxation time θ_j is about 4.3 orders of magnitude in either timescale or strain-rate scale.

Taking into account the above analyses and the experimental results shown in Table 6.1, it is worthwhile to notice the following features of ZWT nonlinear visco-elastic constitutive equation:

1. The constitutive nonlinearity of the tested polymers only comes from the pure elastic equilibrium response $f_e(\varepsilon)$, while all the rate/time-dependent responses are essentially linear. Such a constitutive nonlinearity is a kind of "**rate-independent nonlinearity**", or may be termed **weak nonlinearity**. Thus, if we name such a kind of material as **ZWT material**, it provides a possibility of generalizing what has been established in the linear viscoelasticity to the rate-dependent response of ZWT materials without substantial difficulty.

2. The experimental results for typical engineering plastics show that the ratio of α/E_0 is on the order of 1 to 10, and the ratio of β/E_0 is on the order of 10 to 10^2, as shown in Table 6.1. This means that the nonlinearity should be taken into account when $\varepsilon > 1\%$, while it may be neglected if $\varepsilon < 1\%$, depending on the specific material.

3. The experimental results for typical engineering plastics also show that the relaxation time θ_1 is generally on the order of 10 to 10^2 s, while the relaxation time θ_2 is on the order of 10^{-4} to 10^{-6} s, as shown in Table 6.1. It means that θ_1 is responsible for the visco-elastic behavior at low strain rates, while θ_2 is responsible for the visco-elastic behavior at high strain rates. Since θ_1 is on the order of 10^5–10^8 times higher than θ_2, and each relaxation time θ has its effective influence domain of about 4.3 orders of magnitude, thus θ_1 and θ_2 individually play their respective roles in each effective influence domain.

4. Consequently, under quasi-static loading conditions, where the timescale is on the order of 10^0 to 10^2 s, the high-frequency Maxwell element II (with relaxation time

of 10^0 to 10^2 μs) has already relaxed even at the beginning of loading. The ZWT equation [Eq. (6.61)] is then reduced to

$$\sigma = f_e(\varepsilon) + E_1 \int_0^t \dot{\varepsilon}(t) \exp\left(-\frac{t-\tau}{\theta_1}\right) d\tau \tag{6.65a}$$

5. On the contrary, under impact-loading conditions, where the timescale is on the order of 1 to 10^2 μs, the low-frequency Maxwell element I (with relaxation time θ_1 of 10^{-1} to 1 s) does not have enough time to relax even until the end of loading. Then the Maxwell element I is reduced to a single-spring element with an elastic constant E_1, and, correspondingly, the ZWT equation [Eq. (6.61)] is reduced to

$$\begin{aligned}\sigma &= f_e(\varepsilon) + E_1\varepsilon + E_2 \int_0^t \dot{\varepsilon}(t) \exp\left(-\frac{t-\tau}{\theta_2}\right) d\tau \\ &= \sigma_e(\varepsilon) + E_2 \int_0^t \dot{\varepsilon}(t) \exp\left(-\frac{t-\tau}{\theta_2}\right) d\tau \qquad (6.65b)\\ \sigma_e(\varepsilon) &= f_e(\varepsilon) + E_1\varepsilon\end{aligned}$$

   It means that the nonlinear visco-elastic wave propagation in ZWT polymer bars is actually controlled by Eq. (6.65b).

Theoretically, ZWT equation [Eq. (6.61)] can be deduced from either the Coleman-Noll finite linear visco-elastic theory (Coleman and Noll, 1960; 1961) or the Green-Revlin multiple integral (Green, and Rivlin, 1957). In fact, the one-dimensional form of the Green-Revlin multiple integral can be expressed as:

$$\begin{aligned}\sigma(t) = &\int_{-\infty}^t \phi_1(t-\tau_1)\dot{\varepsilon}(\tau_1)d\tau_1 \\ &+ \iint_{-\infty}^t \phi_2(t-\tau_1, t-\tau_2)\dot{\varepsilon}(\tau_1)\dot{\varepsilon}(\tau_2)d\tau_1 d\tau_2 \\ &+ \iiint_{-\infty}^t \phi_3(t-\tau_1, t-\tau_2, t-\tau_3)\dot{\varepsilon}(\tau_1)\dot{\varepsilon}(\tau_2)\dot{\varepsilon}(\tau_3)d\tau_1 d\tau_2 d\tau_3 + \cdots\end{aligned} \tag{6.66}$$

where ϕ_i is relaxation function, the first term on the right-hand side of the equation is the linear term that obeys the Boltzmann superposition principle, and the rest are nonlinear terms representing the joint contributions of the strain history to the final stress.

However, Eq. (6.66) is obviously mathematically intractable, particularly for engineering applications. In fact, even if we only take the first three terms, at least 28 groups of experiments at different conditions should be completed in order to determine those three relaxation functions ϕ_1, ϕ_2, ϕ_3 (Lockett, 1972).

Practically, from the hint given by the experimental results mentioned above, it is reasonable to assume that (Wang and Yang, 1992)

$$\phi_1(t-\tau_1) = E_0 + E_1 \exp\left(-\frac{t-\tau_1}{\theta_1}\right) + E_2 \exp\left(-\frac{t-\tau_1}{\theta_2}\right)$$

$$\phi_2(t-\tau_1, t-\tau_2) = const. = \alpha$$

$$\phi_3(t-\tau_1, t-\tau_2, t-\tau_3) = const. = \beta$$

Then, the Green-Revlin equation [Eq. (6.66)] immediately reduces to the ZWT equation [Eq. (6.61)].

6.4 Stress Waves Propagating in Nonlinear Visco-Elastic Bars

Now, let us discuss the stress waves propagating in nonlinear visco-elastic bodies, particularly to see how the nonlinear terms in the constitutive relation influence the visco-elastic wave propagation. For convenience, we will mainly discuss the longitudinal propagation in ZWT bars.

The control equations are composed of the movement equation

$$\rho_0 \frac{\partial v}{\partial t} = \frac{\partial \sigma}{\partial x} \tag{6.67}$$

the continuous equation

$$\frac{\partial v}{\partial x} = \frac{\partial \varepsilon}{\partial t} \tag{6.68}$$

and the nonlinear visco-elastic constitutive equation. With regard to ZWT materials, under impact-loading conditions where the timescale is on the order of 1 to 10^2 μs, the ZWT nonlinear visco-elastic constitutive equation [Eq. (6.61)] is reduced to Eq. (6.65)

$$\sigma = (E_0 + E_1)\,\varepsilon + \alpha\varepsilon^2 + \beta\varepsilon^3 + E_2 \int_0^t \dot{\varepsilon}(t) \exp\left(-\frac{t-\tau}{\theta_2}\right) d\tau$$

or equivalently in the differential form:

$$\frac{\partial \sigma}{\partial t} + \frac{\sigma}{\theta_2} = \left[\sigma'_{eff}(\varepsilon) + E_2\right] \frac{\partial \varepsilon}{\partial t} + \frac{\sigma_{eff}(\varepsilon)}{\theta_2} \tag{6.69a}$$

where $\sigma_{eff}(\varepsilon)$ is effective nonlinear pure elastic response:

$$\sigma_{eff}(\varepsilon) = f_e(\varepsilon) + E_1\varepsilon, \qquad \sigma'_{eff}(\varepsilon) = \frac{df_e(\varepsilon)}{d\varepsilon} + E_1 \tag{6.69b}$$

Similar to the characteristics method previously discussed for linear visco-elastic waves propagating in bars, the above partial differential control equations are equivalent to the following three groups of ordinary differential equations. The first and second groups are

$$dx = \pm C_V dt \tag{6.70a}$$

$$dv = \pm\frac{1}{\rho_0 C_V}d\sigma \pm \frac{\sigma - \sigma_{eff}(\varepsilon)}{\rho_0 C_V \theta_2}dt = \pm\frac{1}{\rho_0 C_V}d\sigma + \left[\frac{\sigma - \sigma_{eff}(\varepsilon)}{(\sigma'_{eff} + E_2)\theta_2}\right]dx \tag{6.70b}$$

where the plus and minus signs are for rightward-propagating and leftward-propagating waves, respectively, and the propagation velocity along the characteristics, C_V, is

$$C_V = \sqrt{\frac{1}{\rho_0}\frac{d(\sigma_{eff} + E_2)}{d\varepsilon}} = \sqrt{\frac{\sigma'_e(\varepsilon) + E_1 + E_2}{\rho_0}} \tag{6.71}$$

The third group corresponds to the particle motion loci and the material derivative of the constitutive relation of material [Eq. (6.69)] along the particle motion locus:

$$dx = 0 \tag{6.72a}$$

$$d\varepsilon - \frac{d\sigma}{\sigma'_{eff} + E_2} - \frac{\sigma - \sigma_{eff}(\varepsilon)}{\sigma'_{eff} + E_2}\frac{dt}{\theta_2} = 0 \tag{6.72b}$$

Thus, if the material parameters are known, the nonlinear visco-elastic waves propagating in the bar can be solved by the characteristics method.

Note the terms having *dt* or *dX* in the compatibility conditions, Eqs. (6.70b) and (6.72b). They are the terms that describe the dispersion and dissipation characters of nonlinear visco-elastic waves. Moreover, it is worthwhile to note that the high-frequency relaxation time θ_2 in Eqs. (6.70b) and (6.72b) always appears in the terms with *dt* or *dX*, in the form $(\sigma - \sigma_e)dt/\theta_2$ or $(\sigma - \sigma_e)dX/(C_v\theta_2)$. It means that the dispersion and attenuation of visco-elastic waves, in fact, mainly depend on θ_2 and the stress difference between the overall visco-elastic stress and the pure elastic equilibrium stress $[\sigma(\varepsilon) - \sigma_e(\varepsilon)]$, called "over stress".

In the following, the numerical results by characteristics method are compared between two typical polymers, namely, a thermoplastic one, PMMA, and a thermoset one, epoxy. The high-frequency relaxation time θ_2 for PMMA is 95.4 μs, while for epoxy is 8.57 μs; the former is larger than the latter by about one order. The rest material parameters for numerical calculation are listed in Table 6.1. Under the boundary condition of constant velocity impact (80 m/s) applied at the end ($X = 0$) of a semi-infinite bar, the stress profiles and strain profiles at the different distances from the impact boundary ($X = 0$, 0.2, 0.4, 0.6, 0.8, 1.0 m) are calculated and given respectively in Figs. 6.13(a and b) for PMMA, and in Figs. 6.14(a, b) for epoxy (Wang *et al.*, 1995; Wang, 2003). In those figures, in addition to the wave profiles for nonlinear visco-elastic waves (as shown in solid lines), the profiles for linear approximation are given too for comparison (as shown by dash lines), which can be obtained once the nonlinear elastic coefficients vanish ($\alpha = \beta = 0$).

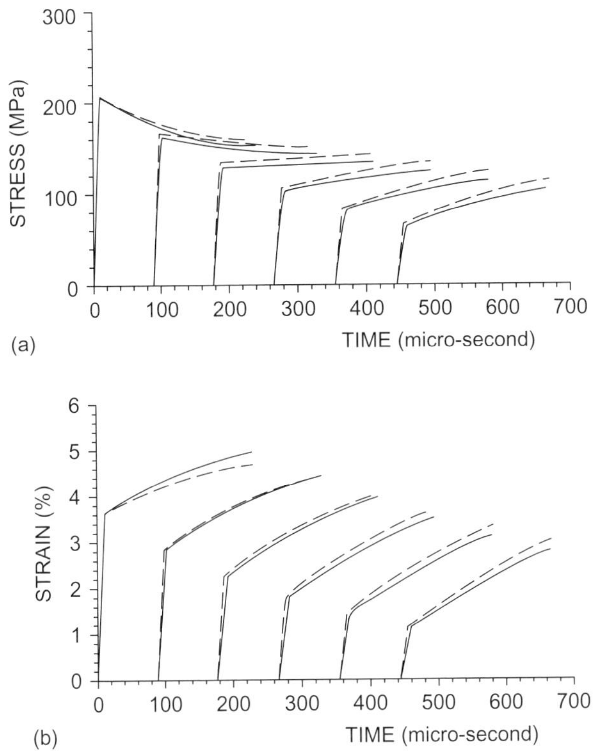

Fig. 6.13. Nonlinear visco-elastic waves in PMMA bar under constant velocity impact condition (80 m/s), showing (a) stress profiles and (b) strain profiles, at $X = 0, 0.2, 0.4, 0.6, 0.8, 1.0$ m, respectively, (solid lines: nonlinear, dash lines: linear).

As can be seen from Figs. 6.13 and 6.14, either for PMMA or for epoxy, the influences of constitutive nonlinear elasticity on wave profile shape and wave attenuation cannot be neglected in the case of high-velocity large deformation; even the visco-elastic response is linear.

Moreover, from Figs. 6.13 and 6.14, it is worthwhile to point out the following basic common characters for visco-elastic waves:

First, in contrast to linear elastic waves, no proportional relation exists between the visco-elastic stress wave, strain wave, and particle velocity wave [see Eqs. (6.70b) and (6.72b)]. One of the most distinct characters of visco-elastic waves is that for the position near and at the impact boundary, the visco-elastic stress decreases with time, showing a "stress-relaxation-like" character, while the visco-elastic strain, on the contrary, increases with time, showing a "strain-creep-like" character. It implies that the visco-elastic dynamic stress $\sigma(X, t)$ at an arbitrary position X cannot be determined directly from an experimentally measured dynamic strain $\varepsilon(X, t)$ at that position by simply multiplying with the apparent elastic modulus.

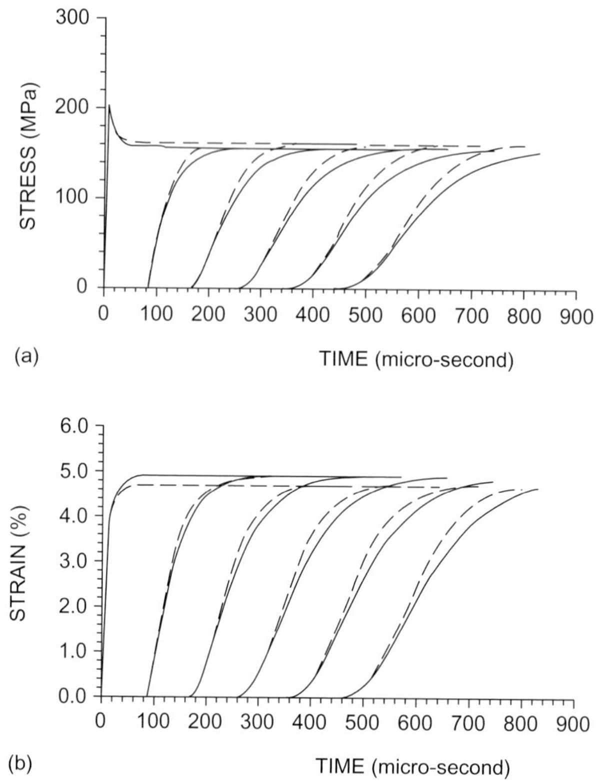

Fig. 6.14. Nonlinear visco-elastic waves in epoxy bar under constant velocity impact condition (80 m/s), showing (a) stress profiles and (b) strain profiles, at $X = 0, 0.2, 0.4, 0.6, 0.8, 1.0$ m, respectively, (solid lines: nonlinear, dash lines: linear).

Secondly, all visco-elastic waves attenuate with distance X, i.e., the wave amplitude decreases with the increase of propagating distance. Such attenuation, according to Eqs. (6.70b) and (6.72b), is mainly dominated by the following θ_2-related term:

$$\frac{\sigma - \sigma_{eff}(\varepsilon)}{(\sigma'_{eff} + E_2)\theta_2}$$

In other words, wave attenuation becomes weaker with increasing θ_2.

Finally, recall that each θ has its "effective influence domain" about 4.3 orders of magnitude in either timescale or strain-rate scale [Eq. (6.64)]. Then, with regard to the propagation of visco-elastic waves, an "effective influence time" (EIT), $t_{eff} = \theta_2$, or an "effective influence distance" (EID), $X_{eff} = C_{v0}/\theta_2$, should correspondingly exist, and both are effectively dominated by θ_2. Once $t > t_{eff}$ or $x > X_{eff}$, θ_2 will no longer play a marked influence. For example, from the material parameters listed in Table 6.1, it can be estimated that $X_{eff} = 0.214$ m for PMMA, and $X_{eff} = 0.0205$ m for epoxy. Noticing such a

θ_2-related character is significant either for the studies on visco-elastic wave propagation (direct problems) or for the studies on visco-elastic constitutive relationships of materials by means of visco-elastic wave propagation (inverse problems).

6.5 Elastic-Visco-Plastic Constitutive Relationship

It has long been known that the plastic behavior of materials under impact loading is markedly different from that under quasi-static loading, and it has been understood that this is caused by the influence of strain rate on plastic deformation. Among researches of great historical significance, the impact tensile experiments of steel wire (Fig. 6.15), conducted in succession by Hopkinson (1872), the father, and Hopkinson (1905), the son, should be specially mentioned. The three most important results obtained from their investigations are: (1) The tensile rupture of steel wire under impact loading is controlled by the height of the falling weight, i.e. the impact velocity, but is independent of its mass. (2) The position at which the steel wire breaks under impact tension is at the fixed suspension end B, but not at the impact end A. (3) The dynamic yield stress measured, Y_d, is as high as about twice the static yield stress, Y_s, and the steel wire can withstand a duration of about 100 μs under a stress level of 1.5 Y_s without any evident yield (**retard yield phenomenon**). The first two results result from stress wave propagation, which can be understood from the discussions given in the previous chapters, while the third is attributed to the strain-rate-dependent mechanical behavior of materials.

Recall what we have discussed in the rate-independent plastic wave theory: on the one hand, it is emphasized that the dynamic stress–strain curves be distinguished from the static stress–strain curves, which means that the strain rate effect is recognized; on the other hand, for convenience, it is assumed that the dynamic stress is a single-valued function of dynamic strain, and thus the constitutive equation in the rate-independent form is adopted. Although this theory has succeeded in solving a great deal of engineering

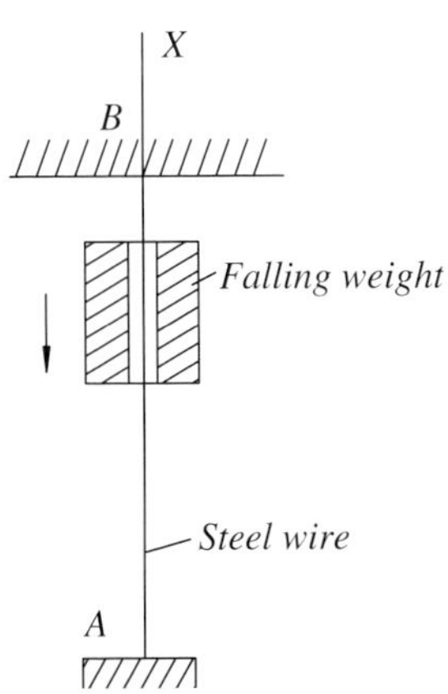

Fig. 6.15. Impact tensile experiment of steel wire.

problems, people found later that the theory did not completely coincide with experimental facts in some aspects. For example, when a bar suffers a constant velocity impact, according to the rate-independent plastic wave theory, a constant plastic strain zone as shown in Fig. 2.10 will be formed around the impacted bar end. However, it was not consistently verified by all the known experiments (Lee, 1956; Riparbelli, 1953; Sternglass and Stuart, 1953). In another example, when an elastic–plastic bar, preloaded to a certain plastic deformation ε_p, suffers an infinite impact disturbance, according to the rate-independent plastic wave theory, an incremental wave will propagate with the plastic wave velocity C_p, depending on the slope of the $\sigma \sim \varepsilon$ curve at $\varepsilon = \varepsilon_p$ (Eq. 2.15). However, several experimental results have shown that the incremental wave in fact propagates, but with elastic wave velocity C_0 (Bell, 1951; Alter and Cautis, 1956; Bianchi, 1964). These facts all indicate that the rate-independent plastic wave theory has some limitations. To remove such limitations, the constitutive relation should take account of the rate-dependent plastic deformation. Thus, the **elastic-visco-plastic wave theory** was developed toward the end of 1940s.

For this purpose, first of all, the elastic-visco-plastic constitutive relation should be established. A number of experiments have been conducted to study the influence of strain rate on the yield stress or flow stress at a given strain (Campbell, 1973). The earliest experimental research can be traced back to the work of Lüdwik (1909). Based on the experimental data reported by a number of researchers, two kinds of empirical equations under one-dimensional stress state were proposed, namely, the power function law and the logarithmic function law expressed, respectively, as:

$$\frac{\sigma}{\sigma_0} = \left(\frac{\dot{\varepsilon}}{\dot{\varepsilon}_0}\right)^n \tag{6.73}$$

$$\frac{\sigma}{\sigma_0} = 1 + \lambda \ln\left(\frac{\dot{\varepsilon}}{\dot{\varepsilon}_0}\right) \tag{6.74}$$

where σ_0 is the yield stress or flow stress measured at quasi-static test in strain rate $\dot{\varepsilon} = \dot{\varepsilon}_0$, and n and λ are material parameters characterizing the strain rate sensitivity of material. Equation (6.73) is geographically represented by a straight line in the $\ln\sigma$–$\ln\dot{\varepsilon}$ plot, while Eq. (6.74) is represented by a straight line in the σ–$\ln\dot{\varepsilon}$ plot, as shown in Fig. 6.16. This means that strain rate displays marked influence only when its value varies in orders.

Equation (6.74) can be rewritten as

$$\dot{\varepsilon} = \dot{\varepsilon}_0 \exp\left[\frac{1}{\lambda}\left(\frac{\sigma - \sigma_0}{\sigma_0}\right)\right] \tag{6.75}$$

Note that $(\sigma - \sigma_0)$ is the difference between dynamic stress and static stress, usually called **over stress** or **extra stress**, and, consequently, $(\sigma - \sigma_0)/\sigma_0$ is the dimensionless over stress. Thus, Eq. (6.75) means that strain rate is a function of over stress.

Based on these experimental results, the so-called **elastic-visco-plastic theory** was developed, which is a generalization of the classical plastic theory to take account of the

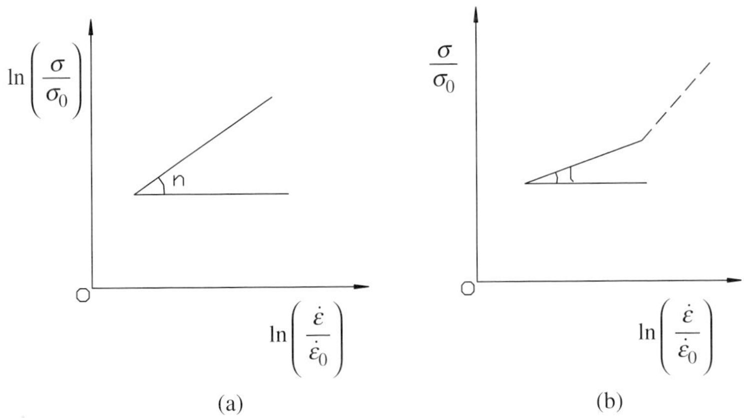

Fig. 6.16. Two kinds of strain rate dependent equation, (a) the power function law and (b) the logarithmic function law.

strain rate effect of plastic deformation, but neglecting the strain rate effect of elastic deformation. In other words, assume that the total strain rate $\dot{\varepsilon}$ is composed of the elastic component (instantaneous response) $\dot{\varepsilon}^e$ and the nonelastic component (noninstantaneous response) $\dot{\varepsilon}^p$:

$$\dot{\varepsilon} = \dot{\varepsilon}^e + \dot{\varepsilon}^p \tag{6.76}$$

where the nonelastic component $\dot{\varepsilon}^p$ is a function of over stress, representing the plastic response coupled with the viscous response.

It was Соколовский (Sokolovskii)(1948) who, based on the perfect plastic model, first proposed the following elastic-visco-plastic constitutive relationship:

$$\dot{\varepsilon} = \frac{\dot{\sigma}}{E} + Sign\,\sigma \cdot g\left[\frac{|\sigma|}{Y_0} - 1\right] \tag{6.77}$$

where the $\dot{\varepsilon}^p$ is regarded as a function of such an over stress, defined as the difference between the dynamic stress σ and the static yield limit of perfect plastic body Y_0, i.e. $g[(|\sigma|/Y_0) - 1]$. Under different constant strain rates, Eq. (6.77) is represented by a set of stress–strain curves, of which the visco-plastic deformation parts are parallel horizontal straight lines, as shown in Fig. 6.17(a). The function g is generally a nonlinear function. If g is a linear function of over stress, then Eq. (6.77) is reduced to the following form:

$$\dot{\varepsilon} = \frac{\dot{\sigma}}{E} + \frac{|\sigma| - Y_0}{\eta} = \frac{\dot{\sigma}}{E} + Sign\,\sigma \cdot \gamma^* \frac{|\sigma| - Y_0}{Y_0} \tag{6.78}$$

which is similar to the visco-elastic constitutive equation for Maxwell body, except the viscous term in Eq. (6.4) is now replaced by the over stress viscous term. If the perfect

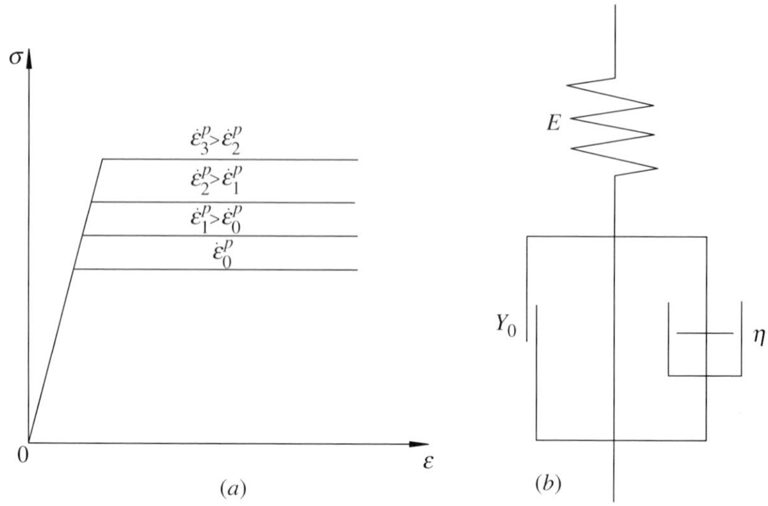

Fig. 6.17. The elastic visco-plastic constitutive model described by Eq. (6.78).

plastic body is rheologically represented by a couple of sliding plates with constant friction force Y_0, then Eq. (6.78) is correspondingly represented by such a three-elements model as shown in Fig. 6.17(b), which is composed of a spring with a visco-plastic model (the so-called Binham model) connected in series, and the latter is again composed of a dashpot and a sliding-plate element connected in parallel. The γ $(=Y_0/\eta)$ in Eq. (6.78) is a material parameter characterizing the viscous character of the visco-plastic model.

In order to take account of strain hardening, Malvern (1951) suggested replacing the Y_0 in Eq. (6.77) with the static stress–strain curve $\sigma_0(\varepsilon)$ and, thus, the following elastic-visco-plastic constitutive equation was proposed:

$$\dot{\varepsilon} = \frac{\dot{\sigma}}{E} + Sign\,\sigma \cdot g\left[\frac{\sigma}{\sigma_0(\varepsilon)} - 1\right] \tag{6.79}$$

which is equal to regarding the friction force Y_0 in the rheological model shown in Fig. 6.17(b) as a function of plastic strain ε^p, $Y = Y(\varepsilon^p)$. Under different constant strain rates, Eq. (6.79) is represented in the σ–ε plane by such a set of stress–strain curves that the visco-plastic deformation parts are all parallel to the static stress–strain curve, namely the distance between any two visco-plastic parts of curves is kept as constant, as shown in Fig. 6.18.

According to Соколовский -Malvern model,only when the over stress is greater than zero can the visco-plastic deformation happen. When the over stress is less than or equal to zero, only elastic deformation happens. Mathematically, it can be expressed by the following

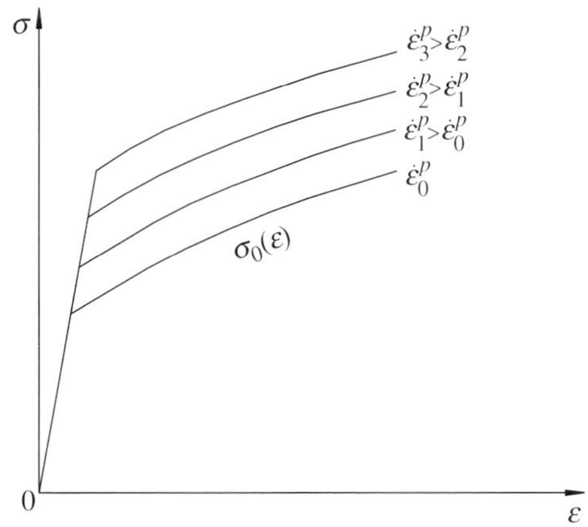

Fig. 6.18. The elastic visco-plastic constitutive model described by Eq. (6.79).

unified form of constitutive equation:

$$\left.\begin{aligned} \dot{\varepsilon} &= \frac{\dot{\sigma}}{E} + \left\langle g\left[\frac{\sigma}{\sigma_0(\varepsilon)} - 1\right]\right\rangle \\ \langle g \rangle &= \begin{cases} 0, & for \dfrac{\sigma}{\sigma_0(\varepsilon)} - 1 \geq 0 \\ g, & for \dfrac{\sigma}{\sigma_0(\varepsilon)} - 1 > 0 \end{cases} \end{aligned}\right\} \tag{6.80}$$

Or, similar to Eq. (6.78), introduce a viscous coefficient γ^*, defined by the following equation:

$$g\,\langle \xi \rangle = \gamma^* \phi(\xi)$$

then Eq. (6.80) can be rewritten in another form as:

$$\left.\begin{aligned} \dot{\varepsilon} &= \frac{\dot{\sigma}}{E} + \gamma^* \langle \phi(F) \rangle \\ \langle \phi(F) \rangle &= \begin{cases} 0, & F \leq 0 \\ \phi(F) & F > 0 \end{cases} \\ F &= \frac{\sigma}{\sigma_0(E)} - 1 \end{aligned}\right\} \tag{6.81}$$

For the Соколовский-Malvern elastic-visco-plastic body, it should be noted that its entering into the visco-plastic region from the elastic region or returning the elastic region from the visco-plastic region completely depends on weather the over stress is greater than zero. As long as the over stress is greater than zero,i.e. either the stress is increasing with

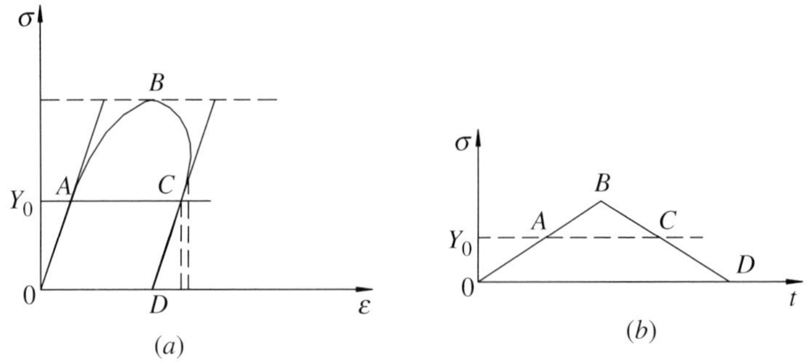

Fig. 6.19. The σ–ε curve for the elastic visco-plastic materials described by Eq. (6.78) in the case of linearly loading and then linearly unloading.

time ($\partial\sigma/\partial t > 0$) or decreasing with time ($\partial\sigma/\partial t < 0$), it follows the same constitutive relationship, i.e. the visco-plastic deformation still continuously develops, even when the stress is decreasing with time. For example, if the stress is first linearly increasing with time, and then linearly decreasing with time, as shown in Fig. 6.19(b), the σ–ε curve calculated according to Eq. (6.78) is shown in Fig. 6.19(a). As can be seen from the figure, the visco-plastic deformation still continuously develops when stress begins to decrease. Only when the stress is unloaded to the value indicated by the point C, showing that over stress equals to zero, a transition from the visco-plastic state to the elastic unloading state happens. Note that point C neither corresponds to the beginning of stress decrease or the beginning of strain decrease. In such aspects, elastic-visco-plastic materials are fully different from elastic–plastic materials.

6.6 Stress Waves Propagating in Elastic-Visco-Plastic Bars

Let us now discuss the stress waves propagating in the Соколовский–Malvern elastic-visco-plastic bars. The control equations are composed of the continuous equation [Eq. (2.12)], the motion equation [Eq. (2.13)], and the elastic-visco-plastic constitutive equation [Eq. (6.81)], namely

$$\left.\begin{aligned} \frac{\partial v}{\partial X} &= \frac{\partial \varepsilon}{\partial t} \\ \rho_0 \frac{\partial v}{\partial t} &= \frac{\partial \sigma}{\partial X} \\ \frac{\partial \varepsilon}{\partial t} &= \frac{1}{E}\frac{\partial \sigma}{\partial t} + \gamma^* \langle \phi(F) \rangle \end{aligned}\right\} \tag{6.82}$$

which are one-order partial differential equations and can be solved by the characteristics method. Similar to what we have discussed on visco-elastic wave propagation, in order to find the characteristics equations and the corresponding characteristic compatibility equations, multiplying each of them with undetermined coefficients N, M, and L, respectively, and then adding together, we have:

$$(L+N)\frac{\partial \varepsilon}{\partial t}+\left(M\rho_0\frac{\partial}{\partial t}-L\frac{\partial}{\partial X}\right)v-\left(\frac{N}{E}\frac{\partial}{\partial t}+M\frac{\partial}{\partial X}\right)\sigma-N\gamma^*\langle\phi(F)\rangle=0 \tag{6.83}$$

These undetermined coefficients should satisfy the following conditions:

$$\frac{dX}{dt}=\frac{0}{L+N}=-\frac{L}{M\rho_0}=\frac{ME}{N} \tag{6.84}$$

which are fully the same as Eq. (6.38), and, thus, similar to Eq. (6.40) and Eq. (6.41), three real characteristics differential equations and three corresponding compatibility differential equations can be obtained, respectively, as:

$$dX=\pm\sqrt{\frac{E}{\rho_0}}dt=\pm C_0 dt \tag{6.85a}$$

$$dv=\pm\frac{1}{\rho_0 C_0}d\sigma\pm C_0\gamma^*\langle\phi(F)\rangle dt=\pm\frac{1}{\rho_0 C_0}d\sigma+\gamma^*\langle\phi(F)\rangle dX \tag{6.85b}$$

and

$$dX=0 \tag{6.86a}$$

$$d\varepsilon=\frac{d\sigma}{E}+\gamma^*\langle\phi(F)\rangle dt \tag{6.86b}$$

The characteristics described by Eq. (6.85a) represent the propagating loci of wave front on the X–t plane, while the characteristics described by Eq. (6.86a) represents the motion loci of particles. Since all these are similar to the situations previously discussed for visco-elastic waves propagating in Maxwell bars, no details need be repeated again, except that the original σ/η should be correspondingly replaced by $\gamma^*\langle\phi(F)\rangle$ now.

As an example, let us discuss a semi-infinite bar, which is originally at rest and in a stress-free state ($v_0=\sigma_0=\varepsilon_0=0$) and, at the bar end ($X=0$), suddenly suffers a constant stress σ^*. Similar to the situation shown in Fig. 6.9, since the relation between stress discontinuity $[\sigma]$ and strain discontinuity $[\varepsilon]$ across the strong discontinuity wave front is determined by the instantaneous response of Eq. (6.81), the strong discontinuous wave propagates along *OA* with elastic wave velocity C_0. Note that, on the one hand, the dynamic compatibility condition and the kinematics compatibility condition across the strong discontinuous wave locus *OA*, namely Eq. (6.42), should be satisfied; and, on the other hand, along the characteristics *OA*, the following characteristics compatibility condition along the rightward characteristics, namely Eq. (6.85b), should be satisfied too:

$$dv=\frac{1}{\rho_0 C_0}d\sigma+\gamma^*\langle\phi(F)\rangle dX$$

Instituting Eq. (6.42a) into the above equation, we have

$$\frac{d\sigma}{dX} = -\frac{1}{2}\rho_0 C_0 \gamma^* \langle \phi(F) \rangle = -\phi[X, \sigma(X)] \tag{6.87}$$

Thus, the problem is attributed to solve the following integral equation:

$$\sigma(X) = \sigma^* - \int_0^X \psi[\xi, \sigma(\xi)]\, d\xi$$

where σ^* has been given by the boundary condition at the bar end. If $\dot{\varepsilon}^p$ is a linear function of the over stress with respect to the yield stress Y_0 of perfect plastic material, namely the constitutive relation is expressed by Eq. (6.78), Then, from Eq. (6.87), the solution along *OA* can be obtained as

$$\sigma = Y_0 + (\sigma^* - Y_0)\exp\left[-\frac{\rho_0 C_0}{2\eta} X\right] \tag{6.88}$$

As can be seen, the attenuation factor in Eq. (6.88) is the same as that in Eq. (6.43a). In other words, the over stress of the strong discontinuous elastic-visco-plastic wave exponentially attenuates with propagation distance X – the same as the attenuation of strong discontinuous Maxwell visco-elastic wave. Once the solution along *OA* is determined, the solutions in the region *AOt* are attributed to solve a solvable characteristics boundary problem, the same as has been discussed for Fig. 6.9.

If the Y_0 in Eq. (6.78) is replaced by the static stress–strain curve $\sigma_0(\varepsilon)$ to take account of strain hardening effect, the problem is still solvable by a similar method (Malvern, 1951). Some results are given in Fig. 6.20 (plotted by solid lines). For comparison, the results calculated according to the rate-independent elastic plastic theory are also given in the same figure (plotted by dash lines).

It is worthwhile to notice the following important differences between the rate-dependent elastic-visco-plastic waves and the rate-independent elastic plastic waves:

1. Because the instantaneous response of elastic-visco-plastic material is elastic, under the same suddenly applied constant velocity impact condition, the initial stress amplitude of an elastic-visco-plastic strong discontinuous wave is higher than that of an elastic-plastic wave. This is the so-called "over stress effect". However, due to the viscous effect, the stress amplitude of an elastic-visco-plastic strong discontinuous wave will continuously attenuate and approach the amplitude of strong discontinuous elastic precursor wave in the elastic–plastic wave theory, i.e. the over stress will gradually decrease until zero, as shown in Fig. 6.20(a). But the propagating velocity of the elastic-visco-plastic wave is kept the same as the elastic wave velocity C_0, regardless of what the value of stress discontinuity is. This coincides with the experimentally observed fact that the incremental wave velocity in a preplastically deformed bar is C_0.

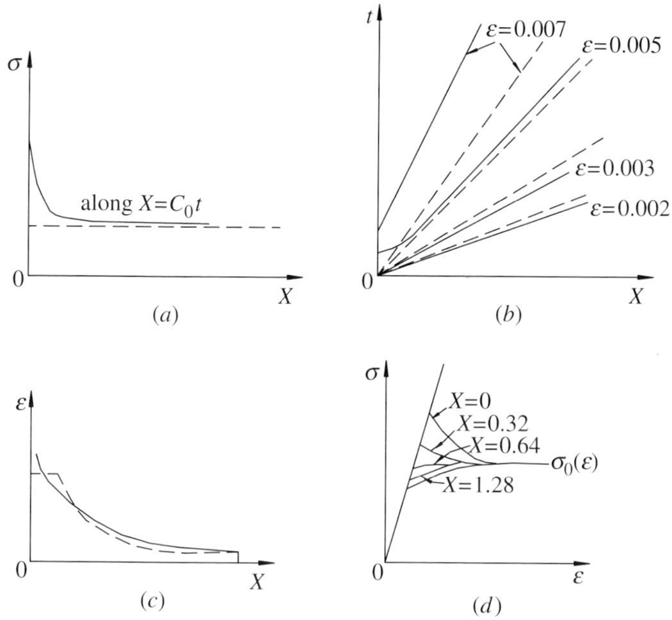

Fig. 6.20. Dynamic response of a semi-infinite bar under constant velocity impact, wherein solid lines indicate the elastic-visco-plastic solution, and dash lines indicate the elastic–plastic solution. (a) the stress distribution along the strong discontinuous wave OA, (b) iso-strain lines on X–t plane, (c) the maximum strain distribution along the bar, (d) the stress–strain curves at different X of the bar.

2. With regard to the weak discontinuous wave region behind the strong discontinuous wave front propagating in a semi-infinite bar under constant initial condition, according to the elastic-plastic wave theory, it is a simple wave region. Consequently, the σ, ε, and v are all invariable along the characteristics, but, according to the elastic-visco-plastic wave theory, it is no longer a simple wave region and, consequently, the iso-strain line no longer coincides with the characteristics, as shown in Fig. 6.20(b). If the slope of iso-strain line $\left.\frac{dX}{dt}\right|_{\varepsilon=const}$ is defined as the phase velocity of strain propagation, then in the case of small strains, the strain propagates quicker in the elastic-visco-plastic bar than in the elastic–plastic bar, because in such case the instantaneous response plays a dominating role. However in the case of large strains, the strain propagates slower in the elastic-visco-plastic bar than in the elastic–plastic bar, because in such case the viscous response with a delayed effect plays a dominating role.

3. When a bar suffers a constant velocity impact, according to the rate-independent plastic wave theory, a constant plastic strain zone as shown in Fig. 2.10 will be formed around the impacted bar end. However, such a constant plastic strain zone no longer exists according to the elastic-visco-plastic wave theory due to

the viscous effect. Moreover, the maximum strain at the bar end predicted by the elastic-visco-plastic wave theory is larger than that predicted by the elastic–plastic wave theory, as shown in Fig. 6.20(c), which is more in conformity with some known experimental facts.

4. At a different cross-section of the bar, the situation of how the strain rate varies with time is different, and, thus, the stress–strain curve calculated at the different cross-section is different, as shown in Fig. 6.20(d). However, the strain rate at every cross-section gradually decreases to zero, so that the stress–strain curve at every cross-section finally approaches the quasi-static stress–strain curve $\sigma_0(\varepsilon)$.

Other examples of elastic-visco-plastic waves propagating in bars (including the problems for finite bars, for the determination of loading–unloading boundaries, i.e. the boundaries of the elastic zone and the visco-plastic zone, and when the instantaneous response contains instantaneous plastic response, etc.) all can be studied on the basis of the previous discussions, or some related monographs can be referred to (e.g. Cristescu, 1967; Nowaski, 1978). In the next two chapters, we will further discuss the one-dimensional strain elastic-visco-plastic waves (in Section 7.11) and the elastic-visco-plastic spherical and cylindrical waves (in Section 8.5), respectively.

CHAPTER 7

One-Dimensional Strain Plane Waves

7.1 Governing Equations

In the elementary theory of stress wave propagating in bars, it is assumed that plane cross sections normal to the longitudinal axis of the bar remain plane in deformation, and on the cross section, there are only uniformly distributing axial stresses σ_X. Therefore, the problem is reduced to a one-dimensional (1D) stress problem. As previously discussed in Section 2.8, this assumption is valid approximately only when the transverse size of the bar is much smaller than the wavelength.

As another extreme case, if the transverse size perpendicular to the propagating direction of longitudinal wave (X axis) is large enough to constrain any transverse motion, namely if

$$u_Y = u_Z = 0 \tag{7.1a}$$

$$\varepsilon_Y = \frac{\partial u_Y}{\partial Y} = 0, \quad \varepsilon_Z = \frac{\partial u_Z}{\partial Z} = 0 \tag{7.1b}$$

$$v_Y = \frac{\partial u_Y}{\partial t} = 0, \quad v_Z = \frac{\partial u_Z}{\partial t} = 0 \tag{7.1c}$$

where symbols denote the same variables as those in previous chapters, and the subscripts Y and Z denote the components along Y and Z direction respectively, then there is only disturbance of axial strain ε_X, propagating in the media, which is called **1D strain plane wave**. In such a case, strictly speaking, there is no transverse inertia effect, but there is constraint from the corresponding lateral normal stress σ_Y and σ_z, so the media is in the state of 3D stress. Assuming that the acting σ_z are uniformly distributed on cross sections, then all nonzero stress components and strain components are functions of X and t only. Due to symmetry, there is

$$\sigma_Y(X, t) = \sigma_Z(X, t) \tag{7.2}$$

Strictly, only when a semi-infinite body is subjected to uniformly distributing normal impact, can it be a problem of 1D strain plane longitudinal wave propagation. However, for a practical problem, during the early period when the disturbances reflected from the lateral boundaries have not arrived at the area discussed yet, it can be treated as a 1D strain plane wave problem. For example, in a contact explosion on a plate or a high velocity normal impact onto a plane, if the plate transverse sizes are larger enough than the plane thickness, during the early period of stress wave propagation, we can view these problems as problems of 1D strain plane longitudinal waves.

Similar to the situation discussed on 1D stress longitudinal waves in bars, the governing equations still consist of three aspects: the continuity equation, the momentum conservation equation, and the constitutive equation of material. If the problem is described by Lagrange variables and the axial stress is assumed to be a single-value function of the axial strain, then the governing equations in form are the same as that of 1D stress longitudinal waves in bars discussed in Section 2.2.

$$\frac{\partial v_X}{\partial X} = \frac{\partial \varepsilon_X}{\partial t} \tag{7.3a}$$

$$\frac{\partial v_X}{\partial X} = \frac{\partial \varepsilon_X}{\partial t} \tag{7.3b}$$

$$\sigma_X = \sigma_X(\varepsilon_X) \tag{7.3c}$$

Note that the current relationship between stress and strain, σ–ε, is no longer the relationship under the 1D tension or compression stress state. Equation (7.3) can also be expressed in other forms, for example, by eliminating σ_X, we obtain the following first-order partial differential equations with respect to v_X and ε_X as unknown variables

$$\frac{\partial v_X}{\partial X} = \frac{\partial \varepsilon_X}{\partial t} \tag{7.4a}$$

$$\frac{\partial v_X}{\partial t} = C_L^2 \frac{\partial \varepsilon_X}{\partial X} \tag{7.4b}$$

By eliminating ε_X, we obtain the following first-order partial differential equations with respect to σ_X and v_X as unknown variables

$$\rho_0 \frac{\partial v_X}{\partial t} = \frac{\partial \sigma_X}{\partial X} \tag{7.5a}$$

$$\rho_0 C_L^2 \frac{\partial v_X}{\partial X} = \frac{\partial \sigma_X}{\partial t} \tag{7.5b}$$

If we use the longitudinal displacement $u_X(X, t)$ as an unknown function, then we obtain the following second-order partial differential equation

$$\frac{\partial^2 u_X}{\partial t^2} - C_L^2 \frac{\partial^2 u_X}{\partial X^2} = 0 \tag{7.6}$$

The C_L in the above equations is the wave velocity of 1D strain longitudinal plane wave

$$C_L = \sqrt{\frac{1}{\rho_0}\frac{d\sigma_X}{d\varepsilon_X}} \tag{7.7}$$

Thus it can be seen that most of the methods used to analyze the 1D stress waves can be directly adopted and thus will not be repeated again here. In the following sections, we will focus on some specific characteristics of 1D strain waves.

7.2 One-Dimensional Strain Elastic Waves

The differences between elastic 1D strain waves and 1D stress waves are mainly caused by the difference in the stress/strain states, and consequently the different longitudinal stress–strain relationships. In the former problem, media are in the state of 3D stress, while in the latter problem, media are in the state of 1D stress. To deduce the relationship between longitudinal stress and strain in 1D strain problems, we first review the general Hooke's law below.

It is well known from elastic mechanics, the general Hooke's law for isotropic, homogenous, linear elastic materials can be written as

$$\left.\begin{aligned}
\varepsilon_X &= \frac{1}{E}[\sigma_X - \nu(\sigma_Y + \sigma_Z)] \\
\varepsilon_Y &= \frac{1}{E}[\sigma_Y - \nu(\sigma_Z + \sigma_X)] \\
\varepsilon_Z &= \frac{1}{E}[\sigma_Z - \nu(\sigma_X + \sigma_Y)] \\
\varepsilon_{XY} &= \frac{1}{2G}\tau_{XY} \\
\varepsilon_{YZ} &= \frac{1}{2G}\tau_{YZ} \\
\varepsilon_{ZX} &= \frac{1}{2G}\tau_{ZX}
\end{aligned}\right\} \tag{7.8}$$

where strain components are defined below:

$$\left.\begin{aligned}
\varepsilon_X &= \frac{\partial u_X}{\partial X}, \quad \varepsilon_{XY} = \frac{1}{2}\left(\frac{\partial u_X}{\partial Y} + \frac{\partial u_Y}{\partial X}\right) \\
\varepsilon_Y &= \frac{\partial u_Y}{\partial Y}, \quad \varepsilon_{YZ} = \frac{1}{2}\left(\frac{\partial u_Y}{\partial Z} + \frac{\partial u_Z}{\partial Y}\right) \\
\varepsilon_Z &= \frac{\partial u_Z}{\partial Z}, \quad \varepsilon_{ZX} = \frac{1}{2}\left(\frac{\partial u_Z}{\partial X} + \frac{\partial u_X}{\partial Z}\right)
\end{aligned}\right\} \tag{7.9}$$

where E is the Young's modulus, G is the shear modulus, and ν is the Poisson's ratio. If we regard stresses as functions of strains, the general Hooke's law in the tensor form can be equivalently expressed as

$$\sigma_{ij} = \lambda\delta_{ij}\varepsilon_{kk} + 2\mu\varepsilon_{ij} \tag{7.10}$$

where δ_{ij} is the Kronecker defined by

$$\delta_{ij} = \begin{cases} 0, & i \neq j \\ 1, & i = j \end{cases}$$

λ and μ are Lame's coefficients:

$$\left.\begin{aligned} \lambda &= \frac{E\nu}{(1+\nu)(1-2\nu)} \\ \mu = G &= \frac{E}{2(1+\nu)} \end{aligned}\right. \tag{7.11}$$

If introducing the volume strain Δ and the hydrostatic stress p (the negative of the average normal stress), respectively, defined by

$$\Delta = \varepsilon_X + \varepsilon_Y + \varepsilon_Z \tag{7.12}$$

$$-p = \tfrac{1}{3}\sigma_{kk} = \tfrac{1}{3}(\sigma_X + \sigma_Y + \sigma_Z) \tag{7.13}$$

and introducing the deviator strain tensor e_{ij} characterizing distortion (shape deformation) and the corresponding deviator stress tensor s_{ij}:

$$\left.\begin{aligned} e_{ij} &= \varepsilon_{ij} - \tfrac{1}{3}\Delta\delta_{ij} \\ s_{ij} &= \sigma_{ij} + p\delta_{ij} \end{aligned}\right\} \tag{7.14}$$

then the generalized Hooke's law can be expressed in two parts: the volumetric deformation law and the distortional deformation law:

$$\left.\begin{aligned} -p &= K\Delta \\ s_{ij} &= 2Ge_{ij} \end{aligned}\right\} \tag{7.15}$$

where K is the volumetric modulus, which is related to other elastic coefficients by the following relation:

$$K = \lambda + \frac{2G}{3} = \frac{E}{3(1-2\nu)} \tag{7.16}$$

Under the current 1D strain condition ($\varepsilon_Y = \varepsilon_Z = 0$), we have

$$\left.\begin{aligned} \Delta &= \varepsilon_X \\ e_{XX} &= \varepsilon_X - \tfrac{1}{3}\varepsilon_X = \tfrac{2}{3}\varepsilon_X \end{aligned}\right\} \tag{7.17}$$

Thus the relationship between longitudinal stress and longitudinal strain can be obtained,

$$\sigma_X = -p + s_{XX} = K\Delta + 2Ge_{XX} = \left(K + \tfrac{4}{3}G\right)\varepsilon_X \tag{7.18a}$$

or from Eq. (7.10), we obtain

$$\sigma_X = (\lambda + 2\mu)\varepsilon_X = E_L\varepsilon_X \tag{7.18b}$$

where E_L is called the **laterally constrained elastic modulus**

$$E_L = K + \tfrac{4}{3}G = \lambda + 2\mu = \frac{(1-\nu)E}{(1+\nu)(1-2\nu)} \tag{7.18c}$$

Substituting Eq. (7.18) into Eq. (7.7), the wave velocity of elastic 1D strain longitudinal waves, C_L^e, can be obtained

$$C_L^e = \sqrt{\frac{E_L}{\rho_0}} = \sqrt{\frac{1}{\rho_0}\left(K + \frac{4}{3}G\right)} = \sqrt{\frac{1}{\rho_0}(\lambda + 2\mu)} = \sqrt{\frac{1}{\rho_0}\frac{(1-\nu)E}{(1+\nu)(1-2\nu)}} \tag{7.19}$$

The laterally constrained stresses σ_Y and σ_Z can be obtained from Eqs. (7.8) and (7.10)

$$\sigma_Y = \sigma_Z = \frac{\nu}{1-\nu}\sigma_X = \frac{\lambda}{\lambda + 2\mu}\sigma_X \tag{7.20}$$

Therefore, all the methods used and the results obtained in the previous discussions of elastic 1D stress longitudinal waves can be in principle adopted to the analyses of elastic 1D strain longitudinal waves, only if that the Young's modulus E is replaced by the laterally constrained elastic modulus E_L. However, the following points should be noticed.

(1) Since $\lambda > 0$, $\mu > 0$, then it is known from Eq. (7.20) that the signs of σ_Y and σ_Z are always the same as the sign of σ_X. So, when a 1D strain plane wave propagates in the media, it is either in the state of 3D compression or 3D tension.

(2) Since in general $0 < \nu < 0.5$, then it is known from Eqs. (7.18) and (7.19) that $E_L > E$, $C_L^e > C_0$, namely a 1D strain elastic longitudinal wave propagates faster than a 1D stress elastic longitudinal wave in bars. Data of C_L^e for common materials are shown in Table 7.1 (Kolsky, 1953), referring to the data of C_0 in the Table 2.1 and C_T in the Table 2.2. For common metals, ν is in the range of 1/4–1/3, and correspondingly E_L/E is in the range of 1.2–1.5, C_L^e/C_0 is in the range of 1.1–1.22. When $\nu \to 1/2$ (namely for incompressible materials), $C_L^e \to \infty$.

(3) The 1D strain elastic longitudinal wave is actually the irrotational plane wave in isotropic, homogenous, linear elastic, infinite media, and C_L^e is actually the velocity of irrotational wave in infinite elastic media. These will be discussed further in Chapter 11.

Table 7.1. Lame's coefficients and the values of 1D strain elastic longitudinal wave velocity C_L^e for some common materials.

	Steel	Copper	Aluminum	Glass	Rubber
λ (GPa $= 10^{10}$ dyne/cm^2)	112	95	56	28	10
μ (GPa $= 10^{10}$ dyne/cm^2)	81	45	26	28	7.0×10^{-4}
C_L^e (km/s)	5.94	4.56	6.32	5.80	1.04

7.3 Elastic–Plastic Constitutive Relationship in 1D Strain Condition

First recall that in 1D stress condition, plastic deformation occurs when the stress reaches the yield stress Y determined in 1D tension test of materials. The yielding criterion can be expressed by

$$\sigma_X = Y \tag{7.21}$$

When the strain-hardening effect is taken into consideration, Y is a function of plastic strain ε_X^p:

$$Y = Y_p\left(\varepsilon_X^p\right) \tag{7.22}$$

or a function of plastic work W_p:

$$Y = Y_W\left(W_p\right) \tag{7.23}$$

where plastic work W_p in 1D stress condition is

$$W_p = \int \sigma_X d\varepsilon_X^p \tag{7.24}$$

According to Eq. (7.22), the derivative of stress with respect to plastic strain is

$$\frac{d\sigma_X}{d\varepsilon_X^p} = \frac{dY_p}{d\varepsilon_X^p} = Y_p' \tag{7.25}$$

while according to Eq. (7.23), it is

$$\frac{d\sigma_X}{d\varepsilon_X^p} = \frac{dY_W}{dW_p} \cdot \frac{dW_p}{d\varepsilon_X^p} = \frac{dY_W}{dW_p} . Y_W = Y_W Y_W' \tag{7.26}$$

So we have $Y_p' = Y_W Y_W'$.

Assuming that the total strain ε is the sum of the elastic strain ε^e and the plastic strain ε^p

$$\varepsilon = \varepsilon^e + \varepsilon^p \tag{7.27}$$

and since in the elastic scope

$$\frac{d\varepsilon_X^p}{d\sigma_X} = \frac{1}{E} \tag{7.28}$$

then in the plastic scope, considering Eqs. (7.25) and (7.26), there should be:

$$\frac{d\varepsilon_X}{d\sigma_X} = \frac{d\varepsilon_X^e}{d\sigma_X} + \frac{d\varepsilon_X^p}{d\sigma_X} = \frac{1}{E} + \frac{1}{Y_P'} = \frac{1}{E} + \frac{1}{Y_W Y_W'}$$

Therefore, the slope $d\sigma_X/d\varepsilon_X$ of the plastic σ_X–ε_X curve in 1D stress condition can be expressed by E and Y_p or Y_W as

$$\frac{d\sigma_X}{d\varepsilon_X} = E_p = \frac{EY_P'}{E + Y_W'} = \frac{EY_W Y_W'}{E + Y_W Y_W'} \tag{7.29}$$

The E_P is called the **plastic hardening modulus**. If E_p is a constant, the material is a linear hardening material.

In the general 3D condition, assuming that the material is isotropic, the hydrostatic stress contributes none to yielding, and the Bauschinger's effect is negligible, then there are two yield criteria extensively used: the Mises criterion (the maximum distortion energy criterion) and the Tresca criterion (the maximum shear stress criterion).

According to the Mises criterion, Eq. (7.21) is generalized to the following criterion

$$\sigma_i = \sqrt{3J_2} = \frac{\sqrt{2}}{2}\sqrt{(\sigma_1 - \sigma_2)^2 + (\sigma_2 - \sigma_3)^2 + (\sigma_3 - \sigma_1)^2} = Y \tag{7.30}$$

where $\sigma_1, \sigma_2, \sigma_3$ are the principle stresses, $J_2 = s_{ij}s_{ij}/2$ is the second invariable of deviator stress tensor, and σ_i is called **effective stress** or **stress intensity** ($\sigma_i = \sigma_1$ in 1D condition). The physical meaning of this criterion is that the plastic deformation begins when the elastic distortion energy $W_d = J_2/(2G)$ reaches its critical value for yielding.

The Tresca criteria can be expressed as

$$\sigma_1 - \sigma_3 = \pm Y \tag{7.31}$$

where σ_1 and σ_3 correspond to the maximum principle stress and the minimum principle stress respectively. The physical meaning of this criterion is that plastic deformation begins when the maximum shear stress τ_{max} [$=(\sigma_1 - \sigma_3)/2$] reaches its critical value for yielding.

In the current 1D strain condition, these two criteria have the same form:

$$\sigma_X - \sigma_Y = \pm Y \tag{7.32}$$

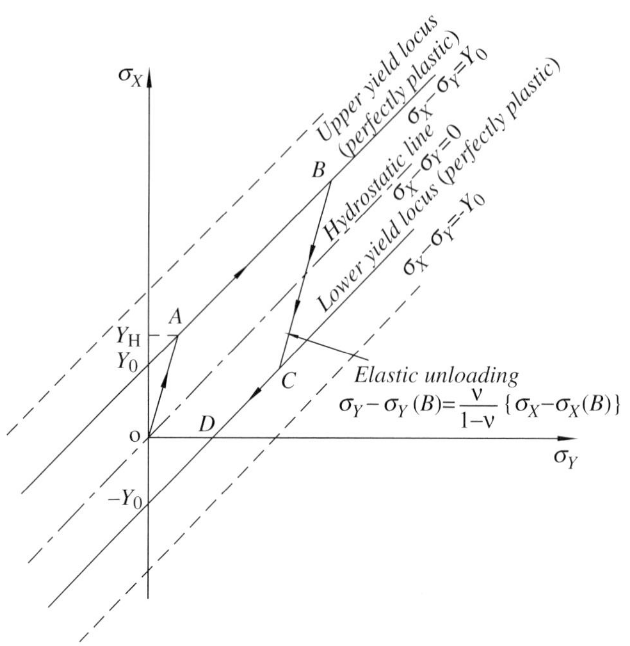

Fig. 7.1. Upper and lower yield loci on stress plane (σ_X, σ_Y).

On the stress plane (σ_X, σ_Y), this equation is graphically represented by two parallel straight lines with slopes equal to 1, which are called the upper and the lower yield loci, respectively, as shown in Fig. 7.1. The range between these two lines is an elastic zone. The two loci are parallel and symmetric to the hydrostatic line $\sigma_X = \sigma_Y (= \sigma_Z)$, in consistence with the assumption that the hydrostatic stress contributes none to yielding. For the perfectly plastic material, $Y = Y_0$, then the yield loci are invariable. As for the isotropic hardening material, since Y is a function of plastic deformation or plastic work, then the distance between the two yield loci increases with increasing the plastic deformation or plastic work, as shown by the two dashed lines in Fig. 7.1.

Noticing the relationship between σ_Y and σ_X in the 1D strain elastic deformation condition [Eq. (7.20)], substituting this relationship into Eq. (7.32), then the initial yield strength Y_H with respect to the axial stress σ_X in 1D strain is

$$Y_H = \frac{1-\nu}{1-2\nu} Y_0 = \frac{\lambda + 2\mu}{2\mu} Y_0 = \frac{K + 4G/3}{2G} Y_0 \tag{7.33}$$

which is called the **limit of laterally constrained yielding** or **Hugoniot elastic limit**, corresponding to the point A in Fig. 7.1. Obviously, there must be $Y_H > Y_0$. For example, if $\nu = 1/3$, then $Y_H = 2Y_0$.

Based on the above discussions, we will further discuss the elastic–plastic relationship, σ_X–ε_X, in the 1D strain condition as follows.

Assume that the plastic deformation contributes none to the volumetric deformation, namely

$$\varepsilon_X^p + \varepsilon_Y^p + \varepsilon_Z^p = 0 \tag{7.34}$$

then the volumetric deformation law is completely a pure elastic law. Moreover, if we notice that in the 1D strain condition we have Eq. (7.17) and the relationship below

$$\sigma_{XX} = \sigma_X + p = \tfrac{2}{3}(\sigma_X - \sigma_Y) = \tfrac{4}{3}\tau \tag{7.35}$$

then the elastic–plastic stress–strain relationship in the 1D strain condition can be written as

$$\left\{\begin{array}{l} \text{Volumetric law}: -p = K\varepsilon_X \\ \text{Distortional law}: \sigma_X - \sigma_Y = \left\{\begin{array}{ll} 2G\varepsilon_X & \text{(Elastic)} \\ \pm Y & \text{(Plastic)} \end{array}\right. \end{array}\right\} \tag{7.36}$$

So, similar to Eq. (7.18), for the loading, the axial relationship of σ_X–ε_X in 1D strain condition can be written as (refer to Fig. 7.2),

$$\sigma_X = -p + s_{XX} = -p + \tfrac{2}{3}(\sigma_X - \sigma_Y) = \left\{\begin{array}{ll} \left(K + \tfrac{4}{3}G\right)\varepsilon_X, & \sigma_X \le Y_H \\ K\varepsilon_X + \tfrac{2}{3}Y, & \sigma_X \ge Y_H \end{array}\right. \tag{7.37}$$

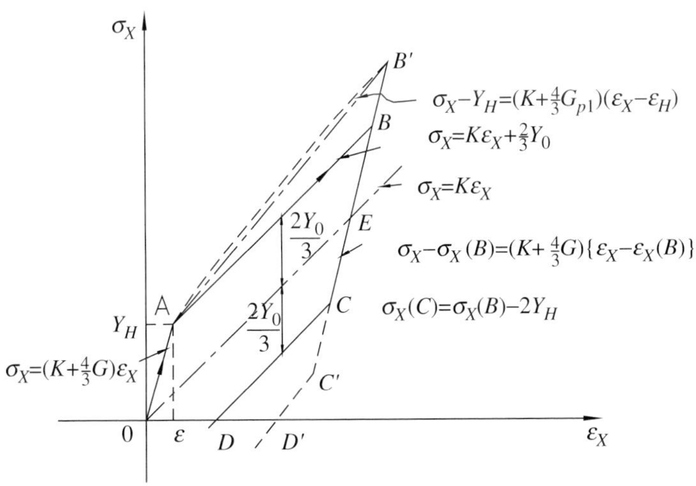

Fig. 7.2. Axial elastic plastic σ_X–ε_X curve in 1D strain condition.

The right side of the above equation is a sum of two items. The first item is related to the volumetric deformation, as graphically represented by the straight line *OE* with slope *K* in Fig. 7.2, while the second item is related to the distortional deformation.

For a perfect plastic material, $Y = Y_0$, then the plastic part of σ_X–ε_X curve parallels to the line *OE* with a perpendicular distance of $2Y_0/3$, as shown by the line *AB* in Fig. 7.2. In such a case, the slopes of σ_X–ε_X curve for both elastic loading and plastic loading are invariable, namely

$$\frac{d\sigma_X}{d\varepsilon_X} = \begin{cases} K + \frac{4}{3}G, & \sigma_X \le Y_H \\ K, & \sigma_X \ge Y_H \end{cases} \tag{7.38}$$

For an isotropic hardening plastic material, the situation is a bit complicated. As can be seen from Eqs. (7.36) and (7.37), the key to determine the slope $d\sigma_X/d\varepsilon_X$ lies on determining $d(\sigma_X - \sigma_Y)/d\varepsilon_X$; namely, how to determine the slope of the distortional stress–strain curve, $d(\sigma_X - \sigma_Y)/d(\varepsilon_X - \varepsilon_Y)$ or $d\tau/d\gamma$, where τ is the maximum shear stress while γ is the maximum shear strain. Based on the symmetry of deformation in the 1D strain condition ($\varepsilon_Y^p = \varepsilon_Z^p$) and the assumption that plastic deformation contributes none to volume change [Eq. (7.34)], it can be seen that there is only one independent among the three principle plastic strains, namely

$$\varepsilon_X^p = \tfrac{2}{3}\left(\varepsilon_X^p - \varepsilon_Y^p\right) = \tfrac{4}{3}\gamma^p \tag{7.39}$$

Then, if *Y* is regarded as a function of axial plastic strain ε_X^p and taken into account of strain hardening, then from the distortional law in Eq. (7.36) it is known that the $d\tau/d\gamma$ in the elastic part and the plastic part can be respectively expressed as

$$\left.\begin{aligned} \frac{d\tau}{d\gamma^e} &= 2G \\ \frac{d\tau}{d\gamma^p} &= \frac{1}{2}\frac{dY_p}{d\varepsilon_X^p}\cdot\frac{d\varepsilon_X^p}{d\gamma^p} = \tfrac{2}{3}Y_p' \end{aligned}\right\} \tag{7.40}$$

Thus, similar to Eq. (7.25) deduced in the 1D stress condition, assuming that the total strain is the sum of an elastic part and a plastic part [Eq. (7.25)], we have

$$\frac{d\gamma}{d\tau} = \frac{d\gamma^e}{d\tau} + \frac{d\gamma^p}{d\tau} = \frac{1}{2G} + \frac{3}{2Y_p'} = \frac{3G + Y_p'}{2GY_p'}$$

Corresponding to the shear modulus *G* in elastic distortional law, the half of the slope of plastic distortional curve is defined as the **plastic shear modulus** G_p

$$G_p = \frac{1}{2}\frac{d\tau}{d\gamma} = \frac{GY_p'}{3G + Y_p'} \tag{7.41a}$$

Similar to deducing Eq. (7.22b), if Y is regarded as a function of the axial plastic work W_p, namely $Y_W(W_p)$, and noting that in the 1D strain condition there is

$$dW_p = \sigma_X d\varepsilon_X^p + 2\sigma_Y d\varepsilon_Y^p = (\sigma_X - \sigma_Y)\, d\varepsilon_X^p = \tfrac{8}{3}\tau d\gamma^p$$

then we have

$$\frac{d\tau}{d\gamma^p} = \frac{1}{2}\frac{dY_W}{dW_p}\cdot\frac{dW_p}{d\gamma^p} = Y_W' \cdot \tfrac{2}{3}(\sigma_X - \sigma_Y) = \tfrac{2}{3}Y_W Y_W'$$

Comparing with Eq. (7.40), it can be seen that $Y_p' = Y_W Y_W'$. This is consistent with the Eq. (7.26) deduced in the 1D stress condition. Then the plastic shear modulus can be also written as

$$G_p = \frac{GY_W Y_W'}{3G + Y_W Y_W'} \tag{7.41b}$$

After introducing the plastic shear modulus, according to Eq. (7.37), the slope of axial stress–strain curve of an isotropic hardening plastic material in the 1D strain condition is

$$\frac{d\sigma_X}{d\varepsilon_X} = \begin{cases} K + \frac{4}{3}G, & \sigma_X \le Y_H \\ K + \frac{4}{3}G_p, & \sigma_X \ge Y_H \end{cases} \tag{7.42}$$

It is obvious that the expression of slope of plastic part is formally similar to that of elastic part, only if the G is replaced by G_p. For a linear hardening plastic material, $G_p = G_{pl}$ (constant), the plastic part of the axial stress–strain curve is a straight line, as graphically presented by the dash-dotted line AB' in Fig. 7.2. This corresponds to

$$Y_p' = Y_W Y_W' = E_Y \text{ (constant)}$$

which indicates Y_p is a linear function of ε_X^p:

$$Y_p = Y_H + E_Y \varepsilon_X^p \tag{7.43a}$$

or equivalently, Y_W^2 is a linear function of plastic work W_p:

$$Y_W^2 = Y_H^2 + 2E_Y W_p \tag{7.43b}$$

where Y_H is the Hugoniot elastic limit of materials [Eq. (7.33)], E_Y is a material constant characterizing the linear hardening, both of which are determined by experiments.

Now, let us look at the loading path and the unloading path on both the plane $(\sigma_X, \varepsilon_X)$ (Fig. 7.2) and the plane (σ_X, σ_Y) (Fig. 7.1), and special attention will be focused on what differs with those in the 1D stress condition. As an example, we discuss the situation for the perfectly plastic material. When it is loaded from zero-state ($\sigma_X = \varepsilon_X = 0$), the loading path first follows the line *OA* representing the linear elastic deformation process. On the σ_X–ε_X plane, according to Eq. (7.37), its slope is $(K + 4/3G)$; while on the σ_X–σ_Y plane, according to Eq. (7.20), its slope is $\nu/(1-\nu)$. After the σ_X reaches the laterally constrained yield limit Y_H Eq. (7.23), the loading path follows the line *AB* representing the plastic deformation process. On the σ_X–ε_X plane, the slope of stress–strain curve is *K* according to Eq. (7.37), while on the σ_X–σ_Y plane, the slope is 1 according to Eqn (7.36). If the material is unloaded at the point *B*, following the assumption of elastic unloading, the unloading path follows the line *BC* which is parallel to the elastic loading line *OA*. On the plane σ_X–ε_X, the unloading stress–strain relation is expressed by

$$\sigma_X = \sigma_X(B) + \left(K + \tfrac{4}{3}G\right)\{\varepsilon_X - \varepsilon_X(B)\} \tag{7.44}$$

while on the plane σ_X–σ_Y, it corresponds to

$$\sigma_Y = \sigma_Y(B) + \frac{\nu}{1-\nu}\{\sigma_X - \sigma_X(B)\} \tag{7.45}$$

Note that the material is only subjected to hydrostatic pressure when it is unloaded to the point *E*. After that, a sign transition of shear stress (and the corresponding distortional deformation) takes place. With a further unloading of σ_X, the shear stress increases in the opposite direction, namely the elastic distortional deformation is loaded in the negative direction following the line *EC*. While at the point *C*, according to Eq. (7.32), the following reverse yield criterion is satisfied,

$$\sigma_X - \sigma_Y = -Y$$

It indicates that a reverse plastic deformation begins in the material. The stress at *C*, $\sigma_X(C)$, can be determined by the above equation and Eq. (7.45)

$$\sigma_X(C) = \sigma_X(B) - \frac{2(1-\nu)}{1-2\nu}Y = \sigma_X(B) - 2Y_H \tag{7.46}$$

After that, with regard to the original axial stress σ_X, it is an unloading process following the line *CD*; while with regard to the yield criteria it is actually a **reverse plastic loading** process following the lower yield locus *CD*, although it is sometimes inadequately called **plastic unloading**.

7.4 One Dimensional Strain Elastic–Plastic Waves

The elastic–plastic σ_X–ε_X relationship in the 1D strain condition discussed in the last section is based on the assumption that the elastic deformation follows Hooke's law. Based on those discussions, now let us analyze the propagation of elasto-plastic 1D strain wave. Research in this area is developed from the works of Wood (1952) and others.

In the loading stage, according to Eq. (7.19), the wave velocity of 1D strain elastic wave is

$$C_L^e = \sqrt{\frac{K + \frac{4}{3}G}{\rho_0}}$$

According to Eqs. (7.7) and (7.42), the wave velocity of 1D strain plastic wave is

$$C_L^p = \sqrt{\frac{K + \frac{4}{3}G_p}{\rho_0}} \tag{7.47}$$

Since $G > 0$ and $Y_p' > 0$ (due to hardening effect), it can be seen from Eq. (7.41) that $G_p/G < 1$, and then consequently we obtain

$$C_L^p = C_L^e$$

This indicates that, same as in the situation of 1D stress waves, during the propagation of a 1D strain elasto-plastic wave, a faster precursor elastic wave propagates ahead, and then is followed by a slower plastic wave. Analyses of the whole problem could follow the same way that was used in analyses of 1D stress waves, including the conservation conditions across the wave front as discussed in Section 2.7. Those will be no longer repeated here. However, there are some points to be specially noticed:

(1) For a perfect plastic material, the plastic portion of its stress–strain curve in the 1D stress condition is a horizontal line $\sigma_X = Y$ (Fig. 7.3). So, there is no one-to-one corresponding, single-valued function relationship between stress and strain. This violates the second assumption of the rate-independent stress–strain relationship [Eq. (2.14)] required in establishing the rate-independent theory of stress waves (Chapter 2). Otherwise, we will deduce that the velocity of a plastic wave is zero, or no propagation of plastic deformation. However, in the 1D strain condition, the rate-independent theory of stress waves can still be adopted for perfect plastic materials. This is because although the distortional δ_X–ε_X curve is graphically a

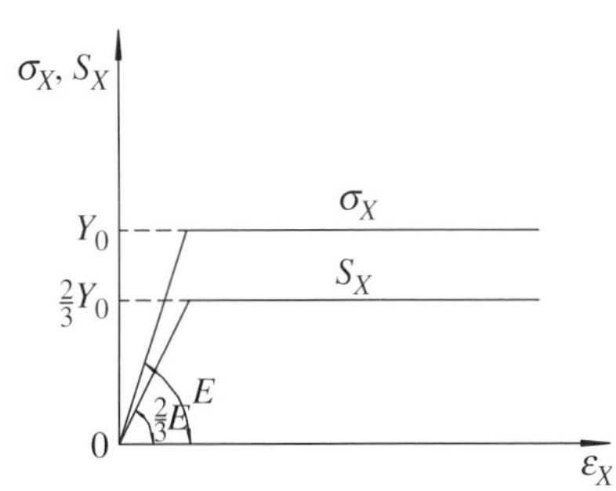

Fig. 7.3. Stress–strain relation of perfect plastic material in 1D stress condition.

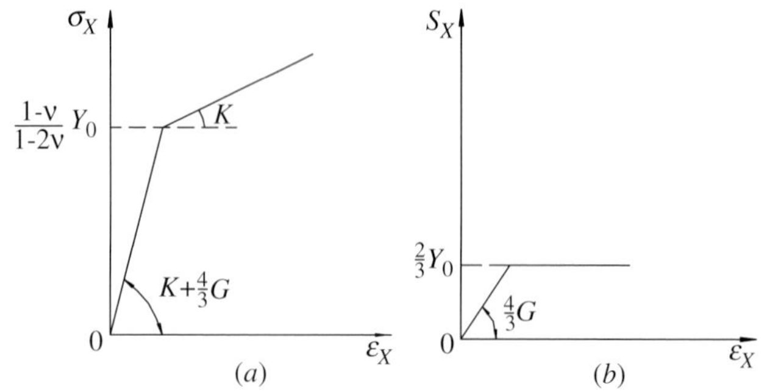

Fig. 7.4. Stress–strain relation of perfect plastic material in 1D strain condition.

horizontal line [Fig. 7.4(b)], the axial stress σ_X now is a one-to-one corresponding single-valued function of the axial strain ε_X, with a linear relation for its plastic portion whose slope depends on the value of K, as shown in Fig. 7.4(a). Thus, formally it takes the same form as the linear hardening stress–strain relationship in the 1D stress condition.

(2) Generally, the plastic wave velocity C_L^p in 1D strain condition is a function of plastic strain ε_X^p or plastic work W_p [Eqs. (7.47) and (7.41)]. For linear hardening and perfect plastic materials, the plastic wave velocity is a constant, and if denoted by C_{L1}^p and C_K respectively, then we have

$$C_{L1}^p = \sqrt{\frac{1}{\rho_0}\left[K + \frac{4GE_Y}{3(3G+E_Y)}\right]} \tag{7.48}$$

$$C_K = \sqrt{\frac{K}{\rho_0}} \tag{7.49}$$

C_K is also called the body wave velocity.

(3) In the 1D stress condition, the plastic wave velocity C_L is much smaller than the elastic wave velocity C_0, about one order less. For example, the ratio E_L/E of metals is about 0.01, then we obtain $(C_L/C_0) \approx 0.1$. While in the 1D strain condition, the plastic wave velocity is relatively high, its lower limit is the body wave velocity C_K under the perfect plastic assumption. Since $K \approx E$, the body wave velocity approaches the velocity of elastic longitudinal wave in bars, $C_K \approx C_0$. For example, if $\nu = 1/3$, then it can be seen from Eqs. (7.16) and (7.11) that $K = E, G = 3E/8$, and consequently $C_K = C_0, C_L^e = \sqrt{2/3}C_0 \approx 1.22C_0$, or $(C_K/C_L^e) \approx 0.82$.

The above discussions are for the elastic–plastic loading process.

During the unloading process, first, there is elastic unloading, and according to Eq. (7.44) the elastic unloading disturbances propagate with C_L^e, the elastic longitudinal wave velocity in infinite media. Because $C_L^e > C_L^p$, namely an elastic unloading wave propagates faster than a plastic loading wave, then similar to those in 1D stress condition, there exist the pursuing unloading and the head-on unloading, etc., depending on different actual situations. However, because C_L^e is not much larger than C_L^p, during the pursuing unloading, the plastic loading wave in the 1D strain condition attenuates slower than that in the 1D stress condition. As to the propagation of elastic–plastic boundaries, the methods used for the 1D stress waves (Section 4.10) can be similarly adopted here. But the difference in yield criterion between those two conditions and the possibility of the appearance of reverse plastic loading boundaries should be noticed (Wang *et al.*, 1983).

Comparing with the 1D stress waves, one unique feature of the 1D strain waves is the propagation of reverse plastic loading waves when σ_X satisfies the reverse yield criterion [Eq. (7.32)] even σ_X has not unloaded to zero yet. During the reverse plastic loading, since $C_L^p < C_L^e$, namely the latter formed reverse plastic wave propagates slower than the earlier formed elastic unloading wave, then the distance between these two wave fronts increases with wave propagation.

7.5 Influence of Reverse Yield on the Propagation of 1D Strain Elastic–Plastic Wave

As an example, let us consider a semi-infinite media subjected to a rectangular impulse (Wang, 1982). Initially, the media is in the natural state. At $t = 0$, the surface of the media ($X = 0$) is suddenly subjected to a uniformly distributed, constant pressure σ_X^*; after $t = T$, it is suddenly unloaded to zero, as shown in Fig. 7.5.

Thus, a loading **dual-wave formation** first occurs at $t = T$, namely, a strong discontinuous elastic 1D strain precursory wave and a strong discontinuous plastic 1D strain wave simultaneously propagate from $X = 0$ with constant velocity C_L^e and C_K; respectively. They are graphically represented by the line *OA* and *OB* in Fig. 7.5(a), respectively. The constant-valued zones 0, 1, and 2 in Fig. 7.5(a) correspond to the points 0, 1, and 2 on the σ_X–v_X plane in Fig. 7.5(b). It is easy to determine

$$\sigma_{X0} = v_{X0} = 0$$

$$\sigma_{X1} = -Y_H, \quad v_{X1} = -\frac{\sigma_{X1}}{\rho_0 C_L^e} = -\frac{Y_H}{\rho_0 C_L^e}$$

$$\sigma_{X1} = \sigma_X^*, \quad v_{X2} = \frac{1}{\rho_0 C_K}\left\{\left(\frac{C_K}{C_L^e} - 1\right) Y_H - \sigma_X^*\right\} = v_X^*$$

since they are totally similar to those for the linear hardening plastic materials in the 1D stress condition.

However, from the beginning of unloading the situation is totally different. At $t = T$, when the stresses on the surface boundary suddenly unload to zero, an unloading dual-wave formation occurs, namely, not only a strong discontinuous elastic 1D strain unloading wave *TB* propagates with a constant velocity C_L^e, but also a strong discontinuous

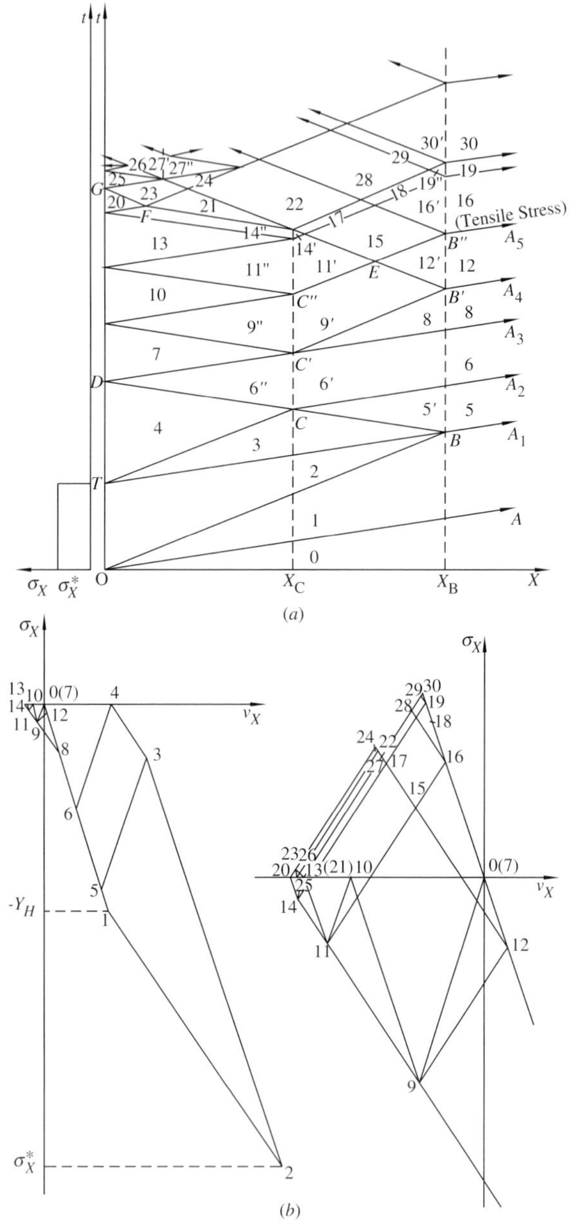

Fig. 7.5. The influence of reverse yield on the propagation of 1D strain wave.

plastic 1D strain reverse loading wave *TC* simultaneously propagates with a constant velocity C_K. The constant-valued zones 3 and 4 behind these two wave fronts correspond to the points 3 and 4 on the σ_X–v_X plane. The states in these two zones can be determined by using the compatibility conditions across discontinuous wave front and the reverse yield condition:

$$\sigma_{X3} = \sigma_X^* + 2Y_H, \quad v_{X3} = v_X^* - \frac{2Y_H}{\rho_0 C_L^e}$$

$$\sigma_{X4} = 0, \quad v_{X4} = v_{X3} + \frac{\sigma_{X3}}{\rho_0 C_K}$$

For the instant before the strong discontinuous elastic unloading wave *TB* catches up with the positive plastic loading wave *OB*, the wave profiles of σ_X–X and ε_X–X are shown as the solid line and the dashed line respectively in Fig. 7.6(a). During the wave propagating process, the distance between the positive plastic loading wave and the positive elastic loading wave increases, while the distance between the elastic unloading wave and the

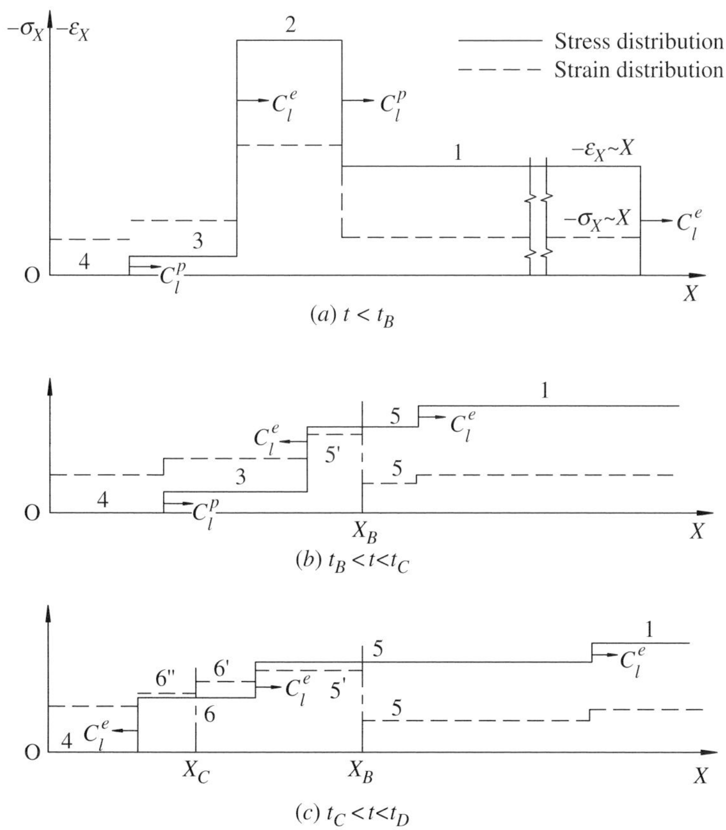

Fig. 7.6. Distribution of stress (solid lines) and strain (dashed lines) at various times.

positive plastic loading wave decreases. At the time $t = t_B$, TB catches up with OB and they interact with each other (pursuing unloading). For convenience, assume that the unloading residual particle velocity v_{X3} is not too high which satisfy the following condition

$$v_{X3} < v_{X1} + \frac{\sigma_{X3} - \sigma_{X1}}{\rho_0 C_L^e}$$

in other words, if

$$\sigma_X^* \geq -\left(\frac{1 + 5\dfrac{C_L^p}{C_L^e}}{1 + \dfrac{C_L^p}{C_L^e}} \right) Y_H$$

then the positive loading wave vanishes after their interaction, and a stationary strong discontinuous strain interface is formed at the point B. The rightward wave from the point B is an elastic unloading wave B_{A1}, while the leftward wave from the point B is an internal reflective elastic wave BC. Note that with regard to the elastic unloading zone 3 the leftward wave BC is an elastic loading wave, reloading the media from σ_{X3} to σ_{X5}; while with regard to the reverse plastic loading wave TC, BC is an elastic unloading wave (head-on unloading wave). In fact, at the time $t = t_C$, BC and TC meet and interact with each other, and consequently a stationary strong discontinuous strain interface is formed at C. The rightward wave from the point C is an elastic internal reflective unloading wave C_{A2}, further unloading the media from σ_{X5} to σ_{X6}. The states in the zones 5 and 6 and in the other further formed zones as well, can be obviously determined by the compatibility conditions across the wave fronts and the continuity requirement that the stresses and particle velocities across the interfaces where waves interact, should be respectively equal to each other, as mentioned previously. The results obtained are shown in Fig. 7.5. It is not difficult for the readers to understand these results by comparing the characteristics in (X, t) plane and the characteristics in (σ_X, v_X) plane. However, the following points should be emphasized:

(1) Corresponding to B and C, two stationary strong discontinuous interfaces with respect to strain are formed at X_B and X_C. It should be noted that they are of different characters. The former is with regard to the positive plastic loading, namely there is a jump of positive plastic strain at X_B; the strain at left side is larger than that at right side, as shown in Fig. 7.6(b). The latter is with respect to the reverse plastic loading, namely there is a jump of reverse plastic strain at X_C; the strain at left side is smaller than that at right side, as shown in Fig. 7.6(c).

(2) Since the media on the left side of the stationary strong discontinuous strain interface X_C have experienced reverse plastic loading up to the value of $\sigma_{X4} = 0$ (zone 4), their reverse succeeded-yield stress due to strain-hardening is zero, then in the interval OX_C, the disturbance $\sigma_X = 0$ propagating rightward from the unloading surface, such as DC, is an elastic wave. On the other hand, in the interval $X_C X_B$ on the right side of the discontinuous interface X_C, the media only experienced positive plastic loading up to the value of $\sigma_{X2} = \sigma_X^*$, then their reverse succeeded-yield

stress is $\sigma_{X3} = \sigma^* + 2Y_H$. Consequently, any unloading disturbance whose absolute value is less than $|\sigma_{X3}|$ will lead to reverse plastic loading. Therefore, when DC' is incident on to the stationary discontinuous strain interface X_C, similar to the situation when wave reflection and transmission occur when an elastic wave propagates from a material with high yield limit into a material with low yield limit (assuming the two materials have identical sound resistance), a rightward reverse plastic loading wave $C'B'$ will be transmitted and a leftward elastic unloading wave will be reflected. On this analogy, all the rightward waves reflected from the unloading surface are elastic waves in the interval OX_C, but when they are incident on to X_C, they will transmit reverse plastic loading waves, only if the stationary discontinuous interface X_C exists. Thereby, the intensity of reverse plastic loading waves on the right side of X_C successively increases as the zone 9′, 11′, and 14′ successively appear in Fig. 7.5 (the absolute value of $|\sigma_X|$ successively decreases and approaches zero).

(3) Since the media in the interval $X_C X_B$ on the left side of the stationary strong discontinuous strain interface X_B experienced positive plastic loading up to the value of $\sigma_{X2} = \sigma_X^*$, while the media on the right side of X_B only experienced elastic deformation, their reverse succeeded-yield stress is $\sigma_{X3} = \sigma_X^* + 2Y_H$ on the left side but is still Y_H on the right side of X_B. Consequently, when a rightward reverse plastic loading wave, such as $C'B'$, propagates to X_B, similar to the situation when a plastic wave propagates from a material with low yield limit into a material with high yield limit (assuming the two materials have identical sound resistance) and the amplitude of the plastic wave is lower than the yield limit of the latter material, there must exist a reflection of a further loaded plastic wave and a transmission of only an elastic wave, then a leftward reverse plastic loading wave $B'E$ will be reflected and a rightward elastic unloading wave $B'A_4$ will be transmitted, where the word unloading is relative to the original positive loading. Thus it can be seen that the reverse plastic wave propagation ends at the stationary strong discontinuous strain interface X_B, unless the intensity of the rightward incident reverse plastic loading wave is higher than the Y_H (tensile) so that the reverse plastic loading could take place on the right side of X_B.

(4) When the leftward reverse plastic loading wave $B'E$ reflected from X_B meets the rightward reverse plastic loading wave $C''E$ transmitted across X_C, a further reverse plastic loading must take place (head-on loading), same as the intensifying effect resulting from the head-on loading of two waves with the same sign. Consequently the tensile stress zone could be formed. Moreover, on the one hand, because of the interaction of the leftward internal reflective tensile waves with a series of successively intensified rightward reverse plastic loading waves transmitted across X_C, enable the magnitude of tensile stress to increase (as in the zone 17). On the other hand, the reflective loading must take place when the rightward internal reflective waves arrive at the stationary discontinuous strain interface X_B, enable the tensile stress to increase further (as in the zone 16). Therefore, for the reverse plastic loading zones in the interval $X_C X_B$, the magnitude of tensile stress in those zones will be successively larger and larger. Meanwhile, the magnitude of tensile stress in the reverse elastic loading zones (namely the tensile elastic unloading zones) formed correspondingly on the right side of X_B will also be successively larger and larger.

(5) Since the media on the left side of the stationary strong discontinuous strain interface X_C only experienced reverse plastic loading up to the value of $\sigma_{X4} = 0$ (zone 4), when the reverse plastic loading waves (tensile stress waves) developed in the interval $X_C X_B$ are leftward incident on to X_C, similar to the situation when a plastic wave propagates from a material with high yield limit into a material with low yield limit (assuming the two materials have identical sound resistance), there will be leftward transmission of an elastic wave and a reverse plastic loading (tensile stress) wave (as in the zones 20 and 22). Along with the media on the left side of X_C getting into the tensile state of reverse plastic loading, the stationary strong discontinuous strain interface X_C disappears. The afterwards leftward reverse plastic loading waves with higher and higher tensile stresses (as in the zone 28) are not influenced by the disappearance of X_C. Finally, the tensile stress zone will be extended to the entire interval OX_B.

(6) If before the leftward plastic loading wave (tensile stress wave) FG is incident on to the unloading surface ($X = 0$), namely at $t < t_G$, the unloading boundary condition ($\sigma_X = 0$) in fact plays a role to promote the development of reverse plastic loading, then after that it contrarily plays an unloading role to the reverse plastic loading, and then the problem can be treated as the same as what was discussed previously for the unloading problem of 1D stress plastic waves.

Thus it can be seen, for the problems of one-dimensional strain elastic plastic wave propagation, only because of the reverse yield effect, the tensile stress zone can be formed near the surface where the impulsive compressive load is subjected to, and the corresponding particle velocity becomes negative. Such a situation neither exists in the problems of one-dimensional stress wave propagation discussed previously, nor exists in the problems of one-dimensional strain wave propagation based on the hydrodynamic model to be discussed in the next section. Once the tensile stress is larger than the critical tensile stress of the concerned material, the so-called **front spalling** may occur. This may be of significance in the study of dynamic fracture, particularly for the brittle materials with lower tensile strength or/and internal pre-cracks.

7.6 State Equation of Solids under High Pressures

The discussions of 1D strain elasto-plastic waves in the above two sections are based on the fact that the elastic deformation of materials follows the Hooke's law. It is clear that it is valid only when strain is small. In problems of high pressure and larger deformation, elastic moduli are not constants any more, but the functions of strain or stress. Then, the elastic components of constitutive equation are **nonlinear**.

Under high pressure, the distortional resistance (shearing strength) of solids can be neglected approximately. Then, the distortion part of constitutive relation can be neglected, and only the volumetric part should be considered. In other words, the constitutive relation is now simplified to a relationship between the hydrostatic pressure p and the volume strain Δ or the specific volume V; which is sometimes called the **state equation of solids under high pressure**. It is equivalent to regarding the solid under high pressure as the compressible nonviscous fluid.

In this section, the state equation of solids under high pressure, obeying the nonlinear volumetric law, will be discussed. In fact, state equations are not only valid in the studies of shock waves in solids under high pressure when shear strength can be neglected, but also valid in studies of nonlinear elasto-plastic waves in solids whose shear strength cannot be neglected.

With regard to the volume strain, according to its engineering definition, if denoted by Δ, is expressed as

$$\Delta = \frac{V - V_0}{V_0} = (1 + \varepsilon_X)(1 + \varepsilon_Y)(1 + \varepsilon_Z) - 1 \tag{7.50}$$

where V_0 is the initial specific volume ($V_0 = 1/\rho_0$), then the corresponding volumetric modulus is:

$$K = -\frac{dp}{d\Delta} = -V_0\frac{dp}{dV} \tag{7.51}$$

This is depicted in Lagrange system, and so is called Lagrange **volumetric modulus**.

While following the definition of logarithmic strain (true strain), if denoted by $\widetilde{\Delta}$, then it is expressed as

$$\widetilde{\Delta} = \ln\frac{V}{V_0} = \tilde{\varepsilon}_X + \tilde{\varepsilon}_Y + \tilde{\varepsilon}_Z, \quad \tilde{\varepsilon} = \ln(1 + \varepsilon) \tag{7.52}$$

then the corresponding volumetric modulus is

$$k = -\frac{dp}{d\widetilde{\Delta}} = -\frac{dp}{d\ln V} = -V\frac{dp}{dV} \tag{7.53}$$

It is depicted in Euler system, and so is called Euler **volumetric modulus**. Obviously, there exists a relationship between k and K

$$k = \frac{V}{V_0}K = (1 + \Delta)K \tag{7.54}$$

In the 1D strain condition, Eqs. (7.51), (7.53), and (7.54) can be reduced respectively to:

$$K = -\frac{dp}{d\varepsilon_X}, \quad k = -\frac{dp}{d\tilde{\varepsilon}} \tag{7.55a}$$

$$k = (1 + \varepsilon_X)K \tag{7.55b}$$

Bridgman equation The studies on state equation of solids under high pressure originated from the experimental studies on the relationship between the volumetric modulus and the hydrostatic pressure under static high pressure. Bridgman (1949) studied such relationships for several tens of elements and compounds under the range of static high pressure p about

10^4–10^5 bars (1–10 GPa). Based on his experimental results, he proposed an empirical equation:

$$\frac{V_0 - V}{V} = -\Delta = ap - bp^2 \tag{7.56}$$

where a and b are material constants. This equation is usually called the **Bridgman equation** or the **isothermal state equation of solids**. For most materials tested, when p is in the unit of bars, a is in the order of 10^{-6}–10^{-7} bar^{-1} and b is in the order of 10^{-12} bar^{-2}. For example, based on the data obtained from more than 400 tests for iron at temperature 24°C, there is:

$$\frac{V_0 - V}{V} = -\Delta = 5.826 \times 10^{-7} p - 0.80 \times 10^{-12} p^2$$

From Eq. (7.56), Lagrange volumetric modulus K as a function of hydrostatic pressure p can be obtained:

$$K(p) = \frac{1}{a - 2bp} = \frac{1}{a\left(1 - \frac{2b}{a}p\right)} \tag{7.57}$$

It is clear that K increases with the increase of p, namely the p–V curve is a concave curve as shown in Fig. 7.7. This indicates that the resistance of volumetric compression of solid materials increases with the increase of compressive deformation; in other words, it is more difficult to compress a solid as its volume is more compressed.

As b/a is in the order of 10^{-5}–10^{-6} bar^{-1}, Eq. (7.57) shows that a pressure in the order of 10^3–10^4 bars (0.1–1 GPa) is needed at least to obtain a change of 1% in K. This shows why the nonlinearity of elastic volumetric law is considered only in problems under high pressure. When $2bp/a < 1$, extending Eq. (7.57) into a power series and neglecting the

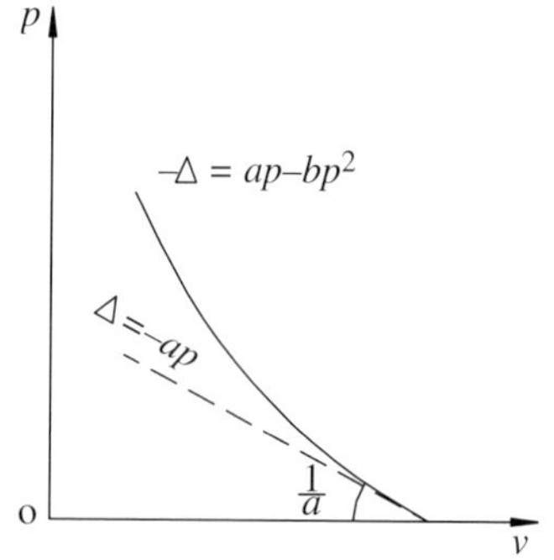

Fig. 7.7. The p–V curve according to the Bridgman equation [Eq. (7.56)].

small quantities of high-orders, we obtain

$$K(p) \approx \frac{1}{a}\left(1+\frac{2b}{a}p\right) \tag{7.58}$$

Obviously, $1/a$ is equivalent to the volumetric modulus K_0 of the linear elastic volumetric law under low pressure; while b is the coefficient depicting the change of K with p when K is approximated by a linear function of p.

The functions in other forms are also used to describe the relationship between p and V; and the material coefficients in these functions are determined by fitting the data form Bridgman's experiments. For instance, based on the theory of solid physics, Pack *et al.* (1948) suggested the following p–V relationship:

$$p = \alpha\left(\frac{V_0}{V}\right)^{\frac{2}{3}}\left\{\exp\left[\beta\left(1-\left(\frac{V}{V_0}\right)^{\frac{1}{3}}\right)\right]-1\right\} \tag{7.59a}$$

In the 1D strain condition, as $V = (1+\varepsilon_X)V_0$, the above equation can be written as:

$$p = \alpha(1+\varepsilon_X)^{-\frac{2}{3}}\left\{\exp\left[\beta\left(1-(1+\varepsilon_X)^{\frac{1}{2}}\right)\right]-1\right\} \tag{7.59b}$$

The values of α and β for several materials were determined based on the Bridgman's experimental data and are listed in Table 7.2 (Broberg, 1956).

Murnagham equation Now let us start from Euler volumetric modulus defined by Eq. (7.52) to discuss the relationship between k and p. If, similar to Eq. (7.58), we assume that k is a linear function of p, namely

$$k = k_0(1+\eta p) \tag{7.60}$$

where k_0 and η are material constants, then the p–V relationship can be written as an ordinary differential equation

$$-V\frac{dp}{dV} = k_0 + np$$

where $n = k_0\eta$. From initial condition $V|_{p=0} = V_0$, we obtain

$$\left(p+\frac{1}{\eta}\right)\left(\frac{V}{V_0}\right)^n = \frac{1}{\eta}\ \text{(constant)} \tag{7.61a}$$

Table 7.2. The values of α and β for several materials [Eq. (7.59b)].

Materials	Iron	Aluminum	Magnesium	Copper	Lead	Silex
ρ_0 (10^3 kg/m^3)	7.86	2.72	1.74	8.93	11.34	2.21
α (10^5 bar)	6.01	1.90	0.639	3.52	1.32	4.7
β	8.4	11.4	15.6	11.7	10.0	2.0

or rewritten as

$$p = \frac{k_0}{n}\left\{\left(\frac{V_0}{V}\right)^n - 1\right\} \tag{7.61b}$$

In the 1D strain condition, it is reduced to

$$p = \frac{k_0}{n}\{(1+\varepsilon_X)^{-n} - 1\} = \frac{k_0}{n}\{\exp(-n\tilde{\varepsilon}_X) - 1\} \tag{7.61c}$$

It should be noticed that Eq. (7.61a) in form is similar to the following isentropic equation for ideal gases

$$p\left(\frac{V}{V_0}\right)^\gamma = \text{constant}$$

And the n in Eq. (7.61a) is similar to the adiabatic exponent γ in the above equation for ideal gases. For metals, n typically takes 4. Equation (7.61) is usually called **Murnagham equation** or the **isentropic state equation of solids**, where material constant K_0 and n can be obtained by experiments on wave propagation.

The Murnagham equation was originated from research on the elastic theory of finite deformation by performing quite complicated mathematical calculations. As shown above, it is actually equivalent to the assumption that k is proportional to p [Eq. (7.60)], namely Euler volumetric modulus increase linearly with the increase of hydrostatic pressure. Obviously, the p–V curve depicted by Eq. (7.61) is also a concave curve, formally similar to that in Fig. 7.7. Or in the 1D strain compressive loading condition, the corresponding axial stress–strain curves σ_X–ε_X is a concave curve as shown in Fig. 7.8, representing the nonlinear elastic volumetric law.

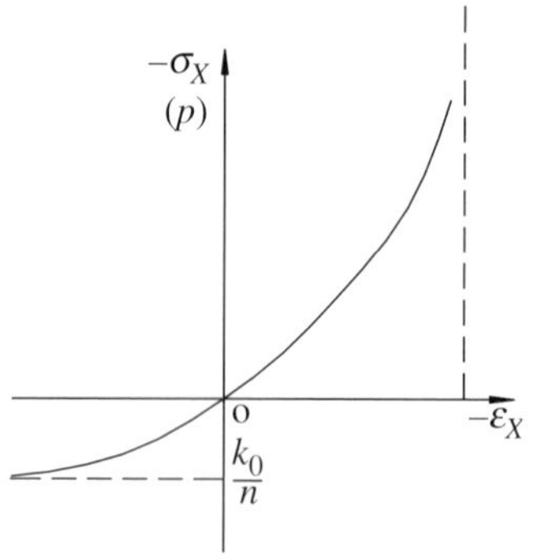

Fig. 7.8. The σ_X–ε_X curve corresponding to the Murnagham equation [Eq. (7.61)] in the 1D strain condition.

Grüneisen equation The Bridgman equation and the Murnagham equation depict the relationship between p and V in an isothermal process and an isentropic process respectively. Thereby, they are the state equations valid only under certain thermodynamics condition, and are insufficient to depict the relationships among state parameters under general thermodynamics conditions involving changes in temperature T and entropy S. In other words, it is inadequate to represent the state equations of solids under general conditions by a relationship relating only two state parameters p and V, and so the relationships among p, V, and other thermodynamics parameters must be considered. For example, we can use a series of Bridgman equations under different temperatures (a series of isothermal p–V curves) or a series of Murnagham equations under different entropies (a series of isentropic p–V curves) to describe the state equations under general conditions. These lead to respectively the so-called **thermal equations of state**:

$$f_T(p, V, T) = 0 \tag{7.62}$$

and the so-called **entropic equations of state:**

$$f_S(p, V, S) = 0 \tag{7.63}$$

They have been usually adopted in the studies on fluid mechanics. Geographically, they are depicted by a surface in a there dimensional space of p, V, T or a surface in a three dimensional space of p, V, S respectively. It is clear that the Bridgman equation represents only the intersection curve of the surface of Eq. (7.62) with a plane of constant temperature; while the Murnagham equation only represents the intersection curve of the surface of Eq. (7.63) and a plane of constant entropy. From the slope of these intersection curves respectively, the isothermal volumetric modulus k can be determined by

$$k_T = -V\left(\frac{\partial p}{\partial V}\right)_T \tag{7.64}$$

and the isentropic volumetric modulus k_S can be determined by

$$k_S = -V\left(\frac{\partial p}{\partial V}\right)_S \tag{7.65}$$

In previous discussions on linear elastic waves, the difference between k_T and k_S has not been emphasized.

However, for the research on shock waves in solids under high pressure, neither Eqs. (7.62) nor (7.63) is sufficient. Because from the equations based on the conservation of mass, momentum, and energy across a shock wave front (refer to Section 2.7), it can be understood that if the distortion of materials is neglected then the parameters in these conservation equations only contain the hydrostatic pressure p, the specific volume V, and the internal energy E (also refer to Section 7.7), while the temperature T and the entropy S are not concerned directly. Therefore, it is convenient to write a state equation in internal energy form which relates p, V, and E:

$$f_E(p, V, E) = 0 \tag{7.66a}$$

Or equivalently, p can be written as a function of E and V

$$p = p(E, V) \tag{7.66b}$$

The problem now is what the actual function for the state equations of solids in internal energy form will be.

To conduct a further discussion, similar to defining the volumetric modulus by Eqs. (7.63) or (7.64), we first introduce a new variable Γ based on Eq. (7.66b):

$$\Gamma \equiv V\left(\frac{\partial p}{\partial E}\right)_V \tag{7.67}$$

which represents the change rate of the pressure with the internal energy change in unit volume under isometric condition, and is usually named **Grüneisen coefficient**.

Considering the first and second thermodynamics law, we have

$$dE = TdS - pdV \tag{7.68}$$

Consequently, in an isometric process, there is

$$\left(\frac{\partial E}{\partial T}\right)_V = T\left(\frac{\partial S}{\partial T}\right)_V \equiv C_V \tag{7.69}$$

where C_V is the isometric specific heat. On the other hand, since it is an isometric process, there should be

$$dV = \left(\frac{\partial V}{\partial p}\right)_T dp + \left(\frac{\partial V}{\partial T}\right)_p dT = 0$$

then we obtain

$$\left(\frac{\partial p}{\partial T}\right)_V = -\left(\frac{\partial p}{\partial V}\right)_T \left(\frac{\partial V}{\partial T}\right)_p = -V\left(\frac{\partial p}{\partial V}\right)_T \cdot \frac{1}{V}\left(\frac{\partial V}{\partial T}\right)_p = k_T \alpha \tag{7.70}$$

where $\alpha \left[\equiv (1/V)(\partial V/\partial T)_p\right]$ is the isobaric expansion coefficient. Therefore, from Eqs. (7.69) and (7.70), we obtain

$$\Gamma = V\frac{\left(\frac{\partial p}{\partial T}\right)_V}{\left(\frac{\partial E}{\partial T}\right)_V} = \frac{k_T \alpha V}{C_V} \tag{7.71}$$

This is a basic formula to determine Grüneisen coefficient by three thermodynamics parameters k_T, α, and C_V, which are easy to measure.

Assume that Γ is a function of V only:

$$\Gamma = \gamma(V) \tag{7.72}$$

it is named Grüneisen assumption. Integrating Eq. (7.67), we obtain

$$p = \frac{\gamma}{V}E + \Pi(V), \tag{7.73a}$$

or rewrite it as

$$p = p_K(V) + \frac{\gamma}{V}\{E - E_K(V)\} \tag{7.73b}$$

This is the well-known **Mie–Grüneisen state equation**, commonly used in internal energy form in the studies of stress waves in solids under high pressure, up till now.

To highlight the physical meaning of $E_K(V)$, integrate Eq. (7.69), to obtain

$$E = E_K(V) + \int_0^T C_V dT = E_K(V) + E_T(V, T) \tag{7.74}$$

This indicates that the internal energy E consists of two parts: the **cold energy** or the energy of cold compression $E_K(V)$ independent of temperature and the **thermal energy** $E_T(V, T)$ dependent on temperature. The former is the internal energy at 0°K, also called the **lattice potential energy** (elastic energy), which includes the interaction energy among ions or neutral molecules (lattice binding energy), the energy of "zero-point oscillation" and the energy of compression of the gas of valence electrons (in a metal), all independent of temperature. The latter is also called **lattice kinetic energy**, which includes the thermal vibration energy of lattices and the thermal excitation energy of electrons, all dependent on temperature.

To highlight the physical meaning of $p_K(V)$, integrate Eq. (7.70), to obtain

$$p = p_K(V) + \int_0^T k_T \alpha dT = p_K(V) + \int_0^T \frac{\Gamma C_V}{V} dT = p_K(V) + p_T(V, T) \tag{7.75}$$

This indicates that pressure p also consists of two parts: the **cold pressure** $p_K(V)$ independent of temperature and the **thermal pressure** $p_T(V, T)$ dependent on temperature, which correspond to the cold energy E_K and the thermal energy E_T, respectively. In fact, applying Eq. (7.68) at 0°K and noting that the entropy keeps zero at 0°K (Nernst theorem), we then have

$$p_K(V) = -\frac{dE_K(V)}{dV} \tag{7.76}$$

while according to Eqs. (7.68) and (7.75), the thermal pressure p_T can be expressed as:

$$p_T(V, T) = -\left(\frac{\partial E_T}{\partial V}\right)_S = \int_0^T \frac{\Gamma C_V}{V} dT$$

according to the Grüneisen assumption [Eq. (7.72)], then we have

$$p_T(V,T) = \frac{\gamma}{V}\int_0^T C_V dT = \frac{\gamma}{T}E_T(V,T) \tag{7.77}$$

This indicates that the ratio of the thermal pressure p_T to the thermal energy E_T is independent of temperature and is a function only of specific volume, $\gamma(V)/V$. This is the right physical meaning of Grüneisen assumption, and is equivalent to the assumption of quasi-harmonic vibration of lattices in the theory of thermal vibration of lattices. In other words, it is assumed that the vibration frequencies of lattices are functions of atomic spacing (i.e. specific volume) and the vibration can be treated as harmonic at each specific volume. Equation (7.77) is actually the Grüneisen equation. In fact, if p_T and E_T in Eq. (7.77) are replaced by $(p - p_K)$ and $(E - E_K)$ respectively, then Eq. (7.73b) is obtained.

7.7 Shock Waves in Solids under High Pressures

Under such high pressure where the shearing strength of solids can be neglected, as discussed previously, solids can be treated as non-viscous compressible fluids. Then, a number of research results of isentropic waves and shock waves in hydrodynamics field can be directly generalized to the studies of stress waves in solids under high pressure; only the differences in the state equations between solids and gases should be considered. Such a kind of approximation is usually called the **hydrodynamics approximation**.

Common solid materials under high pressure show such character that they become more difficult to be compressed with increase in the volumetric compression, namely volumetric modulus increases with the increase of pressure. This is schemed in Figs. 7.7 and 7.8 qualitatively. Thereby, similar to the discussions in Section 2.6, it can be understood that the strong discontinuous shock wave will be formed in the propagation of compression disturbances in the solids under high pressure, while the weak discontinuous rarefaction wave will be formed in the propagation of expansion disturbances.

The propagation of rarefaction waves is similar to the propagation of nonlinear elasto-plastic waves in decreasingly hardening materials discussed previously and so will not be discussed here again. In the following, only the problems related to the shock wave propagating in solids under high pressure will be discussed in detail.

7.7.1 Shock jump conditions

In Section 2.7, by using Lagrange variables we discussed the conditions of conservation of mass, momentum, and energy across a shock wave front in the 1D stress condition, namely the so-called **shock jump conditions**. Here, we will extend those discussions to the 1D strain problems, and the expressions depicted by Euler variables will also be presented.

Lagrange formulation In the material coordinate system, consider a 1D strain plane shock wave propagating with the **material wave velocity** $\mathscr{D}$ in the X-axis direction. Assume that the initial state ahead of the wave front and the final state behind the wave front are all in the thermodynamic equilibrium, then the conservation of mass, momentum, and energy across the wave front of a shock wave can be expressed in the same form as Eqs. (2.55), (2.57), and (2.61) or (2.62), and are summarized below:

$$[v] = -\mathscr{D}[\varepsilon]$$

$$[\sigma] = -\rho_0\mathscr{D}[v]$$

$$[e] = -\frac{[\sigma v]}{\rho_0\mathscr{D}} - \frac{1}{2}[v^2] = \frac{1}{2\rho_0}\left(\sigma^- + \sigma^+\right)[\varepsilon]$$

Here, we only need to regard the σ, ε, and v in these equations as the X-axial stress, strain, and particle velocity in the 1D strain condition.

For solids under high pressure, according to the hydrodynamics approximation, the terms related to the distortion can be neglected. This is equivalent to replacing the σ by $(-p)$ and the ε by $(V_0 - V)/V_0$ in the above three equations. Moreover, consistent with the traditional notations used in the discussion of shock waves in solids under high pressure, we change the notation for the particle velocity by u and for the internal energy by E, and consequently the above equations are reduced to

$$[u] = -\rho_0\mathscr{D}[V] \tag{7.78a}$$

$$[p] = \rho_0\mathscr{D}[u] \tag{7.78b}$$

$$[E] = \frac{[pu]}{\rho_0\mathscr{D}} - \tfrac{1}{2}[u^2] = -\tfrac{1}{2}\left(p^- + p^+\right)[V] \tag{7.78c}$$

These are Lagrange formulations of the so-called **shock jump conditions** or the **Rankine–Hugoniot relations** (**R–H relations** for short) of the plane shock waves in solids under high pressure.

Euler formulation Now we discuss the same problem in the space coordinate system. Consider a plane shock wave that propagates with the **spatial wave velocity** D along the x-axis in the space coordinates, all the state parameters ahead of the wave front are denoted with superscript "+" while those behind the wave front are denoted with superscript "−", as shown in Fig. 7.9.

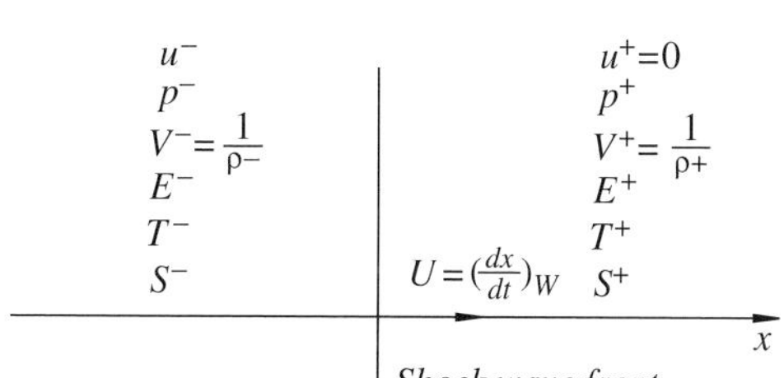

Fig. 7.9. A plane shock wave in a space coordinate system.

First, noticing that the particle velocity relative to the wave front is $(u - D)$, according to the mass conservation, the mass flow rate, m, passing through a wave front should be conservative, namely

$$\rho^-(u^- - D) = \rho^+(u^+ - D) = m$$

or

$$[\rho u] = D[\rho] \tag{7.79a}$$

Second, according to the conservation of momentum, there should be

$$\rho^- u^-(u^- - D) - \rho^+ u^+(u^+ - D) = p^+ - p^-$$

or by using the above mass conservation, we have

$$mu^- + p^- = mu^+ + p^+$$

or

$$[p] = -m[u] = \rho^+(D - u^+)[u] \tag{7.79b}$$

Third, according the energy conservation, there should be

$$\rho^- \left\{\tfrac{1}{2}\left(u^-\right)^2 + E^-\right\}(u^- - D) - \rho^+ \left\{\tfrac{1}{2}\left(u^+\right)^2 + E^+\right\}(u^+ - D) = p^+u^+ - p^-u^-$$

Or by using the above momentum conservation, we have

$$m\left\{\tfrac{1}{2}\left(u^-\right)^2 + E^-\right\} + p^-u^- = m\left\{\tfrac{1}{2}\left(u^+\right)^2 + E^+\right\} + p^+u^+$$

or

$$[E] = -\frac{[pu]}{m} - \frac{1}{2}[u^2] = \frac{[pu]}{\rho^+(D - u^+)} - \frac{1}{2}[u^2] \tag{7.79c}$$

If also using the momentum conservation [Eq. (7.79b)], the above equation can be written in a form independent of the coordinate system,

$$[E] = -\tfrac{1}{2}(p^+ + p^-)[V] \tag{7.79c$'$}$$

which is consistent with Eq. (7.78c). Equations (7.79) are then Euler **formulation** of the **Rankine–Hugoniot relations** in the space coordinate system.

If the coordinate system moves with the particles ahead of a shock wave, then the **relative spatial wave velocity** U of the shock wave, relative to the particles ahead, is

$$U = D - u^+ \tag{7.80}$$

Substituting this into Eq. (7.79) and noticing $u^* = 0$ in the present coordinate system, then Euler formulation of the Rankine–Hugoniot relations in such cases are

$$u^- = -\rho^+ U\,[V] \tag{7.81a}$$

$$[p] = \rho^+ U u^- \tag{7.81b}$$

$$[E] = \frac{\rho^- u^-}{\rho^+ U} - \tfrac{1}{2}\left(u^-\right)^2 \tag{7.81c}$$

Equations (7.78), (7.79), and (7.81) are equivalent to each other. They are different expression forms of the same *R–H* relations in the different coordinate systems. It should be noticed that even in the cases of 1D strain plane shock waves where the distortion is not negligible, since there still exists $\varepsilon_X = (V_0 - V)/V_0$, these relations are still valid, only need to regard the p in those relations as the axial stress $(-\sigma_X)$.

From these three equation sets, it can be seen that the material wave velocity $\mathscr{D}$, the relative spatial wave velocity U, and the absolute spatial wave velocity D of a shock wave can be expressed respectively by

$$\mathscr{D} = V_0\sqrt{-\frac{[p]}{[V]}} \tag{7.82a}$$

$$U = V^+\sqrt{-\frac{[p]}{[V]}} \tag{7.82b}$$

$$D = u^+ + V^+\sqrt{-\frac{[p]}{[V]}} \tag{7.82c}$$

Among these, there obviously exist the following relations

$$D = U + u^+ = \frac{V^+}{V_0}\cdot \mathscr{D} + u^+ \tag{7.83}$$

This is actually a concrete expression of Eq. (2.11) in the present situation. If the state ahead of the shock wave is in an uncompressed natural state ($V^+ = V_0$), then $U = \mathscr{D}$; if moreover it is in a static state ($u^+ = 0$), then $D = U = \mathscr{D}$. Only in this condition, these three wave velocities are identical.

When the magnitude of jump of the state parameters across a wave front approaches infinity, and the discontinuity with the differential under the isentropic condition is replaced, then $\mathscr{D}$, U, and D in Eq. (7.82) are transformed to the corresponding wave velocities of a rarefaction wave (weak continuous wave), namely the material wave velocity C, the relative spatial wave velocity a, and the absolute spatial wave velocity c, respectively:

$$C = C_0\sqrt{-\left(\frac{\partial p}{\partial V}\right)_S} = \frac{\rho}{\rho_0}\sqrt{\left(\frac{\partial p}{\partial \rho}\right)_S} \tag{7.84a}$$

$$a = V\sqrt{-\left(\frac{\partial p}{\partial V}\right)_S} = \sqrt{\left(\frac{\partial p}{\partial \rho}\right)_S} \tag{7.84b}$$

$$c = u + V\sqrt{-\left(\frac{\partial p}{\partial V}\right)_S} = u + \sqrt{\left(\frac{\partial p}{\partial \rho}\right)_S} \tag{7.84c}$$

Be reminded, the above three forms of R–H relations are all deduced for the rightward wave. For the leftward wave, the sign of wave velocity should be reversed, similar to that discussed previously for the strong discontinuous interfaces in bars.

Therefore, there are totally four equations, the three conservation equations for the shock jump conditions, together with a state equation in the internal energy form to characterize material properties. If the initial conditions, namely the initial states before shock loading are given, then there are five unknowns p, V (or ρ), u, E, and the shock wave velocity $\mathscr{D}$ (or U, D) to characterize the final states but only by four equations. Once any of the unknowns is given from the boundary conditions, then the other four unknowns can be determined. Therefore, for a certain material (of which the state equation is known), under the given initial and boundary conditions, the problem of plane shock wave propagation is solvable.

For instance, consider the problem that two plane plates A and B collide at high velocity. There are shock waves generated simultaneously in the two plates, while the state parameters at the impact interface are indeterminate, as shown in Fig. 7.10. Here, there are totally ten unknowns, five each for the plate A and the plate B. We can list four equations each for the plate A and the plate B, together with the continuity conditions of particle velocity

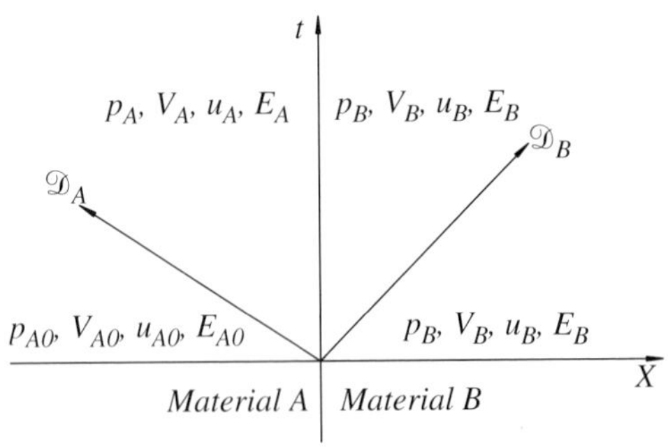

Fig. 7.10. Propagation of shock waves during the collision between two planes A and B.

and pressure, respectively, at the interface:

$$u_A = u_B, \quad p_A = p_B$$

then there are totally ten equations for ten unknowns. Therefore the problem is solvable. Of course, it should be noticed here, the shock wave velocity should take different sign when the *R–H* relations [Eq. (7.78)] or its other equivalent forms) are applied for rightward waves and leftward waves, respectively.

7.7.2 Shock adiabat

If no concrete boundary conditions are specified, in certain initial equilibrium states, the *R–H* relations together with the state equation provide the relationship between any two of the five unknowns, such as *p–V*, *p–u*, etc. The relationships between any couple of unknowns obtained in this way are named **shock adiabat**, or **Rankine–Hugoniot curve**, or briefly **Hugoniot** curves. Because the shock jump process across the shock wave front is an irreversible nonequilibrium process, so the shock adiabat actually only represents the locus of the possible final equilibrium states through a shock jump started at a certain initial equilibrium state (called the center point of Hugoniot); it does not represent the locus of the successive states that the material experienced in the process. This is completely different with those curves in the equilibrium states, such as the isothermal curve which represents the relationship between the state parameters in an isothermal process, or the isentropic adiabatic curve which represents the relationship between the state parameters in an isentropic adiabatic process, or the quasi-equilibrium adiabatic stress–strain curve of a nonincreasingly hardening material discussed previously in the 1D stress elastic–plastic waves. In addition, since Hugoniot curves depend on the initial equilibrium states, so even for the same material, different Hugoniot curves exist corresponding to different initial states. The problem that the shock jump process is a nonequilibrium process with additional entropy-increase will be discussed further, besides it has been discussed a little in Section 2.7.

Among various types of Hugoniot curves, the following three types are commonly used: the *p–V* (or *p–ρ*) curves, the *p–u* curves, and the *U–u* curves.

The p–V Hugoniot curves From the condition of energy conservation [Eq. (7.78c)] and the state equation in internal energy form [Eq. (7.66)], eliminating the internal energy *E*, we obtain the *p–V* relationship in the shock adiabatic condition, namely the *p–V* Hugoniot curves. The shape of *p–V* Hugoniot curves for ordinary materials are concave upwards. Figure 7.11 shows the typical *p–V* Hugoniot curves of materials determined experimentally by shock wave tests (Jones, 1972).

The chord connecting the start point *A* and the final point *B* of a *p–V* Hugoniot curve, as shown in Fig. 7.12, is of specific significance, called the **Rayleigh line**. In fact, it is known from Eq. (7.82) that the velocity of a shock wave depends on the slope of this Rayleigh line. Moreover, it is known from Eq. (7.78c) that the area below a Rayleigh line represents the jump of internal energy across the shock wave front. It should be noticed that the material velocity $\mathscr{D}$ is completely determined by the slope of the Rayleigh line for a given

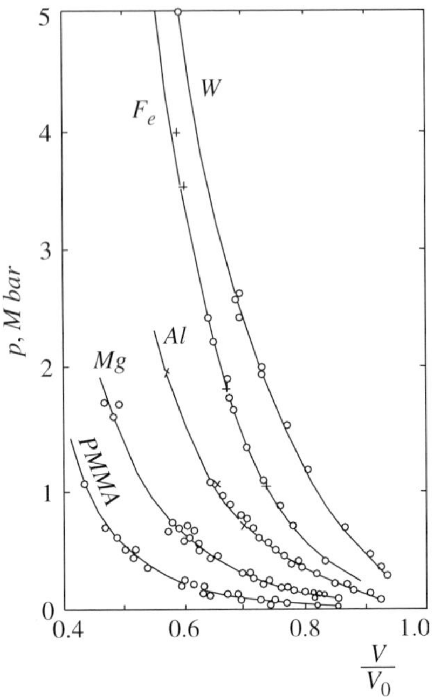

Fig. 7.11. The p–V Hugoniot curves of some materials.

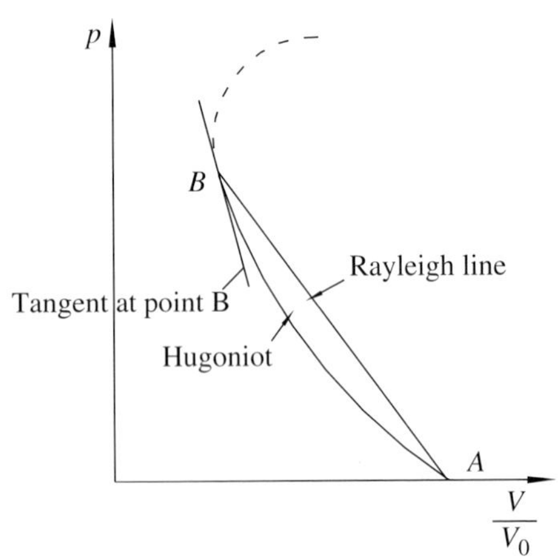

Fig. 7.12. p–V Hugoniot curve and the corresponding Rayleigh line.

material (namely its density is given), but the special velocity additionally depends on the initial density of material, ρ^+, ahead of the shock wave front.

For a stable shock wave, since every part of wave front propagates with an identical wave velocity and the wave velocity depends on the slope of the Rayleigh line, this indicates that the locus of all states that the shock jump experiences is just the Rayleigh line. In the previous content, it has been emphasized that a Hugoniot curve is a locus of the possible final equilibrium states corresponding to a certain initial equilibrium state, but not a locus of the states experienced in a shock jump process. Now we are clear that the locus of the states experienced in a shock jump process is just the Rayleigh line. Certainly, the states of the points on a Rayleigh line except the initial one and the final one are not at thermodynamic equilibrium. In a shock jump process, if the p–V relationship representing an equilibrium process can be approximated by a Hugoniot curve, then the difference of pressure between a Hugoniot curve and the corresponding Rayleigh line for a given specific volume approximately represents the nonequilibrium force, namely the irreversible viscous force (dissipation force) which results in the irreversible entropy-increase in a shock jump. Based on such a mechanism of dissipation, a stable shock wave is realized by adjusting the actual profile shape of shock wave front to make the sum of the equilibrium forces and the nonequilibrium forces just equal to the corresponding pressure of Rayleigh line.

Let us discuss the p–V Hugoniot curve from another angle to illustrate again that a shock jump process is a process having irreversible entropy-increase. Differentiating Eq. (7.78c), and denoting the initial state by subscript 0 and final state without subscript, we obtain

$$dE = \tfrac{1}{2}(V - V_0)dp - \tfrac{1}{2}(p + p_0)dV$$

This shows the differential relationship between two neighboring possible final points on a p–V Hugoniot curve. On the other hand, since any final state is at thermodynamic equilibrium, the first and second law of thermodynamics [Eq. (7.68)] should also be satisfied. Eliminating dE from these two equations, we obtain

$$TdS = \tfrac{1}{2}(V_0 - V)dp + \tfrac{1}{2}(p - p_0)dV = \frac{1}{2}\left\{1 - \frac{\dfrac{p - p_0}{V - V_0}}{\left(-\dfrac{dp}{dV}\right)}\right\}(V_0 - V)dp \tag{7.85}$$

Referring to Fig. 7.12, it is known that the $(p - p_0)/(V_0 - V)$ is the slope of the Rayleigh line; while the $(-dp/dV)$ is the tangent slope of the p–V Hugoniot curve at the final state B. Since the p–V Hugoniot curve for a common material is concave upwards, it means

$$\frac{\left(\dfrac{p - p_0}{V_0 - V}\right)}{\left(-\dfrac{dp}{dV}\right)} < 1$$

then Eq. (7.85) shows that the entropy increases with the increase of pressure along the Hugoniot curves, $dS/dp > 0$. Even under the abnormal condition that the Hugoniot curve is bent back as shown by the dashed line in Fig. 7.12, because $dp/dV > 0$, we still obtain from Eq. (7.85) the conclusion that $dS/dp > 0$.

Now, let us discuss the relationship between the Hugoniot curve, the isentropic adiabatic curve, and the isothermal curve. If we conduct a total differential of $p = p(S, V)$ with respect to V along the p–V Hugoniot curve, then the relationship between the slope dp/dV of the Hugoniot curve and the slope $(\partial p/\partial V)_S$ of the isentropic p–V curve is obtained:

$$\frac{dp}{dV} = \left(\frac{\partial p}{\partial V}\right)_S + \left(\frac{\partial p}{\partial S}\right)_V \left(\frac{dS}{dV}\right) \tag{7.86}$$

According to Eqs. (7.69) and (7.70), in the normal cases, we have

$$\left(\frac{\partial p}{\partial S}\right)_V = \left(\frac{\partial p}{\partial T}\right)_V \left(\frac{\partial T}{\partial S}\right)_V = \frac{k_T \alpha T}{C_V} > 0$$

Since Eq. (7.85) indicates that dS/dV and dp/dV have the same sign during a compressive process ($V < V_0$), then Eq. (7.86) indicates that if $dp/dV < 0$, then

$$\left(-\frac{dp}{dV}\right) > -\left(\frac{\partial p}{\partial V}\right)_S$$

This means that the Hugoniot curve AB is above the isentropic curve AS_1 passing through the initial point A; while it is below the isentropic curve BC passing through the final point B, as shown in Fig. 7.13. If $dp/dV > 0$, then the difference between the slope of the Hugoniot curve and the slope of the corresponding isentropic curve further increases, this conclusion is still valid. In addition, since the area below an isentropic p–V curve represents

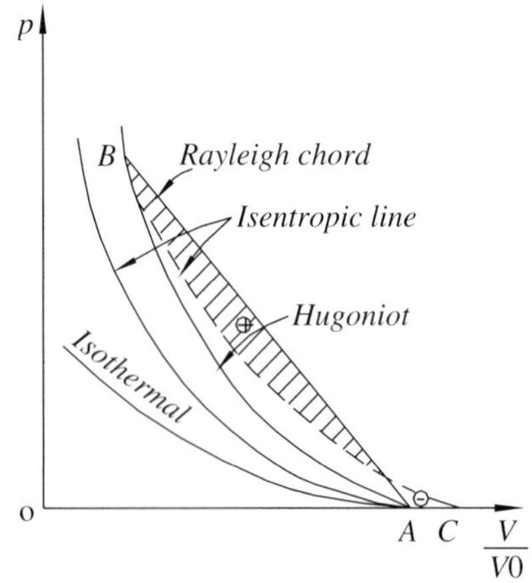

Fig. 7.13. The Hugoniot curve, Rayleigh line, isentropic curve, and isothermal curve.

the reversible change of internal energy during an isentropic process [see Eq. (7.68)], then the area encompassed by the Rayleigh line AB and the expansion isentropic curve BC (as the shadowed area shown in Fig. 7.13) represents the irreversible dissipation of energy during the process of shock jump, which corresponds to the irreversible entropy-increase given by Eq. (7.85).

As for the relation between the slope $(\partial p/\partial V)_S$ of an isentropic curve and the slope $(\partial p/\partial V)_T$ of an isothermal curve, conducting a total differentiation of $p = p(V, T)$ with respect to V under the isentropic condition, we have

$$\left(\frac{\partial p}{\partial V}\right)_S = \left(\frac{\partial p}{\partial V}\right)_T + \left(\frac{\partial p}{\partial T}\right)_V \left(\frac{dT}{dV}\right)_S \tag{7.87}$$

Noting that for an isothermal process, there is

$$dT = \left(\frac{\partial T}{\partial V}\right)_S dV + \left(\frac{\partial T}{\partial S}\right)_V dS = 0$$

and considering Eq. (7.69) and the following Maxwell formula[1]

$$\left(\frac{\partial p}{\partial T}\right)_V = \left(\frac{\partial S}{\partial V}\right)_T$$

we obtain

$$-\left(\frac{\partial T}{\partial V}\right)_S = \left(\frac{\partial T}{\partial S}\right)_V \left(\frac{\partial S}{\partial V}\right)_T = \frac{T}{C_V}\left(\frac{\partial p}{\partial T}\right)_V$$

Thereby, Eq. (7.87) can be rewritten as

$$-\left(\frac{\partial p}{\partial V}\right)_S = -\left(\frac{\partial p}{\partial V}\right)_T + \frac{T}{C_V}\left(\frac{\partial p}{\partial T}\right)_V^2$$

Since $T > 0$, $C_V > 0$, the above equation indicates

$$-\left(\frac{\partial p}{\partial V}\right)_S > -\left(\frac{\partial p}{\partial V}\right)_T$$

which means that an isentropic curve is above an isothermal curve (Fig. 7.13).

[1] Introducing the Helmholtz free energy $A = E - TS$, then the first and second thermodynamics law [Eq. (7.68)] is converted to $dA = -SdT - pdV$; from this, we have $S = -(\partial A/\partial T)_V$ and $p = -(\partial A/\partial V)_T$, and consequently $(\partial p/\partial T)_V = -\partial^2 A/\partial T\partial V = (\partial S/\partial V)_T$.

Since at the initial state (point A), according to Eq. (7.85) there is $\left.\frac{dS}{dV}\right|_A = 0$, then according to Eq. (7.86) there is

$$\left.\frac{dp}{dV}\right|_A = \left.\left(\frac{\partial p}{\partial V}\right)_S\right|_A$$

namely, the slope of a Hugoniot curve and the slope of the corresponding isentropic curve are identical at the initial state point A. Moreover, after a total differentiation of Eq. (7.86) and the successive mathematical operations, it can be proved that

$$\left.\frac{d^2p}{dV^2}\right|_A = \left.\left(\frac{\partial^2 p}{\partial V^2}\right)_S\right|_A$$

$$\left.\frac{d^3p}{dV^3}\right|_A \neq \left.\left(\frac{\partial^3 p}{\partial V^3}\right)_S\right|_A$$

namely, the curvature of a Hugoniot curve and the curvature of the corresponding isentropic curve are also identical, and the difference between these two curves only appear in their third-order derivatives.

In the above discussion, the influence of the phase-transition of materials on the Hugoniot curve has not been taken into consideration, and it has been implicitly assumed that the following relationship is valid,

$$\frac{p_2 - p_1}{V_1 - V_2} > \frac{p_1 - p_0}{V_0 - V_1} \tag{7.88}$$

where the subscript 0 denotes the initial state, the subscript 2 denotes the final state, while the subscript 1 denotes a state between the initial state and the final state on a Hugoniot curve. It is clear that a single stable shock wave front is formed if Eq. (7.88) is satisfied. However, since shock waves are usually formed under high pressure and the irreversible entropy-increase induced in a shock jump process must lead to a dramatic increase in temperature, then the phase-transition may occur under such conditions of high pressure and high temperature. Consequently, Eq. (7.88) is no longer satisfied. For example, it has been found that when iron is under an impact pressure of 13 GPa, a phase-transition from the phase α body-centred cubic (BCC) to the phase ε hexagonal close-packed (HCP) takes place, etc.

In Fig. 7.14, if the point 1 is corresponding to a phase-transition, because of the different material properties before and after the phase-transition, the state equations and the shock adiabatic curves are correspondingly different. For the p–V Hugoniot curve, its phase-transition may be displayed as a discontinuity in the curve shape itself (called the first-order phase-transition) or a discontinuity in the slope of the curve (called the second-order phase-transition).

As an example, let us discuss a second-order phase-transition as shown in Fig. 7.14. If the shock pressure is not much larger than the phase-transition pressure p_1, such as if the final shock state is corresponding to the point 2 in Fig. 7.14, then the absolute value of the slope

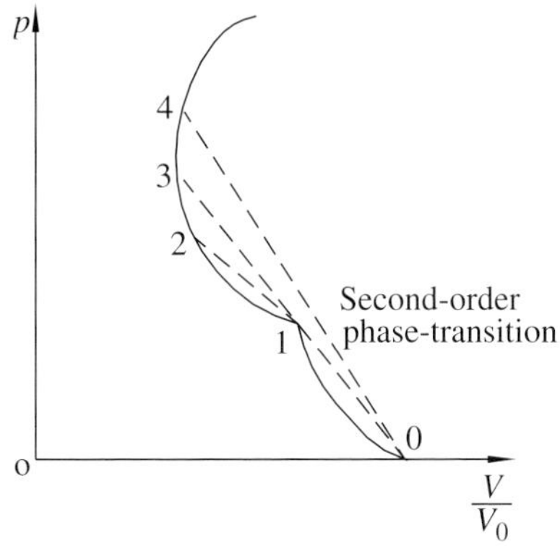

Fig. 7.14. The p–V Hugoniot curve with a phase-transition.

of Rayleigh line 1–2 is smaller than those of Rayleigh line 0–1; consequently, the shock wave velocity in the high-pressure phase is smaller than that in the low-pressure phase. Then, Eq. (7.88) is no longer valid, which means that a single shock wave is unstable and a dual-wave construction will be formed. Only when the shock pressure is high enough (e.g. as high as p_4 at the point 4) so that the slope of Rayleigh line 0–4 is larger than those of Rayleigh line 0–1, a single shock wave is then stable again. It is clear that the point 3 in the figure corresponds to the critical condition for the transition from the dual-wave construction to the stable one-wave structure, which satisfies the following condition

$$\frac{p_3 - p_1}{V_1 - V_3} = \frac{p_1 - p_0}{V_0 - V_1}$$

The critical pressure p_3 (corresponding to the point 3) is called the **overdrive pressure**.

The influence of phase-transition on the propagation of shock waves has been utilized as an important method to study the phase transition, particularly under high pressure. On the other side, effects of phase-transition have been utilized to realize the required phase-transition, such as to synthesize diamond from graphite, etc.

Although the above discussion was with respect to the Hugoniot curves in p–V form, it in fact shows the common properties of Hugoniot curves in other forms, because we can convert the Hugoniot curves in p–V form into other forms through the R–H relationship. Therefore, in the following, the Hugoniot curves in other forms will not be discussed in detail, but only some additional features will be briefly analyzed for the Hugoniot curves in another two commonly used forms, namely the p–u and the U–u Hugoniot curves.

The p–u Hugoniot curves From the R–H relationship [Eq. (7.78b)], it is known that the line connecting the initial point A and the final point B on the Hugoniot curve in p–u forms, namely the slope $[p]/[u]$ of the Rayleigh line represents the shock wave resistance $\rho_0\mathscr{D}$, as shown in Fig. 7.15.

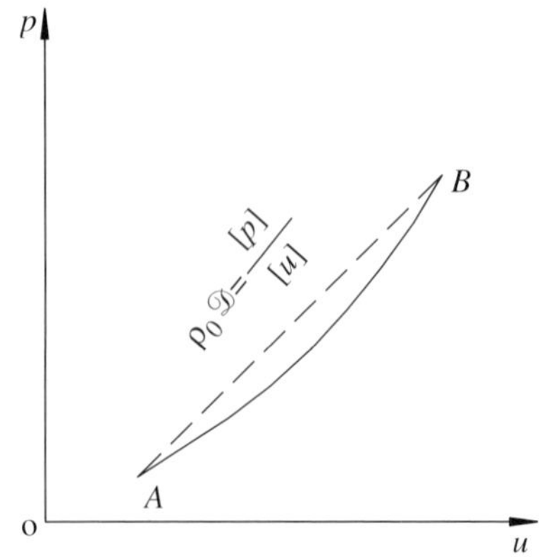

Fig. 7.15. The p–u Hugoniot curve.

In the analyses of the high-velocity collision between two objects or the interaction of two shock waves, and the reflection or transmission of shock waves propagating from one material to another material, the continuity of p and u respectively at the interface must be satisfied and will be utilized to solve the problem. Therefore, it is convenient to use the Hugoniot curves in p–u forms, which is similar to the stressparticle velocity curve (σ–v) in the theory of 1D stress elasto-plastic waves in bars. This point will be clear in the discussion on the reflection or transmission of shock waves in the next section.

The U–u Hugoniot curves In view of the experimental investigation of shock waves and the related measuring technique, the parameters with a dimension of velocity such as the particle velocity u and the shock wave velocity U usually are much easy to be directly measured. In fact, the measurements of these parameters can be decomposed into the measurements of the distance and the time, and both are easy to conduct now. As to the other parameters such as the pressure, specific volume and internal energy, etc., they are relatively difficult to be measured directly in the dynamic conditions. Therefore, in the experimental determination of Hugoniot curves and in the experimental studies of the state equations of materials under high pressure by means of the measurements of the Hugoniot curves, the U–u Hugoniot curve is most commonly used.

A number of experimental results for a lot of materials in a wide range of experimental pressures show that if no phase-transition takes place then the U–u Hugoniot curves display a linear relationship as shown in Fig. 7.16.

$$U = a_0 + su \tag{7.89a}$$

where a_0 and s are material constants. The material which satisfies this relationship sometimes is named the "a, s material". For some materials such as iron, if its experimental U–u Hugoniot curve is a departure from the linear relationship, then an item of square of

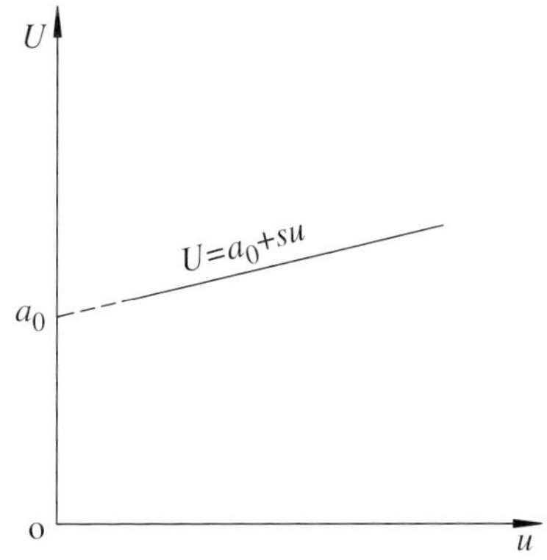

Fig. 7.16. The U–u Hugoniot curve.

u is added, namely

$$U = a_0 + su + qu^2 \tag{7.89b}$$

where q is also a material constant.

Since the nonisentropic shock wave approaches the isentropic sound wave when the particle velocity u approaches zero, it seems that the a_0 in Eq. (7.89) should represent the sound velocity of materials. In fact, the data of a_0 determined in the experiments of shock waves are indeed much closer to the sound velocity of body wave $(k_0/\rho_0)^{1/2}$. For some typical materials, the data of a_0 and s determined by the experiments of shock waves are shown in Table 7.3 (McQueen, *et al.*, 1970). In addition, the density ρ_0 and the Grüneisen coefficient Γ_0 under the normal pressure and temperature are also listed in the table. For the metallic materials, the Γ_0 is usually close to 2.

Table 7.3. Parameters of U–u Hugoniot curves for some typical materials [Eq. (7.89)].

Materials	a_0 (km/s)	s	ρ_0 (g/cm^3)	Γ_0
2024 Al	5.328	1.338	2.785	2.00
921-T Al	5.041	1.420	2.833	2.10
Copper	3.940	1.489	8.93	1.96
Iron (C: 0.18%)*	3.574	1.920	7.85	1.69
Stainless Steel	4.569	1.490	7.896	2.17
U–Mo Alloy	2.565	1.531	18.45	2.03
WC	4.920	1.339	15.02	1.50
Polyethylene (low density)	2.901	1.481	0.915	1.644
Polystyrene	2.746	1.319	1.044	1.18
Plexiglass	2.572	1.536	1.185	0.97
Epoxy	2.678	1.520	1.198	1.13
Durite	2.847	1.404	1.370	0.50

* For mild steel (C: 0.18%), $U = a_0 + su + qu^2$, while $q = -0.068$.

If we introduce $\eta = (V_0 - V)/V_0$, then the parameters u, p, U, E of the final state for the "a, s materials" can be expressed explicitly by the material constant a_0, s, and η. Thus, the mass conservation of the R–H relationship [Eq. (7.81a)] is written as

$$\eta = \frac{u}{U}$$

Combining with Eq. (7.89), we obtain

$$u = \frac{a_0 \eta}{1 - s\eta} \tag{7.90a}$$

$$U = \frac{a_0}{1 - s\eta} \tag{7.90b}$$

while the momentum conservation [Eq. (7.81b)], when $p_0 = 0$, is written as

$$p = \frac{\rho a_0^2 \eta}{(1 - s\eta)^2} \tag{7.90c}$$

and energy conservation [Eq. (7.81c)], when $E_0 = 0$, is written as

$$E = \frac{a_0^2 \eta^2}{2\,(1 - s\eta)^2} \tag{7.90d}$$

Sometimes, the following shock Mach number M_S is introduced

$$M_S = \frac{U}{a_0}$$

then from the R–H relationship [Eq. (7.81)] and the linear U–u Hugoniot relationship [Eq. (7.89)], we obtain

$$\frac{u}{a_0} = \frac{M_s - 1}{s} \tag{7.91a}$$

$$\frac{\rho}{\rho_0} = \frac{s M_s}{1 + (s - 1)\, M_s} \tag{7.91b}$$

$$\frac{p - p_0}{\rho_0 \sigma_0^2} = \frac{M_s\,(M_s - 1)}{s} \tag{7.91c}$$

$$\frac{E - E_0}{a_0^2} = \frac{(M_s - 1)^2}{2s^2} \tag{7.91d}$$

Before ending the discussion on shock adiabat, it should be pointed out that there are two types of problems in relation to the shock adiabat. One is to determine the Hugoniot curves from the given state equations; the other is to determine the state equations from the given

Hugoniot curves. The former is a determinate problem. So far as the problems of stress wave propagation in solids are discussed, the problems involved are mainly the problems of this sort. The latter is an indeterminate problem, namely the problems of this sort can be solved only if a concrete form of state equation has been assumed. When we study the state equations of solids by means of shock wave experiments, the problems involved are mainly the problems of this sort. In the case of Grüneisen state equations, the key is how to determine the Grüneisen coefficients $\gamma(V)$, which is the most difficult to be determined. Fortunately, for the analysis of shock waves the error due to the indetermination of $\gamma(V)$ is not large, and in the engineering application, the $\gamma(V)$ can be approximated by

$$\frac{\gamma(V)}{V} = \rho_0\Gamma_0 \text{ (constant)} \tag{7.92}$$

Or even $\gamma = \Gamma_0$ (constant) in a first approximate analysis. The constant Γ_0 can be determined by other known thermodynamics parameters according to Eq. (7.71). The data of Γ_0 for some typical materials are also listed in Table 7.3.

For the so-called weak shock waves whose intensity is rather weak, the shock jump can be approximately regarded as an isentropic process, and then the corresponding state equations of solids can be expressed in the form of Murnagham equation [Eq. (7.61)] approximately:

$$p = \frac{k_0}{n}\left\{\left(\frac{V_0}{V}\right)^n - 1\right\} = \frac{k_0}{n}\left\{\left(\frac{\rho}{\rho_0}\right)^n - 1\right\}$$

If we have obtained the linear U–u shock adiabat [Eq. (7.89)] from shock experiments, it is easy to determine k_0 and n from a_0 and s. In fact, substituting the Murnagham equation into the equation for the local sound velocity, namely the definition of the relative spatial wave velocity a [Eq. (7.84b)], we obtain

$$a^2 = \left(\frac{\partial p}{\partial \rho}\right)_s = \frac{k_0}{\rho_0}\left(\frac{\rho}{\rho_0}\right)^{n-1}$$

after differentiating, we have

$$\frac{2}{n-1}da = \frac{a}{\rho}d\rho \tag{7.93}$$

On the other hand, since the weak shock wave is approximately regarded as isentropic, then the Riemann invariable across a weak shock wave is constant. This is equivalent to replacing the jump terms [ ] in the conservation condition across a shock wave front [Eq. (7.79)] by the differential terms, and correspondingly replacing the shock wave velocity D by the weak discontinuous wave velocity c. Thus, from Eq. (7.79a) we obtain

$$du = (c-u)\frac{d\rho}{\rho} = \frac{a}{\rho}d\rho$$

Substituting this into Eq. (7.93) and integrating the equation, we have

$$a = a_0 + \frac{n-1}{2}(u - u_0)$$

where the a_0 and u_0 with the subscript 0 represent the initial state quantities ahead of a shock wave respectively, while the symbols without a subscript denote the final state quantities behind a shock wave. Consequently, considering this relationship, the absolute spatial wave velocity D of weak shock waves can be expressed by

$$D = a_0 + u_0 = a + u = \tfrac{1}{2}(a + u + a_0 + u_0) = a_0 + \frac{n+1}{4}u + \frac{n-3}{4}u_0$$

This indicates that theoretically a linear relationship between D and u can be deduced according to the Murnagham equation. If $u_0 = 0$, the above equation is reduced to

$$D = U = a_0 + \frac{n+1}{4}u \tag{7.94}$$

Comparing with the linear U–u Hugoniot curves [Eq. (7.89)] obtained from the experimental measurements, n and k_0 can be determined by s and a_0 according to the following relationship

$$\begin{cases} n = 4s - 1 \\ k_0 = \rho_0 a_0^2 \end{cases} \tag{7.95}$$

7.8 Interaction of Shock Waves in Solids under High Pressure

The general principles to deal with the interaction, reflection, and transmission of shock waves in solids under high pressure are similar to those used in analyzing the shock waves in 1D bars discussed previously. The continuous requirement in pressure and particle velocity respectively at the interface should be satisfied. However, it should be noticed: (i) when the reflective wave is a reloading shock wave, its final state should be on a new Hugoniot curve whose central point is the initial state point ahead of the reflected shock wave. This new Hugoniot curve is different from the original Hugoniot curve whose central point is the initial state point ahead of the incident shock wave; (ii) when the reflective wave is an unloading expansion rarefaction wave, its states are determined by the unloading isentropic curve, namely, the successive states experienced by the media during the rarefaction wave passing through are all on the expansion isentropic curve whose start point is the final state point behind the incident shock wave.

Indeed, as mentioned previously, if the state equation is known, the Hugoniot curve and the isentropic curve passing through any point are determinate. In the case when the Grüneisen state equation is applied, namely the Grüneisen coefficient $\gamma(V)$ is known, it is easy from a known Hugoniot curve to figure out a new Hugoniot curve or a new isentropic curve with an arbitrary point on the known Hugoniot curve as its new central point, as shown in Fig. 7.17. This procedure includes the following steps.

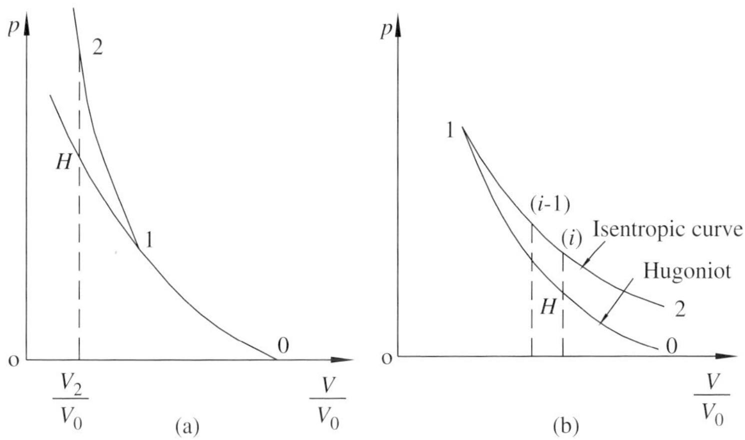

Fig. 7.17. Determining the loading Hugoniot curve (a) and the unloading isentropic curve (b) from a known Hugoniot curve.

Assuming point 0 is the initial state point and point 1 is the final state point of an incidence shock wave. The point 1 is also the initial state point of the reflected shock wave. For an arbitrary point 2 on the Hugoniot curve of the reflected shock wave as shown in Fig. 7.17(a), Eq. (7.78c) should be satisfied, namely,

$$E_2 = E_1 + \tfrac{1}{2}(p_2 + p_1)(V_1 - V_2) \tag{7.96}$$

Because both points 2 and H on the original Hugoniot curve having the same specific volume should satisfy the Grüneisen equation [Eq. (7.73b)], there is

$$E_2 = E_H + \frac{(p_2 - p_H)}{\left(\frac{\gamma}{V}\right)_2} \tag{7.97}$$

where subscript 2 and H denote the corresponding quantities at points 2 and H, respectively. Eliminating E_2 from the above two equations, we obtain

$$E_1 - E_H = \frac{p_2 - p_H}{\left(\frac{\gamma}{V}\right)_2} - \tfrac{1}{2}(p_2 + p_1)(V_1 - V_2) \tag{7.98}$$

On the other hand, both points 1 and H are the points on the original Hugoniot curve with the point 0 being the central point. According to Eq. (7.78c), they should satisfy:

$$E_1 - E_0 = \tfrac{1}{2}(p_1 + p_0)(V_0 - V_1)$$
$$E_H - E_0 = \tfrac{1}{2}(p_H + p_0)(V_0 - V_2)$$

Eliminating E_0 from the above two equations, we obtain

$$E_1 - E_H = \tfrac{1}{2}(p_1 + p_0)(V_0 - V_1) - \tfrac{1}{2}(p_h + p_0)(V_0 - V_2) \tag{7.99}$$

Then eliminating $(E_1 - E_H)$ from Eqs. (7.98) and (7.99), and assuming $p_0 = 0$, we obtain

$$p_2 = \frac{p_H - \left(\dfrac{\gamma}{V}\right)_2 \cdot \dfrac{(p_H - p_1)(V_0 - V_2)}{2}}{1 - \left(\dfrac{\gamma}{V}\right)_2 \cdot \dfrac{(V_1 - V_2)}{2}} \tag{7.100}$$

This is the p–V Hugoniot curve with point 1 as the central point, which is required in analyzing the reflected shock waves.

For the isentropic curve passing through point 1, it can be determined in the following way [Fig. 7.17(b)]: since along an isentropic curve there must be $dS = 0$, then from the thermodynamics law [Eq. (7.68)], there should be

$$dE = -pdV$$

or expressed in a different form, we have

$$E_i = E_{i-1} - \tfrac{1}{2}(p_i + p_{i-1})\,\Delta V \tag{7.101}$$

On the other hand, both the point i and the point H on the original Hugoniot curve having the identical specific volume V_i should satisfy the Grüneisen equation [Eq. (7.73b)], so there is

$$E_i = E_H + \frac{p_i - p_H}{\left(\dfrac{\gamma}{V}\right)_i} \tag{7.102}$$

Eliminating E_i from the above two equations, we obtain

$$p_i = \frac{p_H - \left(\dfrac{\gamma}{V}\right)_i \left[p_{i-1} \cdot \dfrac{\Delta V}{2} + E_H - E_{i-1}\right]}{1 + \left(\dfrac{\gamma}{V}\right)_i \cdot \dfrac{\Delta V}{2}} \tag{7.103}$$

Therefore, the states at the point i can be determined from that at the point $(i - 1)$ by a numerical method, and other points can also be determined point by point in the same way. Thus the expansion isentropic curve 1–2 is determined.

After the Hugoniot curves of reflected shock waves and the expansion isentropic curves of reflected unloading waves can be determined, it is easy to deal with the interaction, reflection, and transmission of shock waves. As mentioned previously, it is convenient to analyze this sort of problem by using p–u curves. In the following, we highlight this by analyzing the reflection and transmission of shock waves at the interface between two materials.

First, we discuss the problem when a plane shock wave $\mathscr{S}_i$ with pressure p_1 is normally incident from a material A with low wave impedance into a material B with high wave impedance, as shown in Fig. 7.18. Assume that both materials were undisturbed initially, which corresponds to the point 0 in the p–u plane. According to Eq. (7.78b), it is known that the slope of the Rayleigh line connecting the initial state point 0 and the final state point 1 represents the shock wave impedance $(\rho_0\mathscr{D})_A$ of material A. Since the impedances of these two materials are different, reflection and transmission simultaneously occur when an incident shock wave arrives at the interface of the two materials. Since the shock wave impedance $(\rho_0\mathscr{D})_B$ of material B is larger than that $(\rho_0\mathscr{D})_A$ of material A, from the

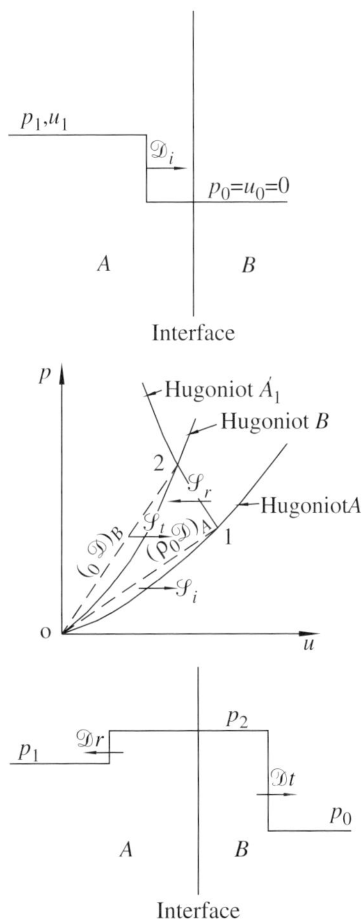

Fig. 7.18. Reflection and transmission of a shock wave which is incident into a high wave impedance material from a low wave impedance material.

previous discussion on the reflection of elastic waves at the interface of different materials (Section 3.5), it is clear that the reflected wave $\mathscr{S}_r$ is a shock wave that further loads the material A from state 1 to state 2, and the transmitted wave $\mathscr{S}_t$ is also a shock wave that compresses the material B from the undisturbed state 0 to state 2. According to the continuity condition for p and u at the interface respectively, and noting the negative sign of $\mathscr{D}$ in the R–H relationship [Eq. (7.78)] for leftward waves, then point 2 is the intersection of the positive Hugoniot curve of material B, whose center point is point 0, and the negative Hugoniot curve of material A, whose center point is point 1. Obviously, the intensity of transmitted wave $\mathscr{S}_t$ is larger than that of incident wave $\mathscr{S}_i$. For instance, when a shock wave with pressure amplitude of 24 GPa propagates from aluminum with low shock wave impedance to iron with high shock wave impedance, the intensity of transmitted shock wave is about 34 GPa.

Now, we discuss the problem that a rightward plane shock wave $\mathscr{S}_i$ with pressure p_1 is normally incident from a material A with high wave impedance into a material B with low wave impedance, as shown in Fig. 7.19. Similar to the previous discussion, it can be deduced that an unloading reflection occurs at the interface of the two materials, namely, the reflected wave $\mathscr{S}_r$ is a expansion rarefaction wave that unloads the material A from state 1 to state 2 and the transmitted wave is a shock wave that loads the material B from state 0 to state 2. Point 2 is the intersection point of the negative isentropic curve of material A passing point 1 and the positive Hugoniot curve of material B, whose center point is point 0. Obviously, the intensity of transmitted wave $\mathscr{S}_t$ is smaller than that of incident wave $\mathscr{S}_i$. For instance, when a shock wave with pressure amplitude of 24 GPa propagates from aluminum with high shock wave impedance to polyethylene with low shock wave impedance, the intensity of transmitted shock wave is about 9.5 GPa.

Figure 7.20 schematically shows the positive Hugoniot curves for several common materials, as well as the negative Hugoniot curves and the negative isentropic curves for several typical explosives, the initial state point for both negative curves is the C–J (Chapman–Jouget) detonation state (the dark points in the figure). The intersection point of any positive curve and any negative curve determine the wave intensity at the interface when the detonation wave of the corresponding explosive is normally incident into the relevant material. If the intersection point is located at a position higher than the C–J point (namely the pressure of the intersection point is higher than the C–J detonation pressure), the wave reflected into the detonation-product is a shock wave; otherwise is a rarefaction wave.

As pointed out in the discussion on the reflection and transmission of elastic waves in 1D bars (Section 3.5), the loading reflection at a rigid wall and the unloading reflection at a free surface can be viewed as examples of transmission into materials with shock wave impedance of ∞ and 0 respectively. However, due to the difference between the Hugoniot curve and the isentropic curve as well as the difference between the Hugoniot curves with different initial state points, the pressure amplitude of wave reflected at a rigid wall is no longer twice that of incident wave, and the particle velocity of wave reflected at a free surface is no longer twice that of incident wave.

For nonporous solid materials, especially metals, when the shock pressure is not very high (e.g. in the order of 10 GPa for metals), the differences between Hugoniot curves and isentropic curves can often be neglected. This is equivalent to the situation of weak shock

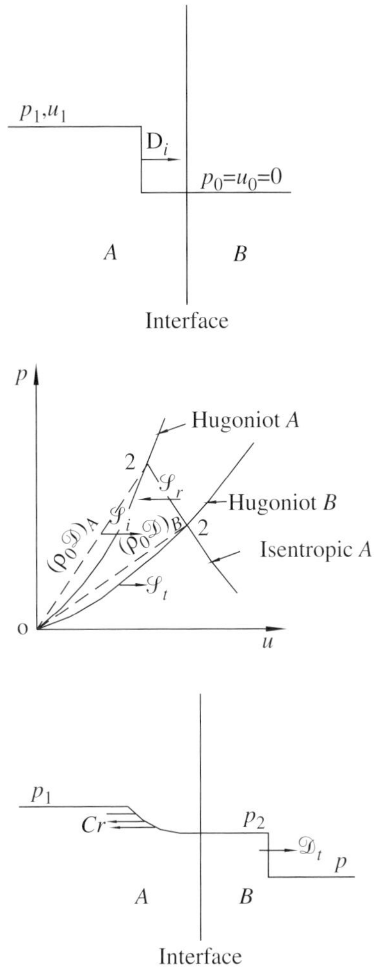

Fig. 7.19. Reflection and transmission of a shock wave which is incident into a low wave impedance material from a high wave impedance material.

waves, for which the entropy-increase due to shock jump can be neglected. In such cases, both the Hugoniot curve of reflected shock wave and the isentropic curve of reflected rarefaction wave can be approximately regarded as the image of incident shock wave about the perpendicular line *ab* passing the final state point (point 1) of the incident shock wave, as shown in Fig. 7.21. Then the problem can be simplified, and consequently the discussions on the reflection of shock waves either at the rigid wall or at the free surface are consistent with the discussions for elastic waves.

It should be noticed that, for elastic waves, the wave impedance $\rho_0 C_0$ is a constant; while for shock waves, the shock wave impedance $\rho_0 \mathscr{D}$ increases with pressure. Therefore, the relative value of wave impedance for two materials also changes with pressure. A special

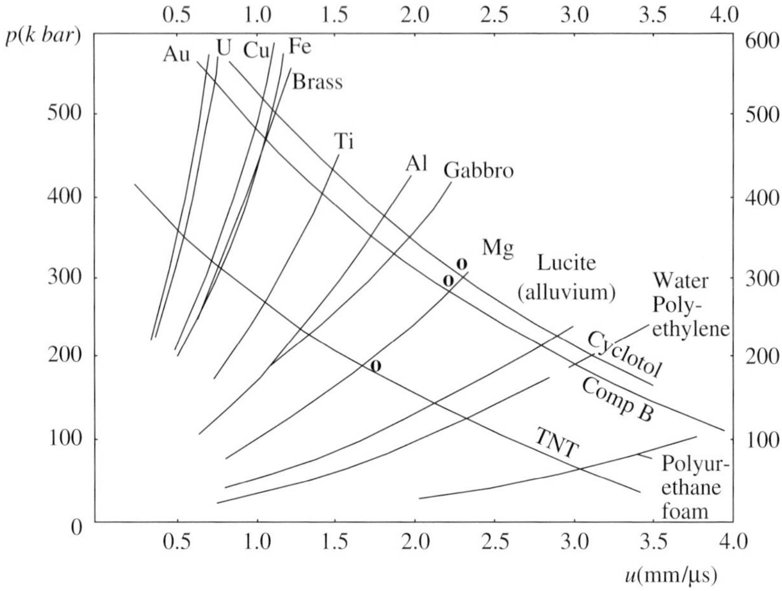

Fig. 7.20. The positive Hugoniot curves for several materials, and the negative Hugoniot curves and negative isentropic curves for several explosives.

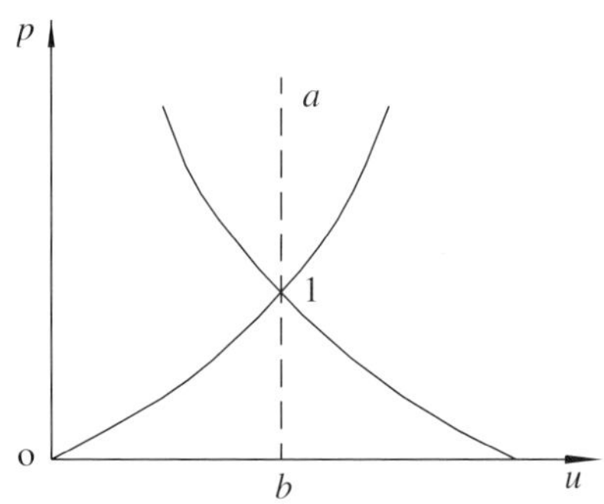

Fig. 7.21. Simplification for weak shock waves.

situation may even occur when the pressure is smaller than a critical pressure p_K, namely $p < p_K$, the shock wave impedance of material A is larger than that of material B; while when $p > p_K$, the shock wave impedance of material A is smaller than that of material B, as shown in Fig. 7.22, where the critical pressure p_K is the intersection point of two Hugoniot curves of the two materials.

To give an example, let us analyze the process of explosive expansion of an explosive device with dispensing liquids (like an air fuel explosive bomb). For convenience,

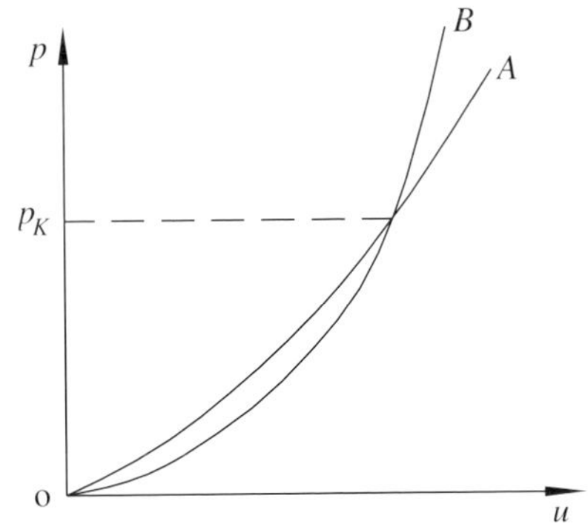

Fig. 7.22. Shock wave impedance increases with pressure.

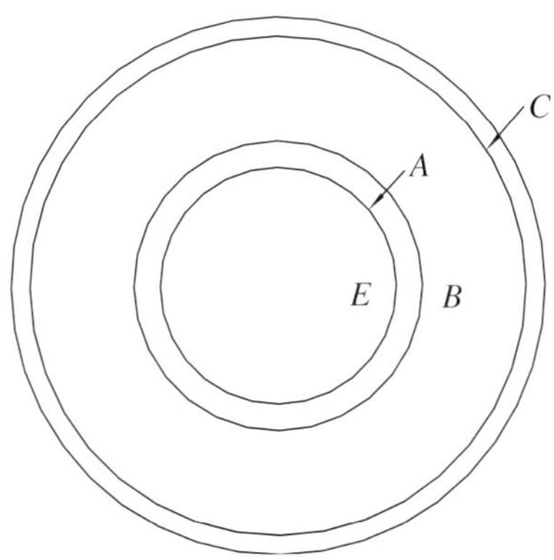

Fig. 7.23. Schematic explosive device for liquid dispensing.

the problem is approximately analyzed in the following by replacing spherical waves with plane waves (Duvall, 1972b), and Lagrange coordinate system will be used here. Figure 7.23 schemes the explosive device. The explosive region *E* (the center sphere) and the liquid to be dispensed *B* are separated by the inner spherical shell *A*; all of them are encased by the outer spherical shell *C*.

Assume that $(\rho_0\mathscr{D})_E > (\rho_0\mathscr{D})_A > (\rho_0\mathscr{D})_B < (\rho_0\mathscr{D})_C$. First, as shown in Fig. 7.24, when the detonation wave whose state is denoted by point 1 arrives at shell *A*, an unloading reflection occurs, whose state corresponds to point 2. Then, when the transmitted shock wave in *A* arrives at the interface of *A* and *B*, an unloading reflection occurs again, whose state corresponds to point 3. However, when the reflected wave returns to the interface of *A* and *E*, a loading reflection occurs, whose state corresponds to point 4. On this analogy, there are a series of leftward rarefaction waves reflected back and a

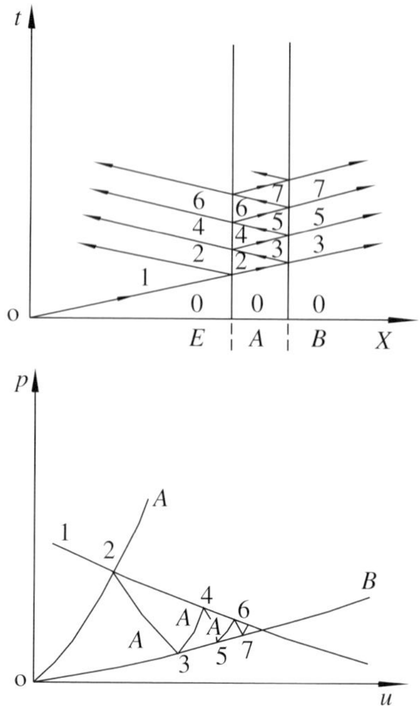

Fig. 7.24. Waves between A and liquid B.

series of rightward shock waves reflected forward in the inner shell A, respectively corresponding to the negative isentropic curves 2–3, 4–5, ..., and the positive Hugoniot curves 3–4, 5–6, ... Simultaneously there are a series of rightward waves transmitted into the liquid B, corresponding to the positive Hugoniot curves 0–3, 3–5, 5–7, and so on.

Second, as shown in Fig. 7.25, when the transmitted shock wave in liquid B arrives at the outer shell C, a loading reflection occurs, corresponding to the negative Hugoniot curve 1–2. Then, when the transmitted shock wave in C arrives at its free surface, an unloading reflection occurs, corresponding to the negative isentropic curve 2–3. This indicates that the expanding particle velocity of the outer shell increases, accelerating the outer shell outward. However, when the reflected wave returns to the interface of B and C, a rightward shock wave is reflected, corresponding to the positive Hugoniot curve 3–4. On this analogy, there are a series of leftward rarefaction waves reflected back and a series of rightward shock waves reflected forward in the outer shell C, respectively corresponding to the negative isentropic curves 4–5, 6–7, ..., and the positive Hugoniot curves 5–6, 7–8, ... With each reflection at the free surface, the expanding velocity of outer shell increases once again, corresponding to the state 3, 5, 7 and so on. As a result, the outer shell successively and acceleratively expands outward until the break of the outer shell, and the liquid is spread to fog-like drops over a large volume in the order of thousands of its original volume.

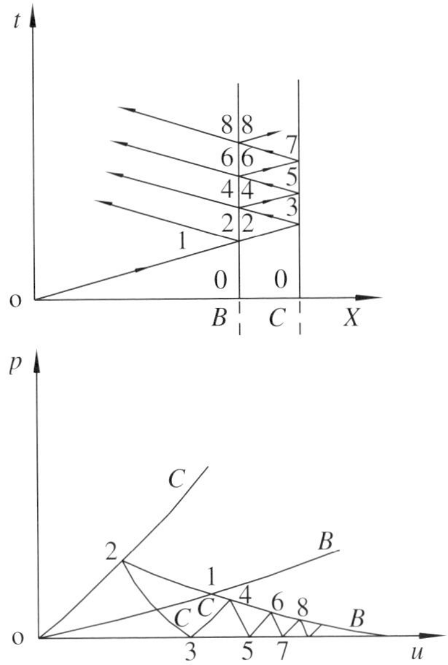

Fig. 7.25. Waves between liquid *B* and *C*.

7.9 Plane Waves in Hydro-Elasto-Plastic Media

Historically, the theory of nonlinear elasto-plastic waves was developed from two different approaches, though the common conclusion was finally reached. On the one hand, based on the theory of linear elastic waves under relative small loading, with the gradual increase of loading, the plastic deformation is taken into consideration (Section 7.4), and furthermore the influence of nonlinear elasticity is taken into consideration (Morland, 1959). This is the approach extending the theory of linear elastic waves under lower stresses to the theory of nonlinear elasto-plastic waves under higher stresses. On the other hand, based on the theory of shock waves in solids under higher pressure neglecting its shear strength (Rice *et al.*, 1958), with the decrease of loading, the shear strength of materials is taken into consideration. This is the approach extending the theory of shock waves of solids under very high pressure (Section 7.7) to the theory of nonlinear elasto-plastic waves under sub-high pressure. Although the two approaches are different, a common conclusion has been reached.

First, we discuss the propagation of plane waves in nonlinear elasto-plastic media following the first approach. Reviewing the discussion on the 1D strain elasto-plastic waves in section 7.4, it was assumed that the elastic deformation follows Hooke's law. Obviously, this assumption is valid only in the problems involving small strain. In the cases of high pressure and large deformation, the elastic moduli of materials are no longer invariables, namely, the elastic constitutive relationship is actually nonlinear. However, if the nonlinear elastic constitutive relationship is written in differential form, it is similar to the differential form of Hooke's law [Eq. (7.15)]

$$\begin{aligned} &\text{volumetric law:} \quad -dp = Kd\Delta \\ &\text{distortional law:} \quad dS_{ij} = 2Gde_{ij} \end{aligned} \tag{7.104}$$

only that the K and G in these equations are generally the functions of strain or stress, representing the local tangent slope of the corresponding stress–strain curve respectively. Correspondingly, if the axial stress–strain relationship σ_X–ε_X in the 1D strain problems is written in differential form, similar to Eq. (7.42), then we obtain

$$d\sigma_X = -dp + dS_{XX} = \begin{cases} \left(K + \frac{4}{3}G\right) d\varepsilon_X, & \sigma_X < Y_H \\ \left(K + \frac{4}{3}G_p\right) d\varepsilon_X, & \sigma_X > Y_H \end{cases} \tag{7.105}$$

Noting $\varepsilon_X = \Delta$ in the 1D strain problems, K and G generally are the functions of volumetric strain Δ or hydrostatic pressure p.

With regard to the nonlinear elastic volumetric modulus $K(p)$, it has actually been discussed in Section 7.6. If the state equation of solid under high pressure is given, then $K(p)$ can be determined. In the cases when the pressure is not high enough to neglect the shear strength of solids, the Bridgman equation [Eq. (7.57)] or Murnagham equation [Eq. (7.60)] are usually used to determine the nonlinear volumetric modulus for engineering application.

With regard to the nonlinear elastic shear modulus $G(p)$, the related experimental data are not as much as that of $K(p)$. Based on the experimental data obtained from the supersonic measurements under the pressure of around 1 GPa, Воронов (Voronov) and Верещакин (Vereshchakin)(1961) proposed, similar to Eq. (7.60), the following linear relationship:

$$G = G_0 (1 + \eta p) \tag{7.106}$$

where η $(=k_0 n)$ is just the material parameter in Murnagham equation [Eq. (7.60)]. Without enough data, the influence of p on G is usually neglected and G is viewed as a constant for engineering applications.

If the plastic shear modulus G_p is also determined [Eq. (7.41)], then the nonlinear elasto-plastic stress–strain relationship in the 1D strain problems is completely determined. Based on this, similar to the discussions in Section 7.4, the problems of nonlinear elasto-plastic plane waves can be solved. Obviously, according to Eq. (7.7), both the nonlinear elastic wave velocity C_L^e and the plastic wave velocity C_L^p

$$C_L^e = \sqrt{\frac{K(p) + \frac{4}{3}G(p)}{\rho_0}} \tag{7.107}$$

$$C_L^p = \sqrt{\frac{K + \frac{4}{3}G_p}{\rho_0}} \tag{7.108}$$

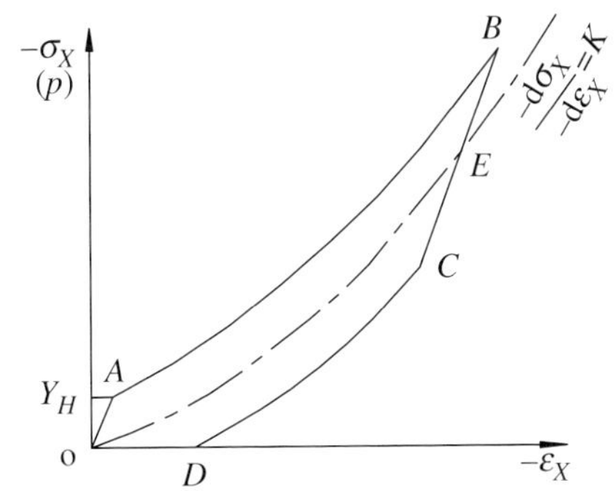

Fig. 7.26. Perfect plastic stress–strain curve in the 1D strain problems.

depend on the axial strain ε_X or volumetric strain Δ; or equivalently, vary with the hydrostatic pressure p.

For perfect plastic materials, $G_p = 0$. In such cases, the stress–strain curves corresponding to the compression loading and the expansion unloading are shown in Fig. 7.26. In the figure, the dash-dotted curve *OE* corresponds to the nonlinear elastic volumetric law; the solid curves *OA* and *AB* correspond to the elastic compression loading and the plastic compression loading respectively, while the solid curves *BC* and *CD* correspond to the elastic expansion unloading and the reverse plastic loading. In the situation of perfect plasticity discussed here, if the pressure is not very high, the curves *AB* and *CD* can be regarded as to keep a constant distance $(2Y_0/3)$ in the σ_X-axis direction with the hydrostatic pressure curve *OE*. Different from the stress–strain curve of linear elasto-perfect plastic material (Fig. 7.2), during the compression loading in the present case, the stress–strain curves of both the elastic loading segment *OA* and the plastic loading segment *AB* are normally concave upward $\left(\left(d^2\sigma_X/d\varepsilon_X^2\right) > 0\right)$. Therefore, the corresponding elastic shock wave and plastic shock waves are formed respectively. Their wave velocities are determined by the tangent slope of Rayleigh line $\overline{OA}$ and $\overline{OB}$ respectively, namely the elastic shock wave velocity $\mathscr{D}_e$ is

$$\mathscr{D}_e = \sqrt{\frac{1}{\rho_0} \cdot \frac{\sigma_X(A)}{\varepsilon_X(A)}}$$

and the plastic shock wave velocity $\mathscr{D}_p$ is

$$\mathscr{D}_p = \sqrt{\frac{1}{\rho_0} \cdot \frac{\sigma_X(B) - \sigma_X(A)}{\varepsilon_X(B) - \varepsilon_X(A)}}$$

In the unloading process, either for the elastic unloading segment *BC* or for the reverse plastic loading segment *CD*, because their slope $d\sigma_X/d\varepsilon_X$ decreases with the decrease of σ_X, both the elastic unloading waves and the reverse plastic loading waves must be the dispersive weak-discontinuous waves, and their wave velocities are determined by Eqs. (7.107) and (7.108) respectively.

It should be pointed out that a process of shock jump is neither isothermal nor isentropic but an irreversible adiabatic process with extra entropy increase. Thereby the segment *OA* and *AB* which are corresponding to the formation of shock waves actually should be the shock adiabatic curves taking account of distortion. Strictly, they are not parallel to the isentropic hydrostatic pressure curve *OE*. In the process of unloading, the elastic unloading segment *BC* is an isentropic adiabatic curve, while the reverse plastic loading segment *CD* is an adiabatic curve paralleling the isentropic hydrostatic pressure curve *OE* but having entropy increase in relation to the irreversible plastic deformation. Therefore, even in the cases of perfect plasticity, the segment *AB* and the segment *CD* are strictly not parallel. Only if the pressure is not very high, so that the differences among the isothermal p–V curve, the isentropic p–V curve, and the shock adiabatic p–V curve are small enough to be negligible, then the shock adiabatic curve and the isentropic adiabatic curve can be viewed equivalent approximately. When their differences should be strictly taken into consideration, the hydrodynamics approximation neglecting the distortion of solids as discussed in Section7.7, should be extended to take account of distortion. This is the second approach introduced at the beginning of this section. A further discussion will be conducted below following this approach.

Reviewing the discussion on shock waves in solids under high pressure in Section 7.7, it was assumed that the shear strength of solids can be neglected, and then materials can be viewed as fluids. It is clear that this is valid only when the shock wave pressure is much higher than the shear strength of materials. In fact, in the 1D strain problems, the axial stress can be decomposed into two parts: the hydrostatic pressure p and the maximum shear stress τ,

$$\sigma_X = -p + s_{XX} = -p + \tfrac{2}{3}(\sigma_X - \sigma_Y) = -p + \tfrac{4}{3}\tau$$

The p increases unlimitedly with increase in the volumetric compression, and it can reach the order of 10^3–10^8 GPa in a wide range of practical problems. The τ increases with increase in the shear deformation, and its limit is the shear strength of materials, in the order of 10 Mpa–1 GPa for most engineering materials. Therefore, if the shock pressure is two orders or more larger than the shear strength of materials, namely the ratio $\tau/p \leq 0.01$, then the term of τ in the above equation can be neglected, and we have approximately $\sigma_X \approx -p$. Only under such condition, the hydrodynamics model provides satisfactory approximate solutions. In contrast, when the shock pressure approaches the shear strength of materials, the hydrodynamics approximation is not valid anymore, and the modification should be made to take the shear strength of materials into consideration.

The modification actually includes two aspects:

(i) Since the state equation of solids under high pressure adopted in the hydrodynamics approximation such as the state equation in the internal energy [Eq. (7.66)], only depicts the nonlinear volumetric deformation law, it should now be replaced by the nonlinear elasto-plastic constitutive relationship including both the volumetric deformation law and the distortional deformation law. If it is assumed that the plastic deformation contributes none to the volumetric deformation, then Eq. (7.66) is still valid and an elasto-plastic distortion law should also be added. In the 1D strain condition, according to Mises criteria or Tresca criteria, similar to Eq. (7.42), the elasto-plastic distortion law in the differential form can be expressed as

$$\frac{ds_{XX}}{d\varepsilon_X} = \begin{cases} \frac{4}{3}G, & |s_{XX}| < \frac{2}{3}Y \\ \frac{4}{3}G_p, & |s_{XX}| = \frac{2}{3}Y \end{cases} \tag{7.109}$$

Such nonlinear elasto-plastic constitutive model is usually called the **hydro-elasto-plastic model**, and the media following such constitutive relationship are usually called the **hydro-elasto-plastic media**. In the 1D strain condition, such constitutive relationship can be written in the form of axial stress versus axial strain, σ_x–ε_x, which is completely consistent with the Eq. (7.105) deduced following the first approach mentioned previously. Thus it can be seen that the results obtained following the two approaches mentioned above are exactly consistent.

(ii) The shock jump condition Eq. (7.78) deduced in the hydrodynamics approximation should be correspondingly modified, namely the hydrostatic pressure $(-p)$ in the original equation now should be replaced by the axial stress σ_X.[2]

Therefore, the governing equation set of the plane shock waves in the hydro-elasto-plastic media includes: three conservation conditions contained in the modified shock jump conditions, two equations contained in the constitutive relationship, an equation connecting σ_X, p, and s_{XX}, an equation connecting the axial strain ε_X and the specific volume V. These seven equations in total are listed below,

$$[u] = \rho_0 \mathscr{D}[V] \tag{7.110a}$$

$$[\sigma_X] = -\rho_0 \mathscr{D}[u] \tag{7.110b}$$

$$[E] = \frac{1}{2}(\sigma_X + \sigma_{X0})\,[V] \tag{7.110c}$$

$$p = p\,(V, E) \tag{7.110d}$$

$$\frac{ds_{XX}}{d\varepsilon_X} = \begin{cases} \frac{4}{3}G, & |s_{XX}| < \frac{2}{3}Y \\ \frac{4}{3}G_p, & |s_{XX}| = \frac{2}{3}Y \end{cases} \tag{7.110e}$$

$$\sigma_X = -p + s_{XX} \tag{7.110f}$$

$$\varepsilon_X = \frac{V}{V_0} - 1 \tag{7.110g}$$

When the initial states (the variables with subscript 0) ahead of the shock wave front are known, there are eight unknowns in the above seven equations, namely variables σ_X, p, s_{XX}, ε_X, V, u, E, and the shock wave velocity $\mathscr{D}$, which depicts the final states behind the

[2]In the deduction of Eq. (7.78), it has been set that the positive sign of p corresponds to the pressure, and correspondingly $\varepsilon_X = 1 - V/V_0$, namely the positive sign of ε_X temporarily corresponds to the compressive strain. Here, consistent with the denotation used in other parts of this book, the positive sign of σ_X and ε_X corresponds to the tensile stress and strain. So, please note the difference of signs in Eqs. (7.110) and (7.78).

shock wave front. If any unknown variable among these is given in the boundary condition, the other seven unknown variables can be determined. In other words, the above seven equations depict the relationships between any two of these unknown variables. These relationships are the shock adiabatic curves or the Hugoniot curves in the case of hydro-elasto-plastic model. It should be emphasized again, these curves only represent the loci of the possible final equilibrium states through a shock jump started at a certain initial equilibrium state; they do not represent the locus of the successive states that the material experienced in the shock jump process.

Figure 7.27 shows three forms of Hugoniot curves: the σ_X–ε_X, the p–ε_X, and the s_{XX}–ε_X, of which the initial state is the undisturbed state 0. The $p-\varepsilon_X$ curve reflects the influence of the volumetric law of solids on the shock jump process, the $s_{XX}-\varepsilon_X$ curve reflects the influence of the distortion law of solids, and the σ_X–ε_X curve reflects the composite influence of the volumetric law and the distortion law. The difference between the σ_X–ε_X curve and the $p-\varepsilon_X$ curve reflects the difference between a hydro-elasto-plastic model and a hydrodynamics model. The point A in the figure corresponds to the Hugoniot elastic limit, $|\sigma_X(A)| = Y_H$. If $|\sigma_X| \le Y_H$, then single elastic shock wave is formed. If $|\sigma_X| > Y_H$, then a dual-wave structure is formed, namely a plastic shock wave follows an elastic precursor shock wave. With increase in pressure, the plastic wave velocity increases. When $|\sigma_X| \ge \sigma_X(K)$, a stable single shock wave is again formed; here, $\sigma_X(K)$ is the overdrive stress, which satisfies the condition that the slope of the line connecting the point K and A just equals the slope of the Rayleigh line OA. If $|\sigma_X| \gg Y_H$, the difference between $\sigma_X-\varepsilon_X$ curve and $p-\varepsilon_X$ curve can be neglected, then the hydrodynamics approximation can be applied.

Comparing with the analysis based on the hydrodynamics, an important feature of the analysis based on the hydro-elasto-plastic theory is that the irreversible plastic deformation is taken into consideration. Therefore, across a shock jump of a plastic shock wave, the irreversible entropy-increase actually consists of two parts:

$$dS = dS_q + dS_p \tag{7.111}$$

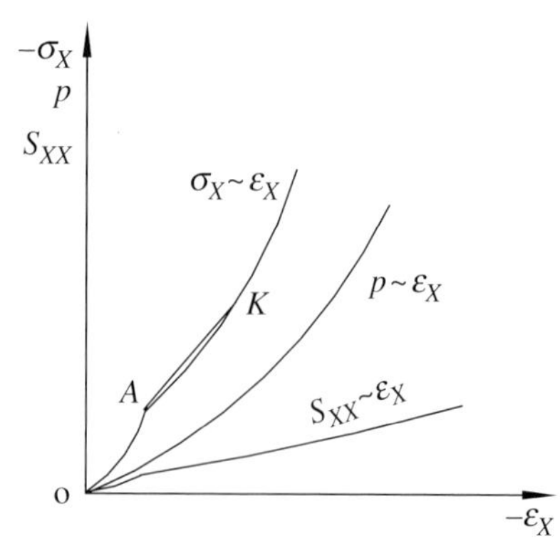

Fig. 7.27. Three forms of Hugoniot curves for hydro-elasto-plastic model.

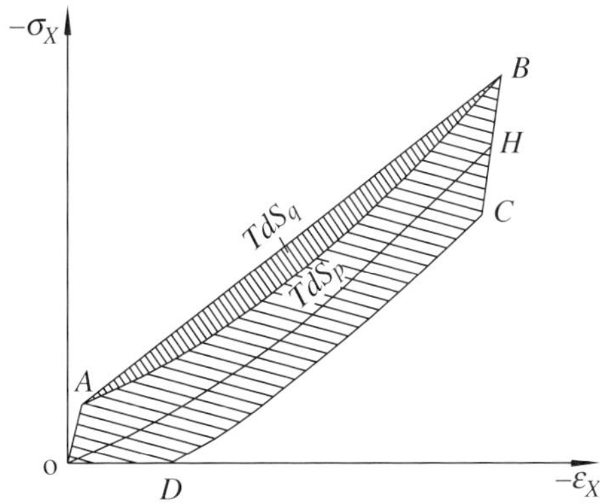

Fig. 7.28. The entropy-increase dS_q and dS_p of a plastic shock wave in the case of hydro-elasto-plastic model.

where dS_q is the extra entropy-increase caused by the formation of a shock wave as discussed in Section 7.7, which can approximately be represented by the area enclosed by the σ_X–ε_X Hugoniot curve AB and the Rayleigh line $\overline{AB}$, as shown in Fig. 7.28; dS_p is the entropy-increase caused by the irreversible plastic deformation (quasi-equilibrium), namely,

$$TdS_p = dW_d^p \tag{7.112}$$

where W_d^p denotes the plastic part of distortional work W_d. Since it is assumed that the plastic deformation contributes none to the volumetric deformation, W_d^p is also the total plastic work.

For the dS_q, similar to the entropy-increase calculated by Eq. (7.85) based on the hydrodynamics approximation, in the present situation, the p and V in Eq. (7.85) are replaced by σ_X and ε_X respectively, and so it can be calculated by

$$TdS_q = \frac{1}{2}V_0(\varepsilon_X - \varepsilon_A)d\sigma_X \cdot \left\{ 1 - \frac{\dfrac{\sigma_X - \sigma_A}{\varepsilon_X - \varepsilon_A}}{\dfrac{d\sigma_X}{d\varepsilon_X}} \right\} \tag{7.113}$$

As for the dS_p, it can be calculated through the plastic distortional work dW_d^p. In the case of 1D strain, the calculation of the plastic work can be simplified by

$$dW_d = V_0 S_{XX} d\varepsilon_X = \frac{4}{3}V_0 \tau d\varepsilon_X \tag{7.114}$$

where the elastic part can be calculated according to the linear elastic distortional law ($S_{XX} = 4G\varepsilon_X/3$), we have

$$dW_d^e = V_0 \cdot \frac{4}{3} G\varepsilon_X d\varepsilon_X = \frac{4}{3}\frac{V_0\tau}{G} d\tau \tag{7.115}$$

then the plastic part is

$$dW_d^p = dW - dW_d^e = \frac{4}{3} V_0\tau \left(d\varepsilon_X - \frac{d\tau}{G} \right)$$

Substituting it into Eq. (7.112), we obtain

$$TdS_p = \frac{4}{3} V_0\tau \left(d\varepsilon_X - \frac{d\tau}{G} \right) \tag{7.116}$$

Under the condition of perfect plasticity, $d\tau \equiv 0$, there is

$$TdS_p = dW_d^p = dW_d = \tfrac{4}{3} V_0\tau d\varepsilon_X$$

Table 7.4 lists the ratios of dS_q and total entropy-increase dS under various pressures (Duvall, 1972a). It can be seen that under high pressures the plastic deformation contributes little to the entropy-increase. In this sense, the hydrodynamics approximation is acceptable. However, under sub-high pressure or the so-called medium pressure, the plastic deformation cannot be neglected. Since the plastic work causes energy dissipation, it is clear that the energy dissipation induced by the plastic shock wave propagation based on the hydro-elasto-plastic analysis is larger, corresponding to a larger increase in temperature.

Comparing with the analysis based on the hydrodynamics approximation, another important feature of the hydro-elasto-plastic analysis lies in the process of unloading. If the state before unloading corresponds to point B in Fig. 7.28, the unloading process begins in an elastic unloading, so it proceeds first following the isentropic curve BC in the figure. Correspondingly, the elastic unloading waves are continuous waves (rarefaction waves), and their wave velocities C_L^e are determined by the tangent slope of the isentropic curve BC [Eq. (7.107)]. Generally, the elastic unloading wave velocity C_L^e are larger than the plastic loading wave velocity C_L^p determined by the slope of Rayleigh line $\overline{AB}$. Therefore, the elastic unloading waves will catch up with the plastic shock waves ahead and interact with them (pursuing unloading). With the decrease of σ_X, when the reverse yield condition

Table 7.4. The ratios of dS_q and total entropy-increase dS under various pressures.

	Aluminum			Copper		
Y (GPa)	0.72	1.74	2.30	0.51	1.28	2.53
σ_X (GPa)	9.0	20.2	37.5	16.6	41.3	81.6
dS_q/dS	0.75	0.87	0.92	0.88	0.95	0.97

is satisfied (point C in the figure), the reverse plastic deformation occurs following the curve CD in the figure, which is determined based on the isentropic expansion curve taking account of distortion. The reverse plastic loading waves are continuous waves, whose wave velocities are determined by the tangent slope of CD [Eq. (7.108)] and generally smaller than the elastic unloading wave velocities.

With regard to the determination of elasto-plastic boundaries, in principle, it is similar to that discussed on the 1D stress elasto-plastic waves in Section 4.10. There are two points that should be noticed. First, for the σ_X–ε_X curve, it should distinguish the loading σ_X–ε_X curve and the unloading σ_X–ε_X curve. In the case of compressive loading when a shock wave is formed, the corresponding σ_X–ε_X curve is a shock adiabatic curve; while in the case of unloading, the σ_X–ε_X curve is either an isentropic adiabatic curve (during elastic unloading), or a quasi-equilibrium adiabatic curve (during reverse plastic loading), which is a curve based on the isentropic p–V curve but considering plastic deformation. And they all change with different initial states. These make the determination of elasto-plastic boundaries much more complicated.

Second, since the reverse plastic loading may occur in the unloading process, the reverse plastic loading boundaries may appear. Moreover, the internal reflective elastic waves which propagate back-and-forth between the positive plastic loading boundary and the reverse plastic loading boundary, not only play an unloading role for the positive plastic loading waves (pursuing unloading), but also play an unloading role for the reverse plastic loading waves (head-on unloading). Thereby, the determination of the reverse plastic loading boundaries and the determination of the positive plastic loading boundaries are connected and dependent on each other. Such coupling character brings new problems to determine the elasto-plastic boundaries. This will become clear in the discussions below.

7.10 Attenuation of Shock Waves in Hydro-Elasto-Plastic Media

Consider a semi-infinite hydro-elasto-plastic body originally at rest and natural state (no initial stress/strain). At the time $t = 0$, on its surface ($X = 0$), it is suddenly subjected to a constant load σ^* and then gradually unloaded, which is expressed by $\sigma_0(\tau)$. We now discuss how the shock waves attenuate due to the pursuing unloading disturbances (Wang *et al.*, 1983). For convenience, the subscript X of the axial stress σ_X is temporarily omitted in this section.

As shown in Fig. 7.29, at the beginning of sudden loading, an elastic precursor shock wave OH with an amplitude Y_H (Hugoniot elastic limit) and a plastic shock wave OA_2 with an initial amplitude of stress $(\sigma^* - Y_H)$ simultaneously propagate rightward. Since the plastic shock wave is continuously unloaded by the pursuing unloading disturbances from the beginning, its amplitude attenuates continuously and its wave velocity $\mathscr{D}_p$ also decreases continuously. In the present example, the unloading boundary by chance coincides with the propagating trajectory of the plastic shock wave. Then, the determination of the unloading boundaries is actually the same problem as to determine how the plastic shock wave attenuates due to the pursuing unloading disturbances.

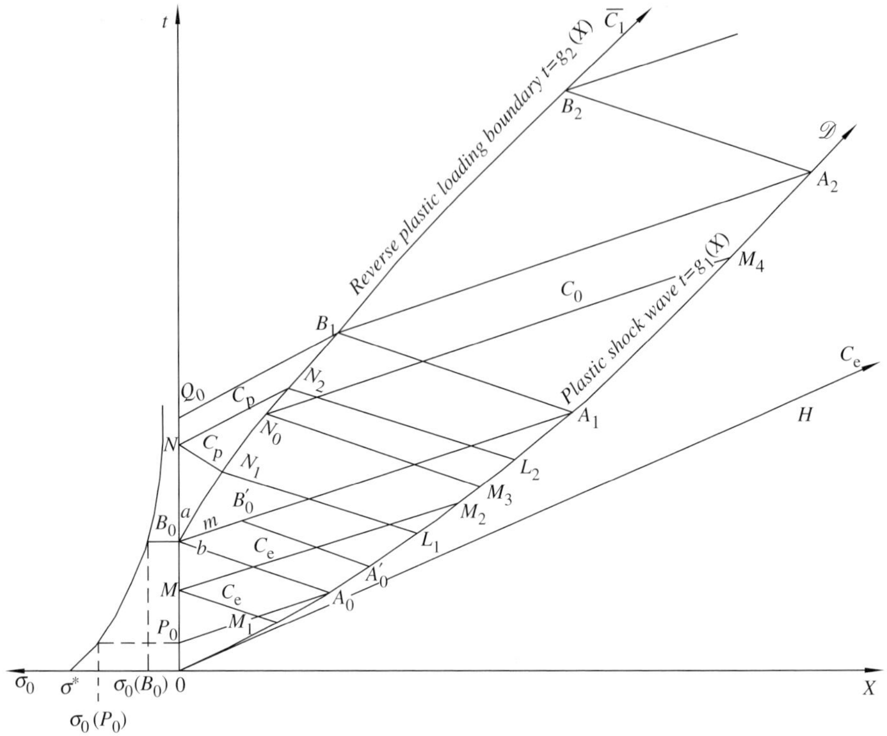

Fig. 7.29. Interaction between a positive plastic loading boundary and a reverse plastic loading boundary.

When the surface load is unloaded to satisfy the condition of reverse yield (the point B_0 in Fig. 7.29), the reverse plastic loading boundary B_0B_2 is formed. First, we discuss the inter-influence between the development of the reverse plastic loading boundary and the unloading boundary A_0A_2. For this purpose, we draw the rightward and leftward characteristics in the elastic unloading zone through the beginning point B_0 of the reverse plastic loading boundary; they intersect the unloading boundary at A_0 and A_1 respectively. And we further draw the successive characteristics A_1B_1, B_1A_2, A_2B_2, and so on in the elastic unloading zone and the characteristics B_1Q_0 in the reverse plastic loading zone.

The following points can be seen from Fig. 7.29 (1) The rightward characteristic MM_2 passing an arbitrary point M_2 on the segment OA_1 in the elastic unloading zone intersects with the axis t, which indicates that the attenuation of shock wave on the segment OA_1 is determined by the surface unload condition for the duration OB_0. (2) the rightward characteristic N_0M_4 passing an arbitrary point M_4 on the segment A_1A_2 in elastic unloading zone intersects with the segment B_0B_1 of the reverse plastic loading boundary, which indicates that the attenuation of unloading boundary A_1A_2 or the attenuation of plastic shock waves on the segment A_1A_2 is determined by the states on the segment B_0B_1 of reverse loading boundary. (3) The leftward characteristic N_2L_2 passing an arbitrary point N_2 on the segment B_0B_1 in elastic unloading zone intersects with the segment A_0A_1 and the rightward characteristic NN_2 passing an arbitrary point N_2 on the reverse plastic loading zone intersects with the axis t, which indicates that the states on the segment B_0B_1 are determined by both the attenuation of plastic shock waves on the segment A_0A_1 and the surface unloading condition for the duration B_0Q_0.

Thus, in order to determine the segment A_1A_2 of the unloading boundary, it is first needed to determine the segment B_0B_1 on the reverse plastic loading boundary. However in order to determine the segment B_0B_1, it is again needed to determine the segment A_0A_1 on the unloading boundary. Therefore, the determination of the whole unloading boundary or the whole propagating trajectory of the plastic shock waves $OA_0A_1A_2A_3$ is a complicated problem coupled with the determination of the reverse plastic loading boundary $OB_0B_1B_2B_3$.

From the above discussion, it is also known that the whole problem can be dealt with by dividing the whole elastic–plastic boundary into some segments, namely we can first determine the segment OA_1 independent of the reverse plastic loading boundary, then we try to determine the segment B_0B_1; after this, similar to the procedures discussed in Chapter 4, by utilizing the conjugate relationship [Eq. (4.37)], from the known elasto-plastic boundary, the corresponding conjugate boundary can be determined. In this war, the segment A_1A_2 on the unloading boundary, the segment B_1B_2 on the reverse plastic loading boundary, etc., can be determined alternately, and then the whole boundary can be determined. Thus, the main problem remaining is how to determine the initial segment OA_1 of the unloading boundary and the initial segment B_0B_1 of the reverse plastic loading boundary.

In the actual treatment, the elastic unloading zone and the reverse plastic loading zone are divided into some subzones, and assume that the characteristics in those subzones can be regarded as straight lines respectively, namely assume that the elastic wave velocity C_L^e and the reverse plastic loading wave velocity C_L^p are constants in the subzones respectively. Note that for an arbitrary point $M_2(X_2, t_2)$ on the segment OA_1, there must exist two associated points connected by two associated characteristics, namely point $M(0, \tau)$ connected by the positive characteristics MM_2 and the so-called conjugate point $M_1(X_1, t_1)$ connected by the negative characteristics MM_1. Then, similar to deducing the conjugate relationship Eq. (4.37), among these three points, the following conjugate relationship can be obtained:

$$\langle \mu_D(M_2) + 1\rangle \sigma(M_2) - \mu_D(M_2)Y_H = \langle \mu_D(M_1) - 1\rangle \sigma(M_1) - \mu_D(M_1)Y_H + 2\sigma_0(M) \qquad (7.117)$$

where $\mu_D = C_L^e/D(\sigma)$ is a known function of σ for a given material. Equation (7.117) shows that if the $\sigma(M_1)$ has been known, then $\sigma(M_2)$ is entirely determined by the unloading surface condition $\sigma_0(M)$.

For an arbitrary point $M_4(X_4, t_4)$ on the segment A_1A_2, there exist a relevant point $N_0(X_{N0}, t_{N0})$ on the reverse plastic loading boundary and a conjugate point $M_3(X_3, t_3)$ connected by the characteristics N_0M_4 and N_0M_3, respectively. Similarly, the following conjugate relationship can be obtained:

$$\begin{aligned}\langle\mu_D(M_4)+1\rangle\,\sigma(M_4)-\mu_D(M_4)Y_H &= \langle\mu_D(M_3)-1\rangle\,\sigma(M_3)\\ &\quad-\mu_D(M_3)Y_H+2\sigma_B(M_0)\end{aligned}\tag{7.118}$$

This equation shows that if the $\sigma(M_3)$ has been known, then the $\sigma(M_4)$ is entirely determined by the stress $\sigma_D(N_0)$ at the relevant point N_0 on the reverse plastic loading boundary.

For an arbitrary point N_2 on the segment B_0B_1, there exist four relevant points N, N_1, L_1, and L_2 connected by the characteristics NN_2, NN_1, N_2L_2, and N_1L_1, respectively, and the following conjugate relationship can be obtained similarly:

$$\begin{aligned}\left(\mu_p+1\right)\sigma\left(N_2\right) &= 2\mu_p\sigma_0(N)-\left(\mu_p-1\right)\sigma(N_1)-\langle\mu_D(L_2)-1\rangle\,\sigma(L_2)\\ &\quad+\{\mu_D(L_1)-1\}\,\sigma(L_1)+\{\mu_D(L_2)-\mu_D(L_1)\}\,Y_H\end{aligned}\tag{7.119}$$

where $\mu_p=C_L^e/C_L^p$. The above equation shows that if $\sigma(N_1)$ has been known then $\sigma(N_2)$ is determined by the free surface condition, the stress $\sigma(L_1)$ and $\sigma(L_2)$ at the relevant points L_1 and L_2 on the unloading boundary A_0A_1.

Thereby, if a small initial segment of OA_1 and a small initial segment of B_0B_1 are known, using the conjugate relationships Eqs. (7.117)–(7.119), the whole unloading boundary and reverse plastic loading boundary can be determined alternately segment by segment. As for the determination of the initial segments of OA_1 and B_0B_1, similar to the discussion in Chapter 4, we can use the "**successive approximations method**" based on the conjugate relationships Eqs. (7.117)–(7.119) or the "**local linearization method**" when the initial propagating velocity of elasto-plastic boundary has been determined.

When the "successive approximations method" is used, it should be pointed out that, although how the zero-order approximation is chosen does not affect the final result, the number of iteration is affected. In the discussion of the attenuation of shock waves in bars in Section 4.6, we took the surface load to be the instantaneous stress of shock wave at each time as the zero-order approximation [Eq. (4.63)]. This is equivalent to making a rigid unloading assumption ($C_e\to\infty$) and the inertia of rigid body after unloading is neglected. In the current situation of 1D strain plastic wave, because the ratio of elastic wave velocity and the plastic wave velocity is not as large as that in the case of 1D stress wave ($C_L^e/C_L^p\approx 0.8$ or larger), the error due to the rigid body assumption obviously becomes larger, and the corresponding attenuation rate of plastic shock wave will be overestimated. If considering that in the case of 1D strain wave the change of $\mathscr{D}_p$ due to the attenuation of shock waves along OA_1 is actually not large, so we can take $\mu_D(M_2)=\mu_D(M_1)$ approximately, then the conjugate relationship [Eq. (7.117)] can be simplified to:

$$\sigma(M_2)=\sigma_0(M)\left\{1+\frac{\mu_D-1}{\mu_D+1}\cdot\frac{\sigma(M_1)-\sigma_0(M)}{\sigma_0(M)}\right\}=\sigma_0(M)\left(1+s_D\right)$$

Since for the 1D strain elasto-plastic wave $\mu_D \approx 1$, and consequently $s_D \ll 1$, then a zero-order assumption better than the rigid unloading assumption can be made:

$$\sigma^{(0)}(M_2) = \sigma_0(M) \tag{7.120}$$

This is equivalent to that when the unloading disturbance catches up with the plastic shock wave at the elastic wave velocity, the stress amplitude of the shock wave immediately decreases to this value, and the internal reflection due to the interaction of these two waves is neglected. By the same argument, for the reverse plastic loading boundary B_0B_1, when we approximately take $\mu_D(L_1) = \mu_D(L_2)$, the conjugate relationship Eq. (7.119) is simplified to:

$$\begin{aligned}\sigma(N_2) &= \sigma_0(N)\left\{1 - \frac{\mu_p - 1}{\mu_p + 1}\cdot\frac{\sigma(N_1) - \sigma_0(N)}{\sigma_0(N)} - \frac{\mu_D - 1}{\mu_D + 1}\cdot\frac{\sigma(L_1) - \sigma(L_2)}{\sigma_0(N)}\right\}\\ &= \sigma_0(N)(1 + S_p)\end{aligned}$$

then similarly, the following zero-order approximation can be made:

$$\sigma^{(0)}(N_2) = \sigma_0(N) \tag{7.121}$$

The remaining actual steps of iteration computation are basically similar to those discussed in Chapter 4. Only for the reverse plastic loading boundary $t = g_2(X)$, once the stress distribution along this boundary $\sigma_B[X) = \sigma(X, g_2(X)]$ is approximately obtained by iteration computation, the reverse yield condition

$$\sigma_B(X) = \sigma_A(X) - 2Y_H \tag{7.122}$$

should be applied so that the trajectory of this boundary $t = g_2(X)$ can be determined, where $\sigma_A(X) = \sigma(X, g_1[X)]$ is the stress distribution along the unloading boundary $t = g_1(X)$.

It should be pointed out that, in the case of $\mu_D \approx 1$ and $\mu_p \approx 1$, Eqs. (7.120) and (7.121) as zero-order approximation often directly provide, although approximate but accurate enough solutions, and no further iteration computation is needed. Such an approximation is called the "**method of simple waves in subzones**". This is equivalent to, when solving the segment OA_1, the zone OA_1B_0 is approximately regarded as a zone of elastic simple waves, and the internal reflection which is generated when the unloading disturbances catch up with the plastic shock waves, is neglected; also when solving the segment B_0B_1, the zone $B_0B_1Q_0$ is approximately regarded as a zone of plastic simple waves, and the internal reflection which is generated when the reverse plastic loading disturbances head-on meet the unloading disturbances, is neglected. On this analogy, for the segment A_1A_2 on the unloading boundary, we approximately set

$$\sigma(M_4) = \sigma_B(N_0)$$

and so on, which is equivalent to, when solving the segment A_1A_2, the zone $A_1B_0B_1A_2$ is approximately regarded as a zone of elastic simple waves, and the relevant internal reflection is neglected.

When the local linearization method is used, the initial propagating velocity $\mathscr{D}_p(\sigma^*)$ of the strong discontinuous unloading boundary can be easily determined by the slope of Rayleigh line of σ_X–ε_X Hugoniot curve. Difficulties lie in the determination of the initial propagating velocity $\overline{C}_L$ of the weak discontinuous reverse plastic loading boundary. It should be first noticed that, for the first-order weak discontinuous interface B_0B_1, same as that shown in Fig. 4.35(a), the $\partial\sigma/\partial t$ which is given by the boundary condition for the state just before the reverse yielding (denoted as $\partial\sigma^b/\partial t$) and the $\partial\sigma/\partial t$ which is given by the boundary condition for the state just after the reverse yielding (denoted as $\partial\sigma^a/\partial t$) cannot be simply regarded as $\partial\sigma^e/\partial t$ and $\partial\sigma^p/\partial t$ respectively, and substituted into Eq. (4.111) for the determination of $\overline{C}_L$. Actually, $\partial\sigma^e/\partial t$ ahead of the reverse plastic loading boundary is $\partial\sigma^m/\partial t$ in the zone m, which is to be solved. Thereby, according to Eq. (4.111), there should be:

$$\frac{\dfrac{\partial\sigma^a}{\partial t}}{\dfrac{\partial\sigma^m}{\partial t}} = \frac{1-\left(\dfrac{\overline{C}_L}{C_L^e}\right)^2}{1-\left(\dfrac{\overline{C}_L}{C_L^p}\right)^2}$$

While along the elastic characteristics B_0A_1, there is:

$$\frac{\partial\sigma^m}{\partial t} - \frac{\partial\sigma^b}{\partial t} = -C_L^e\left(\frac{\partial\sigma^m}{\partial X} - \frac{\partial\sigma^b}{\partial X}\right)$$

Meanwhile B_0B_1 should also satisfy the reverse yield condition Eq. (7.122), of which the differential form in the case of perfect plasticity is:

$$\frac{d\sigma_B(X)}{dX} = \frac{\partial\sigma^m}{\partial X} + \frac{1}{\overline{C}_L}\frac{\partial\sigma^m}{\partial t} = \frac{d\sigma_A(X)}{dX}$$

Eliminating $\partial\sigma^m/\partial X$ and $\partial\sigma^m/\partial t$ from the above three equations, and noticing $\partial\sigma^b/\partial t = \partial\sigma^a/\partial t$ at point B_0, the following formula about $\overline{C}_L$ can be obtained,

$$\frac{1}{\overline{C}_L} = \frac{A + \sqrt{A^2 + 4\left\{\dfrac{1}{\left(C_L^p\right)^2} + \dfrac{A}{C_L^e}\right\}}}{2} \tag{7.123}$$

where $A = (1/\eta) - (1/C_L^e)$, $\eta = \left(\partial\sigma^b/\partial t\right)/\left[(d\sigma_A/dX) - \left(\partial\sigma^b/\partial X\right)\right]$. In actual computation, to avoid the computation of $\partial\sigma^b/\partial X$, we try to express the η by other known variables. Therefore using the compatibility relationships along the elastic characteristics B_0B_0', B_0A_0, and $B_0'A_0'$, we obtain

$$\sigma(B_0') - \sigma(B_0) = \frac{1}{2}\left(1 - \frac{C_L^e}{\mathscr{D}(A_0)}\right)\left\langle\sigma(A_0') - \sigma(A_0)\right\rangle$$

while from the geometric relationship of quadrangle $B_0B_0'A_0A_0'$, it is known:

$$X(B_0') - X(B_0) = \frac{1}{2}\left(1 + \frac{C_L^e}{\mathscr{D}(A_0)}\right)\langle X(A_0') - X(A_0)\rangle$$

Dividing the first equation above by the second one, and let $B_0' \to B_0$, then the relationship between the total differential $\left.\frac{d\sigma(B_0)}{dX}\right|_{B_0A_1}$ of σ along the elastic characteristics B_0A_1 at point B_0 and the total differential $d\sigma_A(A_0)/dX$ of σ along the unloading boundary A_0A_2 at point A_0 is obtained

$$\left.\frac{d\sigma(B_0)}{dX}\right|_{B_0A_1} = -\frac{\mu_D - 1}{\mu_D + 1}\frac{d\sigma_A(A_0)}{dX}$$

while according to the definition of total differential along the characteristics, there is

$$\left.\frac{d\sigma}{dX}\right|_{B_0A_1} = \frac{\partial\sigma}{\partial X} + \frac{1}{C_L^e}\frac{\partial\sigma}{\partial t}$$

Substituting the above results into the definition of η in Eq. (7.123), then the η at point B_0 is

$$\frac{1}{\eta(B_0)} = \frac{\dfrac{d\sigma_A(0)}{dX} + \dfrac{\mu_D(A_0) - 1}{\mu_D(A_0) + 1}\cdot\dfrac{d\sigma_A(A_0)}{dX} + \dfrac{\sigma_0'(B_0)}{C_L^e}}{\sigma_0'(B_0)} \tag{7.124}$$

where $\sigma_0' = d\sigma_0(t)/dt$. If further we consider that the change of the slope of initial segment OA_0 of the unloading boundary is small, there is approximately $\mu_D = \mu_s =$ constant, and we consider that the initial unloading segment OP_0 of the surface load can be dealt with as local linear attenuation, namely approximately $\sigma_0' =$ constant, then referring to Eq. (4.50) it is known:

$$\frac{d\sigma_A(0)}{dX} = \frac{d\sigma_A(A_0)}{dX} = \frac{\mu_D^2 - 1}{2\mu_D}\cdot\frac{\sigma_0'(0)}{C_L^e}$$

Substituting it into Eq. (7.124), we obtain:

$$\frac{1}{\eta(B_0)} = \left(\frac{1}{\mathscr{D}} - \frac{1}{C_L^e}\right)\frac{\sigma_0'(0)}{\sigma_0'(B_0)} + \frac{1}{C_L^e} \tag{7.124$'$}$$

or according to the definition of A in Eq. (7.123), there is

$$A = \left(\frac{1}{\mathscr{D}} - \frac{1}{C_L^e}\right)\frac{\sigma_0'(0)}{\sigma_0'(B_0)}$$

For a given material, the C_L^e, C_L^p, and $\mathscr{D}(\sigma)$ are all known. Therefore, from the $\sigma_0'(0)$ at point 0 and the $\sigma_0'(B_0)$ at point B_0 given by the surface unloading condition, and according to Eqs. (7.123) and (7.124′), the initial propagating velocity of the reverse plastic loading boundary at point B_0, $\overline{C}_L(B_0)$, can be determined.

If the surface load $\sigma_0(t)$ is a linearly attenuating function, then $\sigma_0'(0) = \sigma_0'(B_0)$, and $A = (1/\mathscr{D} - 1/C_L^e)$, therefore Eq. (7.123) is reduced to the result given by Lee and Liu (1964):

$$\frac{1}{\overline{C}_L} = \frac{\left(\frac{1}{\mathscr{D}} - \frac{1}{C_L^e}\right) + \sqrt{\left(\frac{1}{\mathscr{D}} - \frac{1}{C_L^e}\right)^2 + \left(\frac{2}{C_L^p}\right)^2 + \left(\frac{1}{\mathscr{D}} - \frac{1}{C_L^e}\right)\frac{4}{C_L^e}}}{2} \tag{7.125}$$

Once the initial propagating velocity of the elasto-plastic boundary is known, the whole elasto-plastic boundary can be determined by using the corresponding conjugate relationships according to the local linearization approximation method as discussed in Chapter 4. The details will not be repeated here.

As an example, the computational results of attenuation of plastic shock waves in the steel plate subjected to a contact explosion by using the above successive iteration method are shown as the solid line in Fig. 7.30 (Wang *et al.*, 1983). The computation iterations were conducted until the calculated stress reached to an accuracy of three valid digitals. The dotted line near the solid line are the results obtained by the "method of simple waves in subzones" mentioned above, which almost coincide with the solid line with an error less than 1%. The results calculated by the hydrodynamics approximation, namely neglecting the shearing strength of material, are shown as the dashed line in the figure. Also the results calculated by the finite difference numerical method based on the hydro-elasto-plastic model are shown as the dash-dotted line in the figure (Chu *et al.*, 1981). It is clear that the result obtained by the approximation methods discussed in this section agrees better with the result obtained by the finite difference numerical method than the result obtained by the hydrodynamics approximation. It illustrates that the elasto-plastic distortion of solid influences the shock waves to attenuate more quickly. In this example, the maximum stress amplitude of plastic shock waves is 39.3 GPa and the Hugoniot elastic limit of the

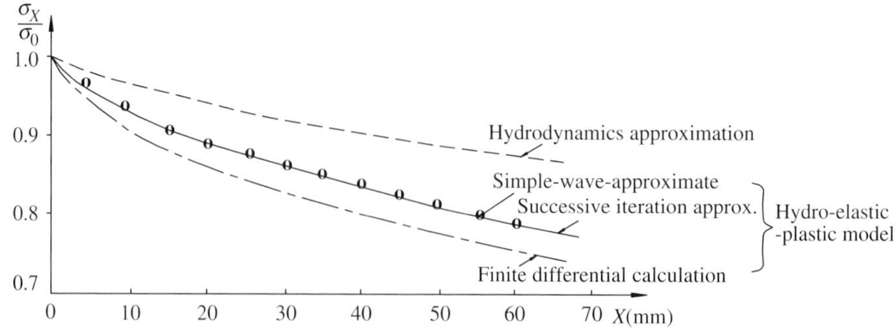

Fig. 7.30. Attenuation of plastic shock waves in steel plate under contact explosion.

material is 1.21 GPa, so the influence of shear strength of material on the attenuation of shock waves is no longer negligible.

7.11 One-Dimensional Strain Elasto-Visco-Plastic Waves

Now, we will generalize the one-dimensional stress Соколовский (Sokolovskii)–Malvern elasto-visco-plastic theory discussed in Section 6.5 to the general 3D stress case, based on the classical plastic-flow theory, and furthermore the 1D strain elasto-visco-plastic waves will be discussed.

It is well known that the general form of the plastic flow theory (increment theory) is expressed as

$$\dot{\varepsilon}_{ij}^{p} = \lambda \frac{\partial F}{\partial \sigma_{ij}} \tag{7.126}$$

where $\dot{\varepsilon}_{ij}^{p} = \partial \varepsilon_{ij}^{p}/\partial t$ is the plastic strain rate, λ is the nonnegative scalar function, F is the plastic potential function. In the frame of the so-called associate theory, F is regarded as the yield function. If the Mises yield condition is applied, namely the plastic potential function is in the form below:

$$F = J_2 - \mathscr{K}^2(W_p) = 0$$

then Eq. (7.126) is reduced to the form below:

$$\dot{\varepsilon}_{ij}^{p} = \lambda s_{ij} \tag{7.127}$$

the J_2 in the above equations is the second invariable of stress deviator s_{ij},

$$J_2 = \tfrac{1}{2} s_{ij} s_{ij} = \tfrac{1}{6}\left\{(\sigma_1 - \sigma_2)^2 + (\sigma_2 - \sigma_3)^2 + (\sigma_3 - \sigma_1)^2\right\}$$

the $\mathscr{K}(W_p)$ is the strain hardening parameter of the material, a function of plastic work W_p. For the perfect plastic materials, $\mathscr{K}$ equals the yield limit in pure shear, k; while $k = Y_0/\sqrt{3}$, where Y_0 is the yield limit in simple tension.

If introducing the second invariable I_2^p of the plastic strain rate tensor $\dot{\varepsilon}_{ij}^{p}$:

$$I_2^p = \tfrac{1}{2} \dot{\varepsilon}_{ij}^{p} \dot{\varepsilon}_{ij}^{p}$$

and defining the following plastic strain rate intensity $\dot{\varepsilon}_i^p$ and stress intensity σ_i respectively as[3] :

$$\dot{\varepsilon}_i^p = \sqrt{I_2^p}$$

$$\sigma_i = \sqrt{J_2}$$

[3]The σ_i defined here is different from the definition of Eq. (7.30) by a factor $\sqrt{3}$.

then from Eq. (7.127), we obtain

$$\tfrac{1}{2}\dot{\varepsilon}^p_{ij}\dot{\varepsilon}^p_{ij} = \lambda^2\left(\tfrac{1}{2}s_{ij}s_{ij}\right)$$

therefore the λ can be expressed as:

$$\lambda = \frac{\dot{\varepsilon}^p_i}{\sigma_i} = \frac{\sqrt{I^p_2}}{\sqrt{J_2}}$$

and finally Eq. (7.127) is reduced to:

$$\dot{\varepsilon}^p_{ij} = \dot{\varepsilon}^p_i \frac{s_{ij}}{\sqrt{J_2}} \tag{7.128}$$

According to the Соколовский (Sokolovskii)–Malvern model (Section 6.5), the inelastic strain rate $\dot{\varepsilon}^p$ is a function of over-stress $[\sigma - Y(W_p)]$. Generalizing this to the 3D general situation, the inelastic strain rate intensity $\dot{\varepsilon}^p_i$ should be a function of "over-stress" $\left(\sqrt{J_2} - \mathscr{K}(W_p)\right)$. Thereby, similar to Eq. (6.81), in the 3D situation, there is

$$\dot{\varepsilon}^p_i = \gamma\left\langle\phi\left(\frac{\sqrt{J_2}}{\mathscr{K}}\right) - 1\right\rangle \tag{7.129}$$

where γ is the viscous coefficient. Thus, the constitutive relationship Eq. (7.128), when the general plastic flow theory is extended to the general visco-plastic theory, becomes

$$\dot{\varepsilon}^p_{ij} = \gamma\left\langle\phi\left(\frac{\sqrt{J_2}}{\mathscr{K}}\right) - 1\right\rangle \frac{s_{ij}}{\sqrt{J_2}} \tag{7.130}$$

If further we assume that the inelastic strain $\dot{\varepsilon}^p_{ij}$ contributes none to the volumetric deformation

$$\dot{\varepsilon}^p_{kk} = 0$$

$$e^p_{ij} = \varepsilon^p_{ij} - \tfrac{1}{3}\varepsilon_{kk}\delta_{ij} = \varepsilon^p_{ij}$$

where e_{ij} is deviator strain, then the general form of elasto-visco-plastic constitutive relationship is:

$$\begin{cases} \dot{\varepsilon}_{kk} = \dfrac{1}{3K}\dot{\sigma}_{kk} \\ \dot{\varepsilon}_{ij} = \dot{\varepsilon}^e_{ij} + \dot{\varepsilon}^p_{ij} = \dfrac{1}{2\mu}\dot{s}_{ij} + \gamma\left\langle\phi\left(\dfrac{\sqrt{J_2}}{\mathscr{K}} - 1\right)\right\rangle \dfrac{s_{ij}}{\sqrt{J_2}} \end{cases} \tag{7.131}$$

This constitutive relationship was first presented by Perzyna (1963) through a generalization of the Соколовский (Sokolovskii)–Malvern's theory.

In the 1D strain condition, Eq. (7.131) is reduced to

$$\begin{cases} \dot{\varepsilon}_X = \dfrac{1}{3K}(\dot{\sigma}_X + 2\dot{\sigma}_Y) \\ \dot{\varepsilon}_X = \left(\dot{\varepsilon}_X^e - \dot{\varepsilon}_Y^e\right) + \left(\dot{\varepsilon}_X^p - \dot{\varepsilon}_Y^p\right) = \dfrac{1}{2\mu}\left(\dot{\sigma}_X - \dot{\sigma}_Y\right) + \sqrt{3}\gamma\left\langle \phi\left(F\right)\right\rangle \end{cases} \tag{7.132}$$

where

$$\langle \phi(F)\rangle = \begin{cases} 0, & F \le 0, \\ \phi(F), & F > 0, \end{cases} \qquad F = \frac{\sqrt{J_2}}{\mathcal{K}} - 1$$

Solving the $\dot{\sigma}_Y$ from the first equation of Eq. (7.132) and then substituting it into the second equation of Eq. (7.132), we obtain the axial elasto-visco-plastic constitutive relationship in the 1D strain condition:

$$\dot{\varepsilon}_X = \frac{1}{E_L}\dot{\sigma}_X + \frac{4\sqrt{3}\mu}{3E_L}\gamma\left\langle \phi(F)\right\rangle \tag{7.133}$$

where E_L is the laterally constrained elastic modulus [Eq. (7.18c)]

$$E_L = K + \frac{4}{3}G = \lambda + 2\mu = \frac{(1-\nu)E}{(1+\nu)(1-2\nu)}$$

Therefore, the continuity equation [Eq. (7.3a)], the momentum conservation equation [Eq. (7.3b)], and the elasto-visco-plastic constitutive relationship [Eq. (7.133)] together formulate the governing equation set of the 1D strain elasto-visco-plastic waves, which are collected below:

$$\begin{cases} \dfrac{\partial v_X}{\partial X} = \dfrac{\partial \varepsilon_X}{\partial t} \\ \rho_0 \dfrac{\partial v_X}{\partial t} = \dfrac{\partial \sigma_X}{\partial X} \\ \dfrac{\partial \varepsilon_X}{\partial t} = \dfrac{1}{E_L}\dfrac{\partial \sigma_X}{\partial t} + \dfrac{4\sqrt{3}\mu}{3E_L}\gamma\left\langle \phi\left(F\right)\right\rangle \end{cases}$$

Compared with the governing equation set of the 1D stress elasto-visco-plastic waves [Eq. (6.82)], it is clear that the forms of these two equation sets are almost identical, what is different is only to replace the Young's modulus E by the laterally constrained elastic modulus E_L and to replace the γ^* by the $4\sqrt{3}\mu\gamma/(3E_L)$. Therefore, similar to that, the differential equations of characteristics [Eq. (6.85)] and the compatibility relationship [Eq. (6.86)] along the characteristics can be obtained from Eq. (6.82), the following three sets of differential equations for the characteristics and the compatibility relationship in the 1D strain elasto-visco-plastic wave condition can be obtained, respectively:

$$\begin{cases} dX = \pm\sqrt{\dfrac{K+\frac{4}{3}G}{\rho_0}}dt = \pm C_L^e dt \\ dv_X = \pm\dfrac{1}{\rho_0 C_L^e}d\sigma_X \pm \dfrac{4\sqrt{3}\mu\gamma}{3\rho_0 C_L^e}\langle\phi(F)\rangle\, dt \\ \quad\;\; = \pm\dfrac{1}{\rho_0 C_L^e}d\sigma_X + \dfrac{4\sqrt{3}\mu\gamma}{3E_L}\langle\phi(F)\rangle\, dX \end{cases} \tag{7.134}$$

$$\begin{cases} dX = 0 \\ d\varepsilon_X = \dfrac{d\sigma_X}{E_L} + \dfrac{4\sqrt{3}\mu\gamma}{3E_L}\langle\phi(F)\rangle\, dt \end{cases} \tag{7.135}$$

Under the given initial-boundary conditions, the concrete steps in solving the problem are the same as those discussed in Section 6.6 for the 1D stress elasto-visco-plastic waves, and will not be repeated here again.

CHAPTER 8

Spherical Waves and Cylindrical Waves

In the practical underground explosions and engineering blasts, there are often problems of point explosion, point impact (severely localized impact loading), and spherical or cylindrical cavity-walls subjected to an explosion loading. In these problems, it is necessary to deal with the propagation of spherical or cylindrical waves.

As different from the plane waves discussed, the front of spherical and cylindrical waves disperses during propagation. Even in an elastic situation, their wave profile keeps changing. In addition, in the propagation of a compressive wave, tensile stress and amplitude oscillation can be formed in the region behind the wave front.

Due to the many similar characteristics between a spherical wave and a cylindrical wave, they can be analyzed together. We will mainly study the propagation of a spherical wave. Later, it will be pointed out that the analyses on a cylindrical wave can be obtained following the same way.

8.1 Continuity Equation and Motion Equation

The fronts of a spherical wave are surfaces of concentric spheres. Because of the spherical symmetry of the motion of particles, the system of spherical coordinates, shown in Fig. 8.1, is chosen to describe the motion. Thus, there is only non-zero displacement component at radial direction $u(r, t)$, and all variables are functions of radius of sphere r and time t, independent of angle θ and φ, then

$$\varepsilon_r(r,t) = \frac{\partial u(r,t)}{\partial r}, \quad v(r,t) = \frac{\partial u(r,t)}{\partial t} \tag{8.1a}$$

$$\varepsilon_\theta(r,t) = \varepsilon_\varphi(r,t) = \frac{u(r,t)}{r} \tag{8.1b}$$

$$\sigma_r = \sigma_r(r,t), \quad \sigma_\theta(r,t) = \sigma_\varphi(r,t) \tag{8.1c}$$

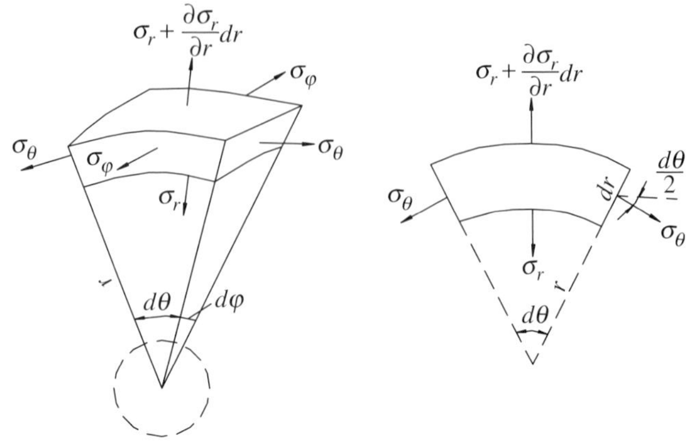

Fig. 8.1. An infinitesimal element in spherical coordinates $(dr, d\theta, d\varphi)$.

where r, θ, and φ are Lagrange (material) coordinates and denote the corresponding components when used as subscripts, other symbols denote the same quantities as in previous chapters. Compared with Eq. (7.2), it can be found that the stress state of a spherical wave is similar to that of a 1D strain plane wave. If the curvature radius approaches infinity ($r \to \infty$), the problem of spherical waves is reduced to that of 1D strain plane waves.

To satisfy the condition that a displacement is a continuous and single value function of r and t, from Eq. (8.1) it is known that the strains ε_r, ε_θ and the radial particle velocity v should follow the compatible condition below:

$$\frac{\partial \varepsilon_r}{\partial t} = \frac{\partial v}{\partial r} \tag{8.2a}$$

$$\frac{\partial \varepsilon_\theta}{\partial t} = \frac{v}{r} \tag{8.2b}$$

This is the continuity equation among the governing equations for a spherical wave, representing the mass conservation.

Considering the momentum conservation in the radial direction (shown in Fig. 8.1), we obtain

$$\left(\sigma_r + \frac{\partial \sigma_r}{\partial r} dr\right)(r + dr)^2 \, d\theta d\varphi - \sigma_r r^2 d\theta d\varphi - 2\sigma_\theta \sin \frac{d\theta}{2} \cdot \left(r + \frac{dr}{2}\right) d\varphi dr$$

$$- 2\sigma_\varphi \sin \frac{d\varphi}{2} \cdot \left(r + \frac{dr}{2}\right) d\theta dr = \rho_0 \left(r + \frac{dr}{2}\right)^2 d\varphi d\theta dr \cdot \frac{\partial v}{\partial t}$$

neglecting high-order small quantities, and noting $\sigma_\theta = \sigma_\varphi$, then

$$\frac{\partial \sigma_r}{\partial r} + \frac{2(\sigma_r - \sigma_\theta)}{r} = \rho_0 \frac{\partial v}{\partial t} \tag{8.3}$$

This is the motion equation of a spherical wave.

Similarly, in analyzing a cylindrical wave whose wave fronts are coaxial cylinders, there is only non-zero displacement component in radial directions $u(r, t)$, and all variables are functions of cylinder radius r and time t, independent of angle θ and φ, then

$$\varepsilon_r = \frac{\partial u}{\partial r}, \; v = \frac{\partial u}{\partial t} \tag{8.4a}$$

$$\varepsilon_\theta = \frac{u}{r}, \; \varepsilon_z = 0 \tag{8.4b}$$

$$\sigma_r = \sigma_r(r, t), \sigma_\theta = \sigma_\theta(r, t), \sigma_z = \sigma_z(r, t) \tag{8.4c}$$

It is shown that the continuity equation of a radial cylindrical wave is totally the same as that of a spherical wave [refer to Eq. (8.2)]. With respect to the motion equation for a cylindrical wave, there is a slight difference in the coefficient of the item $(\sigma_r - \sigma_\theta)/r$ from that for a spherical wave, namely we have

$$\frac{\partial \sigma_r}{\partial r} + \frac{(\sigma_r - \sigma_\theta)}{r} = \rho_0 \frac{\partial v}{\partial t} \tag{8.5}$$

Sometimes Eqs. (8.3) and (8.5) can be written in a unified equation as:

$$\left.\begin{aligned} &\frac{\partial \sigma_r}{\partial r} + n_0 \frac{(\sigma_r - \sigma_\theta)}{r} = \rho_0 \frac{\partial v}{\partial t} \\ &n_0 = \begin{cases} 2 & \text{(spherical waves)} \\ 1 & \text{(radial cylindrical waves)} \end{cases} \end{aligned}\right\} \tag{8.6}$$

Thus, Eqs. (8.2) and (8.6) are the common continuity equation and motion equation of both spherical waves and cylindrical waves. Hence, both the spherical wave and the cylindrical wave can be analyzed jointly as a whole.

8.2 Elastic Spherical Waves and Cylindrical Waves

As discussed earlier for the plane waves, the governing equations consist of continuity equation, motion equation, and material constitutive equation. If the former two have been obtained shown as Eqs. (8.2) and (8.6), then the constitutive relationship can be discussed. The constitutive equation dominates the property of a stress wave, namely, whether a stress wave is elastic, elasto-plastic, or elasto-visco-plastic. At first, we discuss about spherical waves and cylindrical waves in linear elastic media.

Under the condition shown in Eqs. (8.1) and (8.4), the Hooke's law expressed in the form of volumetric deformation law and distortional deformation law Eq. (7.15) is reduced to

$$\sigma_r + n_0\sigma_\theta + n_1\sigma_z = 3K(\varepsilon_r + n_0\varepsilon_\theta) \tag{8.7a}$$

$$\sigma_r - \sigma_\theta = 2G(\varepsilon_r - \varepsilon_\theta) \tag{8.7b}$$

$$(\sigma_r - \sigma_\theta - 2\sigma_z)n_1 = 2G(\varepsilon_r + \varepsilon_\theta)n_1 \tag{8.7c}$$

where

$$n_0 = 2,\ n_1 = 0,\ \text{(for spherical waves)}$$
$$n_0 = 1,\ n_1 = 1,\ \text{(for cylindrical waves)}$$

Then, for a spherical wave, the governing equations consist of five equations, namely Eqs. (8.2a), (8.2b), (8.6), (8.7a), and (8.7b), that can be used to solve five unknown functions σ_r, σ_θ, ε_r, ε_θ, and v. For a cylindrical wave, there is one more unknown function σ_z, but there is also one more equation [Eq. (8.7c)]. Therefore, there are six equations for determining six variables.

As an example, the method of characteristics is used to analyze the propagation of a spherical wave in elastic media. Using Eqs. (8.2) and (8.7) to eliminate ε_r and ε_θ, we obtain

$$\frac{1}{3K}\frac{\partial\sigma_r}{\partial t} + \frac{2}{3K}\frac{\partial\sigma_\theta}{\partial t} - \frac{\partial v}{\partial r} - \frac{2v}{r} = 0 \tag{8.8}$$

$$\frac{1}{2G}\frac{\partial\sigma_r}{\partial t} - \frac{1}{2G}\frac{\partial\sigma_\theta}{\partial t} - \frac{\partial v}{\partial r} + \frac{v}{r} = 0 \tag{8.9}$$

Equations (8.3), (8.8), and (8.9) build up a group of first-order hyperbolic partial differential equations with respect to σ_r, σ_θ, and v. When solving the differential equations by using the method of characteristics, the linear combination of the partial differential equations can be transformed to the equations which only contain the directional derivatives along the characteristics. So, the above three equations are multiplied by the indeterminate coefficients L, M, and N respectively, then the summation of the three equation yields:

$$\left\{L\frac{\partial}{\partial r} + \left(\frac{M}{3K} + \frac{N}{2G}\right)\frac{\partial}{\partial t}\right\}\sigma_r + \left(\frac{2M}{3K} - \frac{N}{2G}\right)\frac{\partial\sigma_\theta}{\partial t} - \left\{(M+N)\frac{\partial}{\partial r} + L\rho_0\frac{\partial}{\partial t}\right\}v$$
$$+ \frac{1}{r}\left\{2L\left(\sigma_r - \sigma_\theta\right) + (N - 2M)v\right\} = 0 \tag{8.10}$$

These coefficients, L, M, and N, should satisfy:

$$\frac{dr}{dt} = \frac{L}{\dfrac{M}{3K} + \dfrac{N}{2G}} = \frac{0}{\dfrac{2M}{3K} - \dfrac{N}{2G}} = \frac{M+N}{L\rho_0} \tag{8.11}$$

Then we obtain either

$$L = 0, \ \ M + N = 0 \tag{8.12}$$

or

$$\frac{2M}{3K} = \frac{N}{2G}, \ \left(\frac{L}{M}\right)^2 = \frac{K + \frac{4}{3}G}{\rho_0 K^2} \tag{8.13}$$

Substitution of the above results into Eqs. (8.10) and (8.11) yields three groups of characteristic equations and the corresponding characteristic relationships. One is resulted from Eq. (8.12),

$$dr = 0 \tag{8.14}$$

$$\left(\frac{1}{3K} - \frac{1}{2G}\right)d\sigma_r + \left(\frac{2}{3K} + \frac{1}{2G}\right)d\sigma_\theta = \frac{3v}{r}dt \tag{8.15}$$

and other two from Eq. (8.13)

$$dr = \pm C_L dt \tag{8.16}$$

$$d\sigma_r = \pm\rho_0 C_L dv - 2\left[(\sigma_r - \sigma_\theta) \pm \left(K - \frac{2G}{3}\right)\frac{v}{C_L}\right]\frac{dr}{r} \tag{8.17}$$

The first family of characteristics [Eq. (8.14)] depicts the motion loci of particles (the coordinates of particles are fixed when Lagrange coordinate system is adopted), while the compatibility condition along this characteristics [Eq. (8.15)] is the differential form of the constitutive equation [Eq. (8.7)]. Actually, Eq. (8.15) can be obtained by differentiating the difference between Eq. (8.7a) and (8.7b) with respect to t and noticing that the partial derivative with respect to t here equals the total derivative with respect to t along the characteristics. The other two families of characteristics then depict the loci of wave fronts that propagate in the positive (rightward) and negative (leftward) directions. The corresponding compatibility condition, Eq. (8.17) specifies the constraint relationship among σ_r, σ_θ, and v in order to satisfy the continuity equation, motion equation, and material constitutive equation. But, comparing Eq. (8.17) with the characteristic compatibility relation for the 1D plane strain waves, it is found that there is an additional term which relates to dr/r. This term represents the dispersive character of spherical waves. When $r \to \infty$, this term tends to vanish, then the problem is reduced to that of 1D plane strain waves.

Therefore, when solving elastic spherical wave by using the method of characteristics, it is necessary to solve simultaneously three equations of characteristic compatibility [Eqs. (8.15) and (8.17)], where three variables σ_r, σ_θ, and v are included. This is equivalent to that by using the governing equations [Eqs. (8.3), (8.8), and (8.9)] to solve the three variables σ_r, σ_θ, and v. The details of solving the problem by the method of characteristics are similar to that used in analyzing visco-elastic waves in Chapter 6.

For example, consider an elastic media with a spherical cavity of radius a, and assume that the initial conditions, namely the σ_r, σ_θ, and v along r axis are known, and the boundary

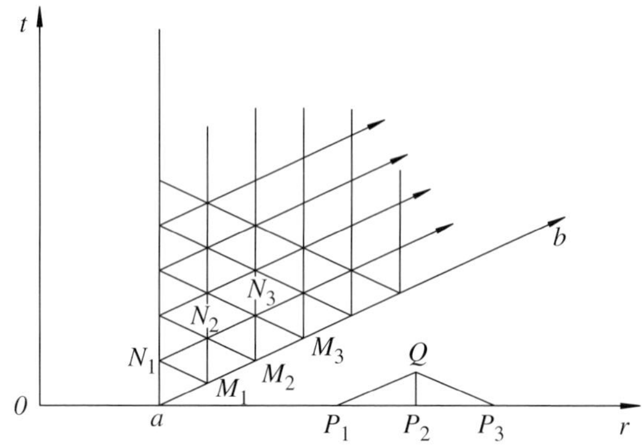

Fig. 8.2. The method of characteristics for solving the problem of elastic spherical waves.

condition, namely the σ_r on $r = a$ is known. It is also assumed that there is temporarily no strong discontinuity involved, namely the spherical wave is a kind of continuous wave. Then the problem is summed up to solve the Cauchy initial-value problem in the *rab* region and the characteristic boundary-value problem in the *bat* region as shown in Fig. 8.2.

For the *bat* region, considering an arbitrary point Q near r axis, there must be three characteristics QP_1, QP_2, and QP_3 that pass point Q and intersect r axis as shown in Fig. 8.2. If the distance from Q to three points P_1, P_2, P_3 is small enough, we can solve the differential by the difference in Eqs. (8.15) and (8.17), then we have

$$\sigma_r(Q) - \sigma_r(P_1) = \rho_0 C_L\{v(Q) - v(P_1)\} - 2\left\{(\sigma_r - \sigma_\theta) - \left(K - \frac{2G}{3}\right)\frac{v}{C_L}\right\}_{P_1} \cdot \frac{r(Q) - r(P_1)}{r(P_1)}$$

$$\sigma_r(Q) - \sigma_r(P_3) = -\rho_0 C_L\{v(Q) - v(P_3)\} - 2\left\{(\sigma_r - \sigma_\theta) + \left(K - \frac{2G}{3}\right)\frac{v}{C_L}\right\}_{P_3} \cdot \frac{r(Q) - r(P_3)}{r(P_3)}$$

$$\left(\frac{1}{3K} - \frac{1}{2G}\right)\{\sigma_r(Q) - \sigma_r(P_2)\} + \left(\frac{2}{3K} + \frac{1}{2G}\right)\{\sigma_\theta(Q) - \sigma_\theta(P_2)\} = \frac{3v(P_2)}{r(P_2)}\{t(Q) - t(P_2)\}$$

As the position of Q, r_Q, and t_Q, can be determined by Eqs. (8.14) and (8.16) in advance and σ_r, σ_θ, and v at points P_1, P_2, and P_3 have been known based on the initial condition, $\sigma_r(Q)$, $\sigma_\theta(Q)$, and $v(Q)$ can be obtained from the above three equations. On this analogy, the solutions in the whole region *rab* can be determined. If the initial condition is a nil-constant condition (undisturbed state), it is easy to prove that the values of the variables in the *rab* region are always zero.

For the *bat* region, to solve a mixed problem by the method of characteristics for 1D stress waves in bars, we should first figure out the values of all variables on the so-called boundary points on the boundary $r = a$, shown as the point N_1 in Fig. 8.2. As long as the distance between N_1 and M_1 (and a) is small enough, according to Eqs. (8.15) and (8.17), there are difference equations along the characteristics M_1N_1 and aN_1 respectively,

$$\sigma_r(N_1) - \sigma_r(M_1) = -\rho_0 C_L\{v(N_1) - v(M_1)\} - 2\left\{(\sigma_r - \sigma_\theta) + \left(K - \frac{2G}{3}\right)\frac{v}{C_L}\right\}_{M_1} \cdot \frac{r(N_1) - r(M_1)}{r(M_1)}$$

$$\left(\frac{1}{3K} - \frac{1}{2G}\right)\{\sigma_r(N_1) - \sigma_r(a)\} + \left(\frac{2}{3K} + \frac{1}{2G}\right)\{\sigma_\theta(N_1) - \sigma_\theta(a)\} = \frac{3v(a)}{r(a)}\{t(N_1) - t(a)\}$$

As the solutions along the characteristics *ab* have just been obtained in solving the *rab* region, while σ_r at point N_1 has been specified based on the boundary condition, so $\sigma_\theta(N_1)$ and $v(N_1)$ can be figured out from the above two equations. After obtaining the solutions at N_1, following the same steps used to solve the Cauchy initial value problems, we can obtain the solutions at point N_2 according to the known data at the points M_1, M_2, and N_1. Solutions at the other so-called interior points can be obtained successively. On this analogy, the solutions of the whole *bat* region can be determined.

With regard to the strong discontinuous spherical wave, according to the displacement continuity condition and the momentum conservation condition across the wave front, noting that for a spherical wave ε_θ must be continuous if u is continuous, then the kinetics compatibility condition [Eq. (2.55)] and the dynamics compatibility condition [Eq. (2.57)] originally deduced for the strong discontinuous plane waves are still valid. In other words, the discontinuous jump across the wave front, with respect to the particle velocity v, stress component σ_r, and strain component ε_r in the direction perpendicular to the spherical wave front, satisfy the following relationships:

$$\left.\begin{aligned} [v] &= \pm D[\varepsilon_r] \\ [\sigma_r] &= \pm\rho_0 D[v] \end{aligned}\right\} \tag{8.18}$$

In the case of elastic wave discussed, $D = C_L$. It can be found that its mathematic expression is completely same as that of the 1D strain plane waves. The dispersive effect of spherical wave is mainly exhibited by the changes in values of the discontinuities $[\sigma_r]$, $[v]$, and $[\varepsilon_r]$ during the propagation of a strong discontinuous spherical wave, even if the region ahead of the wave front is undisturbed.

In Fig. 8.2, if the initial conditions and boundary conditions are assumed to be:

$$\left.\begin{array}{ll} u(r,0) = v(r,0) = \sigma_r(r,0) = \sigma_\theta(r,0) = 0, & a < r \le \infty \\ \sigma_r(a,t) = -p_0, & t \ge 0 \end{array}\right\} \tag{8.19}$$

These initial and boundary conditions correspond to a constant pressure p_0 which is suddenly applied on the inner wall of a spherical cavity $r = a$ in an undisturbed infinite body. A spherical wave is caused by the sudden loading and propagates outwards in the infinite body. Assuming p_0 is not high enough to induce plastic deformation, an elastic strong discontinuous spherical wave front propagates along the line *ab*. On the one hand, Eq. (8.18) should be valid along *ab*. On the other hand, *ab* is also a characteristic so that Eq. (8.17) should be satisfied along *ab*. Considering again that from the displacement continuity condition along *ab* there is $\varepsilon_\theta = u/r = 0$, then from Eq. (8.17) it is known that along *ab* there is

$$\sigma_r - \sigma_\theta = 2G\varepsilon_r = \frac{2G}{K + \frac{4}{3}G}\sigma_r = \frac{2G}{\rho_0 C_L^2}\sigma_r \tag{8.20a}$$

Thus, the dispersion term of spherical wave in the compatibility relationship [Eq. (8.17)] along the characteristics *ab* can be written as

$$\frac{2}{r}\left\{(\sigma_r - \sigma_\theta) - \left(K - \frac{2G}{3}\right)\frac{u}{C_L}\right\}dr = \frac{2\sigma_r}{r}dr$$

while Eq. (8.17) is reduced to:

$$\frac{d\sigma_r}{\sigma_r} = -\frac{dr}{r} \tag{8.20b}$$

After determining the integral constant by using boundary condition $\sigma_r|_a = p_0$, the attenuation of σ_r along *ab* is obtained

$$\sigma_r = -\frac{p_0 a}{r} \tag{8.21a}$$

Substituting this result into Eqs. (8.18) and (8.20), we obtain the following relationships for v and σ_θ along *ab*

$$v = -\frac{\sigma_r}{\rho_0 C_L} = \frac{p_0 a}{\rho_0 C_L r} \tag{8.21b}$$

$$\sigma_\theta = \frac{K - \frac{2}{3}G}{K + \frac{4}{3}G}\sigma_r = \frac{\lambda}{\lambda + 2\mu}\sigma_r = \frac{\nu}{1-\nu}\sigma_r = \frac{-\nu}{1-\nu}\frac{p_0 a}{r} \tag{8.21c}$$

It can be seen from Eq. (8.21c) that the stress amplitude σ_r of a strong discontinuous spherical wave decreases with the increase of its distance r from the center of the sphere in an inversely proportional relation. This is due to the effect of spherical dispersion.

Contrarily, the wave intensity will increase if a spherical wave propagates towards the center point of the sphere (**implosion**). It can also be seen that the proportional relationships between σ_r, σ_θ, and v on the front of a strong discontinuous spherical elastic wave expressed in Eqs. (8.21b) and (8.21c) are the same as the relationship between those on the front of a 1D strain plane elastic wave. Therefore, on the one hand, a strong discontinuous spherical wave differs from a 1D strain plane wave with respect to the attenuation of its intensity under the initial-boundary conditions shown in Eq. (8.19). On the other hand, at each time, the stress state of a strong discontinuous spherical wave is the same as that of a 1D strain plane wave. This means that if σ_r is a compressive stress, σ_θ is also a compressive stress; on the contrary, if σ_r is a tensile stress, σ_θ is also a tensile stress.

The region *bat* swept by a strong discontinuous wave front (behind the front), as shown in Fig. 8.2, is dominated by weak discontinuous waves. After the solutions along *ab* have been obtained by using the relations of a strong discontinuous spherical wave [Eq. (8.21)], as discussed earlier, the problem is summed up to solve a characteristic boundary-value problem (Darboux problem) which is solvable. We will not discuss this in detail again. But it should be emphasized that, as different from the 1D plane strain wave, the spherical waves are no longer simple waves even under the zero initial condition and constant loading boundary conditions [Eq. (8.19)]. Furthermore, due to the dispersion of spherical waves, tensile stress may quickly appear in the region behind the strong discontinuous compressive wave, especially with respect to σ_θ, the magnitude of the tensile stress can, sometimes, even exceed the magnitude of the original compressive stress.

To solve the problem of elastic spherical wave propagation, there are other methods, such as the Fourier transformation (see Miklowitz, 1978). For convenience in connecting and comparing with the previous chapters, we use the characteristics method here. On the other hand, based on the results mentioned above, it is also convenient to further analyze the elasto-plastic spherical waves in the following sections.

Figure 8.3 shows the stresses *vs.* time curves at three points ($r = a, r = 2a$ and $r/a \to \infty$) for the spherical wave propagation under the initial-boundary conditions of Eq. (8.19). For convenience in comparison, the dimensionless stresses $[-(r\sigma_r/ap_0)]$, $[-(r\sigma_\theta/ap_0)]$, and $[-(r\tau/ap_0)]$ are used as vertical coordinates, where τ is the maximum shear stress $|\sigma - \sigma_\theta|/2$. The dimensionless time $\bar{t} = [(C_L t/a) - (r - a/a)]$ is used as horizontal coordinates. These computations are for the case of Poisson ratio $v = 0.25$, namely $\lambda = \mu$ (see Selberg, 1952).

Here, we compare the above results with those in the static analysis. It is known from the analysis of static elasticity (Timoshenko and Goodier, 1951), the well-known Lame solution shows that the stresses in the wall of a thick-wall spherical vessel subjected to a static inner pressure $(-p_0)$ can be expressed by:

$$\left.\begin{aligned}\sigma_r &= \frac{p_0 a^3 (b^3 - r^3)}{r^3 (b^3 - a^3)} \\ \sigma_\theta &= -\frac{p_0 a^3 (b^3 + 2r^3)}{2r^3 (b^3 - a^3)}\end{aligned}\right\} \tag{8.22}$$

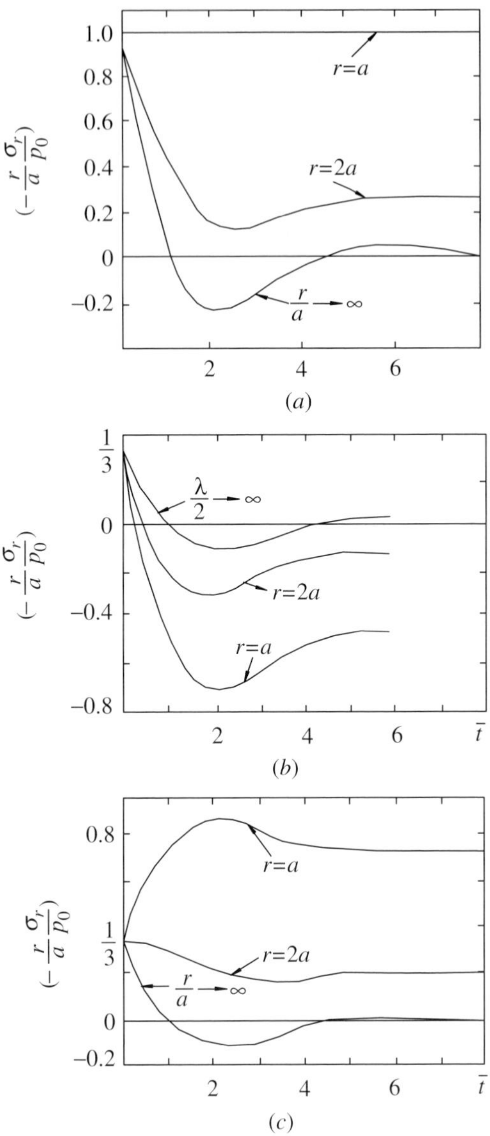

Fig. 8.3. Propagation of elastic spherical waves under initial-boundary condition of Eq. (8.19).

where a and b are the inner and outer radius of the spherical vessel, respectively. For a strong discontinuous spherical wave as just discussed, the dynamic analysis of elastic spherical wave has indicated that both σ_r and σ_θ on the wave front are compressive. However, according to Eq. (8.22) for the static analysis, σ_r is compressive while σ_θ is tensile. Furthermore, it can be seen that the stress amplitude of a strong discontinuous spherical wave is inversely proportional to r according to Eq. (8.21a), while the static stress σ_r is inversely proportional to r^3. With respect to the region dominated by the weak discontinuous spherical waves, the dynamic analysis of elastic spherical wave has shown that, after a strong discontinuous wave has passed, the radial compressive stress σ_r decreases, while the hoop stress σ_θ becomes a tensile stress in a short time. Especially at the inner wall ($r = a$), the magnitude of σ_θ obtained by using wave analysis is larger than that obtained by static analysis, and consequently the maximum shear stress τ obtained by dynamic analysis is also larger than that found in static one.

With regard to the radial cylindrical waves, based on the six equations shown in Eqs. (8.2), (8.5), and (8.7), the problem can be analyzed following the same way. And Eq. (8.18) is still valid for the strong discontinuous cylindrical waves.

Consider an infinite elastic media with a cylindrical cavity, of which the length is infinite and the radius is a. Under the zero initial condition and the boundary condition corresponding to a uniform pressure ($-p_0$) is suddenly loaded on the cavity, similar to Fig. 8.2, a strong discontinuous cylindrical wave propagates at velocity C_L along the line ab, but the amplitude of σ_r attenuates inversely with $\sqrt{r}$, namely, Eq. (8.21) is now replaced by the following relationships for a strong discontinuous cylindrical wave along ab:

$$\sigma_r = -\sqrt{\frac{a}{r}}p_0 \tag{8.23a}$$

$$v = \sqrt{\frac{a}{r}}\frac{p_0}{\rho_0 C_L} \tag{8.23b}$$

$$\sigma_\theta = \sigma_Z = -\frac{\nu}{1-\nu}\sqrt{\frac{a}{r}}p_0 \tag{8.23c}$$

In the region swept by the strong discontinuous wave front (behind the wave front) which is dominated by the weak discontinuous waves, the stresses *vs.* time curves at three points $r/a = 1, 2$, and ∞, similar to that shown in Fig. 8.3, are shown in Fig. 8.4 (see Selberg, 1952).

8.3 Elasto-Plastic Spherical Waves

In the example shown in Fig. 8.2, if the impact loading applied on the inner cavity-wall is intense enough, then plastic deformation will take place. In the following, we will discuss the propagation of elasto-plastic spherical waves.

Assume that the plastic deformation does not contribute to the volumetric deformation [Eq. (7.34)] and the elastic deformation follows the Hooke's law. Obviously, such assumptions are valid under the conditions that the pressure is not too high and the deformation is small. Due to spherical symmetry [Eq. (8.1)], both von Mises yield criterion and Tresca yield criterion are reduced to the form of Eq. (7.32). Therefore, to replace

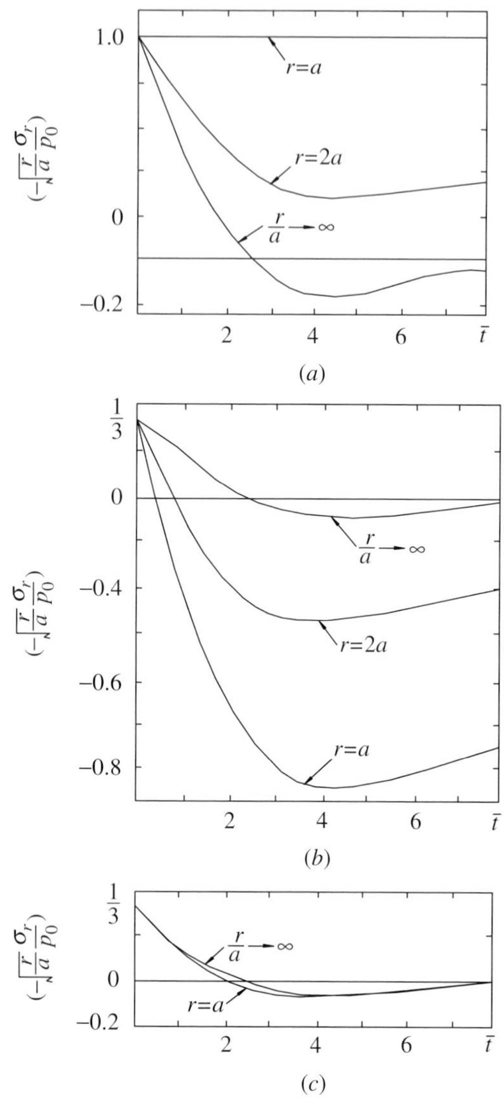

Fig. 8.4. Propagation of an elastic cylindrical wave in an undisturbed body suddenly subjected to a constant loading.

Eq. (8.7), the constitutive equations in the governing equations for elasto-plastic spherical wave are:

$$\text{volumetric deformation law:} \quad \sigma_r + 2\sigma_\theta = 3K(\varepsilon_r + 2\varepsilon_\theta) \tag{8.24a}$$

$$\text{distortional deformation law:} \quad \sigma_r - \sigma_\theta = \begin{cases} 2G(\varepsilon_r - \varepsilon_\theta) & \text{(elastic)} \\ \pm Y & \text{(plastic)} \end{cases} \tag{8.24b}$$

where the yield stress Y can be a function of plastic strain ε_r^p, $Y = Y_p(\varepsilon_r^p)$, alternatively, a function of plastic work W_p, $Y = Y_W(W_p)$. Differentiating Eq. (8.24) with respect to t and after some operations based on the analogy used in deducing of Eq. (7.41), we obtain

$$\frac{\partial \sigma_r}{\partial t} + 2\frac{\partial \sigma_\theta}{\partial t} = 3K\left(\frac{\partial \varepsilon_r}{\partial t} + 2\frac{\partial \varepsilon_\theta}{\partial t}\right) \tag{8.25a}$$

$$\frac{\partial \sigma_r}{\partial t} - \frac{\partial \sigma_\theta}{\partial t} = \begin{cases} 2G\left(\dfrac{\partial \varepsilon_r}{\partial t} - \dfrac{\partial \varepsilon_\theta}{\partial t}\right) & \text{(elastic)} \\ 2G_p\left(\dfrac{\partial \varepsilon_r}{\partial t} - \dfrac{\partial \varepsilon_\theta}{\partial t}\right) & \text{(plastic)} \end{cases} \tag{8.25b}$$

where the plastic shear modulus G_p is the same as that in Eq. (7.41), namely

$$G_p = \frac{GY_p'}{3G + Y_p'} = \frac{GY_W Y_W'}{3G + Y_W Y_W'}$$

Substituting Eq. (8.2) into Eq. (8.25a) and eliminating ε_r and ε_θ, we obtain Eq. (8.8); and substituting Eq. (8.2) into Eq. (8.25b) and eliminating ε_r and ε_θ, we obtain Eq. (8.9). While substituting Eq. (8.2) into Eq. (8.25c) and eliminating ε_r and ε_θ, we obtain

$$\frac{1}{2G_p}\frac{\partial \sigma_r}{\partial t} - \frac{1}{2G_p}\frac{\partial \sigma_\theta}{\partial t} - \frac{\partial v}{\partial r} + \frac{v}{r} = 0 \tag{8.26}$$

The difference between Eq. (8.26) and Eq. (8.9) is due to the shear modulus only, where G_p is used to replace G in Eq. (8.9). Then Eqs. (8.3), (8.8), and (8.26) constitute the governing equations for elasto-plastic spherical waves, that are first-order quasi-linear hyperbolic partial differential equations with respect to σ_r, σ_θ, and v. The difference from the elastic spherical wave equations is only due to G in the third equation being replaced by G_p.

Thus, repeating the steps used in deducing the equations since Eq. (8.10) for the elastic spherical wave, we can obtain three groups of characteristic equations and their corresponding compatibility relationships for solving the elasto-plastic spherical wave by the characteristic method. The results obtained are similar to Eqs. (8.14)–(8.17), except that the G is replaced by G_p and the elastic spherical wave velocity C_L^e

$$C_L^e = \sqrt{\frac{K + \frac{4}{3}G}{\rho_0}}$$

is replaced by the following plastic spherical wave velocity C_L^p

$$C_L^p = \sqrt{\frac{K + \frac{4}{3}G_p}{\rho_0}}$$

Compared to Eq. (7.47), it is known that the plastic spherical wave velocity is the same as the 1D strain plastic wave velocity. Similar to a 1D strain elasto-plastic wave, the plastic spherical wave velocity (C_L^p) is generally smaller than the elastic spherical wave velocity (C_L^e). Consequently, a plastic spherical wave propagates behind an elastic spherical wave. Because C_L^p is generally a function of ε_r^p the two families of characteristics describing the plastic spherical waves on the r–t plane are generally not straight any more. In the special cases of linear hardening materials (G_p = constant) or perfect plastic materials ($G_p = 0$), C_L^p *constant*, and the corresponding characteristics are straight lines. All the above discussions are similar to those of 1D strain elasto-plastic waves. Similar to the situations in the 1D strain wave, once the plastic shear modulus G_p is introduced, the analyses on the elastic spherical wave can be conveniently generalized to study the elasto-plastic spherical wave.

Comparing with the 1D elasto-plastic waves, it should be noted that both the intensity of the elastic precursor spherical wave and the subsequent plastic spherical wave attenuates with the increase in r because of spherical dispersion. This case is similar to that during the propagating process of the elastic precursor spherical wave and the subsequent plastic spherical wave, both of which continuously reflect internal reflective waves and moreover these internal reflective waves interact with each other, thus making the problem more complicated. For example, in the case of linear elasto-perfect plastic materials or linear elasto-linear plastic hardening materials, when the inner wall of a spherical cavity is suddenly subjected to a constant loading, both the elastic precursor spherical wave and the subsequent plastic spherical wave become strong discontinuous waves. The elasto-plastic loading boundary coincides with the propagating locus of strong discontinuous plastic wave until the latter vanishes due to the attenuation effect. In this aspect, it is similar to that of 1D strain elasto-plastic waves. However, the region between an elastic strong discontinuous spherical wave and a plastic strong discontinuous spherical wave is no longer a constant-value region. In this aspect, it differs from that of a 1D strain elasto-plastic wave. This is because that, on the one hand, the stresses on the wave front of an elastic strong discontinuous spherical wave continuously attenuate in inverse proportion to r, and so that the yield criterion is no longer satisfied; and on the other hand, the initial yield condition in a spherical wave is no longer simply expressed by $\sigma_X = Y_H$ [Eq. (7.33)]. In fact, after eliminating σ_θ and ε_r from the three equations of Eq. (8.24), it can be found that the initial yield condition from an elastic state to a plastic state can be expressed as

$$\sigma_r = 3K\varepsilon_\theta + Y_H$$

where

$$Y_H = \frac{K + \frac{4}{3}G}{2G} Y_0$$

is the Hugoniot yield limit. So, the σ_r ahead of the plastic strong discontinuous wave front is not equal to Y_H, but is related to the current value of ε_θ in the elastic region.

Because of the same reason, for a weak discontinuous spherical plastic wave, the spherical dispersion causes the attenuation of stress amplitudes, resulting in the stress originally satisfying the yield condition to no longer satisfy the yield criterion. Consequently, a plastic wave will be reduced or transformed to an elastic wave during its propagation, and the propagating velocity of weak discontinuous elasto-plastic boundary (loading boundary) will be lower than that of a plastic spherical wave. This is another character of spherical waves that differs from that of 1D strain plane waves.

During unloading, an elastic unloading occurs first. Under the assumption of elastic unloading, the propagating velocity of unloading disturbance is C_L^e. When the stress decreases further, a reverse plastic loading occurs and the disturbance propagates at a velocity of C_L^p. These are similar to that of the 1D strain elasto-plastic wave. However, it should be noticed that when an elastic unloading disturbance pursues a plastic loading disturbance, its stress amplitude is going to decrease continuously due to the spherical dispersion, so that the elastic unloading disturbance could be transformed to a reverse plastic loading. Therefore, the propagating velocity of reverse plastic loading boundary will not only be faster than the propagating velocity of plastic loading disturbance itself, but also even faster than the propagating velocity of elastic wave.

Consider an infinite elasto-perfect plastic media ($G_p = 0$) with a spherical cavity of radius a, and the inner wall of the cavity is subjected to a compressive pulse expressed by the following formula

$$\sigma_r(t)\,|_{r=a} = -5.0Y_0 \exp\left(-3.5\frac{C_K t}{a}\right)$$

Figure 8.5 shows the calculated radial stress σ_r profiles at several dimensionless times $\bar{t} = C_K t/a$ (Friedman *et al.*, 1965), where $C_K = \sqrt{K/\rho_0}$ is the 1D strain plastic wave velocity in an elasto-perfect plastic material. From the figure, it can be seen that the strong discontinuous elastic precursor wave continuously attenuates, and the stress σ_r just ahead of the strong discontinuous plastic wave front also keeps changing (although the yield condition is satisfied). The stress amplitude of the strong discontinuous plastic

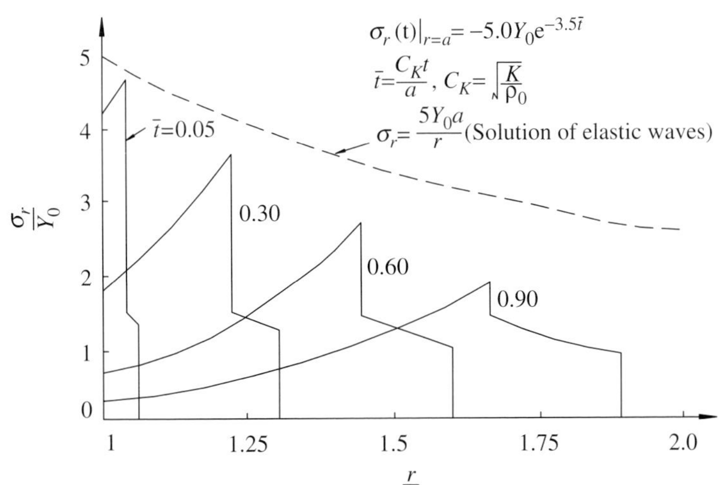

Fig. 8.5. Radial stress profiles of an elasto-perfect plastic spherical wave.

wave attenuates much faster. This is caused by the subsequent unloading of exponential attenuating disturbance on the one hand and the effect of spherical dispersion on the other hand.

With respect to the determination of weak discontinuous boundary of elastic zone and plastic zone for spherical elasto-plastic waves, it is similar to that mentioned in Section 4.10. If it is assumed that the plastic deformation makes no contribution to volumetric deformation, then it is known from Eq. (8.25) that the volumetric modulus is continuous at the elastic–plastic boundary, namely $[K] = 0$; while the shear modulus is discontinuous, namely $[G] \neq 0$, and consequently the wave velocity is discontinuous

$$[C_L] = C_L^p - C_L^e \neq 0 \tag{8.27}$$

This is the essential singularity which dominates the discontinuous properties of elastic–plastic boundary produced for spherical waves. Other singularities along the elastic–plastic boundary depend on this essential singularity and are determined by this singularity. In relation to this, $(\partial\,|\sigma_r - \sigma_\theta|)/\partial t$ or $(\partial\,|\varepsilon_r - \varepsilon_\theta|)/\partial t$ should be considered in judging whether the media is undergoing a plastic loading or an elastic unloading in the study on the problems of spherical elasto-plastic waves, instead of $\partial\sigma_X/\partial t$ or $\partial\varepsilon_X/\partial t$ in the cases of 1D stress wave or 1D strain elasto-plastic waves. Or if introducing the maximum shear stress $\tau = (\sigma_r - \sigma_\theta)/2$ and shear strain $\gamma = (\varepsilon_r - \varepsilon_\theta)/2$, and noticing the continuity equation Eq. (8.2), we have

$$\frac{\partial}{\partial t}(\varepsilon_r - \varepsilon_\theta) = \frac{\partial v}{\partial r} - \frac{v}{r}$$

then we can also equivalently examine $\partial\tau/\partial t$, $\partial\gamma/\partial t$, or $(\partial v/\partial r, v/r)$, etc. The discontinuity of elasto-plastic distortional deformation law [Eq. (8.25b)] across the elastic–plastic boundary can be expressed as

$$\left[\frac{1}{2G}\frac{\partial\tau}{\partial t}\right] = \left[\frac{\partial\gamma}{\partial t}\right] = \frac{1}{2}\left[\frac{\partial v}{\partial r} - \frac{v}{r}\right] \tag{8.28}$$

For the first-order weak discontinuous boundary, namely σ_r, σ_θ, and v are continuous but their first-order partial differentials are discontinuous, similar to Eq. (4.106), we have the following relationships

$$\left[\frac{\partial\sigma_r}{\partial t}\right] = -\overline{C}\left[\frac{\partial\sigma_r}{\partial r}\right] \tag{8.29}$$

$$\left[\frac{\partial v}{\partial t} - \frac{v}{r}\right] = -\overline{C}\left[\frac{\partial v}{\partial r} - \frac{v}{r}\right] \tag{8.30}$$

On the other hand, from the governing equations for an elastic region [Eqs. (8.3), (8.8), and (8.9)] and the governing equations for a plastic region [Eqs. (8.3), (8.8), and (8.26)], it is known that

$$\left[\frac{\partial \sigma_r}{\partial r}\right] = \rho_0 \left[\frac{\partial v}{\partial t} - \frac{v}{r}\right] \tag{8.31}$$

$$\left[\frac{\partial \sigma_r}{\partial t}\right] = \rho_0 \left[C_L^2\left(\frac{\partial v}{\partial t} - \frac{v}{r}\right)\right] \tag{8.32}$$

To simplify the analysis, the items related to $[v/r]$ have been added to Eqs. (8.30) and (8.31). As $[v/r] = 0$, they have no effect on results.

Thus, Eqs. (8.27), (8.29)–(8.32) are the basic equations that the weak discontinuous elasto-plastic boundaries of spherical waves should obey. They are corresponding to Eqs. (4.101), (4.106), and (4.107) for the 1D stress waves. Based on these, we can perform similar discussions mentioned in Section 4.10. For example, for a first-order weak discontinuous boundary, by eliminating $[\partial\sigma_r/\partial t]$, $[\partial\sigma_r/\partial r]$, and $[\partial v/\partial t - v/t]$ from Eqs. (8.29)–(8.32), we obtain the following relationship for the boundary propagating velocity $\overline{C}$

$$\overline{C} = \sqrt{\frac{\left[C_L^2\left(\frac{\partial v}{\partial r} - \frac{v}{r}\right)\right]}{\left[\frac{\partial v}{\partial r} - \frac{v}{r}\right]}}, \text{ or } \frac{\left(\frac{\partial v^p}{\partial r} - \frac{v^p}{r}\right)}{\left(\frac{\partial v^e}{\partial r} - \frac{v^e}{r}\right)} = \frac{1-\left(\frac{C_L^e}{\overline{C}}\right)^2}{1-\left(\frac{C_L^p}{\overline{C}}\right)^2} \tag{8.33}$$

Or following Eq. (8.28), we can rewrite Eq. (8.33) in several equivalent forms, such as:

$$\left.\begin{aligned} &\overline{C} = \sqrt{\frac{\left[C_L^2\frac{\partial\gamma}{\partial t}\right]}{\left[\frac{\partial\gamma}{\partial t}\right]}} = \sqrt{\frac{\left[\left[\frac{C_L^2}{G}\frac{\partial\tau}{\partial t}\right]\right]}{\left[\frac{1}{G}\frac{\partial\tau}{\partial t}\right]}} \\ &\frac{\partial\gamma^p/\partial t}{\partial\gamma^e/\partial t} = \frac{1-\left(\frac{C_L^p}{\overline{C}}\right)^2}{1-\left(\frac{C_L^e}{\overline{C}}\right)^2}, \quad \frac{\partial\tau^p/\partial t}{\partial\tau^e/\partial t} = \frac{G_p}{G}\frac{1-\left(\frac{C_L^p}{\overline{C}}\right)^2}{1-\left(\frac{C_L^e}{\overline{C}}\right)^2} \end{aligned}\right\} \tag{8.33$'$}$$

Referring to Eq. (4.111), Eq. (8.33) can be expressed in terms of $\partial\varepsilon/\partial t$ as [or refer to Eq. (4.73b)]

$$\overline{C} = \sqrt{\frac{\left(C^2\frac{\partial\varepsilon}{\partial t}\right)}{\left(\frac{\partial\varepsilon}{\partial t}\right)}}$$

then replacing only ε in the above equation by γ, we obtain the corresponding equation given in Eq. (8.33′) for spherical waves.

In the above discussions, we assumed that the volumetric modulus K is a constant, which is only valid in situations of small strain and correspondingly the pressure is not too high. So the discussions in this section can be regarded as an extension of the theory of 1D strain elasto-plastic waves in small strain condition (as discussed in Section 7.4) to the spherical waves. In the case of high pressure, it must be considered that the elastic modulus k is not a constant any more, which generally increases with the increase in pressure (Section 7.6), and consequently a shock wave will be formed. In such a case, it is necessary to extend the theory of 1D hydro-elasto-plastic plane wave (Section 7.9) to the spherical waves, so that the analysis becomes much more complicated.

8.4 Approximate Analysis for the Fragmentation of Spherical Shells

As discussed in Chapter 5, if plastic deformation is much larger than elastic deformation and the elastic wave velocity is much faster than the plastic one, especially when the problem we are mainly concerned with is the situation where the elastic wave after having propagated back-and-forth many times in the body studied, then we can approximately let the elastic wave velocity approach infinity. This means that the material is assumed to be rigid-plastic, so that the study will be simplified.

Consider a thick spherical shell with an inner radius $r = a$, which is subjected to an explosion pressure p_a on its inner wall. To simplify the dynamic analysis, let the elastic modulus be approximately infinity. This is equivalent to that in the elasto-plastic constitutive relationship [Eq. (8.24)] discussed in the previous section, the elastic volumetric deformation law is replaced by the volume-incompressible condition ($\varepsilon_{KK} = 0$), while only the yield condition [the second equation in Eq. (8.24b)] is still valid in the distortional deformation law. Then, the governing equations for the problem are composed of the continuous equation [Eq. (8.2)], the motion equation [Eq. (8.3)], and the yield condition [the second equation in Eq. (8.24b)], that are together re-listed below:

$$\text{continuity equation:} \quad \begin{cases} \dfrac{\partial \varepsilon_r}{\partial t} = \dfrac{\partial v}{\partial r} \\ \dfrac{\partial \varepsilon_\theta}{\partial t} = \dfrac{v}{r} \end{cases} \tag{8.34a}$$

$$\text{motion equation:} \quad \frac{\partial \sigma_r}{\partial t} + \frac{2(\sigma_r - \sigma_\theta)}{r} = \rho_0 \frac{\partial v}{\partial t} \tag{8.34b}$$

$$\text{incompressible condition:} \quad \varepsilon_r + 2\varepsilon_\theta = 0 \tag{8.34c}$$

$$\text{yield condition:} \quad \sigma_r - \sigma_\theta = \pm 1 \tag{8.34d}$$

Substituting Eqs. (8.34a) and (8.34b) into the equation obtained by differentiating Eq. (8.34c) with respect to t, the volume incompressible condition becomes

$$\frac{\partial v}{\partial r} + 2\frac{v}{r} = 0$$

or

$$\frac{\partial}{\partial r}\left(vr^2\right) = 0$$

Integrating it, we obtain

$$v = \frac{a^2}{r^2} v_a \tag{8.35}$$

where v_a is the radial velocity of particles on the inner wall ($r = a$). Substituting the above equation and yield condition [Eq. (8.34d)] into the motion equation [Eq. (8.34b)], we obtain

$$\frac{\partial \sigma_r}{\partial r} = \frac{2Y}{r} + \frac{\rho_0 a^2}{r^2}\frac{dv_a}{dt}$$

Integrating this equation and using the boundary condition $\sigma_r|_a = -p_a$ to determine the integral constant, we obtain

$$\sigma_r = -p_a + 2Y \ln\frac{r}{a} + \rho_0 a\left(1 - \frac{a}{r}\right)\frac{dv_a}{dt} \tag{8.36}$$

where the material has been assumed as perfect plastic (Y = constant). Substitution of the above equation into the yield condition yields

$$\sigma_\theta = Y - p_a + 2Y \ln\frac{r}{a} + \rho_0 a\left(1 - \frac{a}{r}\right)\frac{dv_a}{dt} \tag{8.37}$$

In general, the explosion pressure p_a on the inner wall is much larger than the yield stress Y of materials. So unless p_a decreases to Y or even lower, both σ_r and σ_θ on the inner wall are compressive stresses,

$$\sigma_r|_a = -p_a, \quad \sigma_\theta|_a = -\left(p_a - Y\right)$$

This differs from that obtained in static elastic analysis [Eq. (8.22)].

From Eq. (8.37) it is known that the value of compressive stress σ_θ decreases with increasing r. Let $\sigma_\theta = 0$ at $r_h = (b - h)$, then $\sigma_\theta \geq 0$ at $r > r_h$, namely σ_θ is tensile at $r > r_h$. Here, b is the outer radius of the thick spherical shell, and h is the distance from the outer wall.

Taylor (1963) studied the fragmentation of a cylindrical shell. According to the observations recorded by a high-speed camera, it was found that axial cracks firstly appear on the outer wall of the shell, and then grow with the shell expansion. Until the radius of the shell expands to two times its original radius, cracks will penetrate through the shell wall, and lead to shell fragmentation. Obviously, only when the compressive stress zone of σ_θ disappears, the cracks could penetrate up to the inner wall, so that the cylindrical shell could be fragmented.

It is known from Eq. (8.37) that the location r_h or h, where stress σ_θ changes from compressive to tensile, can be determined by the condition of $\sigma_\theta = 0$ at $r = r_h = (b-h)$, namely,

$$p_a - Y = 2Y \ln\left(\frac{b-h}{a}\right) + \rho_0 a \left(1 - \frac{a}{b-h}\right)\frac{dv_a}{dt}$$

The condition that compressive stress zone of σ_θ disappears is $h = b - a$, which results when the right-hand side of the above equation equals zero. Therefore, the shell will be fragmented only till p_a decreases with time to the magnitude of Y.

Assuming that the explosive gas within the spherical shell expands adiabatically, then

$$pV^\gamma = \text{constant}$$

where γ is the exponent of gas adiabatic expansion. The original inner radius of the shell is denoted by a, and the inner radius becomes A at an arbitrary time t due to expansion. Noticing $V = 4\pi A^3/3$, if the initial pressure is p_0, then pressure p_a on inner wall at time t is

$$p_a = p_0\left(\frac{a}{A}\right)^{3\gamma}$$

If the spherical cavity is not fully filled with explosive and the radius of explosive charge is a_c ($<a$), then

$$p_0 = p_c\left(\frac{a_c}{a}\right)^{3\gamma}$$

Because the expansion velocity of a spherical wall is much smaller than that of a detonation wave, it is reasonable to assume that the explosion is instantaneous. So p_c can be regarded as the instantaneous detonation pressure (namely the half of C-J pressure), thus

$$p_a = p_c\left(\frac{a_0}{a}\right)^{3\gamma} \cdot \left(\frac{a}{A}\right)^{3\gamma} = p_c\left(\frac{a_c}{A}\right)^{3\gamma}$$

If the condition that the compressive zone of σ_θ disappears (namely $p_a = Y$) as the criterion of shell fragmentation is taken into consideration, then from the above equation we obtain the following equation for the fragmentation radius A_f:

$$A_f = a_c\left(\frac{p_c}{Y}\right)^{\frac{1}{3\gamma}} \tag{8.38}$$

Al-Hassani and Johnson (1969) analyzed this problem by using Euler variables. In the present section, we use Lagrange variables to analyze the same problem (Wang, 1983).

8.5 Elasto-Visco-Plastic Spherical Waves and Cylindrical Waves

For the rate-dependent materials, if we adopt the Соколовский (Sokolovskii)-Malvern-Perzyna elasto-visco-plastic theory [ref. Eq. (7.131) in Chapter 7], then in solving problems of spherical waves and cylindrical waves, the following constitutive relationship is used to replace Eq. (8.7) in the theory of elastic waves,

$$\dot{\varepsilon}_r + n_0\dot{\varepsilon}_\theta = \frac{1}{3K}\left(\dot{\sigma}_r + n_0\dot{\sigma}_\theta + n_1\dot{\sigma}_z\right) \tag{8.39a}$$

$$\dot{\varepsilon}_r - \dot{\varepsilon}_\theta = \frac{1}{2\mu}\left(\dot{\sigma}_r - \dot{\sigma}_\theta\right) + \gamma\left\langle\phi\left(\frac{\sqrt{J_2}}{\mathrm{K}} - 1\right)\right\rangle\frac{\sigma_r - \sigma_\theta}{\sqrt{J_2}} \tag{8.39b}$$

$$n_1(\dot{\varepsilon}_r + \dot{\varepsilon}_\theta) = -n_1\left\{\frac{1}{2\mu}\left(2\dot{\sigma}_z - \dot{\sigma}_r - \dot{\sigma}_\theta\right) + \gamma\left\langle\phi\left(\frac{\sqrt{J_2}}{\mathrm{K}} - 1\right)\right\rangle\cdot\frac{2\sigma_z - \sigma_r - \sigma_\theta}{\sqrt{J_2}}\right\} \tag{8.39c}$$

Together with the continuity equation [Eq. (8.2)] and the motion equation [Eq. (8.6)], the governing equations can be re-listed below,

$$\left.\begin{aligned}
&\frac{\partial\varepsilon_r}{\partial t} - \frac{\partial v}{\partial r} = 0\\
&\frac{\partial\varepsilon_\theta}{\partial t} - \frac{v}{r} = 0\\
&\rho_0\frac{\partial v}{\partial t} - \frac{\partial\sigma_r}{\partial r} - n_0\frac{(\sigma_r - \sigma_\theta)}{r} = 0\\
&\frac{\partial v}{\partial r} - \frac{1}{3K}\frac{\partial\sigma_r}{\partial t} - \frac{n_0}{3K}\frac{\partial\sigma_\theta}{\partial t} - \frac{n_1}{3K}\frac{\partial\sigma_z}{\partial t} + n_0\frac{v}{r} = 0\\
&\frac{\partial v}{\partial r} - \frac{1}{2\mu}\frac{\partial\sigma_r}{\partial t} - \frac{1}{2\mu}\frac{\partial\sigma_\theta}{\partial t} - \frac{v}{r} - \gamma\left\langle\phi\left(\frac{\sqrt{J_2}}{\mathrm{K}} - 1\right)\right\rangle\cdot\frac{\sigma_r - \sigma_\theta}{\sqrt{J_2}} = 0\\
&n_1\frac{\partial v}{\partial r} - \frac{n_1}{2\mu}\frac{\partial\sigma_r}{\partial t} - \frac{n_1}{2\mu}\frac{\partial\sigma_\theta}{\partial t} + \frac{n_1}{\mu}\frac{\partial\sigma_z}{\partial t} + n_1\frac{v}{r} + n_1\gamma\left\langle\phi\left(\frac{\sqrt{J_2}}{\mathrm{K}} - 1\right)\right\rangle\\
&\quad\cdot\frac{2\sigma_z - \sigma_r - \sigma_\theta}{\sqrt{J_2}} = 0
\end{aligned}\right\} \tag{8.40}$$

where each symbol denotes the same variable used in the previous sections. Thus, for cylindrical waves there are six unknowns $\sigma_r, \sigma_\theta, \sigma_z, \varepsilon_r, \varepsilon_\theta$ and v in six equations. For spherical waves, one unknown σ_z is reduced while one equation (the last one) is also reduced.

When the characteristic method is adopted to solve these hyperbolic first-order partial differential equations, in order to determine the characteristics and the corresponding characteristic compatibility relationships, multiplying the above six equations by six undetermined coefficients L, M, N, P, Q, and R respectively and adding them together, we obtain

$$L\frac{\partial}{\partial t}\varepsilon_r + M\frac{\partial}{\partial t}\varepsilon_\theta + \left\{N\rho_0\frac{\partial}{\partial t} + (P+Q+Rn_1-L)\frac{\partial}{\partial r}\right\}v$$
$$-\left\{n\frac{\partial}{\partial r} + \left(\frac{P}{3K}+\frac{Q}{2\mu}+\frac{Rn_1}{2\mu}\right)\frac{\partial}{\partial t}\right\}\sigma_r - \left(\frac{Pn_0}{3K}-\frac{Q}{2\mu}+\frac{Rn_1}{2\mu}\right)\frac{\partial}{\partial t}\sigma_\theta$$
$$+\left(-\frac{Pn_1}{3K}+\frac{Rn_1}{\mu}\right)\frac{\partial}{\partial t}\sigma_z + (Pn_0 - M - Q + Rn_1)\frac{v}{r} - Nn_0\frac{\sigma_r-\sigma_\theta}{r}$$
$$+\frac{\gamma\langle\phi\rangle}{\sqrt{J_2}}\{2Rn_1\sigma_z - (Rn_1-Q)\sigma_r - (Rn_1-Q)\sigma_\theta\} = 0$$

These coefficients should satisfy

$$\frac{dr}{dt} = \frac{0}{L} = \frac{0}{M} = \frac{P+Q+Rn_1-L}{N\rho_0} = \frac{N}{\dfrac{P}{3K}+\dfrac{Q}{2\mu}+\dfrac{Rn_1}{2\mu}}$$
$$= \frac{0}{\dfrac{Pn_0}{3K}-\dfrac{Q}{2\mu}+\dfrac{Rn_1}{2\mu}} = \frac{0}{-\dfrac{Pn_1}{3K}+\dfrac{Rn_1}{\mu}} \tag{8.41}$$

Then we obtain the following solutions

$$(1) \quad L = M = \frac{Pn_0}{3K} - \frac{\theta}{2\mu} + \frac{Rn_1}{2\mu} = \frac{Rn_1}{\mu} - \frac{Pn_1}{3K} = 0$$

In such a case, we obtain two families of real characteristics. The characteristic equations and the corresponding compatibility relationships are:

$$dr = \pm\sqrt{\frac{K+\frac{4}{3}\mu}{\rho_0}}dt = \pm C_L^e dt, \tag{8.42a}$$

$$d\sigma_r = \pm\rho_0 C_L^e dv + n_0\left\{\frac{1}{r}\left[\left(K-\frac{2\mu}{3}\right)v \pm C_L^e(\sigma_r-\sigma_\theta)\right]\right.$$
$$\left. -\frac{2\mu}{3}\frac{\gamma\langle\phi\rangle}{\sqrt{J_2}}[(1+n_1)\sigma_r - \sigma_\theta - n_1\sigma_z]\right\}dt$$
$$= \pm\rho_0 C_L^e d\upsilon - n_0\left\{\frac{1}{r}(\sigma_r-\sigma_\theta) \pm \left(K-\frac{2\mu}{3}\right)\frac{v}{C_L^e}\right.$$
$$\left. \pm\frac{2\mu}{3C_L^e}\frac{\gamma\langle\phi\rangle}{\sqrt{J_2}}[(1+n_1)\sigma_r - \sigma_\theta - n_1\sigma_z]\right\}dr \tag{8.42b}$$

These two families of characteristics represent the propagating loci of the positive and the negative elasto-visco-plastic waves.

(2) $N = P + Q + Rn_1 - L = 0$

In such a case, we obtain four families of real characteristics in coincidence:

$$dr = 0 \tag{8.43a}$$

The corresponding compatibility relationships are:

(i) $M \neq 0$ and other coefficients are all zero, then

$$d\varepsilon_\theta = \frac{v}{r}dt \tag{8.43b}$$

(ii) $P = L \neq 0$ and other coefficients are all zero, then

$$d\varepsilon_r - \frac{1}{3K}(d\sigma_r - n_0 d\sigma_\theta + n_1 d\sigma_z) + n_0\frac{v}{r}dt = 0 \tag{8.43c}$$

(iii) $Q = L \neq 0$ and other coefficients are all zero, then

$$d\varepsilon_r - \frac{1}{2\mu}(d\sigma_r - d\sigma_\theta) - \left[\frac{v}{r} + \gamma\langle\phi\rangle\frac{\sigma_r - \sigma_\theta}{\sqrt{J_2}}\right]dt = 0 \tag{8.43d}$$

(iv) $Rn_1 = L \neq 0$ and other coefficients are all zero, then

$$n_1 d\varepsilon_r + \frac{n_1}{2\mu}(2d\sigma_z - d\sigma_r - d\sigma_\theta) + n_1\left\{\frac{v}{r} + \gamma\langle\phi\rangle\frac{2\sigma_z - \sigma_r - \sigma_\theta}{\sqrt{J_2}}\right\}dt = 0 \tag{8.43e}$$

The family of characteristics described by Eq. (8.43a) represents the motion loci of particles. The corresponding compatibility relationships [Eqs. (8.43b)–(8.43e)] are actually the total differential forms of the continuity equations [Eq. (8.40b)] and the constitutive equations [Eq. (8.39)]. Because when using Lagrange variables, $\partial/\partial t$ is the same as the total derivative with respect to t along $dr = 0$.

Thus, we have obtained all differential equations of characteristics and the corresponding compatibility relationships for elasto-visco-plastic spherical waves or elasto-visco-plastic cylindrical waves. With given initial and boundary conditions, we can solve the problems of elasto-visco-plastic spherical waves and cylindrical waves by using a numerical characteristics method.

Consider an infinite elasto-visco-plastic media with a spherical cavity of radius $r = a$, which is suddenly subjected to an explosion pressure uniformly distributed on its inner wall, as shown in Fig. 8.6. In the beginning, there is a strong discontinuous spherical wave propagating along the characteristics ab. As discussed in Section 8.2, for the strong discontinuous spherical waves, Eq. (8.18) should be satisfied across the wave front. If the media are in rest and undisturbed state initially, then there exist the following relationships

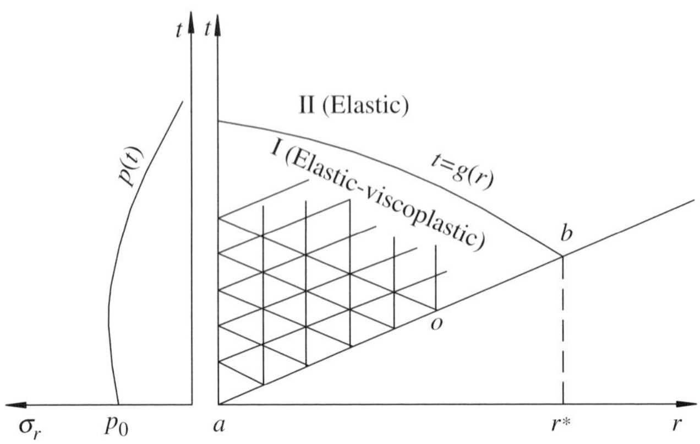

Fig. 8.6. Schematics of a numerical characteristic method to solve a problem of elasto-visco-plastic spherical wave propagation.

to describe the states on the strong discontinuous spherical wave front, which is the same as that for the 1D strain waves,

$$u = \varepsilon_\theta + \varepsilon_\varphi = 0 \tag{8.44a}$$

$$v = -\frac{\sigma_r}{\rho_0 C_L^e}, \qquad \varepsilon_r = \frac{\sigma_r}{K + \frac{4}{3}\mu} \tag{8.44b}$$

$$\sigma_\theta = \sigma_\varphi = \frac{K - \frac{2}{3}\mu}{K + \frac{4}{3}\mu}\sigma_r = \frac{\lambda}{\lambda + 2\mu}\sigma_r = \frac{\nu}{1-\nu}\sigma_r \tag{8.44c}$$

However, as different from the case of 1D strain waves, each variable in the above equations does not remain constant along the characteristics *ab*. In addition, the characteristic compatibility relationship should be satisfied along *ab*. Because, for an elasto-visco-plastic wave, there is only instantaneous response across a strong discontinuous wave front ($\dot{\varepsilon} = \infty$), which implies $W_p = 0$, then the K in the term $\phi\left((\sqrt{J_2}/K) - 1\right)$ of Eq. (8.42b) is reduced as $K = k = Y_0/\sqrt{3}$ (where Y_0 is the yield limit in simple tension). Moreover, consideration of Eq. (8.44) yields [refer to Eq. (8.20a)]

$$\sqrt{J_2} = \frac{\sigma_r - \sigma_\theta}{\sqrt{3}} = \frac{2\mu}{\sqrt{3}\left(K + \frac{4}{3}\mu\right)}\sigma_r$$

Then following similar steps in deducing Eq. (8.20b), along *ab* we have:

$$\frac{d\sigma_r}{dr} = -\frac{n_0\sigma_r}{2r} - \frac{2\mu}{\sqrt{3}C_L^e}\gamma\left\langle\phi\left(\frac{2\mu\sigma_r}{\sqrt{3}\rho_0\left(C_L^e\right)^2 k} - 1\right)\right\rangle \equiv -\psi\left(r, \sigma_r\left(r\right)\right) \tag{8.45}$$

Integrating both the sides of the above equation and using the initial condition $\sigma_r(a, 0) = -p_0$, we obtain the following nonlinear second-kind Volterra integral equation

$$\sigma_r = p_0 - \int_a^r \psi(\xi, \sigma_r(\xi))d\xi \tag{8.46}$$

In general, this equation can be solved by successive approximations method. Once σ_r along *ab* is obtained, the determination of the solutions in region *I* is a solvable characteristic boundary-value problem.

Note that when a strong discontinuous spherical wave propagates along *ab*, its amplitude decreases with increasing r. This is caused by the effect of spherical dispersion on the one hand and the effect of visco-plastic dissipation on the other hand. These two effects correspond to the first and the last items on the right-hand side of the first equality sign in Eq. (8.45) respectively, which defines ψ. Assuming that the over-stress disappears at point $b(r^*)$ shown in Fig. 8.6, namely there is

$$J_2\left(r^*\right) = k^2 \tag{8.47}$$

then the solutions for the region of $r \geq r^*$ are elastic ones [Eq. (8.21)]. By the same argument, the solutions at points in region I (elasto-visco-plastic zone) that satisfy Eq. (8.47) can be solved. Then, we can determine the unloading boundary $t = g(r)$ which is locus of points where the over-stress disappears. In the region II (elastic unloading zone), we can obtain solutions by the method previously used in the analysis on elastic spherical waves. Thus, the whole problem can be solved (Bejda and Perzyna, 1964).

CHAPTER 9

Elastic–Plastic Waves Propagating in Flexible Strings

In the present chapter, the elastic–plastic waves propagating in **flexible strings**, i.e. the strings that only bear stretch force, will be discussed.

In the classic theory of elastic waves propagating in flexible strings, only two simple kinds of elastic wave propagation are dealt with: the longitudinal waves without transverse displacement, which are in fact in the scope of the longitudinal wave propagation theory discussed in Chapter 2, and the transverse waves, with only transverse displacement u but without longitudinal strain.

If it is assumed that the transverse displacement u is always perpendicular to the string (the X-axis) at any time and it is further limited only to study the small-amplitude vibration of strings, [assuming $(\partial u/\partial X)^2 \leq 1$], then the stretch force of string T is unchanged with coordinate X and time t, i.e.:

$$T = T_0 = \text{constant}$$

and the transverse wave velocity C_h is also constant:

$$C_h = \sqrt{\frac{T_0}{\rho_0}} = \text{constant} \tag{9.1}$$

where ρ_0 is the initial density of the string.

The problems of longitudinal and transverse waves simultaneously propagating in strings have been studied by Cole *et al.* (1953) and Smith *et al.* (1958). However, it is limited to the linear elastic strings.

If the loading exerted on the string is high enough, the string will experience plastic deformation, and plastic waves will propagate along the string. In general, both plastic longitudinal waves and plastic transverse waves may simultaneously propagate in the string and then interact. The problem becomes very complicated.

Historically, the studies on elasto-plastic dynamics of strings probably began in the period of World War II, although the research results were published only after the war. The first work can probably be traced to that of Рахматулин (Rakhmatulin) (1945), and then the works of Кристеску (Cristescu) (1954) and Craggs (1954). Compared to the studies on elastic–plastic waves propagating in bars, the studies on elastic–plastic waves propagating in strings are much less satisfactory.

The studies of elastic–plastic dynamics of strings are motivated by a series of practical problems, such as the impact strength of the strings of a parachute, the stress wave propagation in the fibers of long-fiber-reinforced composites, the impact strength of the rope used in mines, the impact deformation of yarn in weaving machines, etc. Moreover, with regard to the stress wave propagation in bars, longitudinal wave propagation is the simplest problem. In more general cases, the bar may be subjected to oblique impact, where upon the problem of elastic–plastic bending wave propagation in beams should be dealt with, as will be discussed in Chapter 10. For slender beams, if the bending stiffness is as small as could be neglected, then the problem is approximately reduced to the study on elastic–plastic wave propagation in flexible strings. Furthermore, based on the studies on stress wave propagation in strings, a further study can be made on the impact problem of membranes, which can be regarded as a generalization of the impact problem of strings. Finally, the study on elastic-plastic wave propagation in strings has been used as a dynamic experimental technique to investigate the dynamic stress–strain relation of materials, as proposed by Рахматулин (Rakhmatulin) (1945b).

Similar to what we discussed on stress wave propagation in bars, the stress wave propagation in strings can be studied either by Lagrange or Euler variables, or a mix of both. In the following, in order to establish a more closed connection between the related chapters in the present book, and also for convenience, the Lagrange variables will be used (Wang, 1964).

9.1 Governing Equations

First of all, it should be made clear again that we are limited to the study of flexible strings. “Flexible” herein means that the stretch force at any point on the string is always acting in the tangential direction of the instantaneous profile of the string, i.e. the string only bears stretch force without any resistance to bending.

In a string that keeps its straight line shape unchanged in the motion throughout, the stretch force acting on the string also keeps its direction invariable throughout. In such a case, only longitudinal waves propagate in the string, and can thus be dealt with completely by the longitudinal wave propagation theory discussed in Chapter 2; no further discussion is needed.

On the other hand, when a transverse wave propagates in a string, a shape change of the string must take place. In other words, only when a direction change of the stretch force vector **T** acting on the string occurs will a transverse disturbance propagate in the string.

If an abrupt change of string shape (break off in shape) takes place, the direction of stretch force vector **T** abruptly changes across this break point of the string. In other words, a strong discontinuity of stretch force direction appears. According to the momentum conservation theorem, there must correspondingly be an abrupt jump of transverse velocity, and *vice versa*. Thus, such a strong discontinuous transverse wave front must propagate in the string in the form of a "break point".

In the following, we will first discuss the propagation of such strong discontinuous transverse waves.

Let us use Lagrange variables S_0 and t to describe the motion of the string, where S_0 is the original arc length of the string, and t is the time. Denote the displacement vector by $\mathbf{u}$, then the particle velocity vector $\mathbf{v}$ is expressed as $\mathbf{v} = \partial\mathbf{u}/\partial t$. Note that there exists, in general, an intersection angle β between the particle velocity vector and the string direction, as shown in Fig. 9.1, where the subscripts 1 and 2 indicate the quantities in front of and behind the strong discontinuity, respectively.

Consider the case in which a strong discontinuous transverse wave propagates an infinitesimal distance along a string segment dS_0 in an infinitesimal time interval dt. The infinitesimal string segment suffers an impulse of

$$(\mathbf{T}_1 + \mathbf{T}_2)dt$$

and the corresponding change of momentum is

$$\rho_0(\mathbf{v}_2 - \mathbf{v}_1)dS_0$$

According to the momentum theorem, we have the following dynamic condition across the strong discontinuous interface:

$$\rho_0 C_h(\mathbf{v}_2 - \mathbf{v}_1) = \mathbf{T}_2 + \mathbf{T}_1 \tag{9.2}$$

where $C_h = dS_0/dt$ is the Lagrange propagating velocity of the strong discontinuous transverse wave and ρ_0 the original line density of the string.

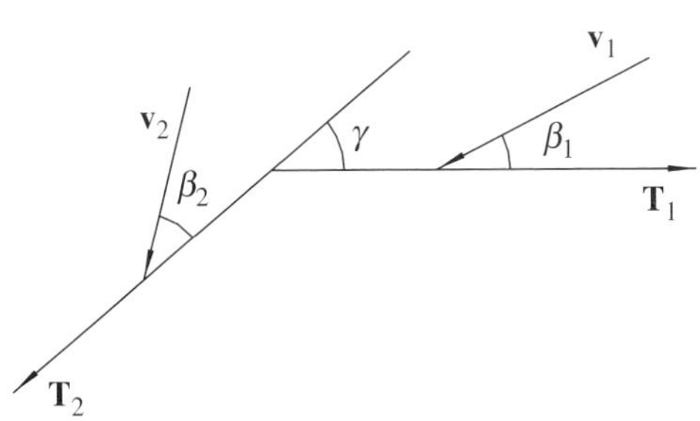

Fig. 9.1. A strong discontinuous transverse wave propagating in the string.

The continuity condition requires the displacements across the strong discontinuous interface to be equal, i.e.:

$$\mathbf{u_2}(S_0, t) = \mathbf{u_1}(S_0, t)$$

This condition should be always held along with the propagation of strong discontinuous interfaces. In other words, the total derivative of $\mathbf{u_2}$ and $\mathbf{u_1}$ following the strong discontinuous interface propagation should be equal, i.e.:

$$\frac{\partial \mathbf{u_2}}{\partial t} + C_h \frac{\partial \mathbf{u_2}}{\partial S_0} = \frac{\partial \mathbf{u_1}}{\partial t} + C_h \frac{\partial \mathbf{u_1}}{\partial S_0} \tag{9.3}$$

which can be rewritten as

$$\mathbf{v_2} - \mathbf{v_1} = -C_h \left(\frac{\partial \mathbf{u_2}}{\partial S_0} - \frac{\partial \mathbf{u_1}}{\partial S_1} \right) \tag{9.4}$$

where $\partial \mathbf{u}/\partial S_0$ is the relative displacement vector. If the unit vectors of infinitesimal string segment before and after deformation are denoted by $\mathbf{S_0}$ and $\mathbf{S_1}$, respectively, as shown in Fig. 9.2.

Then, obviously, we have

$$(1+\varepsilon)\mathbf{S_1} = \mathbf{S_0} + \frac{\partial \mathbf{u}}{\partial S_0} \tag{9.5}$$

where $\varepsilon = \dfrac{dS - dS_0}{dS_0}$ is the tensile strain of the string.

The strain ε and stretch force T are connected to each other by the known constitutive relation of string material in the form of

$$T = T(\varepsilon) \tag{9.6}$$

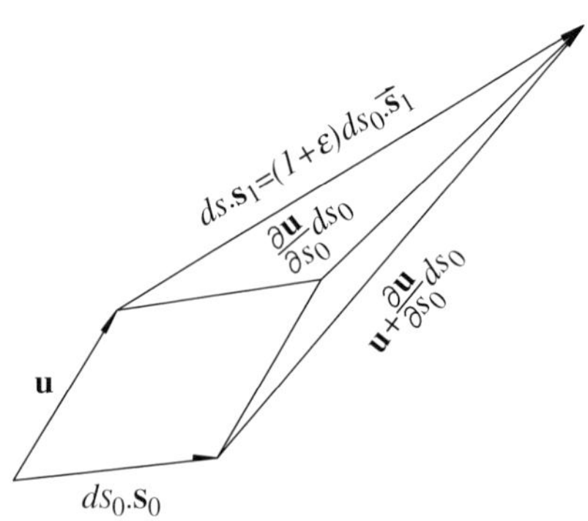

Fig. 9.2. An infinitesimal string segment before and after deformation.

It has been assumed herein and from now on that the stretch force is only the function of strain $T = T(\varepsilon)$, independent of strain rate, and only the situation of $(d^2T/d\varepsilon^2) \leq 0$ will be discussed.

Thus, Eqs. (9.2), (9.4), and (9.6) constitute the basic equations of the current problem. Recalling the Rankine–Hugoniot relations deduced in Section 2.3 for the strong discontinuous longitudinal wave propagating in a bar, it can be found immediately that Eqs. (9.2), (9.4), and (9.6), respectively, correspond to Eqs. (2.57), (2.56), and (2.14).

Without loss of generality, assume that the string is in plane motion, and then the vector equations (9.2) and (9.4) are equivalent to four scalar equations. For example, if we resolve the vector into the tangent component and the normal component with respect to the string segment in front of the strong discontinuous interface, denote the intersection angle between the string segments in front of and behind the "break point" by γ, and take account of Eq. (9.5). Then we have the following four scalar equations (note that the intersection angle between the vectors $\mathbf{S_0}$ and $\mathbf{S_1}$ equals γ behind the strong discontinuity but equals zero in front of the strong discontinuity):

$$\textbf{(9-I)} \begin{cases} \rho_0 C_h[v_2\cos(\beta_2+\gamma) - v_1\cos\beta_1] = T_2\cos\gamma - T_1 & (9.7) \\ \rho_0 C_h[v_2\sin(\beta_2+\gamma) - v_1\sin\beta_1] = T_2\sin\gamma & (9.8) \\ v_2\cos(\beta_2+\gamma) - v_1\cos\beta_1 = C_h[(1+\varepsilon_2)\cos\gamma - 1 - \varepsilon_1] & (9.9) \\ v_2\sin(\beta_2+\gamma) - v_1\sin\beta_1 = C_h(1+\varepsilon_2)\sin\gamma & (9.10) \end{cases}$$

From Eqs. (9.8) and (9.10), we immediately obtain the transverse wave velocity:

$$C_h = \sqrt{\frac{1}{\rho_0}\frac{T_2}{1+\varepsilon_2}} \tag{9.11}$$

On the other hand, from Eqs. (9.7) and (9.9), we have

$$\frac{T_2\cos\gamma - T_1}{(1+\varepsilon_2)\cos\gamma - (1+\varepsilon_1)} = \frac{T_2}{1+\varepsilon_2}$$

which can be reduced to

$$\frac{T_1}{1+\varepsilon_1} = \frac{T_2}{1+\varepsilon_2} \tag{9.12}$$

In order to satisfy this condition, there are two possible solutions, i.e. either

$$T_1 = T_2,\ \varepsilon_1 = \varepsilon_2 \tag{9.13}$$

or

$$\frac{T_1 - T_2}{\varepsilon_1 - \varepsilon_2} = \frac{T_1}{1+\varepsilon_1} = \frac{T_2}{1+\varepsilon_2} \tag{9.14}$$

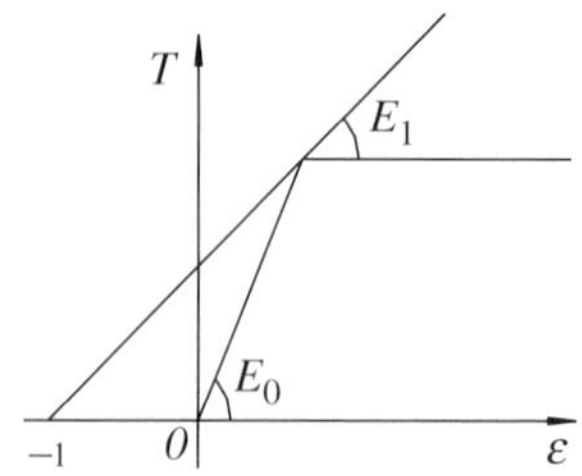

Fig. 9.3. The extension of linear hardening plastic segment just intersects at $\varepsilon = -1$.

Except in the special case of $T = T(\varepsilon)$ shown in Fig. 9.3, only Eq. (9.13) could be the solution of Eq. (9.12). It means that **the transverse wave only brings about shape change of the string, and not strain disturbance**. In the special linear hardening case of $T = T(\varepsilon)$ shown in Fig. 9.3, Eq. (9.14) could be a possible solution of Eq. (9.12). In such cases, the strong discontinuous longitudinal wave and the strong discontinuous transverse wave coincide with each other. The jump of T and ε can be considered as the result induced by the longitudinal wave, while the shape change (break in shape) can be considered as the result induced by the transverse wave.

If $\varepsilon \ll 1$, the stretch force can be approximately regarded as a constant, and Eq. (9.11) is reduced to Eq. (9.1) the elastic transverse wave velocity given by the classic elastic theory under the condition of small-amplitude vibration.

When $\gamma = 0$, Eqs. (9.7)–(9.10), i.e. the equation group (**9-I**), are reduced to:

$$\textbf{(9-II)}\begin{cases} \rho_0 C_h[(v\cos\beta)_2 - (v\cos\beta)_1] = T_2 - T_1 & (9.15) \\ (v\cos\beta)_2 - (v\cos\beta)_1 = C_h(\varepsilon_2 - \varepsilon_1) & (9.16) \\ (v\sin\beta)_2 = (v\sin\beta)_1 & (9.17) \end{cases}$$

They are, in fact, the Rankine–Hugoniot relations for strong discontinuous longitudinal wave propagating in a moving bar, which moves with a transverse particle velocity of $v\sin\beta$ = constant, completely coinciding with Eqs. (2.57) and (2.56), and C_h is reduced to:

$$C_h = \sqrt{\frac{1}{\rho_0}\frac{T_2 - T_1}{\varepsilon_2 - \varepsilon_1}}$$

When the jump across the front of a strong discontinuous transverse wave gradually decreases and approaches infinity, its limit is the weak discontinuous transverse wave. In such cases, use $d\gamma$ instead of γ, and, correspondingly, let

$$T_2 = T_1 dT,\ \varepsilon_2 = \varepsilon_1 + d\varepsilon,\ \beta_2 = \beta_1 + d\beta,\ v_2 = v_1 + dv$$

the equation group (**9-I**) is reduced to the following dynamic condition and kinematics condition across a weak discontinuous transverse wave front:

$$
\textbf{(9-III)}\begin{cases}
\rho_0 C_h[d(v\cos\beta) - v\sin\beta d\gamma] = dT & (9.18)\\
\rho_0 C_h[d(v\sin\beta) + v\cos\beta d\gamma = Td\gamma & (9.19)\\
d(v\cos\beta) - v\sin\beta d\gamma = C_h d\varepsilon & (9.20)\\
d(v\sin\beta) + v\cos\beta d\gamma = C_h(1+\varepsilon)d\gamma & (9.21)
\end{cases}
$$

The wave velocity of a weak discontinuous transverse wave can be obtained immediately from Eqs. (9.19) and (9.21) as

$$C_h(\varepsilon) = \sqrt{\frac{1}{\rho_0}\frac{T(\varepsilon)}{1+\varepsilon}} \tag{9.22}$$

On the other hand, from Eqs. (9.18) and (9.20), we have

$$\frac{T}{1+\varepsilon} = \frac{dT}{d\varepsilon} \tag{9.23}$$

In order to satisfy this condition, there are two possible solutions, i.e. either

$$dT = 0, d\varepsilon = 0 \tag{9.24}$$

or

$$\varepsilon = \varepsilon_A \tag{9.25}$$

where ε_A is the strain at the point A as shown in Fig. 9.4, and the point A is the tangent point of $T = T(\varepsilon)$, with the tangent line starting from $\varepsilon = -1$.

Except in the special case shown in Fig. 9.4, only Eq. (9.24), in general, could be the solution of Eq. (9.23). It shows again that in the case of weak discontinuous transverse wave propagation, **the transverse wave also only brings to shape change** ($d\gamma \neq 0$)

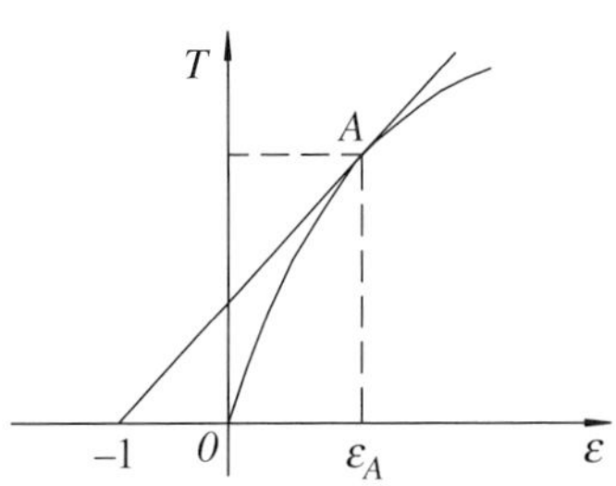

Fig. 9.4. The tangent line at the point A of $T = T(\varepsilon)$ curve just intersects at $\varepsilon = -1$.

of string but not strain disturbance ($d\varepsilon = 0$). In the special case shown in Fig. 9.4, Eq. (9.25) is a solution of Eq. (9.23). In such a case, the weak discontinuous longitudinal wave and the weak discontinuous transverse wave coincide with each other. The change of stretch force and strain, dT and $d\varepsilon$ can be considered as the result induced by the longitudinal wave, while the shape change $d\gamma$ can be considered as the result induced by the transverse wave.

If $d\gamma = 0$, equation group (**9-III**) is reduced to the following familiar relations previously discussed on the weak discontinuous longitudinal wave propagation theory:

$$\textbf{(9-IV)}\begin{cases}\rho_0 C_l d(v\cos\beta) = dT & (9.26)\\ d(v\cos\beta) = C_l d\varepsilon & (9.27)\\ d(v\sin\beta) = 0 & (9.28)\end{cases}$$

where C_h has been rewritten as $C_l = \sqrt{\dfrac{1}{\rho_0}\dfrac{dT}{d\varepsilon}}$. Equation group (**9-IV**) is, in fact, equivalent to Eq. (2.63) deduced in Chapter 2.

The four groups of equations obtained above, (**9-I**) to (**9-IV**), are the basic differential equations for elastic–plastic longitudinal and transverse waves propagating in strings, regardless of strong or weak discontinuous waves. They represent the dynamic conditions and kinematic conditions across the wave front in different cases, respectively.

Obviously, transverse waves propagate with wave velocity $C_h = \sqrt{(1/\rho_0)(T/1+\varepsilon)}$, and longitudinal waves propagate with wave velocity $C_l = \sqrt{(1/\rho_0)(dT/d\varepsilon)}$, which on the Lagrange plane (S_0–t plane) are, respectively, represented by the characteristics families

$$dS_0 = \pm C_h dt \tag{9.29}$$

and

$$dS_0 = \pm C_l dt \tag{9.30}$$

The basic differential relations (**9-III**) and (**9-IV**) give the differential relations across the corresponding positive characteristics (namely along the negative characteristics). Similar to what was discussed in Section 2.1, the differential relations across the negative characteristics (i.e. along the positive characteristics) can be obtained just by exchanging the corresponding positive and negative signs. Starting from these differential relations, the problems of elastic–plastic waves propagating in strings can, in principle, be solved by characteristics method, although they are certainly much complicated than that discussed in Chapter 2.

It should be particularly noticed that the longitudinal wave only results in strain ($d\varepsilon \neq 0$) without shape change ($dr = 0$), whereas the transverse wave only results in shape change ($dr \neq 0$) without strain ($d\varepsilon = 0$); they are interinfluenced. In fact, for the longitudinal wave, according to the equation group (**9-IV**), dT and $d\gamma$ are related with the change of tangential velocity $d(v\cos\beta)$. Therefore, once the transverse wave induces a shape change of string, it must influence the longitudinal wave propagation through the change of tangential velocity. On the other hand, according to Eq. (9.22), the transverse wave velocity C_h is a function of strain ε; therefore, once the longitudinal wave induces a strain change of string, it must also influence the transverse wave propagation.

In different cases, either the longitudinal wave or the transverse wave could be weak discontinuous or strong discontinuous, respectively; and the propagating precedence of the longitudinal wave and the transverse wave could be different, either $C_l > C_h$ or $C_l < C_h$. The situation mainly depends on the stress–strain relation $T = T(\varepsilon)$, the original shape of string, and the initial-boundary conditions. The boundary condition could be given in different forms, such as $\mathbf{T}\,|_{boundary} = \mathbf{T_0}(t)$, $\mathbf{v}\,|_{boundary} = \mathbf{v_0}(t)$ or $\varepsilon\,|_{boundary} = \varepsilon_0(t)$.

With regard to the longitudinal wave, if we are limited to the stress–strain relation $T = T(\varepsilon)$ with the character of $(d^2T/d\varepsilon^2) \le 0$ and without unloading $(\partial\varepsilon/\partial t \ge 0)$, then it has been known from Chapter 2 that the longitudinal wave velocity $C_l = \sqrt{(1/\rho_0)(dT/d\varepsilon)}$ decreases with increasing strain ε. Therefore, under the general loading condition, the longitudinal wave is a kind of continuous wave (weak discontinuous wave). Only if, there is a straight part on the $T = T(\varepsilon)$ curve such as a linear elastic or a linear hardening part, and, moreover, the boundary condition includes strong discontinuity (abrupt loading), then a shock wave (strong discontinuous wave) can form.

With regard to the transverse wave, the situation is a little different from the longitudinal wave. When $\varepsilon \le \varepsilon_A$ (Fig. 9.4), the transverse wave velocity $C_h = \sqrt{(1/\rho_0)(T/1+\varepsilon)}$ increases with increasing ε, so the later disturbances will catch up with the preceding ones, and finally a strong discontinuous transverse wave will form and propagate with the velocity corresponding to the maximum strain ε_m, i.e. $C_h(\varepsilon_m) = \sqrt{(1/\rho_0)[T(\varepsilon_m)/1+\varepsilon_m]}$ Because $C_h(\varepsilon_m) \le C_l(\varepsilon_m)$, the transverse wave will be behind the longitudinal wave if $\varepsilon < \varepsilon_A$, or will coincide with the longitudinal wave if $\varepsilon = \varepsilon_A$. But when $\varepsilon > \varepsilon_A$, the transverse wave velocity C_h decreases with increasing ε. Roughly speaking, it seems that there will be a series of weak discontinuous transverse waves that follow the strong discontinuous transverse wave that corresponding to the ε_A. However, this is not the truth. Because, when $\varepsilon > \varepsilon_A$, we have $[T/(1+\varepsilon)] > (dT/d\varepsilon)$; in other words, when $\varepsilon > \varepsilon_A$, the propagation of transverse waves is faster than the longitudinal wave. Moreover, notice that the strain disturbance ε results from longitudinal propagation, and the smaller the ε is the faster the longitudinal wave propagates. Therefore, along with the transverse waves, successively to catch up with the preceding longitudinal waves, ε ill gradually decrease. Thus, the transverse wave will propagate faster and faster, and a strong discontinuous transverse wave propagating with velocity $C_h(\varepsilon_A)$ will finally form.

It is thus clear that a strong discontinuous transverse wave will finally form. Under the gradually loading boundary condition $[(\partial\sigma_0/\partial t) \ge 0, (\partial\sigma_0/\partial t) \ne \infty]$, the strong discontinuous transverse wave forms gradually. As long as the loading applied on the boundary continuously increases with time $(\partial\sigma_0/\partial t > 0)$, the strong discontinuous wave is an unstable one, i.e. its amplitude and propagating velocity also continuously vary with time. Only when the boundary condition contains a constant loading $(\partial\sigma_0/\partial t = 0)$ can a stable strong discontinuous transverse wave finally form. Obviously, when a constant loading is abruptly applied, then a stable, strong discontinuous transverse wave will immediately form at the beginning.

Finally, it should be noted again that the basic equation group (**9-I**) in the scalar form is obtained by resolving the vector equations (9.2) and (9.4) into the tangent component and the normal component with respect to the string segment in front of the strong discontinuous interface. Alternatively, by resolving the vector equations (9.2) and (9.4) into the tangent component and the normal component with respect to the string segment behind the strong discontinuous interface, we have the following similar equation group:

$$(\mathbf{9\text{-}I'})\begin{cases} \rho_0 C_h[v_2 \cos\beta_2 - v_1\cos(\beta_1-\gamma)] = T_2 - T_1\cos\gamma & (9.31)\\ \rho_0 C_h[v_2 \sin\beta_2 - v_1\sin(\beta_1-\gamma)] = T_1\sin\gamma & (9.32)\\ v_2\cos\beta_2 - v_1\cos(\beta_1-\gamma) = C_h[\varepsilon_2 - (1+\varepsilon_1)\cos\gamma + 1] & (9.33)\\ v_2\sin\beta_2 - v_1\sin(\beta_1-\gamma) = C_h(1+\varepsilon_1)\sin\gamma & (9.34) \end{cases}$$

Obviously, only one of (**9-I**) and (**9-I′**) is independent. For example, Eq. (9.31) can be obtained by first multiplying Eq. (9.7) by $\cos\gamma$, then multiplying Eq. (9.8) by $\sin\gamma$, and then adding them. Moreover, from Eqs. (9.8) and (9.10), and from Eqs. (9.32) and (9.34), we immediately have

$$C_h^2 = \frac{1}{\rho_0}\frac{T_1}{1+\varepsilon_1} = \frac{1}{\rho_0}\frac{T_2}{1+\varepsilon_2}$$

which is just Eq. (9.69) obtained previously.

9.2 Semi-Infinite Straight String under Abrupt Constant-Velocity Oblique Impact

Let us now consider a semi-infinite straight string, which is subjected to an abrupt constant velocity impact, V_0, β_0, at its end S_0. According to the discussions in the preceding section, obviously, a stable, strong discontinuous transverse wave forms immediately at the beginning. In such case, neither characteristic length nor characteristic time is contained, and, consequently, according to the theory of dimensional analysis, only one dimensionless variable S_0/V_0t is contained. This is the so-called "**self-similar problem**". On the S_0–t plot, it corresponds to a bunch of "centered waves", as shown in Figs. 9.5 and 9.7.

If the loading is not too high so that $\varepsilon_m < \varepsilon_A$, then, as shown in Fig. 9.5, first a series of elastic–plastic longitudinal waves propagates in the semi-infinite straight string, behind which a "constant wave region" forms. Within this region, a slower, strong discontinuous transverse wave propagates, accompanied with a break of straight string, but both sides across the break point are still kept in a straight line, as shown in Fig. 9.6. The stretch force T and strain ε are no longer varied in the straight line part of the string behind the break point.

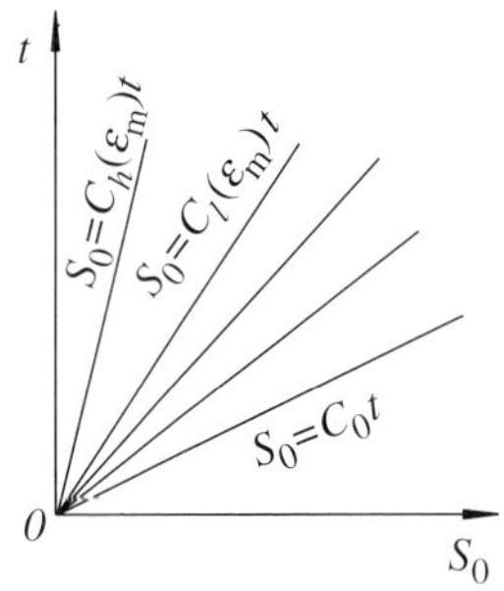

Fig. 9.5. The elastic–plastic longitudinal waves and strong discontinuous transverse wave propagating in a semi-infinite straight string.

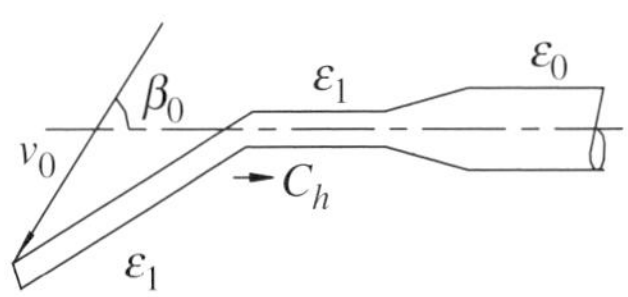

Fig. 9.6. Both sides across the break point of strong discontinuous transverse wave are straight lines.

If the initial strain of string is ε_0 and the initial particle velocity is zero, then for the straight part of string in front of the strong discontinuous transverse wave front, we have

$$v_l = \int_{\varepsilon_0}^{\varepsilon_L} C_l(\varepsilon)d\varepsilon \tag{9.35}$$

According to the equation group (**9-I**), across the strong discontinuous interface we have

$$\rho_0 C_h[v_2\cos(\beta_2+\gamma) - v_1] = T_2\cos\gamma - T_1 \tag{9.36}$$

$$\rho_0 C_h[v_2\sin(\beta_2+\gamma)] = T_2\sin\gamma \tag{9.37}$$

$$v_2\cos(\beta_2+\gamma) - v_1 = C_h[(1+\varepsilon_2)\cos\gamma - (1+\varepsilon_1)] \tag{9.38}$$

$$v_2\sin(\beta_2+\gamma) = C_h(1+\varepsilon_2)\sin\gamma \tag{9.39}$$

For the straight part of the string behind the strong discontinuous transverse wave front, we have

$$v\cos\beta = \text{constant},\ v\sin\beta = \text{constant}$$

By again using the given boundary condition v_0 and β_0 at the impact end, we have

$$v_2 \sin \beta_2 = v_0 \sin(\beta_0 - \gamma) \tag{9.40}$$

$$v_2 \cos \beta_2 = v_0 \cos(\beta_0 - \gamma) \tag{9.41}$$

where β_0 is the intersection angle of the direction of impact velocity v_0 with the initial string, as shown in Fig. 9.6. Since β_0, v_0, ε_0, and $T = T(\varepsilon)$ are known, the seven unknown variables ε_1, v_1, ε_2, v_2, β_2, C_h, and γ can be solved from the seven equations from Eqs. (9.35)–(9.41).

If the loading is high enough so that $\varepsilon_m > \varepsilon_A$, then, as shown in Fig. 9.7, after a series of elastic–plastic longitudinal wave propagation in the semi-infinite straight string, in coincidence with the longitudinal wave of $\varepsilon = \varepsilon_A$, a strong discontinuous transverse wave simultaneously propagates, and, consequently, the string breaks, although both sides of the string across the break point are still kept in a straight line. However, in the straight line part of the string behind the break point, some longitudinal waves still propagate with a velocity slower than that of the transverse wave, and thus the stretch force T and strain ε are still varied. Only after that, finally, there is the "constant wave region", as shown in Fig. 9.8.

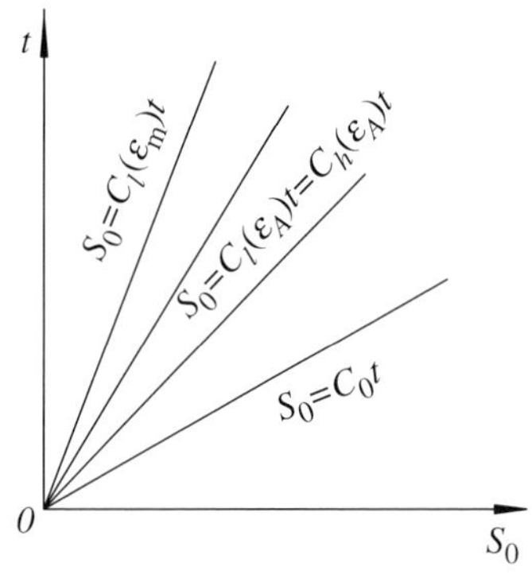

Fig. 9.7. Stress waves propagating in a semi-infinite straight string when $\varepsilon_m > \varepsilon_A$.

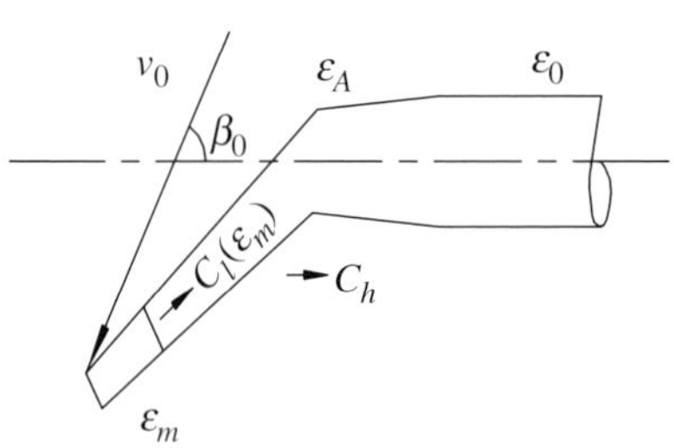

Fig. 9.8. Slower longitudinal waves propagating behind the strong discontinuous transverse wave.

In such cases, Eqs. (9.35)–(9.40) are still held, but Eq. (9.41) should be replaced with the following equation:

$$v_0 \cos(\beta_0 - \gamma) = \int_{\varepsilon_2}^{\varepsilon_m} C_l(\varepsilon) d\varepsilon + v_2 \cos \beta_2 \tag{9.42}$$

Moreover, we have an additional equation to represent the condition that the transverse wave coincides with the longitudinal wave at $\varepsilon = \varepsilon_A$, namely $C_l(\varepsilon_A) = C_h(\varepsilon_A)$, or

$$\left.\frac{dT}{d\varepsilon}\right|_{\varepsilon=\varepsilon_2} = \frac{T(\varepsilon_2)}{1+\varepsilon_2} \tag{9.43}$$

Thus, the eight unknown variables ε_1, v_1, ε_2, v_2, β_2, C_h, γ, and ε_m can be solved by those eight equations, namely Eqs. (9.35)–(9.40), (9.42), and (9.43).

It can be seen, in the self-similar problem, that the transverse wave and longitudinal wave can be dealt with separately, so that solving of the problem becomes much simpler.

In engineering applications, the general $T = T(\varepsilon)$ relation is sometimes simplified as a linear hardening one. In such cases, there are three possible situations:

(1) $E_1 > \dfrac{T_s}{1+\varepsilon_s}$ (Fig. 9.9)

(2) $E_1 < \dfrac{T_s}{1+\varepsilon_s}$ (Fig. 9.10)

(3) $E_1 = \dfrac{T_s}{1+\varepsilon_s}$ (Fig. 9.11)

where T_s and ε_s represent yield stress and yield strain, respectively, and E_1 is the linear hardening modulus.

By using the following relation:

$$c_0 = \sqrt{\frac{E}{\rho_0}} \quad \text{and} \quad c_1 = \sqrt{\frac{E_1}{\rho_0}}$$

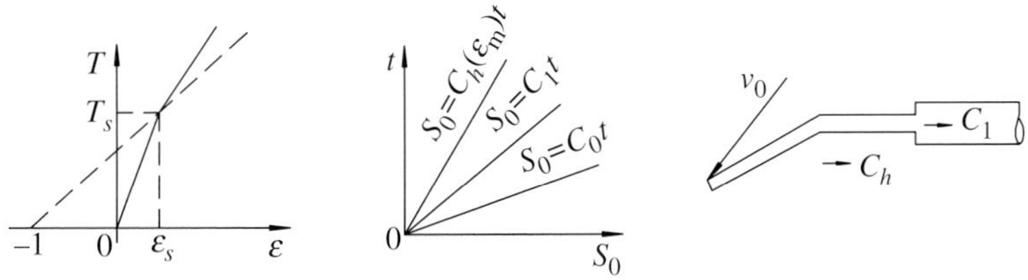

Fig. 9.9. For $E_1 > \dfrac{T_s}{1+\varepsilon_s}$.

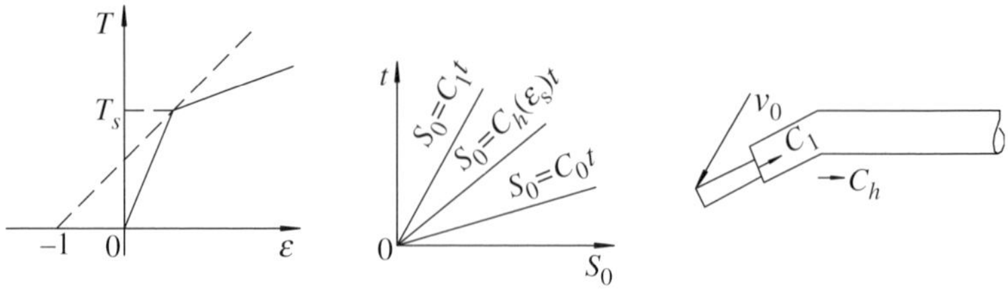

Fig. 9.10. For $E_1 < \dfrac{T_s}{1+\varepsilon_s}$.

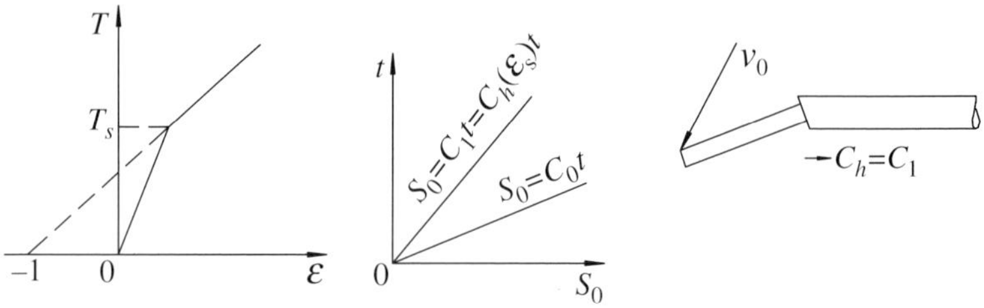

Fig. 9.11. For $E_1 = \dfrac{T_s}{1+\varepsilon_s}$.

the above three situations can also be correspondingly represented as:

(1) $c_0^2\varepsilon_s < c_1^2(1+\varepsilon_s)$

(2) $c_0^2\varepsilon_s > c_1^2(1+\varepsilon_s)$

(3) $c_0^2\varepsilon_s = c_1^2(1+\varepsilon_s)$

where c_0 is the elastic longitudinal wave velocity and c_1 is the plastic longitudinal wave velocity for linear hardening materials, as has been discussed in Chapter 2. Under the boundary condition of abrupt constant loading, both the transverse wave and the longitudinal wave are strong discontinuous waves. Depending on different situations, in the case of (1), the plastic longitudinal wave is faster than the transverse wave, and in the case of (2), the transverse wave is faster than the plastic longitudinal wave, while in the case of (3), both are in coincidence.

For most engineering materials, the situation (1) is usually satisfied. For example, the typical data show that $c_1^2/c_0^2 \approx 0.05$, $\varepsilon_s \approx 0.002$ for steel, and $c_1^2/c_0^2 \approx 0.003$, $\varepsilon_s \approx 0.001$ for copper.

No matter what the situation is, problem solving is much simpler than when including a general $T = T(\varepsilon)$ relation.

9.3 Infinite Straight String under Abrupt Constant-Velocity Oblique Point-Impact

Let us now consider an infinite straight string, which is subjected to an abrupt constant-velocity oblique impact. If the contact arc between the striker and the string is small enough, it could be assumed that the loading is exerted at a point on the string. This is the so-called "**point-impact**". In such cases, as was discussed in the preceding section, it is also a self-similar problem.

However, two different kinds of impact should be distinguished: (1) The impact loading is exerted on a fixed particle of the string, such as $S_0 = 0$, so that there is no relative slide between the striker and the string. Such kind of impact is called "**impact without slide**". (2) The position of impact point on the string is varied with time, namely the striker moves along a certain direction in the space but does not exert the impact loading on a fixed particle of string. So, there is relative slide between the striker and the string. Such kind of impact is called "**impact with slide**".

When an infinite straight string is subjected to an abrupt constant-velocity oblique point-impact, a break point appears at the impact point of string. Such a break point itself is also a strong discontinuous interface. Describing by Lagrange variables, for the impact without slide, the impact point itself does not propagate along the string, so it is a stationary strong discontinuous interface; while for the impact with slide, the impact point is a moving strong discontinuous interface propagating along the string with a certain velocity C_{h0}, which is in general smaller than the transverse wave velocity C_h, as will be seen in the following discussion.

Therefore, for the impact without slide, the problem is equivalent to deal with two semi-infinite straight strings, the left side and the right side of the break point, under impact loading. Such impact problem of semi-infinite straight string is solvable as has been discussed in Section 9.2. However, for the impact with slide, since it is a **varying boundary problem**, the loading boundary itself is undetermined. So, a further discussion about the propagation of impact point itself is required.

The main difference between the impact-point strong discontinuity and the transverse wave strong discontinuity is that there acts external force T_0 on the impact point strong discontinuous interface. Denoting the variables in front of and behind the impact point by subscripts 3 and 4, respectively, as shown in Fig. 9.12, now the dynamic condition

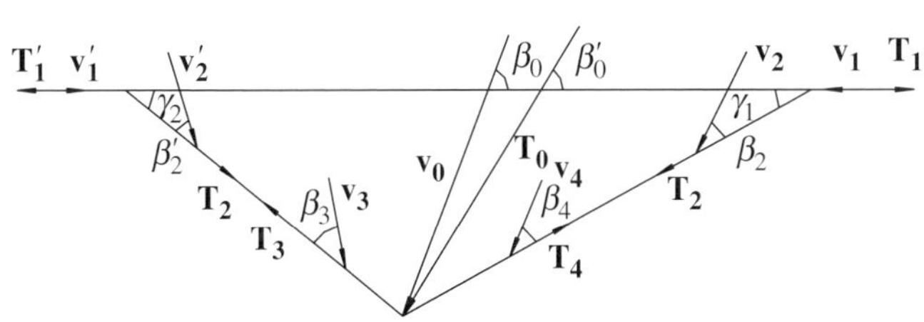

Fig. 9.12. An infinite straight string subjected to abrupt constant-velocity oblique point-impact with slide.

Eq. (9.2) should be changed to:

$$\rho_0 C_{h0}(\mathbf{v_4} - \mathbf{v_3}) = \mathbf{T_4} + \mathbf{T_3} + \mathbf{T_0} \tag{9.44}$$

There is no need for the kinematic condition to change, and, similar to Eq. (9.64), we have

$$\mathbf{v_4} - \mathbf{v_3} = -C_{h0}\left(\frac{\partial \mathbf{u_4}}{\partial s_0} - \frac{\partial \mathbf{u_3}}{\partial s_0}\right) \tag{9.45}$$

Noticing the symbols marked in Fig. 9.12, solving the vector equation (Eqs. (9.44) and (9.45)) into the tangent direction and the normal direction of the vector stretch force $\mathbf{T_3}$, we obtain the following four scalar equations:

$$(\mathbf{9\text{-}V})\begin{cases} \rho_0 C_{h0}[v_4\cos(\pi - \beta_4 - \gamma_1 - \gamma_2) - v_3\cos\beta_3] \\ \qquad = T_4\cos(\gamma_1+\gamma_2) + T_0\cos(\pi - \beta_0' - \gamma_2) - T_3 & (9.46) \\ \rho_0 C_{h0}[v_1\sin(\pi - \beta_4 - \gamma_1 - \gamma_2) - v_3\sin\beta_3] \\ \qquad = -T_4\sin(\gamma_1+\gamma_2) + T_0\sin(\pi - \beta_0' - \gamma_2) & (9.47) \\ v_4\cos(\pi - \beta_4 - \gamma_1 - \gamma_2) - v_3\cos\beta_3 \\ \qquad = C_{h0}[(1+\varepsilon_4)\cos(\gamma_1+\gamma_2) - (1+\varepsilon_3)] & (9.48) \\ v_4\sin(\pi - \beta_4 - \gamma_1 - \gamma_2) - v_3\sin\beta_3 \\ \qquad = -C_{h0}(1+\varepsilon_4)\sin(\gamma_1+\gamma_2) & (9.49) \end{cases}$$

Or solving the vector equation (Eqs. (9.44) and (9.45)) into the tangent direction and the normal direction of the vector stretch force $\mathbf{T_4}$, we obtain

$$(\mathbf{9\text{-}V'})\begin{cases} \rho_0 C_{h0}[v_4\cos\beta_4 - v_3\cos(\pi - \beta_3 - \gamma_1 - \gamma_2)] \\ \qquad = -T_4 + T_0\cos(\beta_0' - \gamma_1) + T_3\cos(\gamma_1+\gamma_2) & (9.50) \\ \rho_0 C_{h0}[v_4\sin\beta_4 - v_3\sin(\pi - \beta_3 - \gamma_1 - \gamma_2)] \\ \qquad = T_0\sin(\beta_0' - \gamma_1) - T_3\sin(\gamma_1+\gamma_2) & (9.51) \\ v_4\cos\beta_4 - v_3\cos(\pi - \beta_3 - \gamma_1 - \gamma_2) \\ \qquad = -C_{h0}[(1+\varepsilon_4) - (1+\varepsilon_3)\cos(\gamma_1+\gamma_2)] & (9.52) \\ v_4\sin\beta_4 - v_3\sin(\pi - \beta_3 - \gamma_1 - \gamma_2) \\ \qquad = -C_{h0}(1+\varepsilon_3)\sin(\gamma_1+\gamma_2) & (9.53) \end{cases}$$

Obviously, among the four equations of dynamic condition, namely Eqs. (9.46), (9.47), (9.50), and (9.51), arbitrary two are independent. Also, among the four equations of kinematic condition, namely Eqs. (9.48), (9.49), (9.52), and (9.53), arbitrary two are independent. Thus, we have totally four independent scalar equations.

Note that, in general, $\beta_0' \neq \beta_0$, namely the $\mathbf{T_0}$ and $\mathbf{v_0}$ are not necessarily in the same direction.

From Eqs. (9.47) and (9.49), and from Eqs. (9.51) and (9.53), the expression of C_{h0} can be obtained as

$$\begin{aligned} C_{h0}^2 &= \frac{1}{\rho_0} \frac{T_4 \sin(\gamma_1+\gamma_2) - T_0 \sin(\beta_0'+\gamma_2)}{(1+\varepsilon_4)\sin(\gamma_1+\gamma_2)} \\ &= \frac{1}{\rho_0} \frac{T_3 \sin(\gamma_1+\gamma_2) - T_0 \sin(\beta_0'-\gamma_1)}{(1+\varepsilon_3)\sin(\gamma_1+\gamma_2)} \end{aligned} \tag{9.54}$$

Assuming $T_3 \neq T_4$, and performing a further mathematical calculation, we obtain

$$C_{h0}^2 = \frac{1}{\rho_0} \frac{T_4 \sin(\beta_0'-\gamma_1) - T_3 \sin(\beta_0'+\gamma_2)}{(1+\varepsilon_4)\sin(\beta_0'-\gamma_1) - (1+\varepsilon_3)\sin(\beta_0'+\gamma_2)} \tag{9.55}$$

$$T_0 = \frac{T_4(1+\varepsilon_3) - T_3(1+\varepsilon_4)}{(1+\varepsilon_3)\sin(\beta_0'+\gamma_2) - (1+\varepsilon_4)\sin(\beta_0'-\gamma_1)} \sin(\gamma_2+\gamma_1) \tag{9.56}$$

In addition, according to Eqs. (9.68) and (9.55), we have

$$C_{h0}^2 = C_h^2(\varepsilon_4) - \frac{T_0}{\rho_0(1+\varepsilon_4)} \frac{\sin(\beta_0'+\gamma_2)}{\sin(\gamma_1+\gamma_2)} = C_h^2(\varepsilon_3) - \frac{T_0}{\rho_0(1+\varepsilon_3)} \frac{\sin(\beta_0'-\gamma_1)}{\sin(\gamma_1+\gamma_2)}$$

Thus, it can be seen, $C_{h0}^2 < C_h^2(\varepsilon_4)$, $C_{h0}^2 < C_h^2(\varepsilon_3)$. It means that the impact point propagates relatively slower in the case of impact with slide.

Eliminating C_{h0} and T_0 from the equation group (**9-V**) and (**9-V′**), two equations to relate other variables can be obtained. For example, from Eqs. (9.49) and (9.53), we can obtain

$$\frac{v_4 \sin(\beta_4+\gamma_1+\gamma_2) - v_3 \sin\beta_3}{(1+\varepsilon_4)} = \frac{v_4 \sin\beta_4 - v_3 \sin(\beta_3+\gamma_1+\gamma_2)}{(1+\varepsilon_3)} \tag{9.57}$$

and eliminating T_0 from Eqs. (9.46) and (9.47), then eliminating C_{h0} from Eq. (9.55), we can obtain

$$\begin{aligned} &\rho_0[v_4 \sin(\beta_4+\gamma_1-\beta_0') - v_3 \sin(\beta_3+\gamma_2+\beta_0')]^2 = [T_4 \sin(\beta_0'-\gamma_1) - T_3 \sin(\beta_0'+\gamma_2)] \\ &\cdot[(1+\varepsilon_4)\sin(\beta_0'-\gamma_1) - (1+\varepsilon_3)\sin(\beta_0'+\gamma_2)] \end{aligned} \tag{9.58}$$

It should be pointed out that the particle velocities across the strong discontinuous impact point are discontinuous $v_3 \neq v_4$, and, moreover v_3 and v_4 are all unequal to v_0. However, the normal component of both impact velocity and particle velocity at the impact point should be equal to each other. This is the so-called **contact condition**. According to this condition, similar to the boundary condition (Eq. (9.40)), we have

$$v_4 \sin\beta_4 = v_0 \sin(\beta_0 - \gamma_1) \tag{9.59}$$

$$v_3 \sin\beta_3 = v_0 \sin(\beta_0 + \gamma_2) \tag{9.60}$$

In fact, $\bar{v}_0$ is also the Euler velocity of impact point propagating in spatial coordinates. Similar to that discussed in Section 2.4, the following relation between Euler wave velocity and Lagrange wave velocity should be satisfied:

$$\mathbf{v_0} = \mathbf{v_3} + (1+\varepsilon_3)C_{h0}\mathbf{S_3} = \mathbf{v_4} + (1+\varepsilon_4)C_{h0}\mathbf{S_4}$$

where $\mathbf{S_3}$ and $\mathbf{S_4}$ are unit vectors in the directions of $\mathbf{T_3}$ and $\mathbf{T_4}$, respectively. By using this relation, Eqs. (9.59) and (9.60) can also be obtained immediately.

Thus, in the case of $\varepsilon_m > \varepsilon_A$, synthesizing all the discussions given above, for the string on the right side of impact point as shown in Fig. 9.12, we must have

$$\textbf{(9-VI)}\begin{cases} v_1 = \displaystyle\int_{\varepsilon_0}^{\varepsilon_1} c(\varepsilon)d\varepsilon & (a) \\ \rho_0 C_h[v_2\cos(\beta_2+\gamma_1) - v_1] = T_2\cos\gamma_1 - T_1 & (b) \\ \rho_0 C_h[v_2\sin(\beta_2+\gamma_1)] = T_2\sin\gamma_1 & (c) \\ v_2\cos(\beta_2+\gamma_1) - v_1 = C_h[(1+\varepsilon_2)\cos\gamma_1 - (1+\varepsilon_1)] & (d) \\ v_2\sin(\beta_2+\gamma_1) = C_h(1+\varepsilon_2)\sin\gamma_1 & (e) \\ v_4\cos\beta_4 = \displaystyle\int_{\varepsilon_2}^{\varepsilon_4} c(\varepsilon)d\varepsilon + v_2\cos\beta_2 & (f) \\ \left.\dfrac{dT}{d\varepsilon}\right|_{\varepsilon=\varepsilon_2} = \dfrac{T(\varepsilon_2)}{1+\varepsilon_2} & (g) \\ v_4\sin\beta_4 = v_2\sin\beta_2 & (h) \end{cases}$$

For the string on the left side of impact point, there are eight similar equations, which can be obtained from **(9-VI)** by using $v_1', \varepsilon_1', v_2', \varepsilon_2', \beta_2', T_1', T_2', \gamma_2, C_h', v_3, \beta_3$, and ε_3 to correspondingly replace $v_1, \varepsilon_1, v_2, \varepsilon_2, \beta_2, T_1, T_2, \gamma_1, C_h, v_4, \beta_4$, and ε_4, respectively . Since the relation $T = T(\varepsilon)$ is a known function, there are 20 unknown variables in the above 16 equations. At the impact point, after eliminating C_{h0} and T_0, there are four more equations, Eqs. (9.57)–(9.60), although a new unknown variable β_0' is introduced. Thus, we have a total of 21 unknown variables but 20 equations, which are established from the dynamic conditions, kinematic conditions, and boundary conditions at the impact point. In order to solve the problem, a supplement equation is required, which should come from the **friction condition** at the impact point in the case of impact with slide.

Note that the equation group **(9-IV)** established is based on a more complicated case of $\varepsilon_m > \varepsilon_A$. If in the case of $\varepsilon_m < \varepsilon_A$, the transverse wave will propagate in a constant wave region, then the last three equations in **(9-IV)** can be replaced by $\varepsilon_4 = \varepsilon_2$, $v_4 = v_2$, and $\beta_4 = \beta_2$, and solving of the problem therefore becomes simpler.

Now, let us discuss the friction condition.

Assume that the friction between the striker and the string follows the Column's friction theorem. In other words, the friction force equals the component of external force $\mathbf{T_0}$ in the normal direction of string multiplied by the friction coefficient f. However, in the case of point impact, the impact point is a break point. Mathematically, neither normal direction nor tangent direction is meaningful with regard to a break point. Therefore, we meet difficulty in applying the Column's friction theorem to the point impact. For this reason, we first regard the impact point as an infinitesimal arc with radius R and included angle γ. Of course, either the normal direction or the tangent direction is determinable at each point of this infinitesimal arc. Meanwhile, we regard the external force $\mathbf{T_0}$ as the resultant force of the external load distributed on this arc. If denote the distributed normal load by N, the distributed friction force by F, and the friction coefficient by f, then, according to the Column's friction theorem, we have

$$F = f \cdot N$$

Let the angular bisector of γ be the Y axis and its normal direction the X axis, the resultant of N as $\mathbf{P} = P_x\mathbf{i} + P_y\mathbf{j}$, and the resultant of F as $\mathbf{Q} = Q_x\mathbf{i} + Q_y\mathbf{j}$. Then we have

$$P_x = \int_{-\gamma/2}^{\gamma/2} RN \sin\alpha d\alpha = 0$$

$$P_y = \int_{-\gamma/2}^{\gamma/2} RN \cos\alpha d\alpha = RN\gamma$$

$$Q_x = \int_{-\gamma/2}^{\gamma/2} RF \cos\alpha d\alpha = RF\gamma = fRN\gamma$$

$$Q_y = \int_{-\gamma/2}^{\gamma/2} RF \sin\alpha d\alpha = 0$$

Thus, it can be seen that: (1) $\mathbf{P} = P_y\mathbf{j} = RN\gamma\mathbf{j}$, i.e. the resultant of distributed normal load is in the direction of Y axis, or the direction of the angular bisector of γ; (2) $\mathbf{Q} = Q_x\mathbf{i} = fRN\gamma\mathbf{i}$, i.e. the resultant of distributed friction force is in the direction of X axis, or the normal direction of the angular bisector of γ; (3) $|\mathbf{Q}| = f\,|\mathbf{P}|$, i.e. when the Column's friction theorem is expressed in the resultant form herein, the normal direction is equivalent to the direction of the angular bisector of γ while the tangent direction is equivalent to its normal direction. When $R \to 0$, the infinitesimal arc approaches a point. This is just the situation of point impact. Thus, from now on, the normal direction and the tangent direction at the break point, in the sense of applying the Column's friction theorem, can be understood as the direction of the angular bisector of the included angle of string at the break point and its normal direction, respectively, as shown in Fig. 9.13. Mathematically, they are just the average of the normal and tangent of the two sides of the included angle.

When the component of external force $\mathbf{T_0}$ in the tangent direction of string is smaller than friction force, the striker cannot slide along the string. This belongs to the problem of impact without slide, which has been solved, as discussed previously. Only when the tangent component of $\mathbf{T_0}$ is large enough to overcome the friction force, the striker will slide along the string. In other words, in the case of impact with slide, the included angle

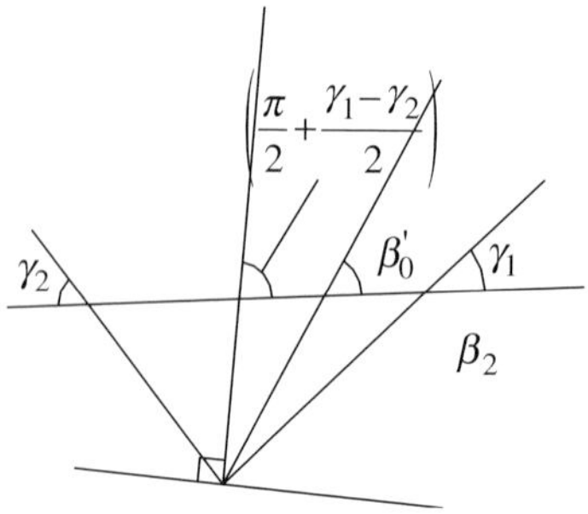

Fig. 9.13. The angular bisector of the included angle of string at the break point and its normal direction.

between the external force $\mathbf{T_0}$ and the normal direction of string $[(\pi/2)+((\gamma_1-\gamma_2)/2)-\beta_0']$ equals the friction angle φ, i.e.

$$\frac{\pi}{2}+\frac{\gamma_1-\gamma_2}{2}-\beta_0'=\varphi=arctg\, f \tag{9.61}$$

Having such a supplement equation, the problem is solvable.

It seems very complicated to solve 21 unknown variables by 21 equations. But, in practice, by using Eqs. (9.59) and (9.60), all the other unknown variables in the equation group (**9-VI**) can be expressed as the functions of only ε_1 and ε_1', and then substituting into Eqs. (9.57), (9.58), and (9.61), the problem is finally reduced to solve three unknown variables ε_1, ε_1', and β_0' by three equations.

The simplest case is, of course, the case without friction. Then, let $f=0$ in Eq. (9.61), and we have

$$\beta_0'=\frac{\pi}{2}+\frac{\gamma_1-\gamma_2}{2} \tag{9.62}$$

It indicates that the direction of external force $\mathbf{T_0}$ is just the direction of angular bisector the included angle at impact point. Substituting Eq. (9.62) into Eq. (9.54), we have

$$T_3=T_4,\quad \varepsilon_3=\varepsilon_4 \tag{9.63}$$

This indicates that without friction, the stretch forces in the left-side and right-side string across the impact point are equal, and the same as for the strains. In addition, it could be imagined that, without friction, any oblique impact under the condition of $\beta_0\neq(\pi/2)$ is a kind of impact with slide. Having Eq. (9.63), the β_0' is not necessary to be taken into consideration in solving the problem. Because, from the equation group (**9-VI**) and, moreover, from Eqs. (9.57), (9.59), (9.60), and (9.63), there are 20 equations, enough to solve the 20 unknown variables.

The problem of infinite straight string subjected to abrupt constant-velocity oblique point impact was first studied by Рахматулин (Rakhmatulin) (1945b). However, as one of the basic equations, Euler equation, which is usually used in mechanical engineering to describe the friction between the pulley and the rope, was adopted by him to describe the dynamic condition at impact point, as

$$T_4 - \rho_0 u_s^2 = e^{f(\gamma_1+\gamma_2)}(T_3 - \rho_0 u_s^2) \tag{9.64}$$

Note that Eq. (9.64) is deduced from the static equilibrium condition, while the problem we discussed herein is a dynamic problem. Only in the case without friction, a result of $T_3 = T_4$, which occasionally coincides with Eq. (9.63), can be obtained from Eq. (9.64). Such a coincidence provides us with a possibility to make use of all results obtained by Рахматулин (Rakhmatulin) under the condition without friction.

In the following, let us discuss more concretely the situation of impact with slide but without friction. For convenience, it is limited to discuss the situation of $\varepsilon_m < \varepsilon_A$. In such cases, the equation group (**9-VI**), together with Eqs. (9.57), (9.59), (9.60), and (9.62), after some mathematical calculations, transform to the following group of equations:

$$v_1 = \int_{\varepsilon_0}^{\varepsilon_1} c(\varepsilon)d\varepsilon$$

$$T_2 = T_1,\ \varepsilon_1 = \varepsilon_2$$

$$C_h = \sqrt{\frac{1}{\rho_0}\frac{T_1}{1+\varepsilon_1}}$$

$$v_2 \cos\beta_2 - v_1 \cos\gamma_2 = C_h(1+\varepsilon_1)(1-\cos\gamma_1)$$

$$v_2 \sin\beta_2 + v_1 \sin\gamma_2 = C_h(1+\varepsilon_1)\sin\gamma_1$$

$$v_4 = v_2$$

$$\beta_4 = \beta_2$$

$$\varepsilon_4 = \varepsilon_2,\ T_4 = T_2$$

$$T_3 = T_4,\ \varepsilon_3 = \varepsilon_4$$

$$v_2 \sin\beta_2 = v_0 \sin(\beta_0 - \gamma_1)$$

$$v_2' \sin\beta_2' = v_0 \sin(\beta_0 + \gamma_2)$$

$$v_2 \sin(\beta_2 + \gamma_1 + \gamma_2) - v_2' \sin\beta_2' = v_2 \sin\beta_2 - v_2' \sin(\beta_2' + \gamma_1 + \gamma_2)$$

The first seven equations are deduced from Eqs. (**9-VI**) for the right-side string. When we consider the left-side string, there are seven other similar equations. After a further arrangement, the unknown variables can be reduced to $\varepsilon(=\varepsilon_4=\varepsilon_2=\varepsilon_1=\varepsilon_3=\varepsilon_2'=\varepsilon_1')$, $v_1(=v_1')$, $b(=b')$, $v_2(=v_4)$, $\beta_2(=\beta_4)$, $v_2'(=v_2)$, $\beta_2'(=\beta_3)$, and γ_2. Those nine unknown variables can be solved by the following nine equations:

$$
\textbf{(9-VII)}\ \begin{cases}
v_1 = \int_{\varepsilon_0}^{\varepsilon} c(\varepsilon)d\varepsilon & (a)\\
C_h = \sqrt{\dfrac{1}{\rho_0}\dfrac{T}{1+\varepsilon}} & (b)\\
v_2\cos\beta_2 - v_1\cos\gamma_2 = C_h(1+\varepsilon)(1-\cos\gamma_1) & (c)\\
v_2\sin\beta_2 + v_1\sin\gamma_1 = C_h(1+\varepsilon)\sin\gamma_1 & (d)\\
v_2'\cos\beta_2' - v_1\cos\gamma_2 = C_h(1+\varepsilon)(1-\cos\gamma_1) & (e)\\
v_2'\sin\beta_2' + v_1\sin\gamma_2 = C_h(1+\varepsilon)\sin\gamma_2 & (f)\\
v_2\sin\beta_2 = v_0\sin(\beta_0-\gamma_1) & (g)\\
v_2'\sin\beta_2' = v_0\sin(\beta_0+\gamma_2) & (h)\\
v_2\sin\beta_2\cos(\gamma_1+\gamma_2) + v_2\cos\beta_2\sin(\gamma_1+\gamma_2) - v_2'\sin\beta_2' & \\
\quad = v_2\sin\beta_2 - v_2'\sin\beta_2'\cos(\gamma_1+\gamma_2) - v_2'\cos\beta_2'\sin(\gamma_1+\gamma_2) & (i)
\end{cases}
$$

Among these equations, from Eqs. (d) and (g) above, we obtain

$$
tg\gamma_1 = \frac{v_0\sin\beta_0}{C_h(1+\varepsilon) - v_1 + v_0\cos\beta_0} = \frac{v_0\sin\beta_0}{c_h + v_0\cos\beta_0} \tag{9.65a}
$$

where

$$
c_h = C_h(1+\varepsilon) - v_1 \tag{9.65b}
$$

Its physical meaning is Euler velocity of transverse wave. Similarly, from Eqs. (e) and (h), we obtain

$$
tg\gamma_2 = \frac{v_0\sin\beta_0}{C_h(1+\varepsilon) - v_1 - v_0\cos\beta_0} = \frac{v_0\sin\beta_0}{c_h - v_0\cos\beta_0} \tag{9.66}
$$

From Eqs. (9.65) and (9.66), the following equations can be solved:

$$
\left\{\begin{aligned}
v_0\cos\beta_0 &= \frac{c_h(tg\gamma_2 - tg\gamma_1)}{tg\gamma_1 + tg\gamma_2}\\
v_0\sin\beta_0 &= \frac{2c_h tg\gamma_1 tg\gamma_2}{tg\gamma_1 + tg\gamma_2}\\
tg\beta_0 &= \frac{2tg\gamma_1 tg\gamma_2}{tg\gamma_2 - tg\gamma_1}\\
ctg\gamma_1 &= 2ctg\beta_0 + ctg\gamma_2
\end{aligned}\right\} \tag{9.67}
$$

In practice, for the certain known ε_0 and β_0, first presume a γ_1, and calculate γ_2 from Eq. (9.67). For example, some results calculated by such way are listed in Table 9.1 [Рахматулин (Rakhmatulin), 1945b].

Table 9.1. Values of γ_1.

γ_2 / β_0	10°	13°	18°	23°	30°	35°	40°	50°	55°
90°	10°00′	13°00′	18°00′	23°00′	30°00′	35°00′	40°00′	50°00′	—
70°	8°50′	11°10′	14°40′	18°00′	22°10′	24°50′	27°30′	32°30′	—
50°	7°40′	9°20′	11°50′	14°00′	16°20′	17°50′	19°10′	21°40′	—
30°	6°10′	7°20′	8°40′	9°40′	10°50′	11°30′	12°10′	13°00′	13°30′

Since γ_1 and γ_2 are known and v_1 and C_h are functions of only ε, then according to Eqs. (*c*)–(*f*), all $v_2 \cos\beta_2$, $v_2 \sin\beta_2$, $v'_2 \cos\beta'_2$, and $v'_2 \sin\beta'_2$ can be expressed as functions of only ε. Substituting them into Eq. (i), ε can be solved, and thus all other unknown variables can be determined. Finally, from Eq. (9.67), the impact velocity v_0, which is required to bring in such value of ε, can be determined. Such a cut-and-trial method was proposed by Рахматулин (Rakhmatulin), (1945b).

For linear hardening materials, the problem becomes much simpler. Let the stretch force–strain relationship be expressed as

$$\begin{cases} T = E\varepsilon, & \varepsilon < \varepsilon_s \\ T = T_s + E_1(\varepsilon - \varepsilon_s) & \varepsilon > \varepsilon_s \end{cases}$$

where T_s and ε_s are yield stretch force and yield strain, respectively, E the elastic modulus, and E_1 the linear hardening modulus. In such case, Eqs. (a) and (b) are deduced to

$$v_1 = c_0(\varepsilon_s - \varepsilon_0) + c_1(\varepsilon - \varepsilon_s)$$

$$C_h = \sqrt{\frac{c_0^2\varepsilon_s + c_1^2(\varepsilon - \varepsilon_s)}{1+\varepsilon}}$$

where $c_0 = \sqrt{E/\rho_0}$ and $c_1 = \sqrt{E_1/\rho_0}$ denote the elastic wave velocity and plastic wave velocity, respectively. Introducing $\bar{c}_1^2 = (c_1/c_0^2)$, $\overline{B}(=C_h/c_0)$, and $\bar{v}_0(=v_0/c_0)$, let $\varepsilon_s = 0.002$, for two different initial strains ε_0 ($\varepsilon_0 = 0$ and $\varepsilon_0 = 0.001$) and four different impact angles β_0 ($\beta_0 = 30°$, $50°$, $70°$ and $90°$), the numerical results calculated for ε, $\overline{B}$, and $\bar{v}_0$ are given in Table 9.2, and the corresponding plot $\varepsilon = \varepsilon(v_0)$ is given in Fig. 9.14 [Рахматулин (Rakhmatulin), 1945b].

In the case of normal impact, we have

$$\beta_0 = \frac{\pi}{2}$$

$$v_0 = v_2 = v'_2$$

$$\gamma_1 = \gamma_2 = \gamma$$

$$\beta_2 = \beta_0 - \gamma_1 = \frac{\pi}{2} - \gamma$$

$$\beta'_2 = \beta_0 - \gamma_2 = \frac{\pi}{2} + \gamma$$

Table 9.2. Values of ε and $\overline{B}$ calculated under different conditions.

$\beta_0 = 90°$, $\varepsilon_0 = 0$, $\bar{c}_1^2 = 0.05$, $\varepsilon_s = 0.002$

$\bar{v}_0$	0.0135	0.0167	0.0192	0.0245	0.0347	0.045	0.0518	0.064	0.0661	0.0742	0.085
ε	0.0020	0.006	0.010	0.020	0.04	0.060	0.0800	0.100	0.120	0.140	0.160
$\overline{B}$	0.0426	0.0441	0.0453	0.0481	0.0530	0.05696	0.06045	0.0656	0.0655	0.068	0.0707

$\beta_0 = 90°$, $\varepsilon_0 = 0.001$, $\bar{c}_1^2 = 0.05$, $\varepsilon_s = 0.002$

$\bar{v}_0$	0.00978	0.1295	0.0164	0.02035	0.0228	0.0283	0.0391	0.0381	0.0423	0.0511
ε	0.002	0.006	0.010	0.016	0.020	0.030	0.040	0.050	0.060	0.080
$\overline{B}$	0.0436	0.0491	0.0463	0.0481	0.0491	0.0518	0.054	0.0563	0.0579	0.0614

$\beta_0 = 70°$, $\varepsilon_0 = 0.001$, $\bar{c}_1^2 = 0.05$, $\varepsilon_s = 0.002$

$\bar{v}_0$	0.0139	0.0182	0.0254	0.0311	0.0376	0.0554
ε	0.006	0.010	0.0219	0.0326	0.0444	0.0787
$\overline{B}$	0.0450	0.0464	0.0499	0.0525	0.0551	0.0612

$\beta_0 = 50°$, $\varepsilon_0 = 0.001$, $\bar{c}_1^2 = 0.05$, $\varepsilon_s = 0.002$

$\bar{v}_0$	0.0113	0.0148	0.0188	0.02426	0.0286	0.0337	0.0423
ε	0.0002	0.0042	0.008	0.0146	0.0207	0.0323	0.0455
$\overline{B}$	0.0431	0.0445	0.0456	0.0478	0.0496	0.00525	0.0555

$\beta_0 = 30°$, $\varepsilon_0 = 0.001$, $\bar{c}_1^2 = 0.05$, $\varepsilon_s = 0.002$

$\bar{v}_0$	0.0142	0.0183	0.0217	0.0263	0.0297	0.0332	0.036	0.0388	0.0425
ε	0.0004	0.0023	0.0058	0.0082	0.0114	0.015	0.0185	0.0223	0.029
$\overline{B}$	0.043	0.0439	0.0446	0.0458	0.0472	0.0487	0.049	0.051	0.0517

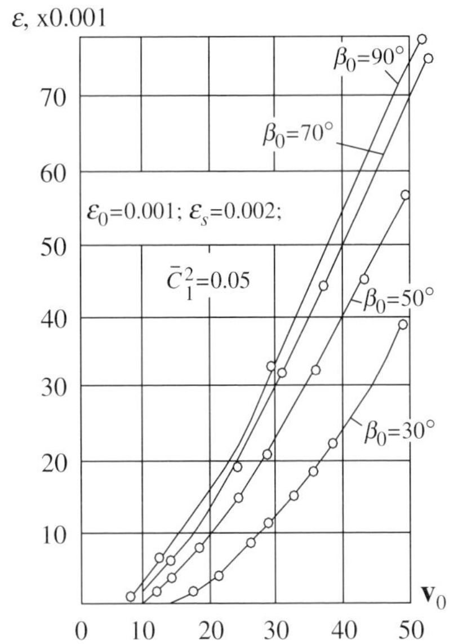

Fig. 9.14. The relation $\varepsilon = \varepsilon(v_0)$ under different impact angle β_0.

Therefore, Eqs. (**9-VII**) are reduced to:

$$
\textbf{(9-VIII)}\quad \begin{cases} v_1 = \displaystyle\int_{\varepsilon_0}^{\varepsilon} c(\varepsilon) d\varepsilon \\ C_h = \sqrt{\dfrac{1}{\rho_0} \dfrac{T}{1+\varepsilon}} \\ v_0 \cos\left(\dfrac{\pi}{2} - \gamma\right) - v_1 \cos\gamma = C_h(1+\varepsilon)(1-\cos\gamma) \\ v_0 \sin\left(\dfrac{\pi}{2} - \gamma\right) + v_1 \sin\gamma = C_h(1+\varepsilon)\sin\gamma \end{cases}
$$

from which four unknown variables ε, v_1, C_h, and γ can be solved. The last two equations, after some mathematical calculations, can be reduced to

$$
\left.\begin{aligned} tg\gamma &= \frac{v_0}{(1+\varepsilon) - v_1} \\ \gamma &= \frac{C_h(1+\varepsilon)}{C_h(1+\varepsilon) - v_1} \end{aligned}\right\} \tag{9.68}
$$

from which, eliminating γ, we have

$$
\left(\frac{C_h(1+\varepsilon)}{C_h(1+\varepsilon) - v_1}\right)^2 - \left(\frac{v_0}{C_h(1+\varepsilon) - v_1}\right)^2 = 1
$$

or

$$v_0^2 = 2C_h(1+\varepsilon)v_1 - v_1^2$$

Substituting the v_1 and C_h of the first two equations in (**9-VIII**) into above equation, for a given v_0, the ε can be solved, and, consequently, C_h, v_1, and γ can also be determined.

For example, as to elastic string, $T = E\varepsilon$ we then have

$$v_1 = c_0(\varepsilon - \varepsilon_0)$$

$$C_h = c_0\sqrt{\frac{\varepsilon}{1+\varepsilon}}$$

$$\bar{v}_0 = \frac{v_0}{c_0} = \sqrt{2(\varepsilon-\varepsilon_0)\sqrt{\varepsilon(1+\varepsilon)} - (\varepsilon-\varepsilon_0)^2} \tag{9.69}$$

When $\varepsilon_0 \approx 0$, $\varepsilon \ll 1$, from Eq. (9.69), we have

$$\varepsilon \approx \frac{\bar{v}_0^{4/3}}{\sqrt[3]{4}} \tag{9.70}$$

Consequently,

$$b \approx c_0\frac{\bar{v}_0^{2/3}}{\sqrt[3]{2}} = 0.8v_0^{2/3}c_0^{1/3} \tag{9.71}$$

Again, from Eq. (9.68), we obtain

$$tg\gamma \approx \frac{v_0}{0.8v_0^{2/3}c_0^{1/3}} = 1.25\cdot\sqrt[3]{\frac{v_0}{c_0}} \tag{9.72}$$

It is thus clear that while v_0 is small, γ is still large. Assume that the elastic limit strain of material is $\varepsilon_s = 0.002$, and it can be seen from Eq. (9.70) that the normal impact velocity required for the beginning of yield in the string is considerably large, $v_0 \approx 55$ m/s.

9.4 Prestretched Strings Subjected to Transverse Impact

Now, let us discuss the problem of prestretched strings subjected to transverse impact, to see how the prestretch force influences the propagating character of both the longitudinal wave and transverse wave in the string (Wang *et al.*, 1992).

Consider an infinite straight string with original line density ρ_0. The string is originally at rest but prestretched by a force T_0 (and a corresponding prestrain ε_0) and subjected to a constant-velocity normal impact, as shown in Fig. 9.15, where the string is coincident with the X axis, and the impact velocity V is coincident with the Y axis (V is positive in the positive direction of Y axis). Obviously, there must be longitudinal and transverse

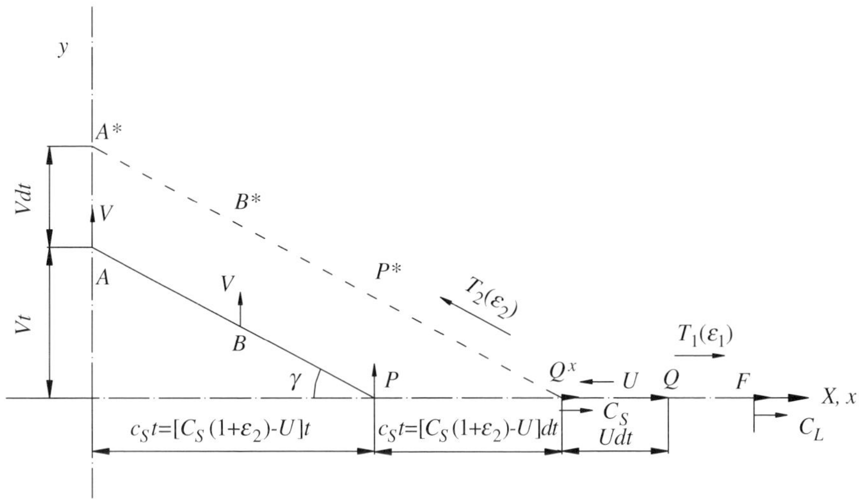

Fig. 9.15. An infinite straight string with prestretched force under transverse impact.

waves in succession, propagating in the string with material velocity (Lagrange velocity) of C_l and C_h, respectively. Due to the symmetry of the problem, we only need to discuss the motion of the half part of the string, $X \geq 0$.

When the problem is described by material coordinates (Lagrange coordinates) for the strong discontinuous longitudinal wave front, we have the following dynamic compatibility condition (momentum conservation) and kinematic compatibility condition (continuity of displacement)

$$[T] = \rho_0 C_l [U]$$

$$[U] = C_l [\varepsilon]$$

where the stretch force T ($=\sigma A_0$) and strain ε are positive for the tensile case, while the particle velocity in the X direction U is positive when it directs to the impact point (i.e. $-X$ direction). Noticing the sign definition for U, it can be seen that the above two equations are actually the dynamic compatibility condition (Eq. 2.57) and kinematic compatibility condition (Eq. 2.55), respectively, deduced in Chapter 2 for the strong discontinuous longitudinal waves in bars.

For the string initially at rest and prestretched by force T_0, if we denote the quantities behind the strong discontinuous longitudinal wave front in string by subscript 1, then the above two equations can be rewritten, respectively, as

$$T_1 - T_0 = \rho_0 C_l U_1 \tag{9.73}$$

$$U_1 = C_l(\varepsilon_1 - \varepsilon_0) \tag{9.74}$$

For the strong discontinuous transverse wave front, the general forms of the dynamic condition and the displacement continuity condition across the wave front have been deduced previously, as expressed by Eq. (**9-I**). In the current case, if we notice $v_1 = U_1$, $v_2 = V$, $b_1 = 0$, $\beta_2 + \gamma = \pi/2$, and denote the quantities behind the strong discontinuous transverse wave front in string by the subscript 2, then Eqs. (**9-I**) can be rewritten as

$$T_2 \cos\gamma - T_1 = -\rho_0 C_h U_1 \tag{9.75}$$

$$T_2 \sin\gamma = \rho_0 C_h V \tag{9.76}$$

$$-U_1 = C_h[(1+\varepsilon_2)\cos\gamma - (1+\varepsilon_1)] \tag{9.77}$$

$$V = C_h(1+\varepsilon_2)\sin\gamma \tag{9.78}$$

Obviously, from Eqs. (9.73) and (9.74), the wave velocity of longitudinal C_l can be immediately obtained:

$$C_l = \sqrt{\frac{1}{\rho_0}\frac{T_1 - T_0}{\varepsilon_1 - \varepsilon_0}} \tag{9.79a}$$

It coincides with the longitudinal wave velocity in the bar (Eq. 2.59).

From Eq. (9.79a), it can be seen that when the constitutive relation of string material $T = T(\varepsilon)$ is linear, the prestretch force T_0 does not influence the longitudinal wave velocity. However, when the constitutive relation of string material $T = T(\varepsilon)$ is nonlinear, the prestretch force T_0 does influence the longitudinal wave velocity. Similar to the situation discussed on the strong discontinuous longitudinal wave in bars, if $(d^2T/d\varepsilon^2) < 0$ [namely, $T = T(\varepsilon)$ is a upwards convex curve], the strong discontinuous longitudinal wave will be transformed into the weak discontinuous longitudinal wave (continuous wave), and, correspondingly, Eq. (9.79) is reduced to

$$C_l = \sqrt{\frac{1}{\rho_0}\frac{dT}{d\varepsilon}} = \sqrt{\frac{A_0}{\rho_0}\frac{d\sigma}{d\varepsilon}} \tag{9.79b}$$

Conversely, if $(d^2T/d\varepsilon^2) > 0$ (namely, $T = T(\varepsilon)$ is a downwards concave curve), then the propagating velocity of strong discontinuous longitudinal wave varies with the prestretch force T_0., since the strong discontinuous wave velocity depends on the slope of Rayleigh line, while the slope of Rayleigh line varies with the position of its initial point, which corresponds to the prestretch force.

For the transverse wave, from Eqs. (9.75) and (9.77), and from Eqs. (9.76) and (9.78), two expression forms of the transverse wave velocity can be obtained:

$$C_h = \sqrt{\frac{1}{\rho_0}\frac{T_2\cos\gamma - T_1}{(1+\varepsilon_2)\cos\gamma - (1+\varepsilon_1)}} \tag{9.80a}$$

$$C_h = \sqrt{\frac{1}{\rho_0}\frac{T_2}{1+\varepsilon_2}} \tag{9.80b}$$

In order to satisfy both these equations, similar to that discussed for Eqs. (9.12)–(9.14), in general they must have

$$T_2 = T_1 \tag{9.81a}$$

$$\varepsilon_2 = \varepsilon_1 \tag{9.81b}$$

It means that the transverse wave only induces string shape change but not strain disturbance.

Thus, the current problem, including all the momentum conservation conditions and displacement continuity conditions for both longitudinal and transverse waves, can finally be summed up as solving the following four independent equations:

$$\rho_0 C_l U = T - T_0 = T_d \tag{9.82a}$$

$$U = C_l(\varepsilon - \varepsilon_0) = C_l \varepsilon_d \tag{9.82b}$$

$$\rho_0 C_h V = T \sin\gamma \tag{9.82c}$$

$$V = C_h(1+\varepsilon)\sin\gamma \tag{9.82d}$$

Herein, for convenience, by noticing Eq. (9.81), the subscripts 1 and 2 are omitted.

Obviously, when the line density ρ_0 and the constitutive relation $T = T(\varepsilon)$ of the string are known, and the prestretch force T_0 (and, correspondingly, the prestrain ε_0) and the transverse impact velocity V are given, the above four independent equations contain five unknown variables: C_l, U, T, C_h, and γ. Therefore, once any one of them can be experimentally determined, the other four can all be solved.

Now, let us discuss an example of how to measure the wave velocity in a prestretched string.

9.4.1 Experimental investigation of wave velocity in a prestretched string

The propagation of longitudinal and transverse waves in the prestretched string was experimentally investigated by Wang (1992). The schematic of experiment is shown in Fig. 9.16.

A gas gun system was used to fire projectiles at vertically suspended yarns with length L. The projectile is composed of a nylon sabot with an attached sharp blade, cut from a razor blade, as shown in Fig. 9.17. A weight is suspended underneath the yarn to apply a prestretch force T_0, which can be adjusted by changing the weight. The impact velocity of projectile V is determined by measuring the time difference Δt_0 passing two laser beams with a fixed distance ΔY_0, namely $V = \Delta Y_0/\Delta t_0$. Two piezoelectric sensors are connected on the top end and lower end of the yarn, respectively. The distance from the impact point to the top sensor L_1 is larger than the distance from the impact point to the lower sensor $L_2 = L - L_1$, so that the longitudinal wave velocity C_l can be directly

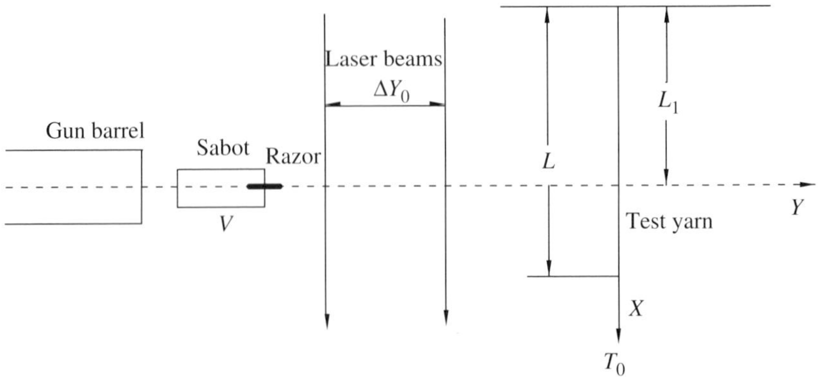

Fig. 9.16. Schematic of the experiment set-up.

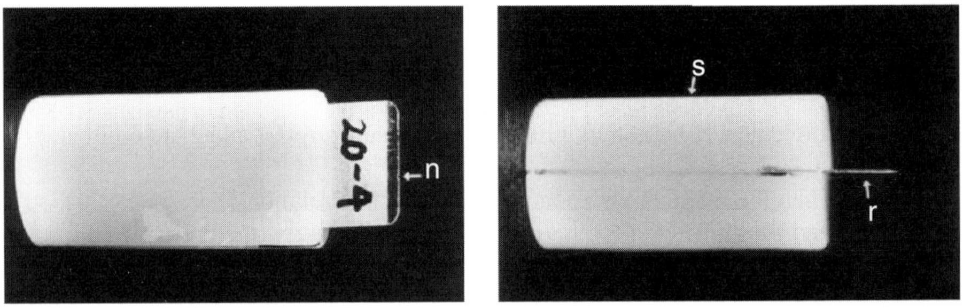

(a) (b)

Fig. 9.17. The projectile attached a sharp razor blade.

measured, $C_l = (L_1 - L_2)/(t_1 - t_2)$, where t_1 and t_2 are the arrival times of longitudinal wave at top sensor and lower sensor, respectively.

A typical oscilloscope record for the longitudinal wave velocity measurement for Kevlar yarn under the transverse impact at $V = 81$ m/s is shown in Fig. 9.18. From the recorded signal jump corresponding to the arrival of the stress wave, it can be deduced that the recorded longitudinal wave is a strong discontinuous one. It means that the constitutive relation $T = T(\varepsilon)$ of the tested material, Kevlar, is either linear or nonlinear and downwards-concave. If it is the former, then the longitudinal wave velocity does not vary with the stretch force; while if it is the latter, the longitudinal wave velocity increases with increasing stretch force.

The experimentally measured longitudinal wave velocities for Kevlar under the conditions of different prestretch forces and different impact velocities are collected in Table 9.3. As can be seen, under the same prestretch force T_0, the longitudinal wave velocity C_l almost does not vary with the impact velocity V, but under the same impact velocity V,

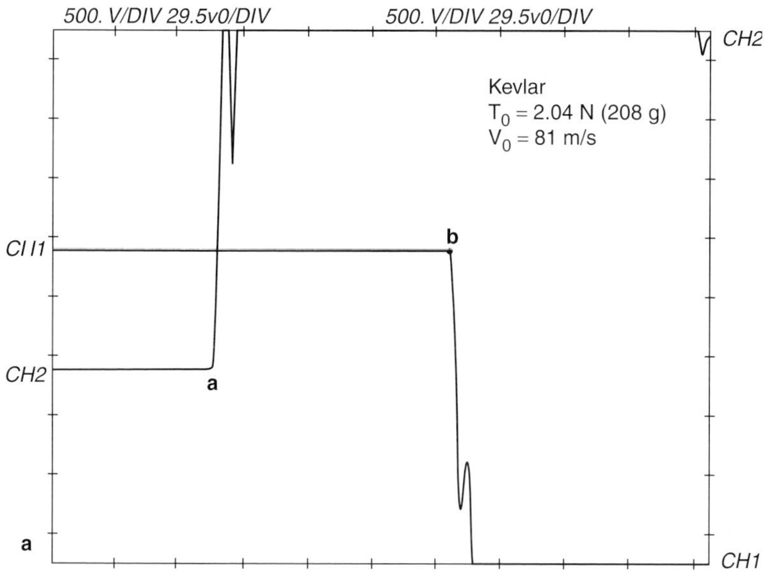

Fig. 9.18. Typical oscilloscope record for the longitudinal wave velocity measurement for Kevlar yarn under the transverse impact at $V = 81$ m/s. (Wang, 1992)

Table 9.3. The longitudinal wave velocities C_l for Kevlar under the conditions of different prestretch forces T_0 (prestress σ_0) and different impact velocities V.

T_0 (N)	2.04	3.02	5.96	10.9	10.9	10.9	10.9	20.7	30.5
σ_0 (MPa)	32	47	93	170	170	170	170	324	478
V (m/s)	81	79	81	54.5	81	138	170	82	81
C_l (km/s)	9.04	9.13	9.22	9.35	9.35	9.36	9.27	9.52	9.79

the C_l of Kevlar increases with increasing prestretch force, showing an approximately linear relation, as shown by the oblique line k in Fig. 9.19. In the same figure, the measured $C_l - T_0$ relations for the polymer yarn Spectra and the Ni-Cr alloy string are also given, shown by the oblique line s and the horizontal line n, respectively. Thus, under the present test conditions, the $T = T(\varepsilon)$ relation for both polymer yarn Kevlar and Spectra are nonlinear and downwards-concave, while for Ni-Cr alloy is linear.

A high-speed image converter camera (Imacon 792), operating at a framing interval of 20 μs, was used to record the break angle γ and the failure processes of the test yarns. Typical high-speed sequences for impacts on Kevlar yarns with prestress of 170 MPa at impact velocity of (a) 55 m/s and (b) 170 m/s are given in Figs. 9.20(a and b), respectively (Wang *et al.*, 1992).

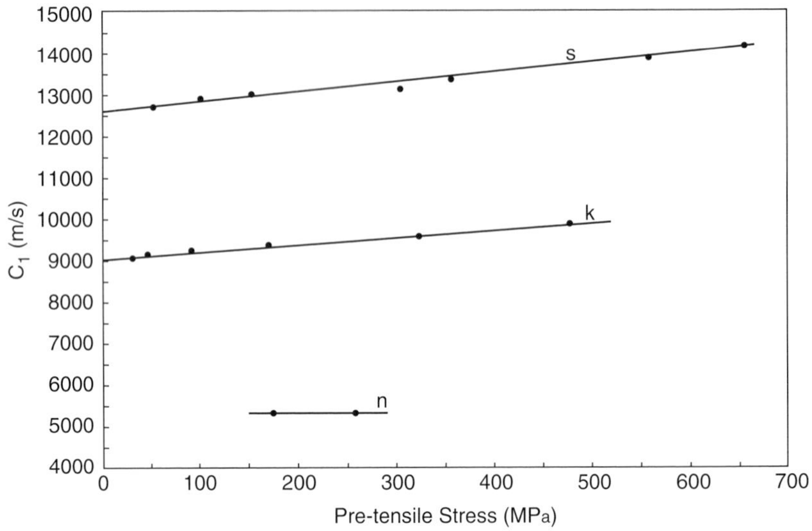

Fig. 9.19. Experimental results for Kevlar (curve k), Spactra (curve s), and Ni-Cr alloy (curve n), showing how C_l varies with T_0.

It is evident that the failure process of a Kevlar yarn under transverse impact is going on by a series of successive broken process of one fiber by one fiber. When a fiber is broken, it springs back, as the stored tension is released. Under a given prestretch force, with increasing the impact velocity, the break angle γ increases, and the yarn is broken in a shorter time.

9.4.2 Experimental study on constitutive relationship of string materials

Once the longitudinal wave velocities of string with different prestretch forces are experimentally determined, the instantaneous stress–strain relation can be deduced. This is similar to solving the second kind of inverse problems previously discussed on the stress wave propagating in bars, i.e. to inversely solve the constitutive relation of materials from the measured wave signals.

For the polymer yarn Kevlar, the measured data of C_l versus T_0 shown in Fig. 9.19, by means of the least square fitting, can be expressed as the following linear relation

$$C_l = 9.10 + 1.56 \times 10^{-3}\sigma_0 \tag{9.83}$$

where the unit of C_l is km/s and the unit of σ_0 is MPa. Substituting the above equation into Eq. (9.79b) and then integrating it, the instantaneous nonlinear $\sigma = \sigma(\varepsilon)$ relation

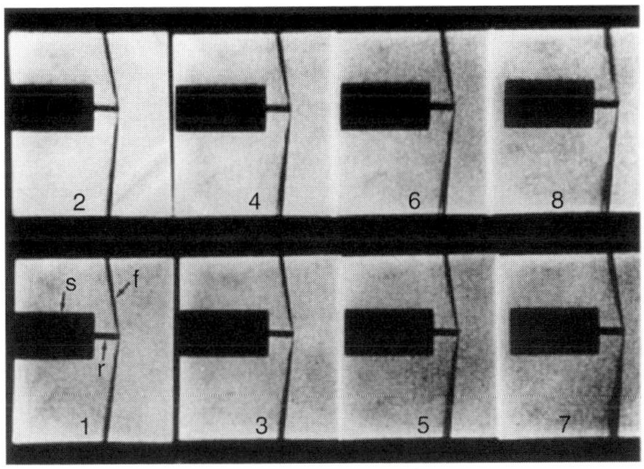

(a)

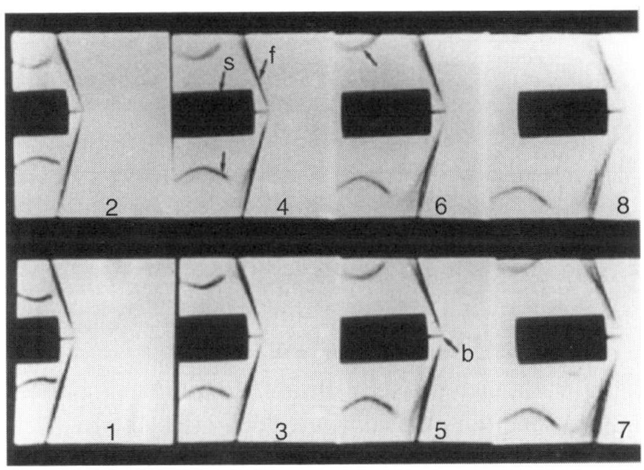

(b)

Fig. 9.20. High-speed sequences for impacts on Kevlar yarns with prestress of 170 MPa at impact velocity of (a) 55 m/s and (b) 170 m/s. Interframe time 20 μs.

can be expressed as

$$\sigma = 5.85 \times 10^3 \left(\frac{1}{1 - 20.8\varepsilon} - 1 \right) \tag{9.84a}$$

or

$$\varepsilon = \frac{\sigma}{20.8\sigma + 1.216 \times 10^5} \tag{9.84b}$$

which is plotted in Fig. 9.21 by the solid line. In the same figure, the instantaneous nonlinear $\sigma = \sigma(\varepsilon)$ relation for polymer yarn Spectra, which is determined by the same method, is plotted by dash line for comparison.

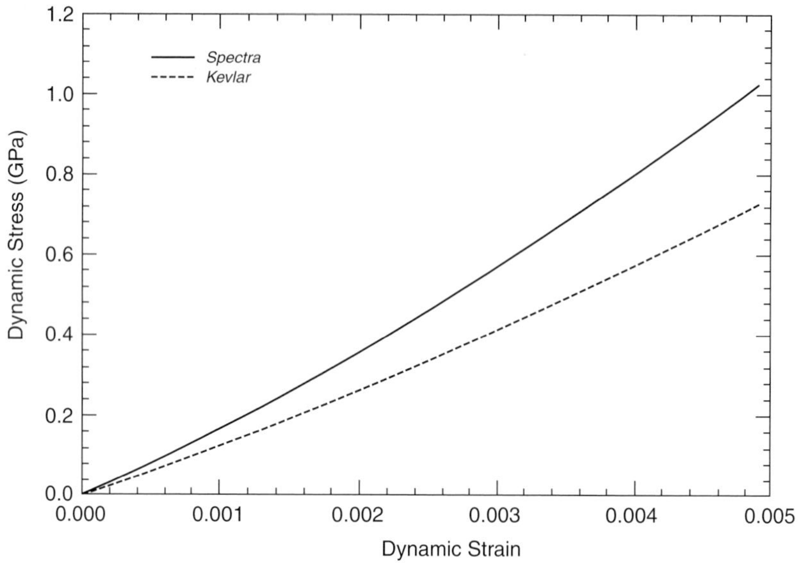

Fig. 9.21. The instantaneous nonlinear $\sigma = \sigma(\varepsilon)$ relations for Kevlar and Spectra.

9.4.3 Determination of transverse wave velocity from longitudinal wave velocity measurements

As can be seen in Fig. 9.15, there exists a simple geographical relation between the transverse impact velocity V and the spatial wave velocity (Euler wave velocity) of transverse wave c_h as

$$V = c_h \tan \gamma \tag{9.85}$$

On the other hand, similar to the derivation of Eq. (2.11), in the current situation, there should exist the following relation between the wave velocity c_h and Lagrange wave velocity C_h (ref. Eq. 9.65b):

$$c_h = C_h(1 + \varepsilon) - U \tag{9.86}$$

The above two equations together with Eqs. (9.82a,b,c,d) are totally six equations, which contain 11 variables: ρ_0, C_l, C_h, c_h, U, V, γ, T, ε, T_0, and ε_0. In general, ρ_0, T_0, and ε_0 are known, and the transverse impact velocity can be measured in the experiment, so the number of remained unknown variables is seven. Therefore, once the longitudinal wave velocity C_l can be measured in the experiment, the other six unknown variables can be determined by the above six equations.

Let us consider Kevlar yarn as an example. According to the measured data of longitudinal wave velocity C_l, all other variables calculated are collected in Table 9.4. Note that for the break angle γ, which is an important quantity characterizing the transverse wave

Table 9.4. The measured longitudinal wave velocity C_l and the calculated other variables for Kevlar under the conditions of different prestretch forces and different transverse impact velocities (Wang, 1992).

T_0 (N)	2.04	3.02	5.96	10.9	10.9	10.9	10.9	20.7	30.5
σ_0 (MPa)	32	47	93	170	170	170	170	324	478
ε_0 (με)	262	383	753	1360	1360	1360	1360	2520	3630
V (m/s)	81	79	81	54.5	81	138	170	82	81
C_l (km/s)	9.04	9.13	9.22	9.35	9.35	9.36	9.27	9.52	9.79
c_h (m/s)	324	333	373	388	424	519	567	527	610
U (m/s)	10	9.2	8.7	3.8	7.48	18	24.9	6.3	5.35
T (N)	10.4	11.0	13.6	14.4	17.5	27.0	32.8	26.3	35.4
σ (MPa)	165	171	210	221	273	414	513	409	555
ε (με)	1350	1400	1710	1780	2160	3320	4030	3220	4180
γ_{cal} (deg.)	14	13.3	12.3	8.0	10.7	14.9	16.7	8.8	7.6
γ_{exp} (deg.)	–	–	13	8.5	11.1	15	17	9.5	8

propagation, both the γ_{cal} calculated from the measured longitudinal wave velocity and the γ_{exp} measured from the high-speed sequences as shown in Fig. 9.20 are given in the same table. As can be seen, both coincide well.

Moreover, as can be seen from Table 9.4, the stress and the strain of string under transverse impact are markedly dependant on the prestretch force and the impact velocity.

Finally, it is worthwhile to notice that the transverse wave velocity of the polymer yarn such as Kevlar is much faster than the transverse wave velocity of metals such as Ni-Cr alloy. A faster transverse wave velocity is favorable to transmit the impact energy out faster and to avoid high localization of strain. Thus, it is of significance in impact-resistant engineering applications.

CHAPTER 10

Elastic–Plastic Waves Propagating in Beams under Transverse Impact (Bending Wave Theory)

Bending motion of bar occurs when bar is subjected to transverse impact or eccentric longitudinal impact. The bar bearing bend load is usually called beam. In the elementary theory of beam which is based on the so-called "plane cross section assumption," the bending motion of beam is attributed to the relative rotation between cross sections and the transverse motion of the neutral axis of beam.

In the previous chapter, we discussed the impact problem of flexible string, neglecting the stiffness of string. The beam can be regarded as a string with un-negligible stiffness.

The dynamic bending problem of beams can be described either by vibration approach or by wave propagation approach. The former is usually discussed in the discipline of Vibration or Dynamics of Structures. In the present chapter, only the latter, namely the bending wave propagation in beams, will be discussed. First, the elastic bending waves will be briefly discussed, and then mainly the plastic bending waves will be discussed. Similar to the situations of plastic longitudinal waves in bars, the plastic bending waves can be dealt with by elastic–plastic analysis, or by rigid-plastic approximate analysis.

It should be noticed that the current chapter is limited to discuss the rate-independent bending wave theory, and also limited to discuss the elemental theory based on the "plane cross section assumption." Obviously, the precise theory without "plane cross section assumption" and taking account of strain rate effect will be much more complicated.

10.1 Basic Assumptions and Governing Equations

Basic equations of the present problem consist of three aspects: the motion equation, the kinematical equation, and the constitutive equation of material. In general cases, the variables used to describe the motion of beam are the functions of spatial coordinates x, y, z, and time t. So, mathematically the problem is very complicated and difficult to solve accurately. For this reason, some limitations and basic assumptions are usually introduced.

First of all, it is limited to discuss the **straight beam** with constant cross-sectional areas, and at least one geometrically symmetrical plane of cross section where the applied loads lie in, exists. Therefore, the deformation of beam axis is always in the symmetrical plane, and no torsion deformation happens. This is the so-called **plane bending** problem. The meaning of such a limitation is to simplify a general three-dimensional problem to a two-dimensional plane problem.

Then let us make **the first basic assumption**: the plane beam cross section perpendicular to the beam axis before deformation remains a plane cross section perpendicular to the beam axis after deformation. This is the so-called **plane cross section assumption**. Thus a beam layer, the length of which is unvaried during the bending, must exist, called **neutral layer** (**surface**). For the beam with rectangular cross section, the neutral layer is another symmetrical plane which is perpendicular to the external load applied to the symmetrical plane. The beam fibers above and below the neutral layer are respectively subjected to tension and compression. The intersection line of the neutral layer with each cross section is called **neutral axis**. We choose the beam axis before deformation as x axis, the neutral axis as y axis, and the external force applied symmetrical to the axis as z axis, as shown in Fig. 10.1. Obviously, the plane cross section assumption attributes the beam bending to two basic motions, namely, the transverse displacement w of beam axis (where w is positive in the positive z direction), and the rotation of each cross section around its neutral axis (where the rotation angle α is positive in the clockwise direction).

Let us denote the transverse shear force and the bending moment by $Q(x, t)$ and $M(x, t)$ respectively, the positive directions of which are defined as shown in Fig. 10.2. Consider an infinitesimal beam element dx. According to the momentum theorem and the moment of momentum theorem, we have the following two dynamic equations

$$\frac{\partial Q}{\partial x} + q(x) = \rho_0 A_0 \frac{\partial v}{\partial t} \tag{10.1}$$

$$\frac{\partial M}{\partial x} - Q = -\rho_0 I \frac{\partial \omega}{\partial t} \tag{10.2}$$

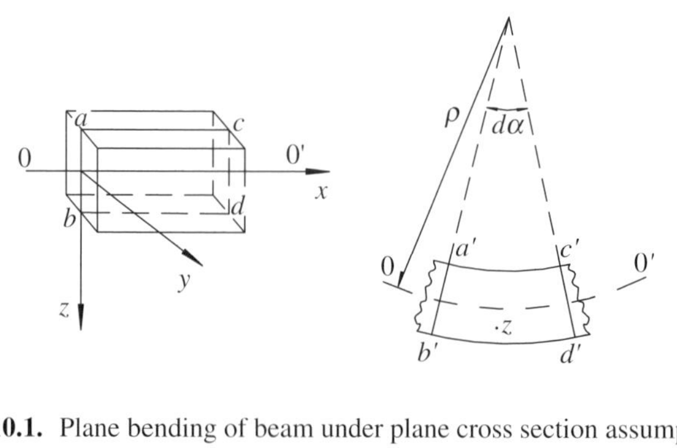

Fig. 10.1. Plane bending of beam under plane cross section assumption.

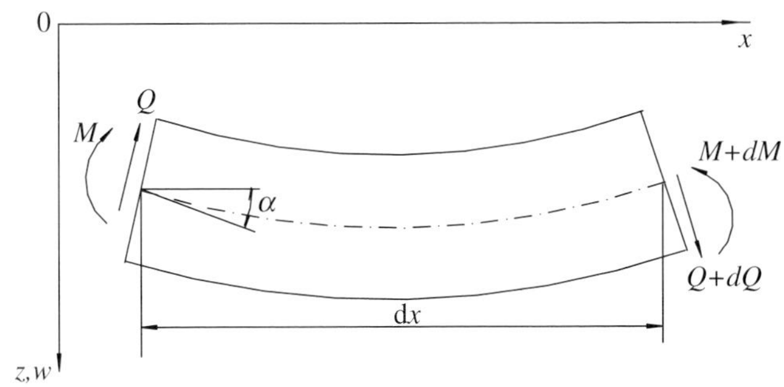

Fig. 10.2. The positive shear force Q and bending moment M acting on an infinitesimal element of beam.

where $q(x)$ is the distributed external load, ρ_0 the initial density of the beam before deformation, A_0 the initial cross section of the beam before deformation, $I = \int_A z^2\,dA$ the moment of inertia of the beam cross section with respect to the neutral axis, v the transverse velocity of the beam axis, and ω the rotating angular velocity of the cross section. As can be seen, the plane cross section assumption is used to simplify the plane bending problem to a one-dimensional problem, namely, the six dynamic equations in the general situation are now reduced to only two, Eqs. (10.1) and (10.2).

The kinematical equations under the plane cross section assumption can be easily expressed as:

$$v = \frac{\partial w}{\partial t} \tag{10.3}$$

$$tg\alpha = \frac{\partial w}{\partial x} \tag{10.4}$$

$$\omega = \frac{\partial \alpha}{\partial t} \tag{10.5}$$

$$k = \frac{1}{\rho} = \frac{\partial \alpha}{\partial s} = -\frac{\dfrac{\partial^2 w}{\partial x^2}}{\left[1 + \left(\dfrac{\partial w}{\partial x}\right)^2\right]^{3/2}} \tag{10.6}$$

where k is the curvature, ρ the radius of curvature, and s the arc length of bending beam axis. Because of the negative sign appearing on the right-hand side of Eq. (10.6), we have defined the positive sign of M and k as shown in Fig. 10.2, wherein the deformed beam axis is shown as a convex curve in the w–x coordinates $(\partial^2 w/\partial x^2) < 0$.

In the bending wave theory to be discussed in the following, we again make the following **second basic assumption: the deformation of beam is sufficiently small (tan $\alpha \ll 1$).** Therefore the difference between Lagrange coordinates and Euler coordinates can be ignored, and then Eqs. (10.4)–(10.6) are reduced to, respectively

$$\alpha = \frac{\partial w}{\partial x} \tag{10.4a}$$

$$\omega = \frac{\partial \alpha}{\partial t} = \frac{\partial^2 w}{\partial t \partial x} \tag{10.5a}$$

$$k = \frac{1}{\rho} = -\frac{\partial^2 w}{\partial x^2} \tag{10.6a}$$

Or, the continuous equations can also be expressed as

$$\frac{\partial v}{\partial x} = \frac{\partial \alpha}{\partial t} = \omega \tag{10.7}$$

$$\frac{\partial k}{\partial t} = -\frac{\partial \omega}{\partial x} \tag{10.8}$$

On the cross section, the normal strain ε_x at an arbitrary point, which is at a distance of z from the neutral axis, can be determined from the geometrical relation as shown in Fig. 10.1, as

$$\varepsilon_x = \frac{(\rho + z)\,d\alpha - \rho\,d\alpha}{\rho d\alpha} = \frac{z}{\rho} = zk \tag{10.9}$$

Under the assumption of small strain, substituting Eq. (10.6a) into the above equation, we have

$$\varepsilon_x = -z\frac{\partial^2 w}{\partial x^2} \tag{10.9a}$$

So the meaning of the plane cross section assumption with respect to the determination of beam deformation is that once the deformation of beam axis is obtained, the deformation at any point of beam can be determined.

Finally, let us discuss the constitutive equation of material. In the cases of longitudinal motion of bars, we are concerned with the simple tensile (compressive) σ–ε relation. In the general cases of rate-independent problems, we need the stress–strain relation to relate the stress tensor with the strain tensor. In the current cases of beam bending, M and Q are used in the dynamic equations instead of stress components, while the variable in relation to the strain ε is the curvature k [Eq. (10.9)]. By the plane cross section assumption, the effect of shear force on the beam deformation is substantially ignored. Thus, with regard to the beam, the constitutive relation of material is transformed to the M–k relationship.

According to the definition of bending moment, M is the resultant moment of the normal stresses on the cross section with respect to the neutral axis, namely

$$M = \int_A \sigma_x z\,dA \tag{10.10}$$

Under the condition of simple tension (compression), the σ–ε relation is easy to determine experimentally. However, in the general situation of three-dimensional stress states, ε_x is not only dependent on σ_x, but also on σ_y and σ_z, namely, σ_x should be a function of $\varepsilon_x, \varepsilon_y$, and ε_z. Usually, at the same time by making the plane cross section assumption, we make the following **third basic assumption**: **there is no interference between each longitudinal beam fiber**, namely, all longitudinal beam fibers are under unidirectional tension (compression), $\sigma_x = \sigma_x(\varepsilon_x)$. If the width and the height of the rectangular cross section are denoted by b and h, respectively, then Eq. (10.10) can be written as

$$M = b\int_{-h/2}^{h/2} \sigma_x(\varepsilon_x) z\, dz$$

In addition, by making the following **fourth basic assumption**: **the σ_x–ε_x curve for tension is the same as that for compression**, the above equation can be written as

$$M = 2b\int_{0}^{h/2} \sigma_x(kz) z\, dz = M(k) \tag{10.11a}$$

or

$$M = \frac{2b}{k^2}\int_{0}^{h/2} \sigma_x(\varepsilon_x)\, \varepsilon_x d\varepsilon_x = M(k) \tag{10.11b}$$

So if σ_x–ε_x curve is known, then $M = M(k)$ curve can be calculated, and M must be a function of k. Wherein, we have in fact implicitly made the following **fifth basic assumption**: **the constitutive equation of material is independent of strain rate**. With regard to this assumption, it should be understood that there approximately exists sole $M = M(k)$ relation under dynamic loading, although the constitutive equation under dynamic loading is surely different from that under static loading.

In the range of linear elastic deformation, $M = M(k)$ is also linear. So, Eq. (10.11) is reduced to

$$M = EIk \tag{10.12}$$

However, it should be noted that the linear hardening elastic–plastic σ_x–ε_x relation such as:

$$\begin{cases} \sigma_x = E\varepsilon_x & \varepsilon_x \le \varepsilon_S \\ \sigma_x = (E - E_1)\varepsilon_S + E_1\varepsilon_x & \varepsilon_x \ge \varepsilon_S \end{cases}$$

does not correspond to a linear hardening elastic–plastic $M = M(k)$ relation. In fact, if the position of the boundary of the elastic layers and the plastic layers is expressed by $z_S = \varepsilon_S/k$, then from Eq. (10.11), we have

$$\begin{aligned} M &= \frac{2bkEz_S^3}{3} + b(E - E_1)\varepsilon_S\left(\frac{h^2}{4} - z_S^2\right) + \frac{2bkE_1}{3}\left(\frac{h^3}{8} - z_S^3\right) \\ &= EIk\left[1 - \lambda - 4\lambda\left(\frac{z_S}{h}\right)^3 + 3\lambda\left(\frac{z_S}{h}\right)\right] \\ &= EIk\left[1 - \lambda - 4\lambda\left(\frac{\varepsilon_S}{hk}\right)^3 + 3\lambda\left(\frac{\varepsilon_S}{hk}\right)\right] \end{aligned} \tag{10.13}$$

where $\lambda = (E - E_1)/E$ and $I = bh^3/12$ (b and h the width and height of the rectangular cross section, respectively). Obviously, the constitutive relation is of linear elasticity till the point of $k_S = 2\varepsilon_S/h$, $M_S = 2EI\varepsilon_S/h$, and then is a nonlinear curve. When $\lambda = 0$, the above equation coincides with Eq. (10.12). When $\lambda = 1$, it corresponds to the perfectly plastic situation, and we have

$$M_S = EI\left[\frac{3\varepsilon_S}{h} - \frac{4\varepsilon_S^3}{h^3k^2}\right]$$

When $k \to \infty$, we have $z_S = \varepsilon_S/k \to 0$ and $M \to 3EI\varepsilon_S/h = 1.5M_S$, as shown in Fig. 10.3. If the full cross section enters into plastic deformation, then curvature k can increase infinitely, and thus the so-called "plastic hinge" is formed.

Hence one can see that, from now on, when we talk about the linear hardening M–k relation or the perfectly plastic M–k relation, they do not correspond to the linear hardening σ–ε relation or perfectly plastic σ–ε relation, respectively.

Thus, we established the total basic equations, including the dynamic equations [Eqs. (10.1) and (10.2)], kinematical equations (10.3), (10.4a), (10.5a), and (10.6a) [or (10.7) and (10.8)], and constitutive equation [Eq. (10.11)]. There are altogether seven equations to solve seven unknown variables M, Q, w, v, α, ω and k. The derivation of these basic equations is based on the above-mentioned five basic assumptions, and is limited to the plane bending problems. Moreover, it has been implicitly assumed that there is no longitudinal tensile (or compressive) loading applied on the beam.

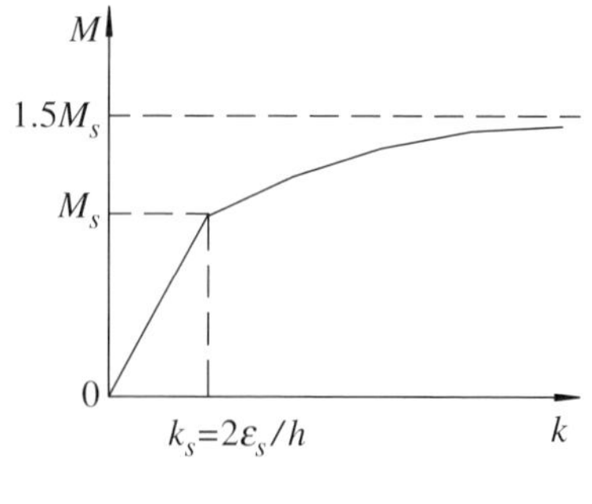

Fig. 10.3. The $M = M(k)$ relation for the elastic-perfectly plastic material.

Because of the introduction of the above-mentioned approximate assumptions, obviously these equations are applicable only under some related conditions. For example, the rotation angle of beam is limited to $\alpha < 10^\circ$ by the small strain assumption. The shear stress effect is disregarded due to the "plane cross section" assumption and the "no interference between each longitudinal beam fiber" assumption. In the case of pure bending, it is true that there is no shear stress. However, in the general cases, there actually exist shear stresses on the cross section of a beam, the result of which is shear force Q. And according to the "paired shear stresses theorem," shear stresses should exist too between the longitudinal beam fibers. Therefore, the beam elements between two original plane cross sections (which are plane before deformation) must experience shear strain, then the original plane cross sections must be distorted and are no longer plane. With regard to the static equilibrium problem, it has been proved in the theory of elasticity that such an approximate "plane cross section" assumption is acceptable only for long beams. This can be interpreted too by an estimation of and comparison between the value-order of each stress component. If the beam length is of order l, the cross-sectional area is of order h^2, and the distributed loading (per unit length) is of order q, then the shear force Q is of order ql, the normal stress between the beam fibers σ_Y is of order $q/l/h$, the shear stress τ is of order Q/h^2 or order ql/h, and the normal stress on the cross section σ_X is of order Ql/h^3 or order ql^2/h^3. So τ/σ_X is of order h/l, and σ_Y/σ_X is of order h^2/l^2. Thus, if l/h is large enough, then by comparing with σ_X, τ is negligible, and σ_Y is more negligible. It is usually recognized that the shear stress effect can be ignored for the beams with the ratio $h{:}l \leq 1{:}5$. However, in the dynamic problems of beams, the key factor is not the ratio of the cross-sectional size of the beam with the length, but the ratio of the cross-sectional size of the beam with the wavelength. As pointed out by Timoshenko, the shear stress effect can be disregarded so long as the wavelength is large enough than the radius of gyration (see Kolsky, 1953). In such cases, the results obtained under the above-mentioned approximate assumptions satisfactorily coincide with the results obtained from the precise theoretical analysis, otherwise corrections are required.

10.2 Elastic Bending Waves

In the simplest theory of elastic bending waves, it is assumed that the motion of each beam element is purely the transverse motion perpendicular to the beam axis; or in other words, the **rotatory inertia of beam cross section is ignored**. In addition, let $q = 0$, the basic equations are then reduced to the following linear equation set:

$$\left\{\begin{aligned}
&\frac{\partial Q}{\partial x} = \rho_0 A_0 \frac{\partial v}{\partial t} && (10.1\text{a})\\
&\frac{\partial M}{\partial x} = Q && (10.2\text{a})\\
&v = \frac{\partial w}{\partial t} && (10.3)\\
&k = -\frac{\partial^2 w}{\partial x^2} && (10.6\text{a})\\
&M = EIk && (10.12)
\end{aligned}\right.$$

Or, the above equation set can be reduced to solve the following fourth-order partial differential equation:

$$C_0^2 R^2 \frac{\partial^4 w}{\partial x^4} + \frac{\partial^2 w}{\partial t^2} = 0 \tag{10.14}$$

where $C_0 = \sqrt{E/\rho_0}$ is the wave velocity of longitudinal elastic wave and $R = \sqrt{I/A_0}$ is the rotating radius of cross section with respect to the neutral axis.

Assume that the bending wave solution of Eq. (10.14) can be expressed in the form of

$$w = D\cos(pt - fx) \tag{10.15}$$

where D is the amplitude, $f = 2\pi/\Lambda$ the wave number, $p = fC$ the circular frequency, while Λ the wavelength, and C the phase velocity of the wave. Substituting Eq. (10.15) into Eq. (10.14), we have

$$p^2 = C_0^2 R^2 f^4$$

or

$$C = \frac{2\pi R C_0}{\Lambda} \tag{10.16}$$

It is thus clear that, in the elementary theory of bending wave propagation, elastic bending wave velocity C is inversely proportional to wavelength Λ. Therefore, as different from the elastic longitudinal waves, where the wave velocity is a constant, a bending disturbance with an arbitrary profile will propagate with a profile-change (the so-called dispersion phenomenon); namely, the disturbances with different wavelengths propagate with different velocities.

However, according to Eq. (10.16), if the wavelength changes infinitesimally, then the wave velocity tends to approach infinity. Physically, an infinite wave velocity is unacceptable. This is because Eq. (10.16) is applicable only when wavelength Λ is much larger than rotating radius R of beam cross section. When wavelength Λ is comparable with rotating radius R, the rotatory inertia $\rho_0 I(\partial\omega/\partial t)$ should not be disregarded again, and the plane cross section assumption should be corrected to take account of shear stress effect.

The correction of taking rotatory inertia into account was at first proposed by Rayleigh (see Kolsky, 1953). In such cases, the basic equation set returns to:

$$\begin{cases} \dfrac{\partial Q}{\partial x} = \rho_0 A_0 \dfrac{\partial v}{\partial t} \\ \dfrac{\partial M}{\partial x} - Q = -\rho_0 I \dfrac{\partial \omega}{\partial t} \\ v = \dfrac{\partial w}{\partial t}, \quad \omega = \dfrac{\partial^2 w}{\partial t \partial x}, \quad \kappa = -\dfrac{\partial^2 w}{\partial x^2} \\ M = EIk \end{cases}$$

or to solve the transverse displacement w from the following partial differential equation:

$$C_0^2 R^2 \frac{\partial^4 w}{\partial x^4} - R^2 \frac{\partial^4 w}{\partial x^2 \partial t^2} + \frac{\partial^2 w}{\partial t^2} = 0 \tag{10.17}$$

As compared to Eq. (10.14), Eq. (10.17) contains an additional term to take account of rotatory inertia. Let Eq. (10.14) be its solution, we obtain

$$C = C_0 \left(1 + \frac{\Lambda^2}{4\pi^2 R^2}\right)^{-1/2} \tag{10.18}$$

The above equation is physically reasonable, since wave velocity C approaches a finite value C_0 when wavelength Λ approaches zero. When ratio R/Λ is small enough, namely either $R/\Lambda \ll 1$ or $\Lambda/R \gg 1$, Eq. (10.18) coincides with Eq. (10.16). It means that the rotatory inertia can be ignored only when Λ is much larger than R.

As pointed out by Timoshenko, the correction of shear stress effect is as important as the correction of rotatory inertia (see Kolsky, 1953). When the shear stress is taken into account, the original plane cross section is distorted due to shear strain, and thus the transverse displacement w of beam axis is actually composed of two parts: w_M which corresponds to rotating angle α induced by applied bending moment M, and w_Q which corresponds to shear strain γ induced by shear force Q:

$$w = w_M + w_Q \tag{10.19}$$

The following elastic constitutive relation exists between the Q and γ:

$$Q = \int_A \tau dA = \int_A G\gamma(z)\, dA = G\gamma A_S \tag{10.20}$$

where A_S is a coefficient which depends only on the shape of beam cross section, $\gamma(z)$ the real shear strain which varies with the cross section, and G the elastic shear modulus. Taking account of such a correction, the basic equations should be correspondingly corrected to:

$$\left.\begin{aligned} &\frac{\partial Q}{\partial x} = \rho_0 A_0 \frac{\partial v}{\partial t} && \text{(translation)} \\ &\frac{\partial M}{\partial x} - Q = -\rho_0 I \frac{\partial \omega}{\partial t} && \text{(rotation)} \end{aligned}\right\} \text{(dynamic equations)}$$

$$\left.\begin{aligned} &w = w_M + w_Q \\ &\omega = \frac{\partial^2 w_M}{\partial t \partial x}, \qquad k = -\frac{\partial^2 w_M}{\partial x^2} && \text{(bending)} \\ &\gamma = \frac{\partial w_Q}{\partial x}, \qquad v = \frac{\partial w}{\partial t} && \text{(shear)} \end{aligned}\right\} \text{[kinematical Eqs. (I)]}$$

or

$$\left.\begin{array}{ll}\dfrac{\partial k}{\partial t}=-\dfrac{\partial \omega}{\partial x} & \text{(bending)}\\ \dfrac{\partial \gamma}{\partial t}+\omega=\dfrac{\partial v}{\partial x} & \text{(shear)}\end{array}\right\}\text{[kinematical Eqs. (II)]}$$

$$\left.\begin{array}{ll}M=EIk & \text{(bending)}\\ Q=GA_s\gamma & \text{(shear)}\end{array}\right\}\text{(constitutive equations)}$$

Substituting the kinematical equations and the constitutive equations into the dynamic equations, respectively, we obtain:

$$GA_S\left(\frac{\partial^2 w}{\partial x^2}-\frac{\partial^2 w_M}{\partial x^2}\right)=\rho_0 A_0\frac{\partial^2 w}{\partial t^2}$$

$$EI\frac{\partial^3 w_M}{\partial x^3}+GA_S\left(\frac{\partial w}{\partial x}-\frac{\partial w_M}{\partial x}\right)=\rho_0 I\frac{\partial^3 w_M}{\partial t^2\partial x}$$

Eliminating the terms containing w_M from the above two equations, finally we obtain the following fourth-order partial differential equation with respect to w:

$$\frac{EI}{\rho_0 A_0}\frac{\partial^4 w}{\partial x^4}-\frac{I}{A_0}\left(1+\frac{E}{\rho_0}\frac{\rho_0 A_0}{GA_S}\right)\frac{\partial^4 w}{\partial x^2\partial t^2}+\frac{I}{A_0}\cdot\frac{\rho_0 A_0}{GA_S}\frac{\partial^4 w}{\partial t^4}+\frac{\partial^2 w}{\partial t^2}=0$$

Noticing $C_0^2=E/\rho_0$ and $R^2=I/A_0$, and introducing $C_Q^2=GA_S/\rho_0 A_0$, the above equation can be rewritten as

$$C_0^2R^2\frac{\partial^4 w}{\partial x^4}-R^2\left(1+\frac{C_0^2}{C_Q^2}\right)\frac{\partial^4 w}{\partial x^2\partial t^2}+\frac{R^2}{C_Q^2}\cdot\frac{\partial^4 w}{\partial t^4}+\frac{\partial^2 w}{\partial t^2}=0 \tag{10.21a}$$

or

$$\frac{\partial^4 w}{\partial x^4}-\left(\frac{1}{C_0^2}+\frac{1}{C_Q^2}\right)\frac{\partial^4 w}{\partial x^2\partial t^2}+\frac{1}{C_Q^2}\frac{1}{C_0^2}\frac{\partial^4 w}{\partial t^4}+\frac{1}{C_0^2R^2}\frac{\partial^2 w}{\partial t^2}=0 \tag{10.21b}$$

which is usually called Timoshenko equation.[1]

[1]If we substitute the constitutive equation into the kinematical equation, together with the dynamic equation, then the following first-order partial differential equation is obtained:

$$\left\{\begin{array}{ll}\dfrac{\partial M}{\partial x}+\rho_0 I\dfrac{\partial \omega}{\partial t}=Q; & \dfrac{1}{EI}\dfrac{\partial M}{\partial t}-\dfrac{\partial \omega}{\partial x}=0\\ \dfrac{\partial Q}{\partial x}-\rho_0 A_0\dfrac{\partial v}{\partial t}=0; & \dfrac{1}{GA_S}\dfrac{\partial Q}{\partial t}-\dfrac{\partial v}{\partial x}=-\omega\end{array}\right\}$$

Let Eq. (10.15) be a solution, we obtain

$$f^4 - \left(\frac{1}{C_0^2} + \frac{1}{C_Q^2}\right) f^2 p^2 + \frac{p^4}{C_0^2 C_Q^2} - \frac{p^2}{C_0^2 R^2} = 0$$

or

$$1 - \left(\frac{C^2}{C_0^2} + \frac{C^2}{C_Q^2}\right) + \frac{C^4}{C_0^2 C_Q^2} - \frac{C^2}{C_0^2 R^2 f^2} = 0 \qquad (10.22)$$

The solution for the above equation provides the relation between bending wave velocity C and wavelength $\Lambda = 2\pi/f$.

For a cylindrical bar with a radius of a, when the material Poisson ratio $\nu = 0.29$, the dimensionless relations between C/C_0 and a/Λ obtained by different theories were given by Davies, as shown in Fig. 10.4 (see Kolsky, 1953).

The results calculated from Timoshenko's theory coincide very well with the results calculated from the precise theory based on the general elastic equations. The different theories show that bending waves display dispersion phenomenon. When $a/\Lambda < 0.1$ (namely, the wavelength is 10 times larger than the radius of the cylindrical bar), the results given by all these theories are very close to each other; otherwise both the elementary theory and the Rayleigh's correction will result in remarkable errors.

10.3 Plastic Bending Waves (Elastic–Plastic Beams)

Studies on the plastic bending waves were carried out later than the plastic longitudinal waves. The earliest theoretical and experimental investigation of plastic deformation of beam under transverse impact may be traced back to the work by Duwez *et al.* (1943, 1950), based on which other research works were developed. In the following, we will mainly discuss their results, the theoretical part of which was established by Bohnenblust.

Consider an infinite beam which is subjected to transverse constant-velocity impact V_1. In the case of elastic deformation, the problem has been solved by Boussinesq in 1885 (see Duwez *et al.*, 1950), and it was shown that w/t is solely dependent on x^2/t, namely the problem possesses similar solution. This result was generalized by Bohnenblust from the elastic bending problem to the plastic bending problem, namely generalized from the linear relation $M = EIk$ to the nonlinear relation $M = M(k)$. They verified that in the

---

If they are solved by the characteristics method, the corresponding characteristics equations are:

$$dM \pm \rho_0 I C_0\, d\omega = \pm C_0 Q\, dt, \text{ along } \quad dx \pm C_0\, dt = 0;$$

$$dQ \pm \rho_0 I C_Q\, dv = \rho_0 A_0 C_Q^2 \omega\, dt, \text{ along } \quad dx \pm C_Q\, dt = 0$$

However, the above two sets of equation are not independent but coupled each other. In fact, M–ω waves depend on Q, while Q–v waves depend on ω.

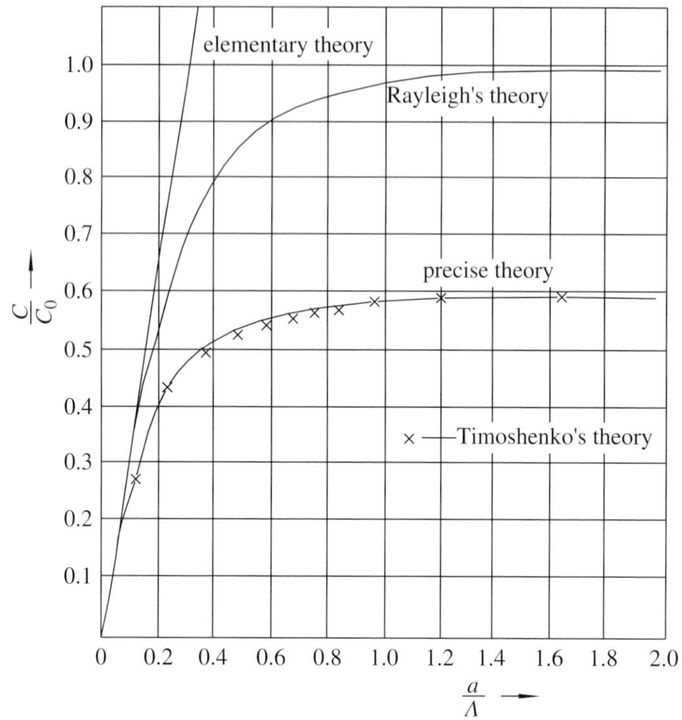

Fig. 10.4. Phase velocities of elastic bending waves for a cylindrical bar with a radius of a ($\nu = 0.29$).

situation of plastic bending, w/t is still solely dependent on x^2/t, thus the problem can be easily solved.

Boussinesq's theory did not provide a precise solution starting from the general equations, but an approximate solution starting from the elementary theory, wherein the shear stress effect and the rotatory inertia are ignored. In such a case, the self-similarity character of solution can be perceived by the dimensional analysis as follows. The basic equation of the current problem is Eq. (10.14). Or by introducing a defined as

$$a^4 = \frac{EI}{\rho_0 A_0} = C_0^2 R^2 \tag{10.23}$$

Equation (10.14) can be rewritten as

$$a^4 \frac{\partial^4 w}{\partial x^4} + \frac{\partial^2 w}{\partial t^2} = 0 \tag{10.14a}$$

Since in the above equation a always appears simultaneously with $\partial/\partial x$, by introducing ξ defined as

$$\xi = \frac{x}{a}$$

Equation (10.14a) can be further reduced to

$$\frac{\partial^4 w}{\partial \xi^4} + \frac{\partial^2 w}{\partial t^2} = 0$$

Note that only constant-velocity V_1 is included in the conditions used to determine the solution for the current problem, while neither characteristic length nor characteristic time are included, then we have

$$w = f(\xi, t, V_1)$$

Noticing that the dimension of $\xi = x/a$ is $t^{1/2}$, choose t and V_1 as the basic dimensions, the problem obviously possesses the following similar solution (see Appendix 3)

$$\frac{w}{V_1 t} = f\left(\frac{\xi^2}{t}\right) = f\left(\frac{x^2}{a^2 t}\right)$$

Or for convenience in practical application, if it is rewritten as

$$w = tf(\eta), \qquad \eta = \frac{x^2}{4a^2 t} \tag{10.24}$$

then we have

$$\frac{\partial w}{\partial t} = f(\eta) - \eta f'(\eta) = v \tag{10.25}$$

$$\frac{\partial^2 w}{\partial t^2} = \frac{\eta^2}{t} f''(\eta) \tag{10.26}$$

$$\frac{\partial w}{\partial x} = \frac{x}{2a^2} f'(\eta) = \frac{\sqrt{t}}{a}\sqrt{\eta} f'(\eta) = \alpha \tag{10.27}$$

$$\frac{\partial^2 w}{\partial x^2} = \frac{1}{2a^2}[f'(\eta) + 2\eta f''(\eta)] = -k \tag{10.28}$$

$$\frac{\partial^2 w}{\partial t \partial x} = -\frac{1}{a\sqrt{t}}\eta^{3/2} f''(\eta) = \omega \tag{10.29}$$

$$\frac{\partial^3 w}{\partial x^3} = \frac{1}{4a^4}\frac{x}{t}[3f''(\eta) + 2\eta f'''(\eta)] \tag{10.30}$$

$$\frac{\partial^4 w}{\partial x^4} = \frac{1}{4a^4 t}\left[3f''(\eta) + 12\eta f'''(\eta) + 4\eta^2 f^{(4)}(\eta)\right] \tag{10.31}$$

Substituting Eqs. (10.26) and (10.31) into Eq. (10.14a), the partial differential equation is then reduced to the following ordinary differential equation

$$4\eta^2 f^{(4)}(\eta) + 12\eta f'''(\eta) + (3 + 4\eta^2) f''(\eta) = 0 \tag{10.32}$$

It is worthwhile to note that $v(=\partial w/\partial t)$ and $\kappa(=-\partial^2 w/\partial x^2)$ are only dependent on η, while $\alpha(=\partial w/\partial x)$, $\omega(=\partial^2 w/\partial t\partial x)$ and $\partial v/\partial t(=\partial^2 w/\partial t^2)$ are also dependent on x or t.

Bohnenblust adopted all the basic assumptions used in the elementary theory of elastic beam. He first presumed that the Boussinesq's self-similar solution [Eq. (10.24)] is still correct for the elastic–plastic beam. After integrating the basic equation, he found that the solution satisfies the initial and boundary conditions, and consequently verified that Eq. (10.24) holds. Actually, we can still use the dimensional analysis to explain it as follows.

Under the basic limitations and assumptions discussed in Section 10.1, and when the rotatory inertia is ignored, the basic equations are

$$\left.\begin{aligned} \frac{\partial Q}{\partial x} &= \rho_0 A_0 \frac{\partial v}{\partial t} &\quad (10.1a)\\ \frac{\partial M}{\partial x} &= Q &\quad (10.2a)\end{aligned}\right\} \text{(dynamic conditions)}$$

$$\left.\begin{aligned} v &= \frac{\partial w}{\partial t} &\quad (10.3)\\ \kappa &= -\frac{\partial^2 w}{\partial x^2} &\quad (10.6a)\end{aligned}\right\} \text{(kinematical conditions)}$$

$$M = M(\kappa) \quad \text{(constitutive equation)} \tag{10.11}$$

Substituting Eqs. (10.3), (10.6a), and (10.11) into Eqs. (10.1a) and (10.2a), we obtain the following equation set, containing only variables Q and w:

$$\frac{\partial Q}{\partial x} = \rho_0 A_0 \frac{\partial^2 w}{\partial t^2} \tag{10.33}$$

$$Q = -\frac{dM}{dk}\frac{\partial^3 w}{\partial x^3} \tag{10.34}$$

which can be further reduced to the following quasi-linear fourth-order partial differential equation with respect to w:

$$\frac{1}{\rho_0 A_0}\frac{dM}{dk}\frac{\partial^4 w}{\partial x^4} - \frac{1}{\rho_0 A_0}\frac{d^2 M}{dk^2}\left(\frac{\partial^3 w}{\partial x^3}\right)^2 + \frac{\partial^2 w}{\partial t^2} = 0 \tag{10.35}$$

It is known that M is a function of k, namely $M = M(k)$ or expressing in the dimensionless form:

$$\frac{M}{M_S} = f\left(\frac{k}{k_S}\right)$$

where M_S and k_S are yield bending moment and yield curvature, respectively. Therefore we have

$$\frac{dM}{dk} = \frac{d(M/M_S)}{d(k/k_S)}\frac{M_S}{k_S} = EI f_1', \quad \frac{d^2M}{dk^2} = \frac{d^2(M/M_S)}{d(k/k_S)^2}\frac{M_S}{k_S^2} = \frac{EI}{k_S} f_1''$$

Recalling the definition of notation a, namely $a^4 = EI/\rho_0 A_0$, Eq. (10.35) can be rewritten as

$$a^4 f_1' \frac{\partial^4 w}{\partial x^4} - a^4 \frac{f_1''}{k_S}\left(\frac{\partial^3 w}{\partial x^3}\right)^2 + \frac{\partial^2 w}{\partial t^2} = 0$$

Introducing $\xi = x/a$ and $Q = w/k_S$, then we have

$$f_1' \frac{\partial^4 Q}{\partial \xi^4} - \frac{f_1''}{a^2}\left(\frac{\partial^3 Q}{\partial \xi^3}\right)^2 + \frac{\partial^2 Q}{\partial t^2} = 0 \tag{10.35a}$$

Noticing that we also have

$$\frac{M}{M_S} = f_1\left(\frac{k}{k_S}\right) = f_1\left(-\frac{1}{k_S}\frac{\partial^2 w}{\partial x^2}\right) = f_1\left(-\frac{\partial^2 Q}{\partial x^2}\right)$$

and once more introducing $\zeta = -Q/a^2$, then Eq. (10.35a) can be further reduced to

$$f_1' \frac{\partial^4 \zeta}{\partial \xi^4} - f_1''\left(\frac{\partial^3 \zeta}{\partial \xi^3}\right)^2 + \frac{\partial^2 \zeta}{\partial t^2} = 0 \tag{10.35b}$$

and

$$\frac{M}{M_S} = f_1\left(-\frac{\partial^2 Q}{\partial x^2}\right) = f_1\left(-\frac{\partial^2 \zeta}{\partial \xi^2}\right)$$

as well. The following original boundary condition

$$\left.\frac{\partial w}{\partial t}\right|_{x=0} = V_1$$

is now reduced to

$$\left.\frac{\partial \zeta}{\partial t}\right|_{\xi=0} = \frac{V_1}{k_S a^2}$$

Under such a boundary condition, the solution of Eq. (10.35b) can be written as

$$\zeta = g_1\left(\xi, t, \frac{V_1}{k_S a^2}\right)$$

The dimensions of each variable in the above equation are, respectively:

$$
\begin{aligned}
[t] &= T \\
[\xi] &= \left[\frac{x}{\sqrt[4]{EI/\rho_0 A_0}}\right] = \frac{L}{LT^{-1/2}} = T^{1/2} \\
[\zeta] &= \left[\frac{w}{k_S\sqrt{EI/\rho_0 A_0}}\right] = \frac{L}{L^{-1}\cdot L^2T^{-1}} = T \\
\left[\frac{V_1}{k_S a^2}\right] &= \left[\frac{V_1}{k_S\sqrt{EI/\rho_0 A_0}}\right] = \frac{LT^{-1}}{L^{-1}\cdot L^2T^{-1}} = L^0T^0
\end{aligned}
$$

Therefore, obviously we have

$$\frac{\zeta}{t} = g\left(\frac{\xi^2}{t}, \frac{V_1}{k_S a^2}\right)$$

or

$$\frac{w}{k_S\cdot\sqrt{EI/\rho_0 A_0}\cdot t} = g\left(\frac{x^2}{\sqrt{EI/\rho_0 A_0}\cdot t}, \frac{V_1}{k_S\cdot\sqrt{EI/\rho_0 A_0}}\right)$$

It shows that Eq. (10.24) still holds for elastic–plastic beam.

Thus, according to Eq. (10.28) k is only dependent on η, and consequently according to Eq. (10.11) M is also only dependent on η, but according to Eq. (10.2a) Q should be

$$Q = \frac{\partial M}{\partial x} = \frac{dM}{d\eta}\frac{\partial \eta}{\partial x} = \frac{1}{a\cdot\sqrt{t}}\sqrt{\eta}\frac{dM}{d\eta} \tag{10.36}$$

In order to deal with the functions which are dependent only on η, S is introduced which is defined as

$$S = \frac{2a^2}{EI}\sqrt{t}Q = \frac{2a^2}{EI}\sqrt{\eta}\frac{dM}{d\eta} \tag{10.37}$$

Since we have

$$\frac{\partial Q}{\partial x} = \frac{EI}{2a^3\sqrt{t}}\frac{ds}{d\eta}\frac{\partial \eta}{\partial x} = \frac{EI}{2a^4 t}\sqrt{\eta}S'$$

and moreover taking into account Eqs. (10.26) and (10.30), Eqs. (10.33) and (10.34) can be correspondingly reduced to

$$S' = 2\eta^{3/2}f''(\eta) \tag{10.38}$$

$$S = -\frac{1}{EI}\sqrt{\eta}\cdot\left[3f''(\eta) + 2\eta f'''(\eta)\right]\cdot\frac{dM}{dk} \tag{10.39}$$

If we eliminate S from these equations, then the following fourth-order ordinary differential equation can be obtained:

$$4\eta^2\frac{dM}{dk}f^{(4)} + \left[12\eta\frac{dM}{dk} - 4\frac{\eta^2}{a^2} - \frac{d^2M}{dk^2}\left(\eta f''' + 3f''\right)\right]f''' + \left[3\frac{dM}{dk} + 4EI\eta^2 - 9\frac{\eta}{a^2}\frac{d^2M}{dk^2}f''\right]f'' = 0 \tag{10.40}$$

In the case of elastic beam, we have $M = EIk$, then Eq. (10.40) coincides with Eq. (10.32). Obviously, substituting Eqs. (10.26), (10.30), and (10.31) directly into Eq. (10.35), we can obtain Eq. (10.40) too.

Alternatively, eliminating $f(\eta)$ from Eqs. (10.38) and (10.39), an ordinary differential equation with respect to S can be obtained. To this end, differentiating Eq. (10.38), we have

$$S'' = \sqrt{\eta}\left[3f''(\eta) + 2\eta f'''(\eta)\right]$$

Comparing it with Eq. (10.39), we obtain

$$S'' + EIS\frac{dk}{dM} = 0 \tag{10.41}$$

Obviously, Eqs. (10.40) and (10.41) are equivalent to each other. From the former, the transverse displacement $w = tf(\eta)$ can be solved and then k, M, and Q can be obtained by differentiating operations. From the latter, S can be solved and then M, k, and w can be obtained by integrating operations. Mathematically, solving Eq. (10.41) is simpler than Eq. (10.40), since Eq. (10.41) is a second-order ordinary differential equation. It was Bohnenblust who just started from this equation to study the plastic bending wave propagating in beams.

Now we discuss the initial and boundary conditions of the current problem, they are

$$t = 0,\ 0 < x \le \infty$$

$$w = v = \alpha = \omega = k = M = 0 \tag{10.42}$$

$$\text{or } w = \frac{\partial w}{\partial t} = \frac{\partial w}{\partial x} = \frac{\partial^2 w}{\partial t\partial x} = \frac{\partial^2 w}{\partial x^2} = M = 0$$

$$t > 0,\quad x = \infty$$

$$w = v = \alpha = \omega = k = M = 0 \tag{10.43}$$

$$\text{or } w = \frac{\partial w}{\partial t} = \frac{\partial w}{\partial x} = \frac{\partial^2 w}{\partial t\partial x} = \frac{\partial^2 w}{\partial x^2} = M = 0$$

$$t > 0,\quad x = 0$$

$$v = \frac{\partial w}{\partial t} = V_1 \tag{10.44}$$

By introducing the dimensionless independent variable defined as $\eta = x^2/4a^2t$, the initial condition (10.42) and the boundary condition (10.43) both correspond to the boundary condition at $\eta = \infty$. From Eqs. (10.24)–(10.29) it can be known that there should be the following boundary conditions:

$$\begin{aligned} &\eta = \infty, \quad w = v = \alpha = \omega = k = M = 0 \\ &\text{or } f(\eta) = \eta f'(\eta) = \eta^{3/2} f''(\eta) = 0 \end{aligned} \tag{10.45}$$

$$\eta = 0, \quad v = \frac{\partial w}{\partial t} = f(\eta) - \eta f'(\eta) = f(\eta) = V_1 \tag{10.46}$$

Note that the following condition has been used in Eq. (10.46).

$$\lim_{\eta \to 0} \eta f'(\eta) = 0$$

This can be explained as follows: at the impact point $x = 0$, in a finite duration the angle of rotation of cross section $\alpha = \partial w/\partial x$ should also be finite. Then according to Eq. (10.27), we must have $\lim_{\eta\to 0} \sqrt{\eta} f'(\eta) = A$, where A is a finite quantity, thus the above equation holds. Alternatively, from the conditions of $w(0, t) = V_1 t = t \cdot f(0)$, Eq. (10.46) can be directly obtained and thus it is required to satisfy $\lim_{\eta\to 0} \eta f'(\eta) = 0$.

In addition, if at $x = 0$, the rotating angle $\alpha = \partial w/\partial x$ is a continuous variable, then because the deformations of beam on the right and left sides of $x = 0$ are anti-symmetric, we must have

$$\alpha(0, t) = \left.\frac{\partial w}{\partial x}\right|_{x=0} = 0$$

Or according to Eq. (10.27), the continuous condition of $\partial w/\partial x$ can be expressed as

$$\lim_{\eta \to 0} \sqrt{\eta} f'(\eta) = A = 0 \tag{10.47}$$

If a strong discontinuity of $\partial w/\partial x$ appears at the impact point (e.g. the so-called plastic hinge appears and then a break in shape of beam axis appears), then in the above equation $A \neq 0$; but because $\alpha(0, \tau)$ is always finite, A is at least still finite. If at any time t, $\partial w/\partial x|_{x=0}$ remains continuous, then we have

$$\left.\frac{\partial \alpha}{\partial t}\right|_{x=0} = \left.\frac{\partial^2 w}{\partial t \partial x}\right|_{x=0} = -\lim_{\eta \to 0} \eta^{3/2} f''(\eta) = 0 \tag{10.47a}$$

When we start from Eq. (10.41) to solve the problem, these boundary conditions should be rewritten to the corresponding expressions in the form as functions of $S(\eta)$. Noticing Eq. (10.38), corresponding to Eqs. (10.45) and (10.46), we have

$$S'(\infty) = 0 \tag{10.48}$$

$$V_1 = \frac{1}{2} \int_{\infty}^{0} d\eta \int_{\infty}^{0} \frac{S'(\eta)}{\eta^{3/2}} d\eta \tag{10.49}$$

According to Eqs. (10.27)–(10.29), Eq. (10.38) can also be written in the following different forms:

$$\begin{aligned} S'(\eta) &= 2\eta^{3/2} f''(\eta) = -2a\sqrt{t}\frac{\partial^2 w}{\partial t \partial x} = -2a\sqrt{t}\omega \\ &= 2a^2\sqrt{\eta}\frac{\partial^2 w}{\partial t^2} - \sqrt{\eta} f'(\eta) = -2a^2\sqrt{\eta}k - \frac{a}{\sqrt{t}}\alpha \end{aligned} \tag{10.50}$$

At $x = 0$, the continuous condition of α, namely Eq. (10.47) or (10.47a), can be expressed as

$$S'(0) = 0 \tag{10.51}$$

It is thus clear that, according to Eq. (10.50), $S'(\eta)$ in its physical meaning can be understood as a function of angular velocity ω and similar to Eq. (10.37) depends only on η. The condition (10.51) can be understood as that of the angular velocity at $x = 0$ equals zero, or at $x = 0$ the rotation angle is continuous (equal to zero) while curvature k is finite. If a plastic hinge is formed at $x = 0$, then a discontinuity of α will take place, and consequently $S'(0) \neq 0$, and a discontinuity of $S'(0)$ will take place. This will be mentioned again in the following discussion.

Once $S(\eta)$ is solved from Eq. (10.41), Q, M, α, k, and w can be obtained from the following equations.

According to Eq. (10.37), we have

$$Q = \frac{EI}{2a^3\sqrt{t}} S(\eta) \tag{10.52}$$

and then the external force $P(t)$ at $x = 0$ can be calculated, $P(t) = 2Q(0, t) = (EIS(0))/(a^3\sqrt{t})$

According to Eq. (10.36), we have

$$M = \frac{EI}{2a^2}\int_{\infty}^{\eta} \frac{S(\eta)}{\sqrt{\eta}}\, d\eta = -\frac{EI}{2a^2}\int_{\eta}^{\infty} \frac{S(\eta)}{\sqrt{\eta}}\, d\eta \tag{10.53}$$

wherein the boundary condition (10.45) has been utilized.

According to Eq. (10.38), we have

$$f'(\eta) = \frac{1}{2}\int_{\infty}^{\eta} \frac{S'}{\eta^{3/2}}\, d\eta = -\frac{S'(\eta)}{\sqrt{\eta}} + \int_{\infty}^{\eta} \frac{S''}{\sqrt{\eta}}\, d\eta;$$

wherein the boundary conditions (10.45) and (10.48) have been utilized. Then according to Eq. (10.27) we have

$$\alpha = \frac{\partial w}{\partial x} = -\frac{\sqrt{t}}{a}S'(\eta) - \frac{\sqrt{t}}{a}\sqrt{\eta}\int_{\eta}^{\infty}\frac{S''}{\sqrt{\eta}}d\eta; \tag{10.54}$$

and according to Eq. (10.50a) we have

$$k = -\frac{S'}{2a^2\sqrt{\eta}} - \frac{a}{2a\sqrt{t}\sqrt{\eta}} = \frac{1}{2a^2}\int_{\eta}^{\infty}\frac{S''}{\sqrt{\eta}}d\eta \tag{10.55}$$

Finally according to Eqs. (10.24) and (10.38) we have

$$w = \frac{t}{2}\int_{\infty}^{\eta}d\eta\int_{\infty}^{\eta}\frac{S'}{\eta^{3/2}}d\eta$$

wherein the boundary conditions (10.45) have been utilized. With regard to the last double integral, by changing the integral sequence as shown in Fig. 10.5, it can be rewritten as

$$w = \frac{t}{2}\int_{\infty}^{\eta}d\zeta\int_{\infty}^{\zeta}\frac{S'(\xi)}{\xi^{3/2}}d\xi = \frac{t}{2}\int_{\eta}^{\infty}\frac{S'(\xi)}{\xi^{3/2}}d\xi\cdot\int_{\eta}^{\xi}d\zeta = \frac{t}{2}\int_{\eta}^{\infty}\frac{S'(\xi)}{\sqrt{\xi}}d\xi - \frac{t}{2}\eta\int_{\eta}^{\infty}\frac{S'(\xi)}{\xi^{3/4}}d\xi$$

and finally it can be written as

$$w = \frac{t}{2}\int_{\infty}^{\eta}\frac{S'}{\sqrt{\eta}}d\eta - \frac{t}{2}\eta\int_{\eta}^{\infty}\frac{S'}{\eta^{3/2}}d\eta \tag{10.56}$$

Noticing that $(1/2)\eta\int_{\eta}^{\infty}(S'/\eta^{3/2})\,d\eta$ is similar to $-\eta f'(\eta)$, and according to Eq. (10.46) we have $\lim_{\eta\to 0}\eta f'(\eta) = 0$, then the boundary condition (10.46) or (10.49) can be expressed as

$$V_1 = \frac{1}{2}\int_{0}^{\infty}\frac{S'}{\sqrt{\eta}}d\eta \tag{10.57}$$

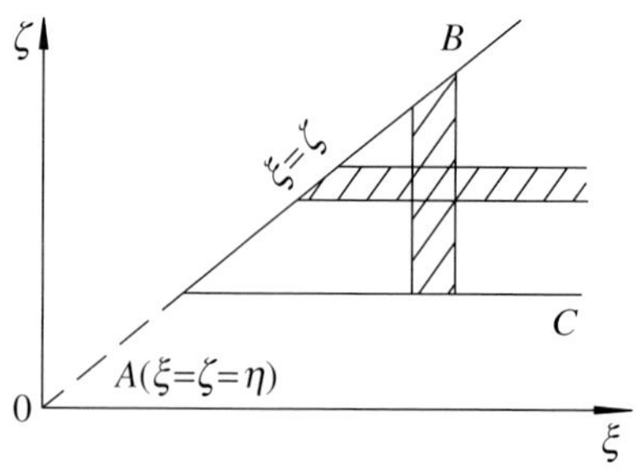

Fig. 10.5. Change of the integral sequence of double integrate.

In fact, Eq. (10.56) has provided the following relation

$$f(\eta) = \frac{1}{2}\int_{\eta}^{\infty} \frac{S'}{\sqrt{\eta}}\,d\eta - \frac{1}{2}\eta \int_{\eta}^{\infty} \frac{S'}{\eta^{3/2}}\,d\eta = \frac{1}{2}\int_{\eta}^{\infty} \frac{S'}{\sqrt{\eta}}\,d\eta + \eta f'(\eta)$$

Comparing it with Eq. (10.25), we obtain

$$v = \frac{\partial w}{\partial t} = \frac{1}{2}\int_{\eta}^{\infty} \frac{S'}{\sqrt{\eta}}\,d\eta \tag{10.58}$$

It means that the boundary condition which should be satisfied is Eq. (10.57).

Thus, the whole problem is solved. The previous discussions can be summed up as follows: when an infinite beam is subjected to a transverse constant-velocity impact V_1, the so-called self-similar solutions (10.24) exist regardless of whether the beam deformation is elastic or elastic–plastic. Then the problem is reduced to solve the ordinary differential equation (10.40) or (10.41). If we start from Eq. (10.41) to solve the function $S(\eta)$ under the boundary conditions (10.48) and (10.57) [namely (10.49) and (10.51)], then Q, M, α, k, w, and v can be obtained respectively from Eqs. (10.52)–(10.56) and (10.58). So long as the impact velocity is the same as V_1, the beam deflection curves at different times are completely the same $w(x, t)$ if these are expressed in the form of w/t and $x/t^{1/2}$; particularly, the point x_0 where $w = 0$ is proportional to $t^{1/2}$. For a certain value of $\eta_1 = x^2/(4a^2t), w/t, v, \alpha/t^{1/2}, k, M$, and $Qt^{1/2}$ have the same values respectively. This means that these quantities propagate along the beam axis with the wave velocity $dx/dt = a\sqrt{\eta_1}/\sqrt{t}$. In other words, the wave velocity is not a constant but decreases with increasing $\sqrt{t}$. Such a phenomenon coincides with the dispersive character of elastic bending waves as pointed out earlier.

It should be noted, both the rotatory inertia and the shear stress effect are disregarded in the above analyses. As pointed out in the analyses on elastic bending waves, such elementary theory is reliable only when the wavelength is much larger than the size of beam cross section. We can imagine that a similar situation exists for the plastic bending waves. Obviously, a more precise solution can be obtained only when the rotatory inertia and the shear stress effect are taken into consideration. In addition, in the above analyses, the reflection and interaction of waves, as well as the more complicated situations in relation to the loading–unloading process, are all not yet discussed.

The key to the whole problem lies in solving Eq. (10.41). The nonlinearity of *dM/dk* makes the problem mathematically complicated. In general, if $M = M(k)$ relation is experimentally determined, the solution is usually obtained by numerical calculation. However, if $M = M(k)$ relation can be approximately described by linear hardening plastic or perfectly plastic model, as shown in Fig. 10.6, then it is possible to solve the problem by analytical calculation. In the following, we will discuss two such special cases respectively.

We will first discuss the corresponding elastic beam, and determine the critical condition for the beginning of plastic yield.

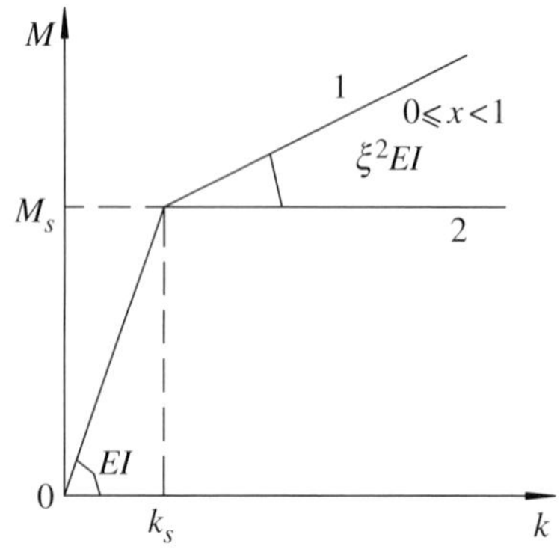

Fig. 10.6. The perfectly plastic and linear hardening plastic $M = M(k)$ relations.

(1) In elastic cases ($M = EIk$)

In the case of elastic deformation, Eq. (10.41) is reduced to the following second-order constant-coefficient linear homogeneous ordinary differential equation:

$$S'' + S = 0$$

Its general solution is $S = C_1 \cos\eta + C_2 \sin\eta$, and the indeterminate coefficients are determined by the boundary conditions (10.51) and (10.57):

$$S'(0) = -C_1 \sin 0 + C_2 \cos 0 = 0,\ C_2 = 0$$

$$V_1 = -\frac{C_1}{2}\int_0^{\infty} \frac{\sin\eta}{\sqrt{\eta}} d\eta = -\frac{C_1}{2}\sqrt{\frac{\pi}{2}},\ C_1 = -2\sqrt{\frac{2}{\pi}}V_1$$

Therefore the solution is

$$S(\eta) = -2\sqrt{\frac{2}{\pi}}V_1 \cos\eta \tag{10.59}$$

According to Eqs. (10.53) and (10.55) we have

$$M(\eta) = \sqrt{\frac{2}{\pi}\frac{EIV_1}{a^2}}\int_{\eta}^{\infty} \frac{\cos\eta}{\sqrt{\eta}} d\eta, \quad k(\eta) = \sqrt{\frac{2}{\pi}\frac{V_1}{a^2}}\int_{\eta}^{\infty} \frac{\cos\eta}{\sqrt{\eta}} d\eta$$

As can be seen, $S(\eta)$ is a cosine function of η and its amplitude is $2V_1\sqrt{2/\pi}$, while the absolute amplitudes of the period functions $M(\eta)$ and $k(\eta)$ decrease with increasing η, as shown in Fig. 10.7.

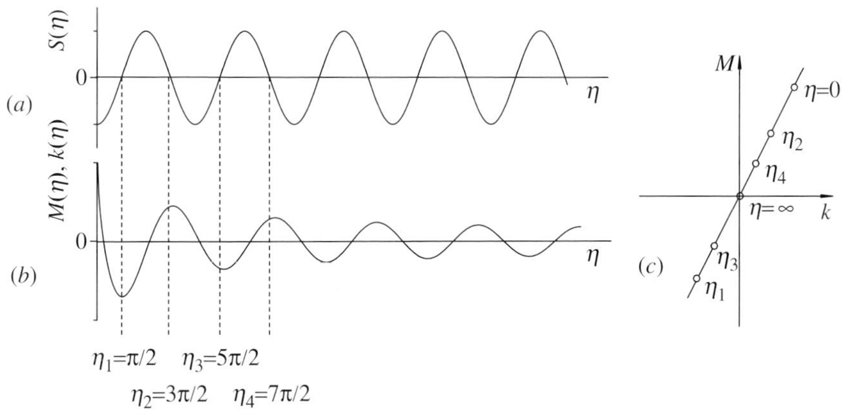

Fig. 10.7. In the case of elastic deformation ($M = EIk$).

The maximum values appear at $\lambda = 0$:

$$S(0) = -2\sqrt{\frac{2}{\pi}}V_1, \quad M(0) = \frac{EIV_1}{a^2}, \quad k(0) = \frac{V_1}{a^2} \tag{10.60}$$

At $\eta = (2n+1)\pi/2$ $(n = 0, 1, 2, \ldots)$, $S(\eta)$ reaches its null point, while according to Eq. (10.36), these are extreme points for $M(\eta)$ and $k(\eta)$, which alternatively reach the positive extreme and the negative extreme. It means that with the propagation of bending wave starting from the impact point $x = 0$ along the beam axis, at each cross section the beam will be successively convex upwards and concave downwards, and the bending deformation will gradually increase after each convex–concave alternative. However, the propagation velocity of bending wave is not a constant. For example, the propagation of the point $w = 0$ is proportional to $t^{1/2}$. Let us estimate how x_0 which is the first point of $w = 0$, varies with time t. Note that now $\eta_0 = x_0^2/4a^2t$ is a small quantity. According to Eq. (10.56), now η_0 should satisfy the following equation:

$$\int_{\eta_0}^{\infty} \frac{\sin\eta}{\sqrt{\eta}}\,d\eta - \eta_0 \int_{\eta_0}^{\infty} \frac{\sin\eta}{\eta^{3/2}}\,d\eta = 0$$

By using the following series expansion relations:

$$\frac{\sin\eta}{\sqrt{\eta}} = \frac{1}{\sqrt{\eta}}\left[\eta - \frac{1}{3!}\eta^3 + \frac{1}{5!}\eta^5 - \cdots\right] = \left[\eta^{1/2} - \frac{1}{3!}\eta^{5/2} + \frac{1}{5!}\eta^{9/2} - \cdots\right]$$

$$\frac{\sin\eta}{\sqrt{\eta^3}} = \left[\eta^{-1/2} - \frac{1}{3!}\eta^{3/2} + 1/5!\eta^{7/2} - \cdots\right]$$

the above equation is reduced to

$$\int_{\eta_0}^{0} \frac{\sin \eta}{\sqrt{\eta}} d\eta + \int_{0}^{\infty} \frac{\sin \eta}{\sqrt{\eta}} d\eta - \eta_0 \int_{\eta_0}^{0} \frac{\sin \eta}{\eta^{3/2}} d\eta - \eta_0 \int_{0}^{\infty} \frac{\sin \eta}{\eta^{3/2}} d\eta$$

$$= -\left[\frac{2}{3}\eta^{3/2} - \frac{2}{7}\cdot\frac{1}{3!}\eta^{7/2} + \frac{2}{11}\cdot\frac{1}{5!}\eta^{11/2} - \cdots\right]_0^{\eta_0} + \sqrt{\frac{\pi}{2}}$$

$$+ \eta_0\left[2\eta^{1/2} - \frac{2}{5}\cdot\frac{1}{3!}\eta^{5/2} + \frac{2}{9}\cdot\frac{1}{5!}\eta^{9/2} - \cdots\right]_0^{\eta_0} - 2\sqrt{\frac{\pi}{2}}\eta_0 = 0$$

Ignoring $\eta_0^{7/2}$ and higher order small quantities, we have

$$2\sqrt{\frac{\pi}{2}}\eta_0 - \frac{4}{3}\eta_0^{3/2} - \sqrt{\frac{\pi}{2}} = 0$$

Thus finally we obtain

$$x_0 = \sqrt{4a^2 t \eta_0} \approx 2.13 a\sqrt{t} = 2.13 \cdot \sqrt[4]{\frac{EI}{\rho_0 A_0}}\sqrt{t} \tag{10.61}$$

It indicates that the relation between x_0 and $\sqrt{t}$ is dependent only on beam size and material properties, but independent of impact velocity V_1.

Since the maximum of M and k occurs at point $x = 0$ and is proportional to V_1, with increase of V_1 the plastic deformation must occur first at point $x = 0$. From Eq. (10.60), let $M(0) = M_S$, the critical impact velocity V_S which is corresponding to the beginning of yield for the elastic beam, is obtained as

$$V_S = \frac{a^2 M_S}{EI} = a^2 k_S = \frac{2a^2 \varepsilon_S}{h} \tag{10.62}$$

where h is the height of the beam cross section and ε_S the tensile (compressive) yield stress.

(2) In linear hardening plastic cases

When $V_1 > V_S$, the plastic deformation takes place in the beam. If we assume that the M–k relation is described by the so-called linear hardening model, namely, comprising two straight lines with the slope of EI and $\xi^2 EI$ ($\xi < 1$) respectively, as shown in Fig. 10.6, then in the elastic deformation region and the plastic deformation region, Eq. (10.41) is respectively reduced to

$$S'' + S = 0$$

$$S'' + \frac{1}{\xi^2} S = 0$$

The general solutions are respectively:

$$S(\eta) = C_1 \cos\eta + C_2 \sin\eta = A_1 \sin(\eta - A_2) \quad (10.63)$$

$$S(\eta) = C_3 \cos\frac{\eta}{\xi} + C_4 \sin\frac{\eta}{\xi} = A_3 \sin(\eta - A_4) \quad (10.64)$$

Thus, $S(\eta)$ comprises these solutions in different regions. At the connecting point of these two kinds of solutions, S as well as S' at both regions should be the same, respectively. As can be seen from Eq. (10.50a), such requirement means that rotating angle α and curvature k should be continuous at the connecting point of elastic region and plastic region.

For example, for $0 \le \eta \le \eta_0$, since $M(\eta) \ge M_S$, the plastic deformation takes place, corresponding to the solution of (10.64); while for $\eta \ge \eta_0$, the deformation is still elastic, corresponding to the solution of (10.63), as shown in Fig. 10.8. Now we have totally five indeterminate coefficients: C_1, C_2, C_3, C_4, and η_0, which can be determined by the following five conditions, namely, the boundary conditions (10.51) and (10.57) at $\eta = 0$, the continuous conditions at $\eta = \eta_0$ for both $S(\eta)$ and $S'(\eta)$ calculated respectively from Eqs. (10.63) and (10.64), and finally $M(\eta_0) = M_S$.

Figure 10.8 displays the situation where at each beam cross section the plastic deformation only takes place once and no unloading happens. So the problem is relatively simple to solve. In such a case, at an arbitrary beam cross section x, at first a series of positive–negative alternative bending elastic waves pass through, and the amplitude successively increases. At time $t = x^2/(4a^2\eta_0)$, the cross section gets into plastic bending under the action of $+M$ (if V_1 is positive). After that the plastic deformation monotonously goes up. When t approaches infinity, M approaches $M(0)$.

If after the first plastic deformation, an unloading and soon afterwards a second plastic loading take place, the problem is more complicated. Let us first have a look at how the cross section comes into the unloading and re-loading.

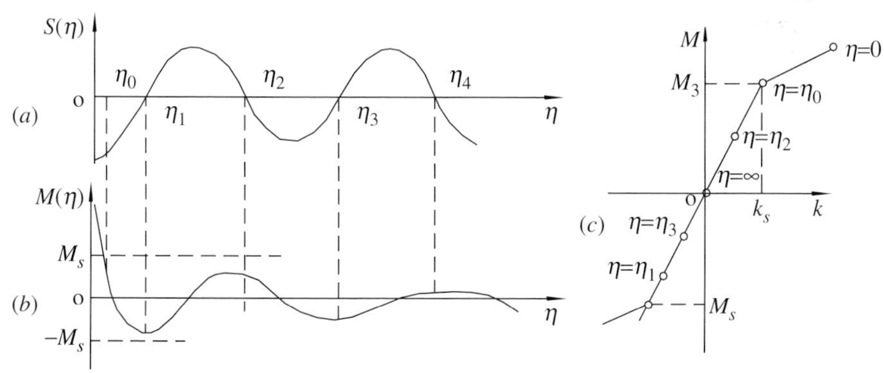

Fig. 10.8. The solutions when the plastic deformation only takes place once at beam cross section and no unloading happens.

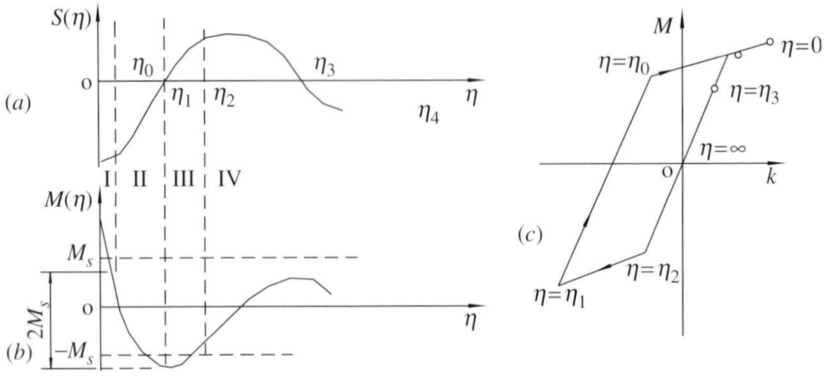

Fig. 10.9. The solutions when the unloading and the secondary plastic loading take place at beam cross section.

With increasing V_1, η_0 in Fig. 10.8 moves rightwards. However, η_0 is unable to reach point η_1, because M reaches its negative extreme at point η_1. So between η_0 and η_1, there must exist a null-point of M, or there must exist an interval $|M(\eta)| < M_S$. However, it should be noted that the absolute value of $M(\eta_1)$ also increases with increasing V_1. Therefore, if V_1 is large enough, an interval $M(\eta_1) \leq -M_S$ will finally appear. Consequently a new plastic region ($\eta_1 \leq \eta \leq \eta_2$) will be formed in both S–η and M–k plots, and an accompanying unloading region will appear between η_0 and η_1, as shown by regions II and III in Fig. 10.9, respectively. In such a case, for an arbitrary cross section x of beam, the motion is in this way: first a series of positive–negative alternative bending elastic waves pass through; at time $t = x^2/(4a^2\eta_2)$, the cross section gets into plastic bending under the action of $-M_S$, and the plastic deformation gradually goes up; then at time $t = x^2/(4a^2\eta_1)$, the negative M reaches its extreme; after that the absolute value of M decreases, namely the unloading of post-plastic deformation takes place. Along with the development of unloading process, M transforms to positive, and then at time $t = x^2/(4a^2\eta_0)$ the cross section gets into the secondary plastic deformation under the action of M. Later on the plastic deformation monotonously goes up. On this analogy, if V_1 is large enough, plastic deformation and the accompanying unloading can take place one more time. The cross section finally always gets into plastic bending under the action of $+M$ provided that V_1 is positive. Before this final monotonously increasing plastic bending, how many times the positive–negative alternative plastic bending and the accompanying unloading will take place, depends on how large the V_1 is. It should be noted that the **interior unloading** discussed here should be distinguished from the unloading of external applied load.

In the following, the situation where an unloading and a secondary plastic loading take place as shown in Fig. 10.9, will be further discussed in detail. For the situations where the plastic loading takes place many times, the problem in principle can be treated in the same way.

First, it is assumed for the M–k relation that the unloading curve is parallel to the elastic straight line, and if the unloading begins at $M(\eta_1)$, then when the inverse reloading reaches

$M(\eta_0)$, which satisfies the condition $|M(\eta_0) - M(\eta_1)| = 2M_S$, the plastic deformation takes place once again, as shown in Fig. 10.9(c). In other words, we have

$$M(\eta_2) = -M_S \tag{10.65}$$

$$|M(\eta_0) - M(\eta_1)| = 2M_S \tag{10.66}$$

This is equivalent to assuming that the unloading process is an elastic one and the plastic reloading process takes into account the Bauschinger's effect.

In the regions I–IV in Fig. 10.9, $S(\eta)$ is obtained respectively as

$$\text{I} \quad S = A\cos\left(\frac{1}{\xi}\right)\eta \qquad 0 \le \eta \le \eta_0; \tag{10.67a}$$

$$\text{II} \quad S = -B\sin(\eta - \eta_1) \qquad \eta_0 \le \eta \le \eta_1; \tag{10.67b}$$

$$\text{III} \quad S = -C\sin\left(\frac{1}{\xi}\right)(\eta - \eta_2) \qquad \eta_1 \le \eta \le \eta; \tag{10.67c}$$

$$\text{IV} \quad S = C\sin(\eta - \eta_3) \qquad \eta_2 \le \eta \tag{10.67d}$$

where the conditions $S'(0) = 0$ and $S(\eta_1) = 0$ have been utilized, and consequently the requirement of $S(\eta)$ in regions II and III should be continuous at point η_1 has been naturally satisfied. The equation set (10.67) contains eight indeterminate coefficients: $A,\ B,\ C,\ D,\ \eta_0,\ \eta_1,\ \eta_2$, and η_3. They can be determined by the following eight equations: two equations to express the requirement that $S(\eta)$ in different regions should be continuous at the connecting points (except η_1), three equations to express the requirement that $S'(\eta)$ in different regions should be continuous at the connecting points, and three equations [Eq. (10.57), (10.65), and (10.66)], namely

$$\left\{\begin{array}{ll}
A\cos\left(\frac{1}{\xi}\right)\eta_0 = -B\sin(\eta_0 - \eta_1) & \text{(a)} \\
\frac{A}{\xi}\sin\frac{\eta_0}{\xi} = B\cos(\eta_0 - \eta_1) & \text{(b)} \\
B = \frac{C}{\xi} & \text{(c)} \\
-C\sin\frac{\eta_2 - \eta_1}{\xi} = D\sin(\eta_2 - \eta_3) & \text{(d)} \\
\frac{C}{\xi}\cos\frac{\eta_2 - \eta_1}{\xi} = -D\cos(\eta_2 - \eta_3) & \text{(e)} \\
\int_0^{\infty}\frac{S'(\eta)}{\sqrt{\eta}}\,d\eta = 2V_1 & \text{(f)} \\
\int_{\eta_2}^{\infty}\frac{S(\eta)}{\sqrt{\eta}}\,d\eta = 2V_S & \text{(g)} \\
\int_{\eta_0}^{\eta_1}\frac{S(\eta)}{\sqrt{\eta}}\,d\eta = -4V_S & \text{(h)}
\end{array}\right\} \tag{10.68}$$

The practical calculation process is as following: choose a pair of values η_2 and η_3 to determine the values of η_0, η_1 and the ratios A:B:C:D from Eqs. (10.68a–c). Then by trial and error, find a pair of values η_2 and η_3 to satisfy Eqs. (10.68g, h). The impact velocity V_1 is then calculated from Eq. (10.68f). This calculation is repeated until the given V_1 is obtained. The integrals involved in the calculation can be easily reduced to the Fresnel integrals, the values of which can be found from the integral table (e.g. see Sparrow, 1934).

(3) In perfectly plastic cases

If the M–k relation of beam is described by the so-called perfect plastic model, as shown by curve 2 in Fig. 10.6, then the calculation will be much simplified. Such a situation can be regarded as a special case of linear hardening plastic model when $\xi = 0$.

Let us first discuss the situation where at each beam cross section the plastic deformation only takes place once, as shown in Fig. 10.8 but now $\xi = 0$. Since $(\eta_0/\xi) < \pi/2$, region I will shrink to a point $\eta = 0$. At this point, $M(0) = M_S$ while k can increase infinitely, namely a plastic hinge is formed. So the solution of the current problem is reduced to

$$S = -B\sin(\eta - \eta_1), \quad 0 \le \eta \tag{10.69}$$

The indeterminate coefficients B and η_1 are determined by the condition of $M(0) = M_S$ and Eq. (10.57):

$$\begin{cases} B\displaystyle\int_0^\infty \frac{\cos(\eta - \eta_1)}{\sqrt{\eta}}\,d\eta = -2V_1 \\ B\displaystyle\int_0^\infty \frac{\sin(\eta - \eta_1)}{\sqrt{\eta}}\,d\eta = 2V_S \end{cases} \tag{10.70}$$

Because $\cos(\eta-\eta_1) = \cos\eta\cos\eta_1 + \sin\eta\sin\eta_1$, $\sin(\eta-\eta_1) = \sin\eta\cos\eta_1 - \cos\eta\sin\eta_1$ and $\int_0^\infty (\cos\eta/\sqrt{\eta})d\eta = \int_0^\infty (\sin\eta/\sqrt{\eta})\,d\eta = \sqrt{\pi/2}$, from the above two equations we obtain

$$\begin{cases} B\sqrt{\dfrac{\pi}{2}}\cos\eta_1 = V_S - V_1 \\ B\sqrt{\dfrac{\pi}{2}}\sin\eta_1 = -(V_S + V_1) \end{cases} \tag{10.70a}$$

and then we obtain

$$\begin{cases} B = 2\sqrt{\dfrac{V_1^2 + V_S^2}{\pi}} \\ \eta_1 = tg^{-1}\left(\dfrac{V_1 + V_S}{V_1 - V_S}\right) = \dfrac{\pi}{4} + tg^{-1}\left(\dfrac{V_S}{V_1}\right) \end{cases} \tag{10.71}$$

Note that since a plastic hinge is formed at point $x = 0$, $(\partial w/\partial x)$ at $x = 0$ is no longer continuous and k at $x = 0$ could increase infinitely. In fact, according to Eq. (10.54), now we have

$$\tan\theta = \alpha|_{x=+0} = \left.\frac{\partial w}{\partial x}\right|_{x\neq+0} = -\frac{\sqrt{t}}{a}S'(\eta)\bigg|_{\eta=0} = \frac{\sqrt{t}}{a}B\cos\eta_1 \tag{10.72a}$$

or from Eq. (10.70a) it can be rewritten as

$$\tan\theta = \alpha|_{x=+0} = \sqrt{\frac{2}{\pi}\frac{1}{a}}(V_S - V_1)\sqrt{t} = \sqrt{\frac{2k_S}{\pi V_S}}(V_S - V_1)\sqrt{t} \tag{10.72b}$$

It indicates that $\alpha(0, t)$ increases in direct proportion to $\sqrt{t}$. Note that with regard to α the deformation of beam is anti-symmetric with respect to the impact point, so the solution of $\alpha|_{x=-0}$ is obviously the reversal of the sign of the solution of $\alpha|_{x=+0}$ expressed by Eq. (10.72). It can be noted from Eq. (10.50a) that a discontinuity of a at $x = 0$ must correspond to a discontinuity of $S(\eta)|_{\eta=0}$. In fact, it can be directly seen from Eq. (10.69) that the condition of $S'(0) = 0$ expressed by Eq. (10.51) no longer holds, and the value of $S'(\eta)_{\eta=+0} = -\frac{a}{\sqrt{t}}\alpha\Big|_{x=+0}$ is finite.

The above solution holds only when $|M(\eta)| < M_S$, $(\eta \neq 0)$. With increasing V_1, we will first have $M(\eta_1) = -M_S$. According to Eq. (10.53) we obtain

$$M(\eta_1) = \frac{EIB}{2a^2}\int_{\eta_1}^{\infty}\frac{\sin(\eta - \eta_1)}{\sqrt{\eta}}\,d\eta = -M_S = -\frac{EIV_S}{a^2}$$

$$B\left[\cos\eta_1\int_{\eta_1}^{\infty}\frac{\sin\eta}{\sqrt{\eta}}\,d\eta - \sin\eta_1\int_{\eta_1}^{\infty}\frac{\cos\eta}{\sqrt{\eta}}\,d\eta\right] = -2V_S$$

From Eq. (10.70a) we know $2V_S = B\sqrt{(\pi/2)}(\cos\eta_1 - \sin\eta)$, and eliminating B from the above equation we obtain

$$\sin\eta_1\frac{1}{\sqrt{2\pi}}\int_0^{\eta_1}\frac{\cos\eta}{\sqrt{\eta}}\,d\eta - \cos\eta_1\frac{1}{\sqrt{2\pi}}\int_0^{\eta_1}\frac{\sin\eta}{\sqrt{\eta}}\,d\eta = \sin\eta_1 - \cos\eta_1$$

The solution of this equation is $\eta_1 = 70.6°$, and correspondingly we have $V_1 = 2.087V_S$, which is the critical velocity of the above solution. In other words, when $V_3 < V_1 < 2.087V_S$, we have the solution (10.69). In such a case, only at $x = 0$ a plastic hinge is formed and the rotating angle of plastic bending at the impact point is determined by Eq. (10.72), while remaining part of the beam is in the state of elastic bending.

When $V_1 = 2.087V_S$, region III as shown in Fig. 10.9 will appear. However, since $(\eta_2 - \eta_1)/\alpha < \pi$, when a approaches zero, region III will shrink at point $\eta = \eta_1$.

The solution of the current problem is:

$$\left\{\begin{array}{ll} S = -B\sin(\eta - \eta_1), & 0 \le \eta \le \eta_1 \\ S = -D\sin(\eta - \eta_1), & \eta_1 \le \eta \end{array}\right\} \tag{10.73}$$

The indeterminate coefficients B, D, and η_1 can be determined by the following three conditions: Eq. (10.57), $M(\eta_1) = -M_S$ and $-M(\eta_1) = 2M_S$, namely

$$\left\{\begin{array}{l} B\displaystyle\int_0^{\eta_1} \frac{\cos(\eta - \eta_1)}{\sqrt{\eta}}\, d\eta + D\int_{\eta_1}^{\infty} \frac{\cos(\eta - \eta_1)}{\sqrt{\eta}}\, d\eta = -2V_1 \\ D\displaystyle\int_{\eta_1}^{\infty} \frac{\sin(\eta - \eta_1)}{\sqrt{\eta}}\, d\eta = -2V_S \\ B\displaystyle\int_0^{\eta_1} \frac{\sin(\eta - \eta_1)}{\sqrt{\eta}}\, d\eta = 4V_S \end{array}\right\} \tag{10.74}$$

For the same reason as was discussed earlier, because regions I and III now shrink to two points, the discontinuity of $S'(\eta)$ appears at point $\eta = 0$ and point $\eta = \eta_1$, as shown in Fig. 10.10.

It can be seen from the basic equation (10.41) that these two points are singular points, $S'' \to \infty$. Thus, corresponding to $\eta = 0$, a stationary plastic hinge is formed at $x = 0$; and corresponding to the discontinuity of $S'(0)$, $\alpha(0, t)$ is discontinuous at $x = 0$, namely the deflection curve of beam possesses a nonzero angle $\alpha \neq 0$ at $x = 0$, the value of which is still calculated according to Eq. (10.72). On the other hand, corresponding to $\eta = \eta_1$, another plastic hinge is formed at $x_1 = 2a\sqrt{\eta_1 t}$. Noting that x_1 is directly proportional to $\sqrt{t}$, so this plastic hinge moves with a velocity of $C_{\eta_1} = dx_1/dt = a\sqrt{\eta_1/t}$ in the direction deviating from the impact point. Corresponding to the discontinuity of $S'(\eta)$, a jump of curvature Δk happens at point x_1, although $\alpha|_{x=x_1} \left(= \frac{\partial w}{\partial x}\Big|_{x=x_1}\right)$ is still continuous. It can be physically explained as follows. Since a constant bending moment M_S acts at the plastic

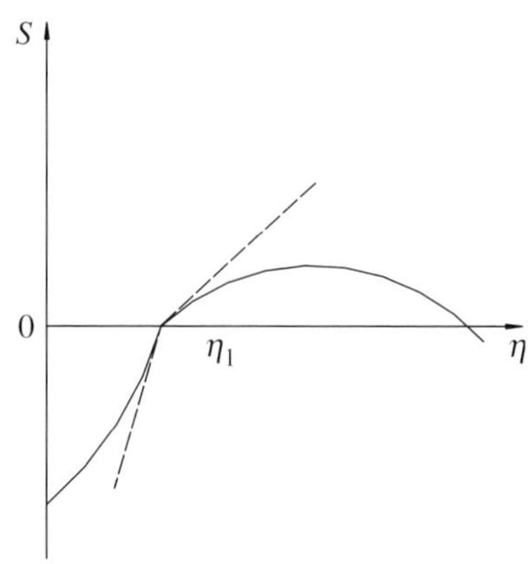

Fig. 10.10. The discontinuity of $S'(\eta)$ appears at $\eta = 0$ and $\eta = \eta_1$.

hinge, the curvature can increase infinitely, although it needs time for such a process. For a stationary plastic hinge, a finite discontinuity of a takes place after a finite interval, and the jump level of discontinuity increases continuously with time, as described by Eq. (10.72). However, for a traveling plastic hinge, since it only passes instantaneously through each cross section, there is not enough time to let the discontinuity of α to take place, so $\alpha|_{x=x_1}$ is still continuous. It can be seen from Eq. (10.50), the discontinuity of $S'(\eta)$ now only means the discontinuity of curvature k or the discontinuity of angular velocity ω, and obviously the jump level of discontinuity of curvature Δk is

$$\Delta k = k(\eta_1 - 0) - k(\eta_1 + 0) = \frac{1}{2a^2\sqrt{\eta_1}}\left[S'(\eta_1+0) - S'(\eta_1-0)\right] = \frac{k_S}{2V_S}\cdot\frac{1}{\sqrt{\eta}}(B-D) \tag{10.75}$$

Once B, D, and η_1 are obtained from Eq. (10.74), θ and Δk can be calculated from Eqs. (10.72) and (10.75). When $V_1 \gg V_S$, we have the following approximate equations given by Duwez *et al.* (1943, 1950)

$$\begin{cases} \eta_1 \approx \dfrac{3V_S}{V_1} \\ tg\theta \approx \sqrt{t}\sqrt{\dfrac{k_S}{3}\dfrac{V_1^{3/2}}{V_S}} \end{cases} \tag{10.76}$$

In Fig. 10.11, the deflection distribution curves and the bending moment distribution curves are given for three different situations, namely (a) $V_1 < V_S$ (elastic beam), (b) $V_S < V_1 < 2.087V_S$ (one stationary plastic hinge), and (c) $V_1 > 2.087V_S$ (stationary and traveling plastic hinges).

Thus, in the case of perfectly plastic M–k relation, the plastic bending of beam is accomplished through the stationary plastic hinge and the traveling plastic hinges which propagate longitudinally along the beam axis.

The deflection curves of rectangular cross section beams under transverse constant-velocity impact were experimentally measured by Duwez *et al.* (1943, 1950) too. The tested beams were made of two different metals, namely, cold-rolled low-carbon steel, and annealed copper. The actual M–k curve experimentally measured (plotted by solid line) and the approximate M–k curve used for computation (plotted by dashed line) are shown in Fig. 10.12 for steel and in Fig. 10.13 for copper, respectively. The comparisons between the experimental and the computational deflection curves for both metals are given in Figs. 10.14 and 10.15, respectively.

As can be seen, the shapes of the theoretical curves and the experimental curves are similar. In the case of the cold-rolled steel beams, the elastic deflection curve agrees rather well with the experimental curve. However, in the case of the annealed copper beams, the elastic–plastic curve is in rather good agreement with the experimental curve. Moreover, it was observed from the experiments that plastic strains are localized in a very small region near the impact point in the steel beam, while plastic strains extend to some distance from the impact point in the copper beam. Thus, as explained by the authors, it seems that the influence of plasticity is greater in copper beam than in steel beam.

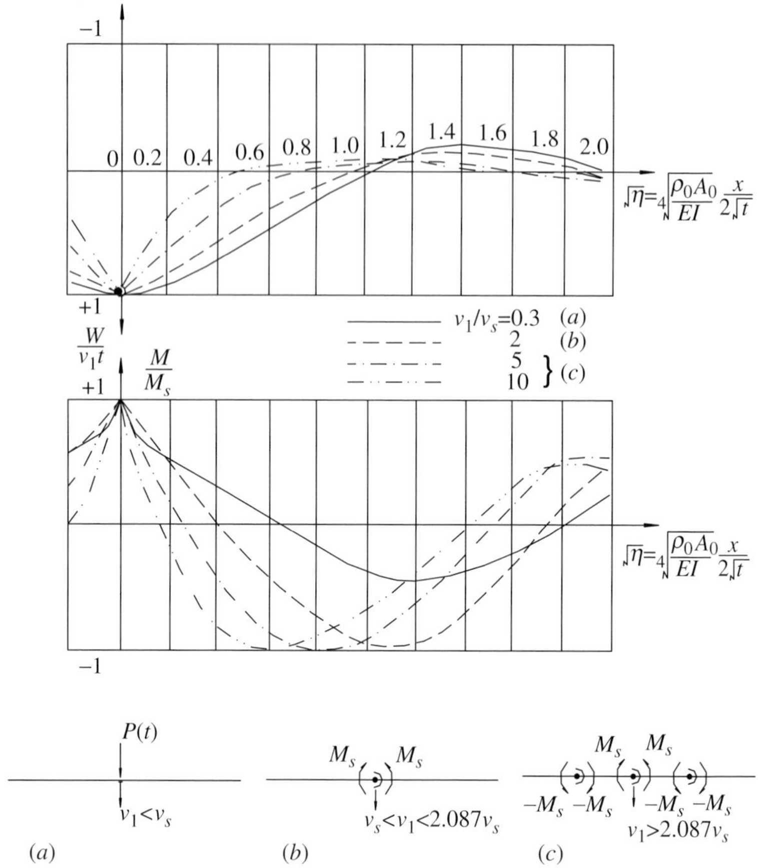

Fig. 10.11. Deflection curves and bending moment curves for an infinite beam under transverse constant-velocity impact V_1 in different situations: (a) $V_1 < V_S$, (b) $V_S < V_1 < 2.087V_S$, and (c) $V_1 > 2.087V_S$.

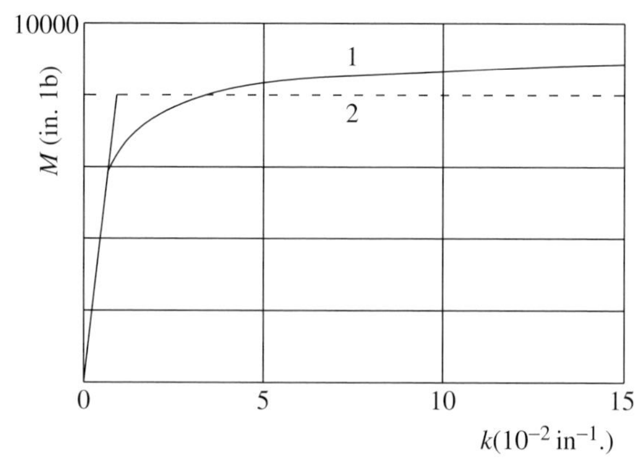

Fig. 10.12. *M*–*k* curve for cold-rolled steel beam (solid line 1 – experimentally measured curve and dashed line 2 – approximate curve). [Reproduced from Duwez *et al.* (1950) by permission of ASTM]

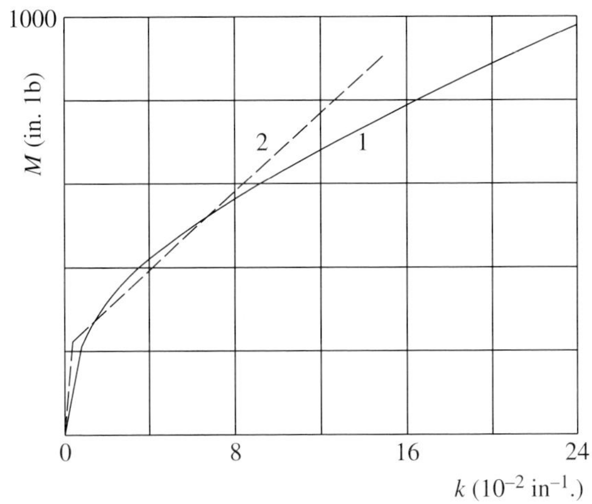

Fig. 10.13. *M–k* curve for annealed copper beam (solid line 1 – experimentally measured curve and dashed line 2 – approximate curve). [Reproduced from Duwez *et al.* (1950) by permission of ASTM]

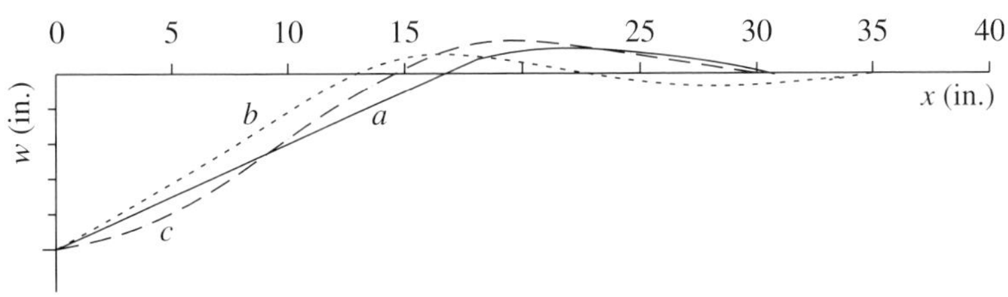

Fig. 10.14. Deflection curves for cold-rolled steel beam (solid line *a* – experimentally measured curve, dotted line *b* – computed elastic–plastic beam curve, and dashed line *c* – computed elastic beam curve). [Reproduced from Duwez *et al.* (1950) by permission of ASTM]

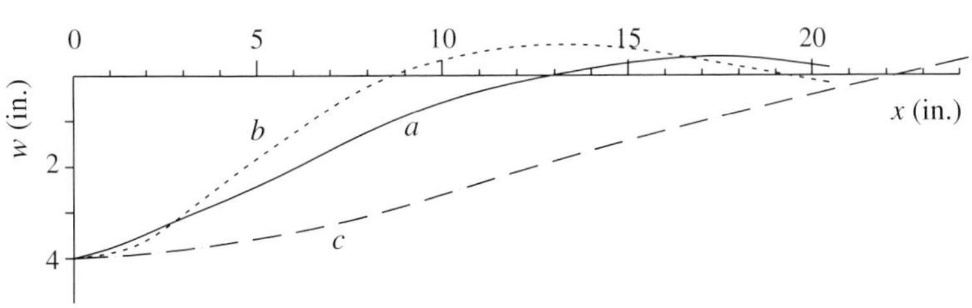

Fig. 10.15. Deflection curves for annealed copper beam (solid line *a* – experimentally measured curve, dotted line *b* – computed elastic–plastic beam curve, and dashed line *c* – computed elastic beam curve). [Reproduced from Duwez *et al.* (1950) by permission of ASTM]

With regard to some appreciable differences between the computational and the experimental curves, it can be attributed to the following main reasons: (1) The M–k curve used for computation is an approximation; (2) The experimental results may be influenced by the bending waves reflected from the beam ends; and (3) The elementary theory itself contains a series of basic assumptions, particularly the rotatory inertia and the shear stress effect are ignored; as well as the computation is not based on the dynamic M–k curve but on the static M–k curve.

In addition, the above experiments verified that the distance x_0 at which $w = 0$, is directly proportional to $t^{1/2}$, namely, when the bending wave propagates along the beam axis its velocity is not a constant but dependent on x^2/t.

10.4 Rigid-Plastic Analysis

The rigid-plastic analysis of beam was developed by Lee and Symonds (1952). Similar to the earlier discussion on the plastic waves in bars, if the plastic deformation is much larger than the elastic deformation, then the elastic deformation could be ignored, and the problem is reduced to an analysis of the so-called **rigid-plastic beam**.

Assume that the M–k relation is simplified as shown in Fig. 10.16. It is equivalent to making such an assumption: when $|M| < M_0$, the beam behaves as a rigid body, while when $|M| = M_0$ at any cross section of the beam, the curvature can increase infinitely, namely, the so-called plastic hinge is formed, as illustrated in Fig. 10.17. Thus, the problem is reduced to the rigid motion of beam segments which are connected to each other by

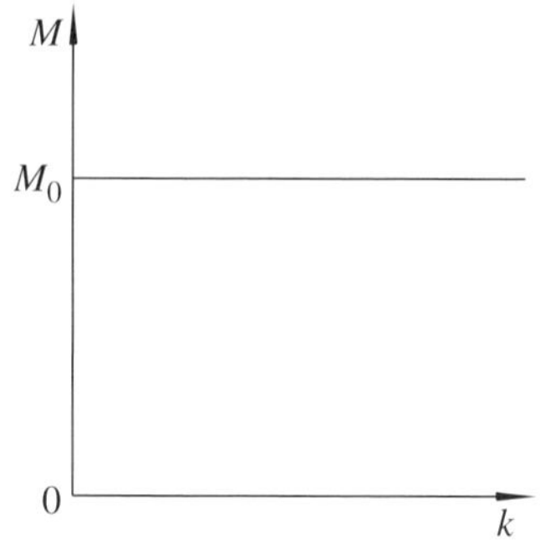

Fig. 10.16. Rigid-plastic M–k curve.

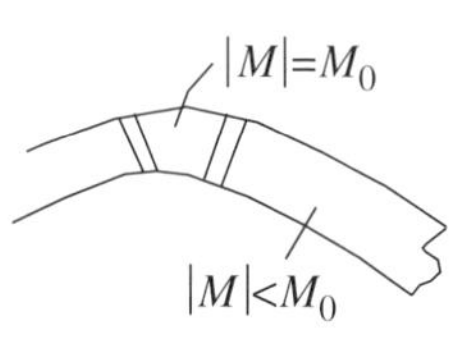

Fig. 10.17. Plastic hinge at $|M| = M_0$.

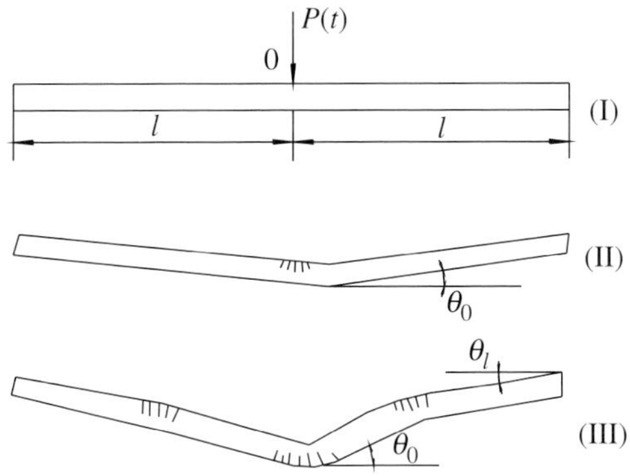

Fig. 10.18. Three phases of movement in the rigid-plastic analysis for a finite straight beam with constant section area under mid-point impact.

plastic hinges. It should be noticed that an actual plastic hinge is gradually formed and occupies a finite plastic deformation region. Only if the plastic deformation is much larger than the elastic deformation, then we can regard the plastic hinge as a singular cross section formed suddenly, and the plastic deformation is concentrated on this cross section. The problem will be much simplified by such an approximation. Not only the problem of finite beam can be solved, including the unloading of external load, but also the external load condition is not limited to the constant-velocity impact, these could not be treated in Section 10.3.

Now let us consider a finite straight beam with a length of $2l$, two stress-free ends and constant cross-sectional area, which is subjected to an impulsive force $P(t)$ at the central point, as shown in Fig. 10.18.

The movement of the beam can be described by dividing it into three stages or three phases. In phase I, $M < M_0$, only translational motion of rigid body is involved; In phase II, at the impact point $M = M_0$, there exists the relative rotation of two halves of beam around the middle plastic hinge; In phase III, at the cross section departing from the impact point at a certain distance $M = -M_0$, in addition to the middle stationary plastic hinge, there are two traveling plastic hinges on the two sides of beam.

In the following, let us discuss each phase in more detail.

Phase I $\left(\mu = \dfrac{Pl}{M_0} < 4\right)$:

When $P(t)$ starts to increase from zero, the beam also begins to move in parallel to itself as a motion of rigid body. In this stage, the acceleration a of each point is the same, which is

$$a = \frac{P}{2ml} \tag{10.77}$$

where m is the density of beam per unit length (line density). This is equivalent to a static beam subjected to a uniformly distributed load $-ma$. So the bending moment distribution is

$$M(x) = \int_0^{l-x} ma\xi \, d\xi = \frac{m(l-x)^2}{2}a = \frac{P(l-x)^2}{4l} \tag{10.78}$$

Obviously, the maximum bending moment exists at the central point of beam, $x = 0$, which is

$$M(0) = \frac{ml^2}{2}a = \frac{Pl}{4}$$

Therefore, when $P = 4M_0/l$, we have $M(0) = M_0$, a plastic hinge is then formed at the central point, and the beam no longer moves in the form of the translational motion of rigid body. By introducing a dimensionless parameter $\mu = Pl/M_0$, when it satisfies the following condition

$$\mu = \frac{Pl}{M_0} = 4 \tag{10.79}$$

Phase I ends and Phase II begins.

Phase II ($4 \le \mu < 22.9$):

In this stage, there is a constant bending moment M_0 acting at the middle plastic hinge. The two halves of the beam still move in the form of rigid motion, but they all relatively rotate around the plastic hinge, as shown in Fig. 10.19. Let a_0 be the linear acceleration at the central point of beam $x = 0$, θ_0 the rotation angle of the right half of the beam around the hinge (which is positive in the anti-clockwise direction), $\alpha_0 = d^2\theta_0/dt^2$ the corresponding angular acceleration, then the momentum equation, and the moment of momentum equation (the moment with respect to $x = l/2$) for the right half of the beam are expressed respectively as

$$\frac{P}{2} = \int_0^l ma_0 \, dx - \int_0^l \frac{d^2\theta_0}{dt^2} x \, dx = ml\left(a_0 - \frac{l\alpha_0}{2}\right) \tag{10.80}$$

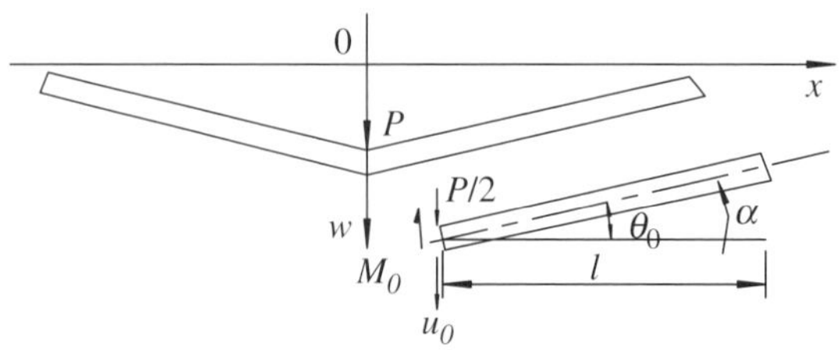

Fig. 10.19. Relative rigid rotations of the two halves of the beam around the hinge.

$$\frac{Pl}{4} - M_0 = 2\int_0^{l/2} m\frac{d^2\theta_0}{dt^2}x^2\,dx = \frac{ml^3}{12}\alpha_0 \tag{10.81}$$

From these equations, a_0 and α_0 can be solved and expressed as

$$a_0 = \frac{M_0}{ml^2}\left(\frac{2Pl}{M_0} - 6\right) = \frac{M_0}{ml^2}(2\mu - 6) \tag{10.82}$$

$$l\alpha_0 = \frac{M_0}{ml^2}\left(\frac{3Pl}{M_0} - 12\right) = \frac{M_0}{ml^2}(3\mu - 12) \tag{10.83}$$

When $P(t)$ is given, both linear acceleration $a_0(t)$ and angular acceleration $\alpha_0(t)$ can be calculated from the above equations. Then by integrating with respect to time t, the linear velocity and the angular velocity can be further calculated, and displacement $w(0, t)$ and angular displacement θ_0 can be finally calculated.

Note that even when $P(t)$ decreases, the above equations still hold as long as the plastic hinge does not disappear. In fact, with decreasing $P(t)$, first the sign of a_0 and α_0 will change from positive to negative, while the sign of angular velocity is still positive, which means that the plastic hinge does not immediately disappear. Even after $P(t)$ has decreased to zero, the plastic hinge can still exist. This is because of the fact that after the unloading process of external load begins, the rigid kinetic energy of beam segments can still transform to the plastic deformation energy, and thus the additional plastic deformation is further induced. As can be seen from this example, the problem concerning the unloading process of external load can be solved by the rigid-plastic analysis.

During Phase II, except the central plastic hinge at $x = 0$, the bending moment at any other cross section of the beam is smaller than M_0. If $P(t)$ increases further, new plastic hinge may appear at other cross sections. The critical condition for the end of Phase II will be discussed as follows.

Let us first look at the bending moment distribution, which can be obviously expressed as (see Fig. 10.19)

$$\begin{aligned} M(x) &= \frac{P}{2}x - M_0 - \int_0^x ma_0 x\,dx + \int_0^x m\frac{d^2\theta_0}{dt^2}(x-\xi)\xi\,d\xi \\ &= \frac{P}{2}x - M_0 - \frac{mx^2}{2}a_0 + \frac{mx^3}{6}\alpha_0 \end{aligned} \tag{10.84a}$$

By substituting Eqs. (10.82) and (10.83) into the above equation, we have

$$\begin{aligned} M(x) &= \frac{\mu M_0}{2}\frac{x}{l} - M_0 - \frac{1}{2}M_0(2\mu - 6)\left(\frac{x}{l}\right)^2 + \frac{1}{6}M_0(3\mu - 12)\left(\frac{x}{l}\right)^3 \\ &= M_0\left(\frac{\mu - 4}{2}\frac{x}{l} - 1\right)\left(1 - \frac{x}{l}\right)^2 \end{aligned} \tag{10.84b}$$

The maximum of bending moment appears at the cross section $x = x_1$, where $(dM/dx)_{x_1} = 0$, that is

$$\frac{P}{2} - ma_0 x_1 + \frac{1}{2} ma_0 x_1^2 = 0$$

or

$$\left(\frac{x_1}{l}\right)^2 - \frac{4}{3}\frac{\mu - 3}{\mu - 4}\left(\frac{x_1}{l}\right) + \frac{\mu}{3(\mu - 4)} = 0$$

The above equation has two roots. This equation and the corresponding $M(x_1)$ can be solved as

$$\begin{cases} \dfrac{x_1}{l} = 1, & M(x_1) = 0 \\ \dfrac{x_1}{l} = \dfrac{\mu}{3(\mu - 4)}, & M(x_1) = M_0\left[1 + \dfrac{9\mu^2 - 2\mu^3}{27(\mu - 4)^2}\right] \end{cases} \tag{10.85}$$

Let us first discuss the second root, which is related to the beginning of Phase II. Due to $x_1/l \leq 1$, the second root exists only when $\mu \geq 6$. In fact, in the range of $4 \leq \mu \leq 6$, the bending moment distribution monotonously decreases with x, and the minimum appears at $x = l$, i.e. $M(l) = 0$, which corresponds to the first root. If by substituting $4 \leq \mu \leq 6$ into Eq. (10.85), the value of $M(x_1)$ is positive, but $M(x_1)$ decreases with increasing μ or $P(t)$; when $\mu = 6$, $M(x_1) = 0$, and when $\mu > 6$ the value of $M(x_1)$ is negative. Thus, the critical condition for the end of Phase II is determined by the following equation:

$$M(x_1) = M_0\left[1 + \frac{9\mu^2 - 2\mu^3}{27(\mu - 4)^2}\right] = -M_0$$

This is a cubic equation with respect to μ:

$$2\mu^3 - 63\mu^2 + 432\mu - 864 = 0$$

Its only real root is $\mu = 22.9$, and corresponding to this real root we have

$$x_1 = \frac{\mu}{3(\mu - 4)} l = 0.404l \tag{10.86}$$

Therefore when $\mu = 22.9$, $M = -M_0$ at $x = \pm 0.404l$ two new plastic hinges appear, so Phase II ends and Phase III begins.

Phase III ($\mu \geq 22.9$):

When $\mu = 22.9$, Phase III begins. If the external load increases continuously, then the limit bending moment M_0 will take place at $x < 0.404l$. In other words, with increasing external load, the plastic hinge originally at x_1 will move (travel) towards the impact point $x = 0$.

This may not be evident immediately, but can be seen from a further discussion by solving the motion equations together with the kinematic conditions across the hinge as follows.

First of all, it should be pointed out that as soon as the plastic hinge moves to a new position, the bending moment M at the original position decreases to a value smaller than M_0, and consequently the relative rotation no longer exists there. In fact, on the left and right sides of plastic hinge, there must be $M < M_0$. Otherwise, a region of constant M must exist. But according to the relation $dM/dx = Q$, the shear force must be zero in this region, and consequently the acceleration of all particles in this region should be zero too. It means that this region moves in the form of rigid motion, no longer as a plastic hinge.

For convenience, let us consider the right half of the beam. Now imagine that at an instance t, the traveling plastic hinge is at position x_h. Since both the left side and the right side of x_h are rigid bodies, the momentum equations and the moment of momentum equations for them can be expressed respectively as (see Fig. 10.20)

$$\frac{P}{2} = \int_0^{x_h} ma_0\,dx - \int_0^{x_h} m\frac{d^2\theta_0}{dt^2}x\,dx = mx_h\left(a_0 - \frac{\alpha_0 x_h}{2}\right) \tag{10.87}$$

$$\frac{Px_h}{4} - 2M_0 = 2\int_0^{x_h/2} m\frac{d^2\theta_0}{dt^2}x^2\,dx = \frac{mx^3\alpha_0}{12} \tag{10.88}$$

$$0 = \int_{x_h}^{l} ma_l\,dx + \int_{x_h}^{l} m\frac{d^2\theta_l}{dt^2}x\,dx = m(l - x_h)\left[a_l + \frac{\alpha_l(l - x_h)}{2}\right] \tag{10.89}$$

$$M_0 = 2\int_0^{(1-x_h)/2} m\frac{d^2\theta_l}{dt^2}\xi^2\,d\xi = \frac{m(l - x_h)^2}{12}\alpha_1 \tag{10.90}$$

where a_l is the linear acceleration at point $x = l$, $a_l = d^2\theta_l/dt^2$ is the angular acceleration of rigid segment on the right side of the traveling plastic hinge. The above dynamic conditions, totally four equations, are not enough to solve the five unknown variables a_0, α_0, a_l, α_l, and x_h. Thus the kinematical condition of the traveling plastic hinge should be further considered as follows.

Now let us consider the kinematical equation for the traveling plastic hinge. Denoting the left side and the right side of the plastic hinge position by x_h^- and x_h^+ respectively, the

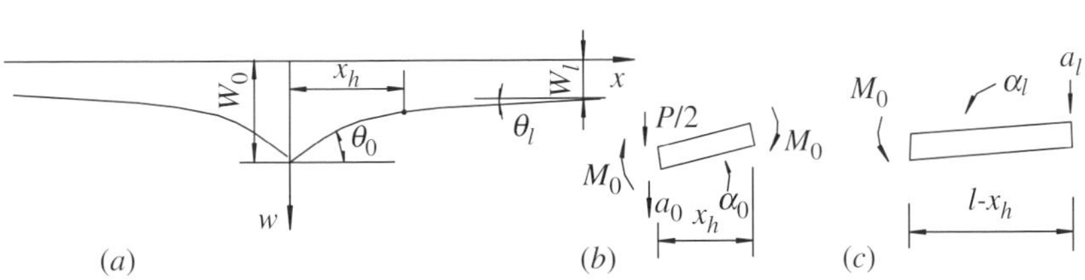

Fig. 10.20. The traveling plastic hinge moves towards the impact point ($\mu \geq 22.9$).

continuity of displacement w at the plastic hinge is expressed as

$$w(x_h^-) = w(x_h^+)$$

which should hold regardless of the time and the position of the plastic hinge. In other words, for the left and right sides of the hinge, the total derivatives of displacement following the traveling plastic hinge should be the same:

$$\frac{d}{dt}w[x_h^-] = \frac{d}{dt}w[x_h^+]$$

$$\frac{\partial w(x_h^-)}{\partial t} + c\frac{\partial w(x_h^-)}{\partial t} = \frac{\partial w(x_h^+)}{\partial t} + c\frac{\partial w(x_h^+)}{\partial x}$$

where $c = dx_h/dt$ is the traveling velocity of the plastic hinge along the beam axis. Therefore, the following relation between the transverse velocity $v = \partial w/\partial t$ and the rotating angle $\theta = -(\partial w/\partial x)$ exists

$$v(x_h^-) - v(x_h^+) = +c\lfloor \theta(x_h^-) - \theta(x_h^+) \rfloor \tag{10.91}$$

As mentioned in Section 10.3, when a traveling plastic hinge moves along the beam axis, it instantaneously (in an infinitesimal time duration) passes through each cross section, there is not enough time to let the discontinuity of rotation angle to take place, so the rotation angle θ across the hinge is continuous. Since $\theta(x_h^-) = \theta(x_h^+)$, by Eq. (10.91) the transverse velocity is also continuous. Similarly, since we have

$$\frac{d}{dt}\theta(x_h^-) = \frac{d}{dt}\theta(x_h^+)$$

$$\frac{d}{dt}v(x_h^-) = \frac{d}{dt}v(x_h^+)$$

correspondingly we can establish the following kinematical relation between angular velocity $\omega = \partial\theta/\partial t$ and curvature $k = \partial\theta/\partial x = -(\partial^2 w/\partial x^2)$:

$$\omega(x_h^-) - \omega(x_h^+) = -c\lfloor k(x_h^-) - k(x_h^+) \rfloor \tag{10.92}$$

and the kinematical equation between angular velocity $\omega = \partial\theta/\partial t = -(\partial^2 w/\partial t\partial x)$ and the transverse acceleration $a = \partial v/\partial t = \partial^2 w/\partial t^2$:

$$a(x_h^-) - a(x_h^+) = c\lfloor \omega(x_h^-) - \omega(x_h^+) \rfloor \tag{10.93}$$

Since angular velocity ω is discontinuous across the plastic hinge, it can be seen from Eqs. (10.92) and (10.93) that transverse acceleration a and curvature k must be discontinuous too.

Having the above results, the velocities and the accelerations for both the sides of the plastic hinge are obtained as following

$$w = w_0 - \int_0^x \theta \, dx \tag{10.94}$$

for $x < x_h$,

$$v = \frac{\partial w}{\partial t} = \frac{\partial w_0}{\partial t} - \int_0^x \frac{\partial \theta}{\partial t} dx = v_0 - \frac{d\theta_0}{dt} x \tag{10.95}$$

$$a = \frac{\partial^2 w}{\partial t^2} = \frac{\partial^2 w_0}{\partial t^2} - \int_0^x \frac{\partial^2 \theta}{\partial t^2} dx = a_0 - \alpha_0 x \tag{10.96}$$

for $x > x_h$,

$$\begin{aligned} v = \frac{\partial w}{\partial t} &= \frac{\partial w_0}{\partial t} - \int_0^{x_h} \frac{\partial \theta}{\partial t} dx - \int_{x_h}^x \frac{\partial \theta}{\partial t} dx - \frac{dx_h}{dt} \left[\theta\left(x_h^-\right) - \theta\left(x_h^+\right)\right] \\ &= v_0 - \frac{d\theta_0}{dt} x_h - \frac{d\theta_l}{dt}(x - x_h) \end{aligned} \tag{10.97}$$

$$\begin{aligned} a = \frac{\partial^2 w}{\partial t^2} &= a_0 - \alpha_0 x_h - \alpha_l (x - x_h) - \frac{dx_h}{dt}\left(\frac{d\theta_0}{dt} - \frac{d\theta_l}{dt}\right) \\ &= a_0 - \alpha_0 x_h - \alpha_l (x - x_h) - \frac{dx_h}{dt}(\omega_0 - \omega_l) \end{aligned} \tag{10.98}$$

From Eqs. (10.95) and (10.97) it is also indicated that the transverse velocity is continuous across the plastic hinge $v\left(x_h^-\right) = v\left(x_h^+\right)$.

From Eqs. (10.95) and (10.97), we have respectively,

$$a\left(x_h^-\right) = a_0 - \alpha_0 x_h \tag{10.99}$$

$$a\left(x_h^+\right) = a_0 - \alpha_0 x_h - \frac{dx_h}{dt}(\omega_0 - \omega_l) \tag{10.100}$$

On the other hand, we have

$$a\left(x_h^+\right) = a_l + \alpha_l (l - x_h) \tag{10.101}$$

then from Eqs. (10.100) and (10.101) we obtain an additional equation (kinematic condition) to relate a_0, α_0, a_l, α_l, and x_h:

$$a_0 - \alpha_0 x_h - a_l - \alpha_l (x - x_h) = \frac{dx_h}{dt}\left(\frac{d\theta_0}{dt} - \frac{d\theta_l}{dt}\right) \tag{10.102}$$

By introducing $\xi = x_h/l$ and $\mu = Pl/M_0$, Eqs. (10.87)–(10.90) and (10.102) can be rewritten as

$$\frac{a_0}{l} = \frac{M_0}{ml^3}\left(\frac{2\mu}{\xi} - \frac{12}{\xi^2}\right) \tag{10.103}$$

$$\alpha_0 = \frac{d^2\theta_0}{dt^2} = \frac{M_0}{ml^3}\left(\frac{3\mu}{\xi^2} - \frac{24}{\xi}\right) \tag{10.104}$$

$$\frac{a_l}{l} = \frac{-M_0}{ml^3}\frac{6}{(1-\xi)^2} \tag{10.105}$$

$$\alpha_l = \frac{d^2\theta_l}{dt^2} = \frac{M_0}{ml^3}\frac{12}{(1-\xi)^3} \tag{10.106}$$

$$\frac{M_0}{ml^3}\left(\frac{\mu}{\xi} - \frac{12}{\xi^2} + \frac{6}{(1-\xi)^2}\right) = -\frac{d\xi}{dt}(\omega_0 - \omega_l) = -\frac{d\xi}{dt}\left(\frac{d\theta_0}{dt} - \frac{d\theta_l}{dt}\right) \tag{10.107}$$

The motion of beam in Phase III will be determined from Eqs. (10.103)–(10.107). They hold regardless of whether the external load increases or decreases from a certain $\mu > 22.9$, only if $d\theta_0/dt > d\theta_l/dt$ (the traveling plastic hinge does not disappear).

The solutions of these equations are complicated due to the presence of the term $(d\xi/dt)(\omega_0 - \omega_l)$. If this term is absent, then a simple relation exists between μ and ξ, and a_0, α_0, a_1, and α_l can be easily expressed as functions of μ. The velocities and the displacements can be obtained by direct integration. At present, since the relation between μ and ξ involves unknown quantities, $d\xi/dt$, $d\theta_0/dt$, and $d\theta_l/dt$, it is impossible to solve the velocities and the displacements by direct integration. In such a case, the problem can be solved by either of the following two methods.

One is the method of successive approximations. First, set $(d\xi/dt)(\omega_0 - \omega_l) = 0$, a first approximate solution can be solved from Eqs. (10.103)–(10.106), and then $(\omega_0 - \omega_l)$ and $d\xi/dt$ can be calculated by integration. Substituting this approximate $(d\xi/dt)(\omega_0 - \omega_l)$ into Eq. (10.107), the second approximate solution can be solved from the equation set. The process should be repeated until the difference $(d\xi/dt)(\omega_0 - \omega_l)$ is calculated from the previous approximation and the present approximation is less than the required error.

Another one is the method of step-by-step integration. First, μ_1 and ξ_1 at some time t_1 is used to calculate the accelerations at this instant from Eqs. (10.103)–(10.106). Then the velocities at time $t_2 = t_1 + \Delta t$ can be calculated, for example

$$(\omega_0)_2 = (\omega_0)_1 + (\alpha_0)_1 \Delta t$$

After $(\omega_0 - \omega_l)$ at time t_2 is calculated, μ_2 and ξ_2 at t_2 can be solved from Eq. (10.107) by repeated trials. Then the accelerations at time t_2 can be evaluated again from Eqs. (10.103)–(10.106), and ω_0 and ω_l at t_2 can be re-calculated, for example

$$(\omega_0)_2 = (\omega_0)_1 + \tfrac{1}{2}\left[(\alpha_0)_1 + (\alpha_0)_2\right]\Delta t$$

After the new ξ_2 and μ_2 is re-calculated, such a process can be repeated to obtain the values at a new time $t_3 = t_2 + \Delta t$ and so forth.

In the motion of Phase III, if the external load begins to decrease (unloading), the problem can be dealt with by the same principle as discussed in Phase II. Equations (10.104)–(10.107) still hold so long as the traveling plastic hinge does not disappear. If P_m, the maximum of $P(t)$, is large enough, there still will be a relative angular velocity $(\omega_0 - \omega_l)$, even $P(t)$ decreases to zero. In this case, plastic hinge continues to move until $\omega_0 = \omega_l$. The final position $\xi_{\max}$ where the plastic hinge just disappears is obtained from Eq. (10.107) with $\mu = 0$ and $(d\xi/dt)\,(\omega_0 - \omega_l) = 0$.

$$\xi_{\max} = 2 - \sqrt{2} = 0.586 \tag{10.108}$$

After the traveling plastic hinge disappears, the action of the stationary plastic hinge at the central point still continues. In this case, Eqs. (10.28) and (10.83) deduced for Phase II should be used to solve the problem in this stage. Eventually the central stationary hinge disappears $(\omega_0 = 0)$, the beam thereafter translates as a rigid body without any further deformation. The final plastic strain is the permanent deformation caused by the impact.

As mentioned earlier, this rigid-plastic analysis is expected to provide satisfactory results when the plastic deformation is much larger than the elastic deformation, in other words the plastic work should be much larger than the elastic strain energy. In the current problem, the plastic work done at the central stationary hinge is $M_0\theta_0$. On the other hand, the elastic bending strain energy for a beam with length l under the action of bending moment M is known as $(1/2)M\theta = Ml/2P = M^2 l/2EI$, and its maximum is $M_0^2 l/2EI$. The requirement that the plastic work should be much larger than the elastic strain energy means

$$M_0\theta_0 \gg \frac{M_0^2 l}{2EI} \tag{10.109}$$

For the beam with two stress-free ends, its fundamental frequency of elastic vibration f_e is known as

$$f_e = \frac{(4.73)^2}{2\pi l^2} \cdot \sqrt{\frac{EI}{\rho_0 A_0}}$$

or its fundamental period τ is

$$\tau = \frac{1}{f_e} = \frac{2\pi l^2}{(4.73)^2} \cdot \sqrt{\frac{\rho_0 A_0}{EI}}$$

Then Eq. (10.109) can be rewritten in a dimensionless form as

$$\frac{\theta_0}{M_0 T^2/\left(ml^3\right)} \gg \frac{1}{2.5}\frac{\tau^2}{T^2} \tag{10.110}$$

where T is the impact duration. The above equation can be regarded as a criterion to evaluate whether the rigid-plastic analysis could be applied. In addition, the present analysis is also limited in the situation where $\theta_0 \leq 10^\circ$, since it has been assumed in the present theory that the deformation is sufficiently small, namely $\theta \approx -(\partial w/\partial x)$.

10.5 Shear Failure of Beams under Transverse Impact

In the previous sections, the bending waves propagating in the beam subjected to transverse impact were mainly discussed. However, it should be emphasized again, as we have pointed out in Section 10.2, that the bending waves are in fact resulted from the interaction of two kinds of waves, M–ω waves and Q–v waves, which are coupled with each other. It implies that the effect of shear force effects have been taken into account. So, shear force plays more important role in the dynamic theory than in the static theory of beams, since the transverse inertia is included in the present governing equation set.

With regard to the dynamic failure modes for beams subjected to transverse impact, a number of experimental and theoretical investigations revealed (Menkes and Opat, 1973; Jones, 1989) that there are at least three basic failure modes, namely, large ductile deformation or large deflection (Mode I), tensile tearing (Mode II), and transverse shearing (Mode III). The former two modes mainly resulted from bending wave propagation, while the third one is closely related to the transverse shear wave propagation.

In the following, the propagation characters of plastic shear wave in beams will be discussed (Wang and Jones, 1996).

In accordance with the discussions in Section 10.1, assume that the plane cross section perpendicular to the beam axis before the deformation remains a plane cross section perpendicular to the beam axis after the deformation (**plane cross section assumption**). Denote the transverse shear force and the transverse displacement on the cross section of the beam by Q and w, then correspondingly, the shear strain is $\gamma = \partial w/\partial X$ and the transverse particle velocity is $v = \dot{w} = \partial w/\partial t$. Introduce $m(= \rho_0 A_0)$ to denote the linear density of the beam, where ρ_0 is the initial density of the beam and A_0 is the initial section area. Similar to the strong discontinuous longitudinal waves propagating in the bar as discussed in Chapter 2 [see Eqs. (2.55) and (2.57)], for the shear disturbance propagating in the form of strong discontinuity (shear hinge), obviously there should exist the following dynamic compatibility condition (momentum conservation)

$$[Q] = -m\dot{\xi}[\dot{w}] \tag{10.111}$$

and the following kinematic compatibility condition (continuity condition)

$$[\dot{w}] = -\dot{\xi}[\gamma] \tag{10.112}$$

where $\dot{\xi} = d\xi/dt$ indicates the propagation velocity of strong discontinuous shear disturbances propagating along the neutral axis of beam (X axis) and ξ the position of shear hinge on the X axis.

From Eqs. (10.111) and (10.112), we have

$$\dot{\xi}^2 = \frac{[Q]}{m[\gamma]} \tag{10.113}$$

It means that the propagation velocity is mainly dependent on Q–γ character of the beam.

Completely similar to the situation discussed in Chapter 2 for the strong discontinuous longitudinal wave propagating in bars (see Section 2.6), only when the plastic part of Q–γ curve shows a linear hardening character ($dQ/d\gamma$ = constant) or an increasing hardening character ($dQ/d\gamma > 0$), can the plastic shear hinge forms.

In the case of rigid-linear hardening plastic beam, as shown in Fig. 10.21, its Q–γ relation is expressed as

$$\gamma = 0 \qquad \text{if } Q < Q_0 \tag{10.114a}$$

$$Q = Q_0 + G_p\gamma \qquad \text{if } Q \geq Q_0 \tag{10.114b}$$

where Q_0 indicates the yield shear force and G_p the linear hardening modulus. Then the propagation velocity of plastic shear hinge is invariable:

$$C_Q = \pm\sqrt{\frac{G_p}{m}} \tag{10.115}$$

where the plus and minus signs correspond to the plastic shear hinge propagating in the positive and negative directions along X axis, respectively.

Note that when G_p approaches zero, Eq. (10.114) is reduced to the rigid-perfectly plastic Q–γ relation, and then $\dot{\xi} = 0$ which is corresponding to a stationary plastic shear hinge. In such a case, it can be seen from the continuity condition across a strong discontinuous interface [Eq. (10.112)] that the transverse velocity across the shear hinge must

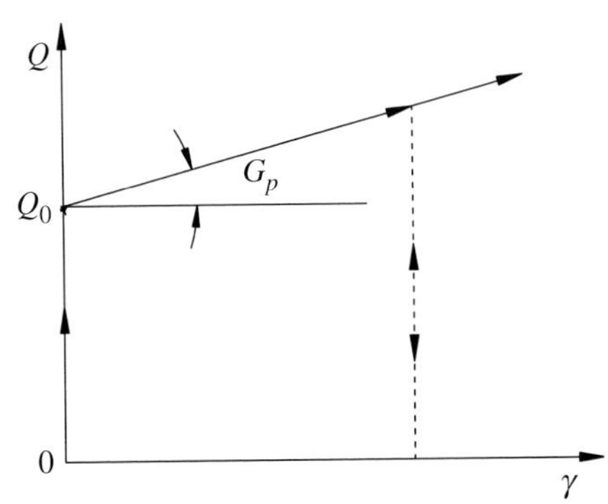

Fig. 10.21. The Q–γ curve for rigid-linear hardening plastic beam.

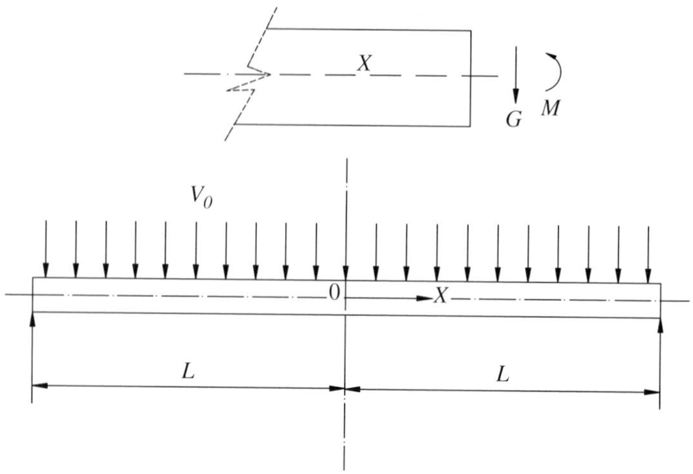

Fig. 10.22. Simple supported beam with length $2L$ subjected to a uniformly distributed impulsive velocity V_0.

be continuous, namely $[\dot{w}] = 0$. Hence the plastic hardening character of Q–γ relation is of significance for the study of shear hinge.

In the following, the shear failure of a simple supported beam subjected to transverse impact is analyzed as an example.

Consider a simple supported beam with length $2L$ subjected to a uniformly distributed impulsive velocity V_0, as shown in Fig. 10.22.

According to the momentum theorem and the moment of momentum theorem [Eqs. (10.1) and (10.2)], when the rotatory inertia effects are disregarded, we have

$$\frac{\partial Q}{\partial X} = m\ddot{w} \tag{10.116a}$$

$$\frac{\partial M}{\partial X} = Q \tag{10.116b}$$

The initial and boundary conditions for the problem shown in Fig. 10.22 are

$$w = 0,\ \dot{w} = V_0, \quad \text{for } t = 0 \tag{10.117a,b}$$

$$Q = 0 \quad \text{for } X = 0 \tag{10.117c}$$

$$M = 0 \quad \text{for } X = \pm L \tag{10.117d}$$

Only one-half of the beam, $0 \leq X \leq L$, is considered henceforth owing to the symmetry about $X = 0$. If a traveling plastic shear hinge moves from $X = L$ in the negative direction of X with velocity $\dot{\xi} = -C_Q$ towards the beam mid-span $X = 0$, then jumps in the transverse shear force, shear strain, and transverse particle velocity across the plastic shear hinge, namely, $[Q]$, $[\gamma]$, and $[\dot{w}]$ are, respectively

$$[Q] = (-Q_1) - (-Q_0) \tag{10.118a}$$

$$[\gamma] = (-\gamma_1) - 0 \tag{10.118b}$$

$$[\dot{w}] = 0 - \dot{w} \tag{10.118c}$$

where the subscript 1 denotes the quantity just behind the traveling hinge. Thus, now Eqs. (10.111) and (10.112) are respectively reduced to

$$Q_1 - Q_0 = mC_Q\dot{w} \tag{10.119a}$$

$$\dot{w} = C_Q\gamma_1 \tag{10.119b}$$

According to the rigid-linearly hardening plastic model, both parts of the beam on either side of the traveling hinge are rigid, so the transverse velocity distribution with respect to X for $t > 0$, as shown in Fig. 10.23(a), is

$$\dot{w} = V = \dot{W} \qquad \text{for } 0 \le X \le \xi \tag{10.120a}$$

$$\dot{w} = 0 \qquad \text{for } \xi \le X \le L \tag{10.120b}$$

where $V = \dot{W}(t)$ denotes the transverse velocity at $X = 0$, ξ is the transient location of the plastic shear hinge and in the present case

$$\xi = L - C_Q t \tag{10.121}$$

The transverse shear force distribution can be obtained from Eq. (10.116a), as shown in Fig. 10.23(b)

$$Q = m\ddot{W}X \qquad \text{for } 0 \le X \le \xi \tag{10.122a}$$

$$Q = -Q_1(t) \qquad \text{for } \xi \le X \le L \tag{10.122b}$$

where $\ddot{W} = d\dot{W}/dt$, which can be obtained from Eq. (10.117c) and by substituting Eqs. (10.118a) and (10.121) into Eq. (10.122)

$$\ddot{W} = -\frac{Q_0}{m(L - C_Q t)} \tag{10.123}$$

Integrating this equation with respect to t, together with the initial condition in Eq. (10.117b), we have

$$\dot{w}(t) = \dot{W}(t) = V_0 + \frac{Q_0}{mC_Q}\ln\left(1 - \frac{C_Q t}{L}\right) \quad \text{for } 0 \le X \le \xi \tag{10.124}$$

Note that $\dot{W}(t)$ decreases with time t, since $\ln(1 - (C_Q t/L)) < 0$ when $\xi < L$. Thus, there exists a time T_S which satisfies the condition $\dot{W}(T_S) = 0$ and consequently $[\dot{w}] = 0$.

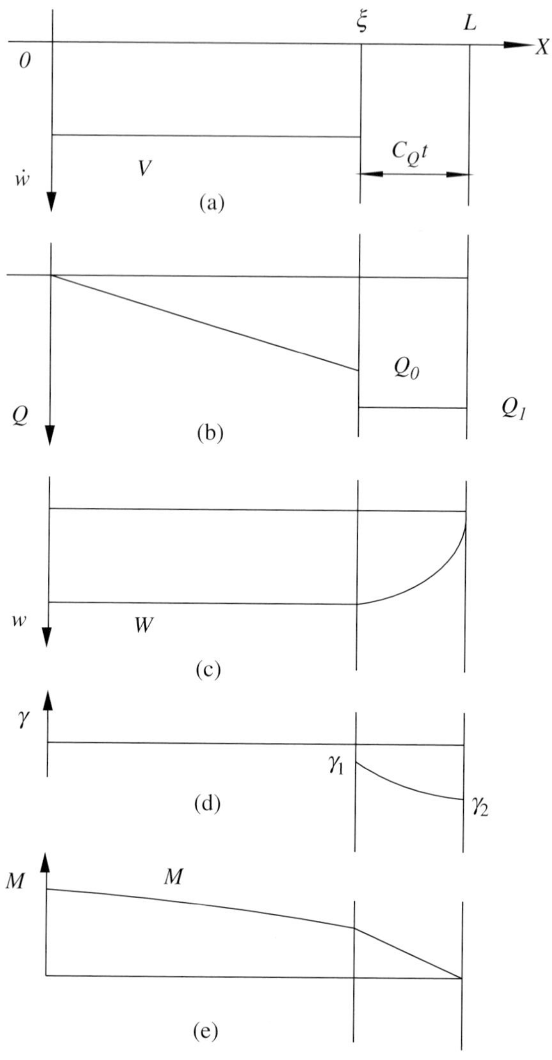

Fig. 10.23. Simply supported beam subjected to impulsive loading at time $t < T_S$. (a) Transverse velocity profile, (b) shear force distribution, (c) transverse displacement profile, (d) shear strain distribution, and (e) bending moment distribution.

In other words, when $t = T_S$, the traveling hinge disappears at the corresponding location $X = \xi_S$, and the motion of the whole beam ceases at the same time. T_S can be determined from Eq. (10.124) setting $\dot{w}(t) = 0$

$$T_S = \frac{L}{C_Q}\left[1 - \exp\left(-\frac{mC_Q V_0}{Q_0}\right)\right] \qquad (10.125a)$$

Note that the beam segment $0 \le X \le \xi_S$ reaches its final state at $t = T_S$, while the motion of the rigid part behind the traveling hinge ($\xi \le X \le L$) ceases immediately upon arrival of the plastic hinge, as indicated by Eq. (10.120b), namely the motion of the outer regions stops at time $T(X)$ which is given by

$$T(X) = \frac{L - X}{C_Q} \quad \text{for } \xi_S \le X \le L \tag{10.125b}$$

where $\xi_S = L - L_S$ and L_S is the final length of the plastic zone

$$L_S = C_Q T_S = L\left[1 - \exp\left(-\frac{mC_Q V_0}{Q_0}\right)\right] \tag{10.126}$$

Now the transverse displacement can be obtained by integrating Eq. (10.124) and utilizing the initial condition in Eq. (10.117a) to give

$$w(t) = V_0 t - \frac{Q_0 L}{mC_Q^2}\left\{\left(1 - \frac{C_Q t}{L}\right)\left[\ln\left(1 - \frac{C_Q t}{L}\right) - 1\right] + 1\right\} \quad \text{for } 0 \le X \le \xi \tag{10.127}$$

This equation together with Eqs. (10.125a,b) predict the final permanent displacement profile, as shown in Fig. 10.23(c)

$$w_f = W_f = \frac{V_0 L}{C_Q}\left\{1 + \frac{Q_0}{mC_Q V_0}\left[\exp\left(-\frac{mC_Q V_0}{Q_0}\right) - 1\right]\right\} \quad \text{for } 0 \le X \le \xi_S \tag{10.128a}$$

$$w_f = w_f(X) = \frac{V_0 L}{C_Q}\left\{\left(1 - \frac{X}{L}\right) - \frac{Q_0}{mC_Q V_0}\left[\frac{X}{L}\left(\ln\frac{X}{L} - 1\right) + 1\right]\right\} \quad \text{for } \xi_S \le X \le L \tag{10.128b}$$

Differentiating $w_f(X)$ with respect to X, the shear strain distribution $\gamma(X) = \partial w/\partial X$ can be obtained, as shown in Fig. 10.23(d)

$$\gamma(X) = 0 \quad \text{for } 0 \le X \le \xi_S \tag{10.129a}$$

$$\gamma(X) = -\gamma_S\left(1 + \frac{Q_0}{mC_Q V_0}\ln\frac{X}{L}\right) \quad \text{for } \xi_S \le X \le L \tag{10.129b}$$

the absolute value of which increases with X up to its maximum value γ_S

$$\gamma_S = \frac{V_0}{C_Q} \tag{10.129c}$$

Note that Eqs. (10.129) can also be derived from Eqs. (10.119b), (10.124), and (10.125a), showing that the kinematic condition across the traveling plastic hinge is satisfied.

On the other hand, in order to satisfy the dynamic condition across the traveling plastic shear hinge, it can be seen from Eqs. (10.119a) and (10.124) that the shear force Q_1 should be the following function of t

$$Q_1(t) = mC_Q V_0 + Q_0\left[1 + \ln\left(1 - \frac{C_Q t}{L}\right)\right] = Q_S + Q_0 \ln\left(1 - \frac{C_Q t}{L}\right) \quad (10.130a)$$

where

$$Q_S = Q_0 + mC_Q V_0 \quad (10.130b)$$

is the maximum value of $Q_1(t)$ at $t = 0$, since Eq. (10.130a) shows that $Q_1(t)$ decreases with t because the length of the plastic zone, $l_S = C_Q t$, is less than L.

The decrease of $Q_1(t)$ with time means that the beam segment behind the plastic shear hinge ($\xi \le X \le L$) is in a rigid-unloading state, which leads to a non-uniform distribution of plastic shear strain described by Eq. (10.129b) as shown in Fig. 10.23(d).

The strain rate across a traveling plastic shear hinge ($C_Q > 0$) is theoretically infinite when $[\gamma] \neq 0$. Therefore, in order to estimate the strain rate in a beam from an engineering viewpoint, the absolute value of average shear strain γ_{av} over the localized plastic zone L_s is first defined as

$$\gamma_{\text{av}} = \frac{\left|\int_{L-L_S}^{L} \gamma(X)\, dX\right|}{L_S} = \frac{W_f}{L_S} \quad (10.131a)$$

Consequently the average shear strain rate can be approximately defined as

$$\dot{\gamma}_{\text{av}} = \frac{\gamma_{\text{av}}}{T_S} = \frac{W_f}{L_S T_S} \quad (10.131b)$$

A further analysis is required to determine the bending moment distribution, to see whether or not the foregoing theoretical analysis satisfies the associated yield condition. From Eqs. (10.116b), (10.121)–(10.123), and (10.130), together with the boundary condition for the bending moment at $X = \xi = (L - C_Q t)$, the bending moment distribution $M(X)$ is expressed as

$$M(X) = Q_1(L - X) = Q_S L\left[1 + \frac{Q_0}{Q_S}\ln\left(1 - \frac{C_Q t}{L}\right)\right]\left(1 - \frac{X}{L}\right) \quad \text{for } (L - C_Q t) \le X \le L \quad (10.132a)$$

$$\begin{aligned} M(X) &= Q_1 C_Q t + \frac{Q_0 L}{2}\left(1 - \frac{C_Q t}{L}\right)\left[1 - \frac{X^2}{(1 - C_Q t/L)^2 L^2}\right] \\ &= Q_S L\left\{\frac{C_Q t}{L} + \frac{Q_0}{Q_S}\left[\frac{C_Q t}{L}\ln\left(1 - \frac{C_Q t}{L}\right) + \frac{1}{2}\left(1 - \frac{C_Q t}{L}\right)\right.\right. \\ &\quad \left.\left. \times \left(1 - \frac{X^2}{(1 - C_Q t/L)^2 L^2}\right)\right]\right\}, \quad \text{for } 0 \le X \le (L - C_Q t) \end{aligned} \quad (10.132b)$$

According to Eq. (10.132b), the maximum bending moment with respect to spatial coordinate X occurs at the mid-span ($X = 0$), as shown in Fig. 10.23(e).

$$M_{x\,\max} = M(0,t) = Q_1 C_Q t + \frac{Q_0 L}{2}\left(1 - \frac{C_Q t}{L}\right) = \frac{Q_0 L}{2} + \left(Q_1 - \frac{Q_0}{2}\right) C_Q t \tag{10.133}$$

which increases monotonously with t.

Equation (10.130) gives $Q_1 = Q_0$ when $t = T_S$, so that the maximum bending moment with respect to both the special coordinate X and the time coordinate t, M_{MAX}, is

$$M_{\mathrm{MAX}} = M(0, T_S) = \frac{Q_0 L}{2}\left(1 + \frac{C_Q T_S}{L}\right) \tag{10.134a}$$

or, in a dimensionless form

$$\overline{M}_{\mathrm{MAX}} = \nu(1 + \overline{L}_S) \tag{10.134b}$$

where the following dimensionless final plastic zone length $\overline{L}_S$, the dimensionless bending moment $\overline{M}$, and the well-known dimensionless strength ratio ν have been introduced and defined as

$$\overline{L}_S = \frac{L_S}{L} = \frac{C_Q T_S}{L}, \quad \overline{M} = \frac{M}{M_0}, \quad \nu = \frac{Q_0 L}{2M_0} \tag{10.135}$$

and M_0 is the fully plastic bending moment at yield.

If a square yield condition relating Q and M is assumed, then it is evident from Eq. (10.134b) that when $\nu(1 + \overline{L}_S) \geq 1$ the maximum bending moment at the mid-span ($X = 0$) satisfies the yield criterion ($M_{\mathrm{MAX}} \geq M_0$) and a new plastic bending hinge is formed there. The problem becomes much more complicated. Therefore, the foregoing theoretical analysis for the pure shear failure of beams is valid only for

$$\nu(1 + \overline{L}_S) \leq 0 \tag{10.136}$$

For a beam having relatively high yield strength but relatively weak strain-hardening character, we have

$$\frac{mC_Q V_0}{Q_0} = \frac{G_p \gamma_S}{Q_0} \ll 1 \tag{10.137}$$

and subsequently from Eq. (10.126), we have

$$\overline{L}_S \cong \frac{mC_Q V_0}{Q_0} \ll 1 \tag{10.138}$$

This means that the final length of a plastic zone L_S is much shorter than L, the one-half span length of the beam. In this circumstance, most of the equations deduced earlier can be simplified to facilitate engineering applications. In fact, Eqs. (10.125a), (10.128), (10.129), and (10.131) now are respectively reduced to:

$$T_S \cong \frac{mLV_0}{Q_0} \tag{10.139}$$

$$w_f = W_f \cong \frac{mV_0^2 L}{2Q_0} \quad \text{for } 0 \le X \le (L - L_S) \tag{10.140a}$$

$$w_f \cong \frac{V_0 L}{C_Q}\left[\left(1 - \frac{X}{L}\right) - \frac{Q_0}{2mC_Q V_0}\left(1 - \frac{X}{L}\right)^2\right] \quad \text{for } (L - L_S) \le X \le L \tag{10.140b}$$

$$\gamma(X) \cong -\gamma_S\left[1 - \frac{Q_0}{mC_Q V_0}\left(1 - \frac{X}{L}\right)\right] \quad \text{for } (L - L_S) \le X \le L \tag{10.141}$$

$$\dot{\gamma}_{\text{av}} \cong \frac{Q_0}{2mC_Q L} \tag{10.142}$$

Equations (10.139) and (10.140) are identical to the well-known Nonaka's solution for the rigid-perfectly plastic beam (Nonaka, 1967). Thus, the Nonaka's solution can be regarded as an approximate solution of the present analysis when $\overline{L}_S \cong 0$.

Finally, it can be seen from the above analyses that there are at least three possible modes of shear failure for the beam subjected to impulsive loading:

(a) Excess transverse deflection failure. It occurs when the transverse displacement w_f predicted by Eq. (10.128) exceeds the critical transverse displacement w_c, which is pre-determined from the design requirements, namely

$$w_f \ge w_c \tag{10.143}$$

(b) Excess transverse shear strain failure. It occurs when the maximum shear strain γ_S predicted by Eq. (10.129) exceeds the critical shear strain γ_c, which is pre-determined from the design requirements or the fracture shear strain of material, namely

$$\gamma_S \ge \gamma_c \tag{10.144}$$

Note that the critical or fracture shear strain γ_c of a material is generally dependent on the shear strain-rate $\dot{\gamma}$.

(c) Adiabatic shear failure. This mode of failure is related to the balance between the stress drop due to thermal softening by adiabatic heating ($\partial\tau/\partial T < 0$, where τ is the shear stress and T the temperature) and the stress rise due to strain hardening ($\partial\tau/\partial\gamma > 0$) and strain-rate hardening ($\partial\tau/\partial\dot{\gamma} > 0$) of a material, namely the so-called thermo-viscoplastic constitutive instability of a material (Wang *et al.*, 1987, 1988). Macroscopically, for a given environmental temperature, the critical thermo-viscoplastic instability criterion can be described, for example, by the following equation (Wang, 1986; Wang and Bao He-sheng, 1991)

$$f(\gamma, \dot{\gamma}) = \left(1 + g\frac{\dot{\gamma}}{\dot{\gamma}_0}\right)\left(A - \frac{\alpha\beta G_1}{T_0\rho C}\right)\left(\frac{\tau_0}{G_1} + \gamma\right) - 1 = 0 \qquad (10.145)$$

where G_1, g, and α characterize the strain-hardening, strain-rate hardening, and thermal-softening properties, respectively, τ_0, $\dot{\gamma}_0$, and T_0 are the characteristic stress, strain-rate, and temperature for a quasi-static test, respectively; ρ and C are the density and the specific heat of the material, respectively; β is the Taylor–Quinney coefficient, namely the fraction of viscoplastic work converted to heat ($\beta = 0.9 \sim 1.0$ in general). The A in the equation is a material parameter which characterizes a certain state of the adiabatic shearing process, for example, the initiation of a deformed shear band, the initiation of a transformed shear band, or the initiation of adiabatic fracture, etc.

The mode which dominates the shear failure of a practical beam will generally depend on which critical condition among Eqs. (10.143)–(10.145) is first satisfied.

For the readers who are interested in the studies on the dynamic elastic–plastic problems of beams, more detailed analyses and examples can be found from the early review literatures or books written by, for example, Abramson *et al.* (1958) and Rakhmatulin and Demyanov (1961); and found from the recent book written by Jones (1989).

CHAPTER 11

General Theory for Linear Elastic Waves

11.1 Linear Elastic Waves in Infinite Media

In the general three-dimensional cases, when body forces are not considered, the motion equations (momentum conservation) are expressed as

$$\left.\begin{aligned}\rho_0\frac{\partial v_X}{\partial t}&=\frac{\partial\sigma_{XX}}{\partial X}+\frac{\partial\sigma_{YX}}{\partial Y}+\frac{\partial\sigma_{ZX}}{\partial Z}\\ \rho_0\frac{\partial v_Y}{\partial t}&=\frac{\partial\sigma_{XY}}{\partial X}+\frac{\partial\sigma_{YY}}{\partial Y}+\frac{\partial\sigma_{ZY}}{\partial Z}\\ \rho_0\frac{\partial v_Z}{\partial t}&=\frac{\partial\sigma_{XZ}}{\partial X}+\frac{\partial\sigma_{YZ}}{\partial Y}+\frac{\partial\sigma_{ZZ}}{\partial Z}\end{aligned}\right\}\tag{11.1}$$

For homogeneous, isotropic, and linearly elastic media, substituting the Hooke's law in the Lame's form [Eq. (7.10) into Eq. (11.1)] to eliminate the stress components, we obtain

$$\left.\begin{aligned}\rho_0\frac{\partial v_X}{\partial t}&=\frac{\partial}{\partial X}(\lambda\Delta+2\mu\varepsilon_{XX})+\frac{\partial}{\partial Y}(2\mu\varepsilon_{XY})+\frac{\partial}{\partial Z}(2\mu\varepsilon_{ZZ})\\ \rho_0\frac{\partial v_Y}{\partial t}&=\frac{\partial}{\partial X}(2\mu\varepsilon_{XY})+\frac{\partial}{\partial Y}(\lambda\Delta+2\mu\varepsilon_{YY})+\frac{\partial}{\partial Z}(2\mu\varepsilon_{YZ})\\ \rho_0\frac{\partial v_Z}{\partial t}&=\frac{\partial}{\partial X}(2\mu\varepsilon_{XZ})+\frac{\partial}{\partial Y}(2\mu\varepsilon_{YZ})+\frac{\partial}{\partial Z}(\lambda\Delta+2\mu\varepsilon_{ZZ})\end{aligned}\right\}\tag{11.2}$$

Further, expressing strain components by displacement components u_X, u_Y, and u_Z [Eq. (7.9)], Eq. (11.1) can be re-written as

$$\left.\begin{aligned}\rho_0\frac{\partial^2 u_X}{\partial t^2}&=(\lambda+\mu)\frac{\partial\Delta}{\partial X}+\mu\left(\frac{\partial^2}{\partial X^2}+\frac{\partial^2}{\partial Y^2}+\frac{\partial^2}{\partial Z^2}\right)u_X\\ \rho_0\frac{\partial^2 u_Y}{\partial t^2}&=(\lambda+\mu)\frac{\partial\Delta}{\partial Y}+\mu\left(\frac{\partial^2}{\partial X^2}+\frac{\partial^2}{\partial Y^2}+\frac{\partial^2}{\partial Z^2}\right)u_Y\\ \rho_0\frac{\partial^2 u_Z}{\partial t^2}&=(\lambda+\mu)\frac{\partial\Delta}{\partial Z}+\mu\left(\frac{\partial^2}{\partial X^2}+\frac{\partial^2}{\partial Y^2}+\frac{\partial^2}{\partial Z^2}\right)u_Z\end{aligned}\right\}\tag{11.3}$$

all symbols in the above equations represent the same variables as those in Chapter 7.

Differentiating the three equations in Eq. (11.2) with respect to X, Y, and Z respectively, and adding together, further utilizing the following continuity conditions

$$\frac{\partial v_X}{\partial X} = \frac{\partial \varepsilon_{XX}}{\partial t}, \quad \frac{\partial v_Y}{\partial Y} = \frac{\partial \varepsilon_{YY}}{\partial t}, \quad \frac{\partial v_Z}{\partial Z} = \frac{\partial \varepsilon_{ZZ}}{\partial t} \tag{11.4}$$

then we obtain

$$\rho_0 \frac{\partial^2 \Delta}{\partial t^2} = (\lambda + 2\mu)\left(\frac{\partial^2}{\partial X^2} + \frac{\partial^2}{\partial Y^2} + \frac{\partial^2}{\partial Z^2}\right)\Delta \tag{11.5}$$

where $\Delta = \varepsilon_{XX} + \varepsilon_{YY} + \varepsilon_{ZZ}$ represents the volume dilatation. Equation (11.5) is a linear hyperbolic partial differential equation with respect to Δ, and it means that volume dilatation Δ propagates with a wave velocity C_L^e

$$C_L^e = \sqrt{\frac{\lambda + 2\mu}{\rho_0}}$$

Therefore, this wave is sometimes called **dilatation wave**. Comparing with Eq. (7.19), it can be seen that the velocity of the dilatation wave is just the velocity of one-dimensional strain elastic longitudinal wave. Note that when the dilatation wave propagates, it is actually accompanied with distortion deformation.

Differentiating the third and second equations of Eq. (11.2) with respect to Y and Z, respectively, and subtracting the latter from the former to eliminate Δ, further utilizing the following continuity conditions

$$\frac{\partial v_Z}{\partial Y} - \frac{\partial v_Y}{\partial Z} = \frac{\partial}{\partial t}\left(\frac{\partial u_Z}{\partial Y} - \frac{\partial u_Y}{\partial Z}\right) = 2\frac{\partial \omega_X}{\partial t}$$

then we obtain

$$\rho_0 \frac{\partial^2 \omega_X}{\partial t^2} = \mu\left(\frac{\partial^2}{\partial X^2} + \frac{\partial^2}{\partial Y^2} + \frac{\partial^2}{\partial Z^2}\right)\omega_X \tag{11.6}$$

where ω_X denotes the rotation with respect to X-axis

$$\omega_X = \frac{1}{2}\left(\frac{\partial u_Z}{\partial Y} - \frac{\partial u_Y}{\partial Z}\right)$$

The other two components of the rotation vector $\boldsymbol{\omega} = (1/2) rot\,\mathbf{u}$ are ω_Y and ω_Z

$$\omega_Y = \frac{1}{2}\left(\frac{\partial u_X}{\partial Z} - \frac{\partial u_Z}{\partial X}\right)$$

$$\omega_Z = \frac{1}{2}\left(\frac{\partial u_Y}{\partial X} - \frac{\partial u_X}{\partial Y}\right)$$

Analogous to the deduction of Eq. (11.6), from the first and second equations, and from the third and first equations of Eq. (11.2), the linear hyperbolic partial differential equations with respect to ω_Y and ω_Z can be obtained.

$$\rho_0 \frac{\partial^2 \omega_Y}{\partial t^2} = \mu \left(\frac{\partial^2}{\partial X^2} + \frac{\partial^2}{\partial Y^2} + \frac{\partial^2}{\partial Z^2} \right) \omega_Y$$

$$\rho_0 \frac{\partial^2 \omega_Z}{\partial t^2} = \mu \left(\frac{\partial^2}{\partial X^2} + \frac{\partial^2}{\partial Y^2} + \frac{\partial^2}{\partial Z^2} \right) \omega_Z$$

This means that rotation $\boldsymbol{\omega}$ propagates with the velocity

$$C_T = \sqrt{\frac{\mu}{\rho_0}} = \sqrt{\frac{G}{\rho_0}}$$

Since the propagation of rotation disturbance induces pure distortion or shear deformation, it is usually named as **distortion wave** or **shear wave**. Comparing the above equation for C_T with Eq. (2.78), it can be seen that the velocity of distortion wave is just the velocity of torsion wave [Eq. (2.78)], which has been discussed in Chapter 2.

It is easy to prove that a dilatation wave is an **irrotational wave** ($\boldsymbol{\omega} = 0$), while a distortion wave is an **isometric wave** ($\Delta = 0$). In fact, the first equation of Eq. (11.3) can be re-written as

$$\rho_0 \frac{\partial^2 u_X}{\partial t^2} = (\lambda + 2\mu) \left(\frac{\partial^2}{\partial X^2} + \frac{\partial^2}{\partial Y^2} + \frac{\partial^2}{\partial Z^2} \right) u_X + 2(\lambda + \mu) \left(\frac{\partial \omega_Z}{\partial Y} - \frac{\partial \omega_Y}{\partial Z} \right)$$

The second and the third equations of Eq. (11.3) can also be re-written in a similar form, and then it can be shown that the displacement in the irrotational case $\mathbf{u_p}$ satisfies the following equation

$$\rho_0 \frac{\partial^2 \mathbf{u_p}}{\partial t^2} = (\lambda + 2\mu) \left(\frac{\partial^2}{\partial X^2} + \frac{\partial^2}{\partial Y^2} + \frac{\partial^2}{\partial Z^2} \right) \mathbf{u_p}, \quad \boldsymbol{\omega}_\mathbf{p} = \frac{1}{2} rot \mathbf{u_p} \tag{11.7}$$

which indicates that $\mathbf{u_p}$ propagates with velocity C_L^e. Meanwhile, from Eq. (11.3), it can be directly deduced that the displacement in the isometric case $\mathbf{u_s}$ satisfies

$$\rho_0 \frac{\partial^2 \mathbf{u_s}}{\partial t^2} = \mu \left(\frac{\partial^2}{\partial X^2} + \frac{\partial^2}{\partial Y^2} + \frac{\partial^2}{\partial Z^2} \right) \mathbf{u_s}, \qquad \Delta_s = div \mathbf{u_s} = 0 \tag{11.8}$$

which indicates that $\mathbf{u_s}$ propagates with wave velocity C_T.

So far, it has been established that an arbitrary displacement can be divided into two parts, an irrotational part and an isometric part.

$$\begin{gathered} \mathbf{u} = \mathbf{u_p} + \mathbf{u_s} \\ rot \mathbf{u_p} = 0, \qquad div \mathbf{u_s} = 0 \end{gathered} \tag{11.9}$$

Each part satisfies Eqs. (11.7) and (11.8) respectively. Substituting Eq. (11.9) into Eq. (11.3) and re-writing to a vector form, we have

$$\begin{aligned}\frac{\partial^2 \mathbf{u}}{\partial t^2} &= \frac{\partial^2 \mathbf{u_p}}{\partial t^2} + \frac{\partial^2 \mathbf{u_s}}{\partial t^2} \\ &= \frac{\lambda + 2\mu}{\rho_0}\left(\frac{\partial^2}{\partial X^2} + \frac{\partial^2}{\partial Y^2} + \frac{\partial^2}{\partial Z^2}\right)\mathbf{u_p} + \frac{\mu}{\rho_0}\left(\frac{\partial^2}{\partial X^2} + \frac{\partial^2}{\partial Y^2} + \frac{\partial^2}{\partial Z^2}\right)\mathbf{u_s}\end{aligned} \tag{11.10}$$

This means that in homogeneous, isotropic, and linearly elastic media, the propagation of an arbitrary displacement disturbance can generally be divided into an irrotational wave and an isometric wave, irrelatively propagating with velocity C_L^e and C_T, respectively. Because these waves propagate in the interior of a body, independent of the boundary effects, they are named as **body waves**, to distinguish from the **surface waves**, which are generated and propagate on surfaces, and will be discussed in Section 11.3.

It is established in vector analysis that there must be a potential in an irrotational field. In other words, $\mathbf{u_p}$ must be a gradient of a scalar φ

$$\mathbf{u_p} = grad\varphi \tag{11.11}$$

φ is called the scalar potential of displacement; while a nondivergent field must be a solenoidal field. In other words, $\mathbf{u_s}$ must be a curl of a vector $\boldsymbol{\psi}$.

$$\mathbf{u_s} = rot\boldsymbol{\psi} \tag{11.12}$$

$\boldsymbol{\psi}$ is called the vector potential of displacement. Then Eq. (11.9) can be re-written as

$$\mathbf{u} = grad\varphi + rot\boldsymbol{\psi} \tag{11.13a}$$

Equations (11.7) and (11.8) can be re-written in the following wave equations with respect to φ and $\boldsymbol{\psi}$, respectively.

$$\frac{\partial^2 \varphi}{\partial t^2} - \frac{\lambda + 2\mu}{\rho_0}\left(\frac{\partial^2}{\partial X^2} + \frac{\partial^2}{\partial Y^2} + \frac{\partial^2}{\partial Z^2}\right)\varphi = 0 \tag{11.13b}$$

$$\frac{\partial^2 \boldsymbol{\psi}}{\partial t^2} - \frac{\mu}{\rho_0}\left(\frac{\partial^2}{\partial X^2} + \frac{\partial^2}{\partial Y^2} + \frac{\partial^2}{\partial Z^2}\right)\boldsymbol{\psi} = 0 \tag{11.13c}$$

Thus the studies on irrotational waves and isometric waves are transferred to the studies on the displacement functions φ and $\boldsymbol{\psi}$, which satisfy Eq. (11.13).

It is easy to prove that an irrotational wave is a longitudinal wave while an isometric wave is a transverse wave. Now let us consider an arbitrary plane wave propagating in homogeneous, isotropic infinite media with velocity C. Without loss of generality, let this plane wave propagate along X-axis, then displacements u_X, u_Y, and u_Z are functions of $\xi = X - Ct$:

$$u_X = u_X(X - Ct), u_Y = u_Y(X - Ct), u_Z = u_Z(X - Ct)$$

Substituting these functions into Eq. (11.3), we obtain

$$\rho_0 C^2 u''_X = (\lambda + 2\mu) u''_X$$
$$\rho_0 C^2 u''_Y = \mu u''_Y$$
$$\rho_0 C^2 u''_Z = \mu u''_Z$$

where ($''$) denotes a second order differential with respect to ξ. To obtain at least one nontrivial solution of u''_X, u''_Y, and u''_Z, it is possible only under one of the two conditions below,

$$C^2 = \frac{\lambda + 2\mu}{\rho_0} = C_L^2, \quad u''_Y = u'' = 0$$

which means that there is disturbance propagating in the X-direction only; or

$$C^2 = \frac{\mu}{\rho_0} = C_T^2, \quad u''_X = 0$$

which means that there is disturbance propagating in the Y- or Z-directions only. These results on the one hand prove that an arbitrary plane wave propagating in homogeneous, isotropic infinite media is either an irrotational wave propagating with velocity C_L^e or an isometric wave propagating with velocity C_T, in accordance with Eq. (11.10), on the other hand, they prove that an irrotational wave is a longitudinal wave while an isometric wave is a transverse wave.

As $(\lambda + 2\mu) > \mu$, a longitudinal wave propagates faster than a transverse one. During seismic, longitudinal waves are always observed first, followed by the transverse waves. Because of this, these two kinds of waves are usually named as P(primary)-waves and S(secondary)-waves. The S-waves are further categorized into SH-waves that propagate parallel to the free surface (the horizon) and *SV*-waves that propagate perpendicular to the free surface.

11.2 Oblique-Incidence, Reflection, and Transmission of Elastic Plane Waves

Although irrotational waves and isometric waves are unrelated during their propagation, they are often correlated when they are reflected or refracted at the body surface or at the interface between two media. Only if the direction of an incident wave is perpendicular to the surface or interface, namely in the case of a normal incidence, then these two kinds of waves are uncoupled and thus can be solved according to the principles discussed in Section 3.5. Otherwise, during an oblique incidence, in order to satisfy the given boundary conditions, whatever be the incident wave i.e. an irrotational wave only or an isometric

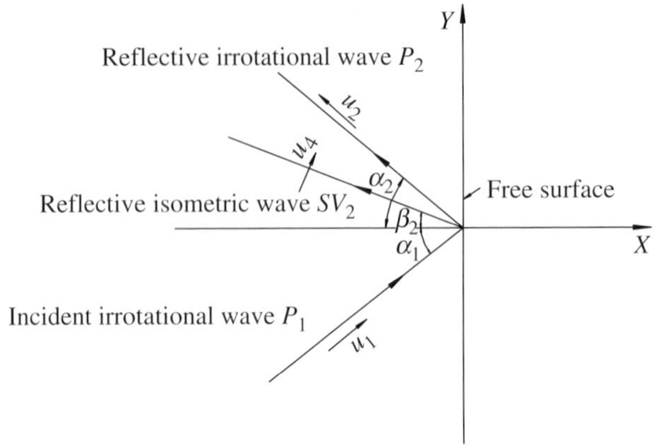

Fig. 11.1. Reflection of an irrotational plane wave obliquely incident at a free surface.

wave only, in general, both an irrotational wave and an isometric wave are simultaneously reflected and refracted. This is called **mode coupling**.

As an example of mode coupling, now let us consider an irrotational plane wave which is obliquely incident onto a free surface. Setting the free surface as Y–Z coordinate plane as shown in Fig. 11.1, assuming that the propagating direction of the incident irrotational wave P_1 is on the X–Y plane, then the displacement components u_X, u_Y are independent of Z and $u_Z = 0$. Regarding the situation of $u_Z \neq 0$, namely, the reflection of SH-wave will be discussed in the ensuing sections. Now consider the general possible situation of an oblique-incidence, namely both an irrotational wave and an isometric wave are simultaneously reflected. Denote the angle between the direction of P_1 and X-axis by α_1, the angle between the direction of the reflected irrotational wave P_2 and X-axis by α_2, the angle between the direction of the reflected isometric wave SV_2 and X-axis by β_2. They are named as the incidence angle, the reflective angle of the irrotational wave, and the reflective angle of the isometric wave, respectively. Thereupon, if we regard the concerned waves as harmonic waves, then the displacements u_1, u_2 of the longitudinal wave P_1, P_2 along their wave-vector direction $\mathbf{k_1}$, $\mathbf{k_2}$, respectively, and the displacement u_4 of the transverse wave SV_2 along the normal of its wave vector $\mathbf{k_4}$ can be expressed respectively as

$$\left.\begin{aligned} u_1 &= A_1 \exp\{i(Xk_1 \cos\alpha_1 + Yk_1 \sin\alpha_1 - \omega_1 t)\} \\ u_2 &= A_2 \exp\{i(-Xk_2 \cos\alpha_2 + Yk_2 \sin\alpha_2 - \omega_2 t)\} \\ u_4 &= A_4 \exp\{i(-Xk_4 \cos\beta_2 + Yk_4 \sin\beta_2 - \omega_4 t)\} \end{aligned}\right\} \tag{11.14}$$

where A is the amplitude, ω the circular frequency, k the wave number, the subscripts used here are identical with the subscripts of u mentioned above. The relationship between ω, k, and wave velocity is

$$\frac{\omega_1}{k_1} = \frac{\omega_2}{k_2} = C_L, \qquad \frac{\omega_4}{k_4} = C_T \tag{11.15}$$

Obviously, the components along X and Y directions of u_1, u_2, and u_4 are respectively:

$$\left.\begin{aligned} u_{1X} &= u_1 \cos\alpha_1, & u_{1Y} &= u_1 \sin\alpha_1 \\ u_{2X} &= -u_2 \cos\alpha_2, & u_{2Y} &= u_2 \sin\alpha_2 \\ u_{4X} &= u_4 \sin\beta_2, & u_{4Y} &= u_4 \cos\beta_2 \end{aligned}\right\} \tag{11.16}$$

The conditions that stress components σ_{XX}, σ_{YY}, σ_{XZ} on the free surface ($X = 0$) are all zero, should be satisfied. Following Hooke' law [Eq. (7.10)], also recalling that in the current situation u_X, u_Y are independent of Z and $u_Z = 0$, then the u_X and u_Y on the free surface should satisfy the conditions below:

$$\left.\begin{aligned} \sigma_{XX}|_{X=0} &= \left\{(\lambda + 2\mu)\frac{\partial u_X}{\partial X} + \lambda\frac{\partial u_Y}{\partial Y}\right\}\bigg|_{X=0} = 0 \\ \sigma_{XY}|_{X=0} &= \left\{\mu\left(\frac{\partial u_Y}{\partial X} + \frac{\partial u_X}{\partial Y}\right)\right\}\bigg|_{X=0} = 0 \end{aligned}\right\} \tag{11.17}$$

where u_X and u_Y include all the contributions of incident waves and reflected waves, namely,

$$u_X = u_{1X} + u_{2X} + u_{4X}$$
$$u_Y = u_{1Y} + u_{2Y} + u_{4Y}$$

Substituting Eq. (9.16) and Eq. (9.14) in Eq. (9.17), we obtain

$$\begin{aligned} &k_1\left(\lambda + 2\mu\cos^2\alpha_1\right)\cdot A_1\exp\{i(Yk_1\sin\alpha_1 - \omega_1 t)\} + k_2\left(\lambda + 2\mu\cos^2\alpha_2\right) \\ &\quad\cdot A_2\exp\{i(Yk_2\sin\alpha_2 - \omega_2 t)\} - 2\mu k_4\sin\beta_2\cos\beta_2 \\ &\quad\cdot A_4\exp\{i(Yk_4\sin\beta_2 - \omega_4 t)\} = 0 \\ &2k_1\sin\alpha_1\cos\alpha_1 A_1\exp\{i(Yk_1\sin\alpha_1 - \omega_1 t)\} - 2k_2\sin\alpha_2\cos\alpha_2 \\ &\quad\cdot A_2\exp\{i(Yk_2\sin\alpha_2 - \omega_2 t)\} - k_4\left(\cos^2\beta_2 - \sin^2\beta_2\right) \\ &\quad\cdot A_4\exp\{i(Yk_4\sin\beta_2 - \omega_4 t)\} = 0 \end{aligned} \tag{11.18}$$

In order for these boundary conditions to be satisfied for any Y and t, there must be:

$$\begin{aligned} k_1\sin\alpha_1 = k_2\sin\alpha_2 = k_4\sin\beta_2 \\ \omega_1 = \omega_2 = \omega_4 = \omega \end{aligned} \tag{11.19}$$

Also considering Eq. (11.15), we obtain the following **law of equal in visual velocities**,

$$\frac{C_L}{\sin\alpha_1} = \frac{C_L}{\sin\alpha_2} = \frac{C_L}{\sin\beta_2} = C \tag{11.20a}$$

where C is the so-called visual velocity. Equation (11.20a) is the Snell's law in optics, and can equivalently be written as

$$\alpha_1 = \alpha_2$$

$$\frac{\sin\alpha_1}{\sin\beta_2} = \frac{k_T}{k_L} = \frac{C_L}{C_T} = \sqrt{\frac{\lambda+2\mu}{\mu}} = \sqrt{\frac{2(1-\nu)}{(1-2\nu)}} \equiv \kappa \tag{11.20b}$$

where $k_T = k_4$ and $k_L = k_1 = k_2$ are the wave numbers of SV-wave and P-wave respectively. Equation (11.20b) indicates that the reflective angle α_2 of a reflected irrotational wave equals the incidence angle α_1; the reflective angle β_2 of a reflected isometric wave is less than the incidence angle α_1. In general $\beta_2 > 0$, except in the case of normal incidence where $\alpha_1 = 0$ and then $\beta_2 = 0$. Therefore, when an irrotational wave is obliquely incident on a free surface, in general, both an irrotational wave and an isometric wave are reflected and coupled together. Readers are suggested to prove by themselves that if only an irrotational wave is reflected, then in case the first equation of the boundary conditions [Eq. (11.17)], namely $\sigma_{XX}|_{X=0} = 0$ (and consequently $\alpha_1 = \alpha_2$) is satisfied, the second equation is no longer satisfied; in order to satisfy simultaneously the second equation, an isometric wave must be simultaneously reflected.

Substituting the above relationship into Eq. (11.18), we obtain

$$k_L(\lambda + 2\mu\cos^2\alpha_1)(A_1 + A_2) - 2\mu k_T A_4 \sin\beta_2\cos\beta_2 = 0$$

$$2k_L\sin\alpha_1\cos\alpha_1(A_1 - A_2) - k_T A_4(\cos^2\beta_2 - \sin^2\beta_2) = 0$$

Moreover, utilizing Eq. (11.20b), there is

$$\frac{\lambda}{\mu} + 2\cos^2\alpha_1 = \frac{\sin^2\alpha_1}{\sin^2\beta_2} - 2\sin^2\alpha_1 = \frac{\sin^2\alpha_1}{\sin^2\beta_2}\cdot\cos 2\beta_2$$

By operating the above two equations, we obtain

$$(C_L\cos 2\beta_2)\frac{A_2}{A_1} - (C_T\sin 2\beta_2)\frac{A_4}{A_1} = -C_L\cos 2\beta_2$$

$$(C_T\sin 2\alpha_1)\frac{A_2}{A_1} + (C_L\sin 2\beta_2)\frac{A_4}{A_1} = C_T\sin 2\alpha_1$$

They are the algebraic equations set with respect to the displacement reflection coefficients A_2/A_1 and A_4/A_1. On solving the equations, we obtain

$$\frac{A_2}{A_1} = \frac{\begin{vmatrix} -C_L \cos 2\beta_2 & -C_T \sin 2\beta_2 \\ C_T \sin 2\alpha_1 & C_L \cos 2\beta_2 \end{vmatrix}}{\begin{vmatrix} C_L \cos 2\beta_2 & -C_T \sin 2\beta_2 \\ C_T \sin 2\alpha_1 & C_L \cos 2\beta_2 \end{vmatrix}}$$

$$\frac{A_4}{A_1} = \frac{\begin{vmatrix} C_L \cos 2\beta_2 & -C_L \cos 2\beta_2 \\ C_T \sin 2\alpha_1 & C_T \sin 2\alpha_1 \end{vmatrix}}{\begin{vmatrix} C_L \cos 2\beta_2 & -C_T \sin 2\beta_2 \\ C_T \sin 2\alpha_1 & C_L \cos 2\beta_2 \end{vmatrix}}$$

Expanding these equations, we have

$$\frac{A_2}{A_1} = \frac{\sin 2\alpha_1 \sin 2\beta_2 - \kappa^2 \cos^2 2\beta_2}{\sin 2\alpha_1 \sin 2\beta_2 + \kappa^2 \cos^2 2\beta_2} = \frac{tg\beta_2 tg^2 2\beta_2 - tg\alpha_1}{tg\beta_2 tg^2 2\beta_2 + tg\alpha_1} \tag{11.21a}$$

$$\frac{A_4}{A_1} = \frac{2\kappa \sin 2\alpha_1 \cos 2\beta_2}{\sin 2\alpha_1 \sin 2\beta_2 + x^2 \cos^2 2\beta_2} = \frac{4 \sin \beta_2 \cos 2\beta_2 \cos \alpha_1}{tg\beta_2 \sin^2 2\beta_2 ctg\alpha_1 + \cos^2 2\beta_2} \tag{11.21b}$$

It can be seen from Eq. (11.21b) that the reflection coefficient is a function of Poisson's ratio ν and the incident angle α_1. When $\nu = 1/3$ (then $C_L/C_T = \sin\alpha_1/\sin\beta_2 = 2$), how the reflection coefficients A_2/A_1 and A_4/A_1 varies with the incident angle α_1 calculated by Eq. (11.21) is shown in Fig. 11.2 (see Kolsky, 1953).

It can be seen from Fig. 11.2 that (a) when the incident angle approaches 48°, the amplitude of the reflective SV-wave approaches its maximum, which is even larger than the amplitude

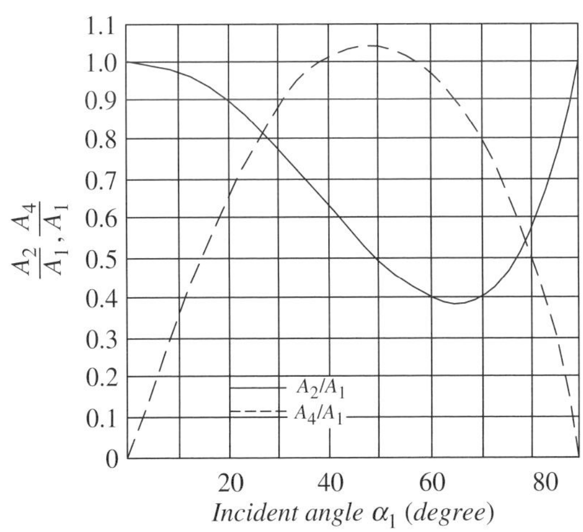

Fig. 11.2. Reflection coefficients A_2/A_1 and A_4/A_1 vary with the incident angle α_1.

of the incident irrotational wave ($A_4/A_1 \approx 1.04$); (b) when the incident angle approaches 65°, the amplitude of the reflective P-wave approaches its maximum ($A_2/A_1 \approx -0.38$); (c) in the case of normal incidence ($\alpha_1 = 0$), only a P-wave ($A_2/A_1 = -1$) but no SV-wave ($A_4 = 0$) is reflected, then this problem is the same as that discussed in Section 3.3; (d) in the case of grazing incidence, namely $\alpha_1 = \pi/2$, there are $A_4 = 0$ and $A_2/A_1 = -1$ again. But Eq. (11.21) shows $u_1 = -u_2$, which indicates a trivial solution, namely waves disappear. However, it should be noticed that such solutions are obtained under the plane wave assumption, namely both incidence waves and reflecting waves are plane waves. If the reflected waves are not limited to homogeneous plane waves only, then nontrivial solutions may exist during the grazing incidence of P-waves at a free surface. In such cases, a nonhomogeneous wave similar to a surface wave will be formed which will be discussed in Section 11.3.

When the oblique incidence of P-waves or SV-waves is discussed in terms of the scalar potential defined by Eq. (11.7) and the vector potential defined by Eq. (11.8), under the current condition of $u_z = (\partial u_X/\partial Z) = (\partial u_Y/\partial Z) = 0$, the displacement components u_{pX}, u_{pY} of P-wave and the displacement components u_{sX}, u_{sY} of SV-wave can be expressed respectively by

$$\begin{aligned} u_{pX} &= \frac{\partial \varphi}{\partial X}, \qquad u_{pY} = \frac{\partial \varphi}{\partial Y} \\ u_{sX} &= \frac{\partial \psi_Z}{\partial Y}, \qquad u_{sY} = -\frac{\partial \psi_Z}{\partial X} \end{aligned} \tag{11.22}$$

then the problem is reduced to the analysis on φ and ψ_Z, and the subscript Z of ψ_Z, can be temporarily neglected here. If the u_1, u_2, and u_4 in Eq. (11.14) are replaced by φ_1, φ_2, and ψ_4, respectively, and the amplitudes of the corresponding harmonic waves are expressed by a_1, a_2, and a_4, namely

$$\begin{aligned} \varphi_1 &= a_1 \exp\{i(Xk_1 \cos\alpha_1 + Yk_1 \sin\alpha_1 - \omega_1 t)\} \\ \varphi_2 &= a_2 \exp\{i(-Xk_2 \cos\alpha_2 + Yk_2 \sin\alpha_2 - \omega_2 t)\} \\ \psi_4 &= a_4 \exp\{i(-Xk_4 \cos\beta_2 + Yk_4 \sin\beta_2 - \omega_4 t)\} \end{aligned} \tag{11.23}$$

then the equal visual velocity law, Eq. (11.20), can be deduced again, and the displacement reflection coefficients a_2/a_1 and a_4/a_1 can also be deduced in the same way as Eq. (11.21). They are related with A_2/A_1 and A_4/A_1 as

$$\begin{aligned} &\frac{u_{2X}}{u_{1X}} = -\frac{u_{2Y}}{u_{1Y}} = -\frac{a_2}{a_1}, \qquad \frac{u_{4X}}{u_{1X}} = \frac{u_{4Y}}{u_{1Y}} = -\kappa\frac{a_4}{a_1} \\ &\left(\frac{A_2}{A_1} = \frac{a_2}{a_1}, \qquad \frac{A_4}{A_1} = -\kappa\frac{a_4}{a_1}\right) \end{aligned} \tag{11.24}$$

Figure 11.3 shows the relationship between a_2/a_1 (and consequently A_2/A_1) and incident angle α_1 with various ν (Arenberg, 1948). When $\nu > 0.26$, A_2/A_1 is always negative; when $\nu < 0.26$, there exist such incident angles that $A_2/A_1 = 0$, which means that there is no P-wave but only SV-wave reflected. In other words, a conversion of wave mode ($P \rightarrow SV$)

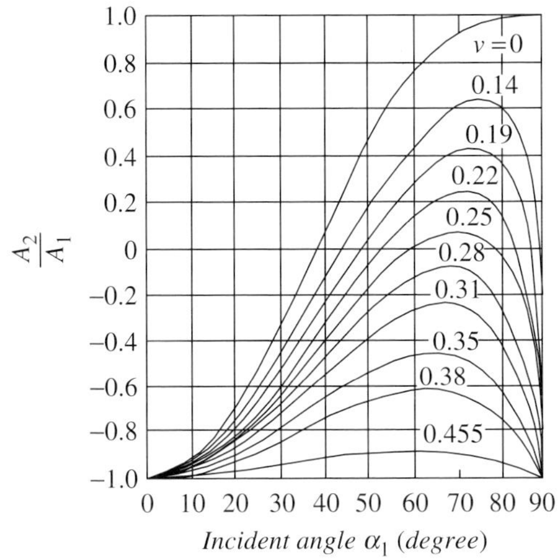

Fig. 11.3. Relationship between A_2/A_1 and incident angle α_1.

occurs through the wave reflection. This phenomenon is called **total mode conversion** or **polarization conversion**. The corresponding incident angle is called **mode conversion angle**, denoted as α $(P \to SV)$, which is a function of ν. For example, if $\nu = 1/4$, then the wave mode conversion of $P \to SV$ occurs when $\alpha_1 = 60°$ and $77.2°$.

In the same manner, it is easy to prove that when a SV_1 wave propagating in the direction parallel to X–Y plane is obliquely incident onto the free surface (Y–Z plane), in general two coupled waves are reflected, a reflected isometric wave P_2 and a reflected irrotational wave SV_2, as shown in Fig 11.4.

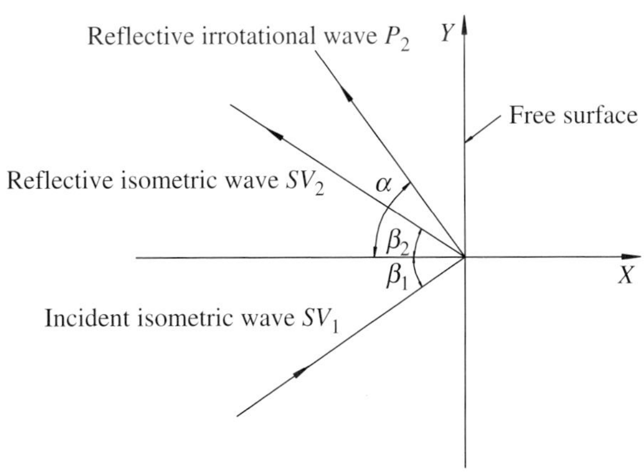

Fig. 11.4. Reflection of an isometric plane wave obliquely incident at a free surface.

Moreover, Snell's law is still valid

$$\frac{C_T}{\sin\beta_1} = \frac{C_T}{\sin\beta_2} = \frac{C_L}{\sin\alpha_2}$$

and the displacement reflection coefficients are:

$$\begin{aligned}\frac{A_4}{A_3} &= \frac{\sin\beta_1 \sin 2\alpha_2 - \kappa^2\cos^2 2\beta_1}{\sin 2\beta_1 \sin 2\alpha_2 + \kappa^2\cos^2 2\beta_1}\\ \frac{A_2}{A_3} &= \frac{2\kappa\sin 2\beta_1 \cos 2\beta_1}{\sin 2\beta_1 \sin 2\alpha_2 + x^2\cos^2 2\beta_1}\end{aligned} \tag{11.25}$$

where A_3 is the displacement amplitude of *SV*-wave, β_1 the incident angle, other symbols denote the same variables as mentioned previously. In the same way, when analyzing with respect to φ and ψ, similar to Eq. (11.24), we obtain

$$\begin{gathered}\frac{u_{2X}}{u_{3X}} = \frac{u_{2Y}}{u_{3Y}} = \frac{1}{\kappa}\frac{a_2}{a_3}, \qquad \frac{u_{4X}}{u_{3X}} = -\frac{u_{4Y}}{u_{3Y}} = \frac{a_4}{a_3}\\ \left(\frac{A_2}{A_3} = \frac{1}{\kappa}\frac{a_2}{a_3}, \qquad \frac{A_4}{A_3} = \frac{a_4}{a_3},\right)\end{gathered} \tag{11.26}$$

Figure 11.5 shows the relationship between a_4/a_3 (and consequently A_4/A_3) and incident angle β_1 with various ν (Arenberg, 1948).

As can be seen from Fig. 11.5 that (a) in the cases of both normal incidence ($\beta_1 = 0$) and 45° angle oblique incidence ($\beta_1 = \pi/4$), no *P*-wave but only *SV*-wave is reflected;

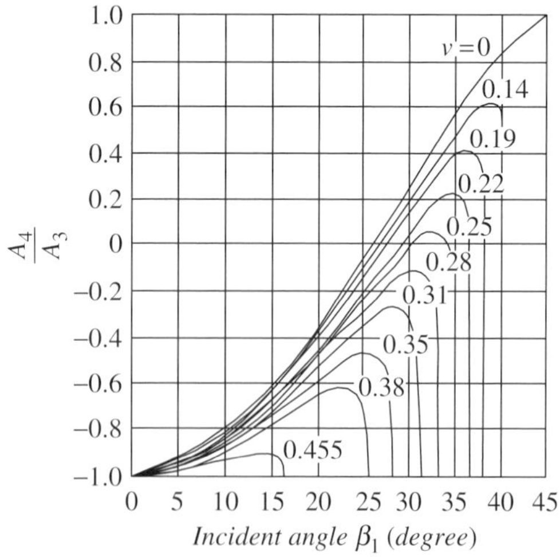

Fig. 11.5. Relationship between A_4/A_3 and incident angle β_1.

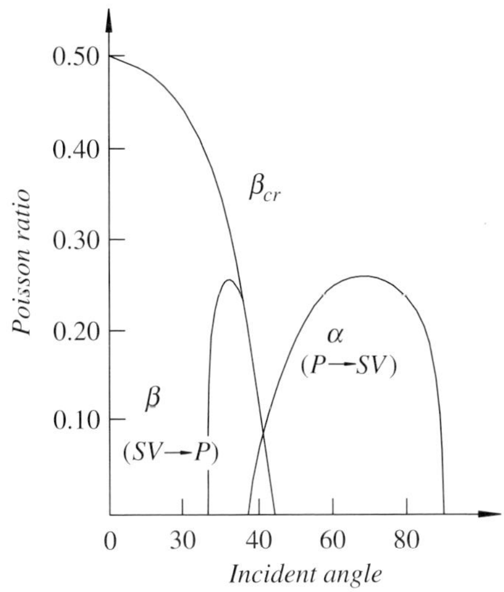

Fig. 11.6. The relationship between β_{cr}, α $(P \rightarrow SV)$, and Poisson's ratio ν.

(b) if $\nu < 0.26$ (except $\nu = 0$), there exist such incident angles that $A_4/A_2 = 0$, which indicates that mode conversion of $SV \rightarrow P$ occurs; the corresponding angle is denoted as $\beta(SV \rightarrow P)$; (c) since $\kappa = (\sin\alpha/\sin\beta) = (C_L/C_T) > 1$, there exists such a critical angle $\beta_{cr} = \sin^{-1}(1/\kappa)$ so that if the incident angle $\beta_1 > \beta_c$, then the value of reflective angle α_2 is no longer a real number. This phenomenon, analogous to optics, is called the **total reflection**.

The critical angle of total reflection β_{cr}, similar to the mode conversion angle $\alpha(P \rightarrow SV)$, is a function of Poisson's ratio ν. The relationship between β_{cr} and ν is shown in Fig. 11.6 (Arenberg, 1948).

So far we have discussed the coupling reflection of P-wave and SV-wave when they are obliquely incident onto the free surface. As for the SH-wave, namely the isometric wave of which the direction of displacement disturbances are parallel to the free surface ($u_X = u_Y = 0$, $u_Z \neq 0$), it can be proved by a similar method used previously i.e. when an SH-wave is obliquely incident onto a free surface, only an equal-amplitude SH-wave, but no other mode wave, is reflected, and the reflective angle equals the incident angle.

Analogous to this, the problems of reflection and refraction can be treated in the same way as that when an elastic wave is obliquely incident on an interface between two different materials A and B. According to the continuity requirement at the interface for the normal stress, shear stress, normal displacement and shear displacement, it can be proved that whatever be the obliquely incident wave i.e. an irrotational wave or an isometric wave, in general, four types of waves are generated at the interface, namely a reflected irrotational wave, a reflected isometric wave, a refracted irrotational wave, and a refracted isometric wave, and Snell's law is still valid. For instance, if the obliquely incident wave is an

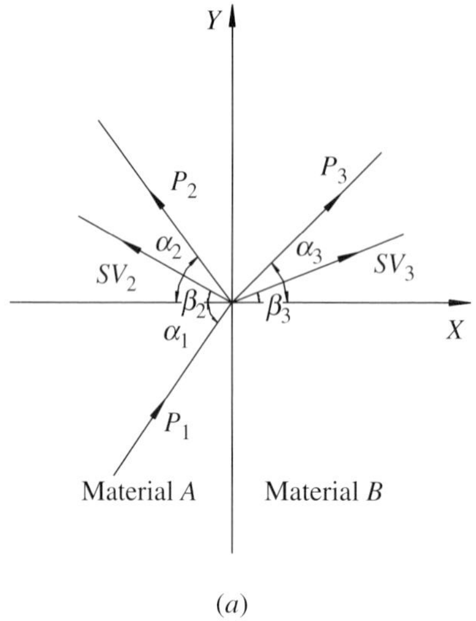

Fig. 11.7. Reflection and refraction when an elastic wave is obliquely incident on an interface between two media.

irrotational wave (P-wave), as shown in Fig. 11.7(a), then we have

$$\frac{\sin\alpha_1}{(C_L)_A}=\frac{\sin\alpha_2}{(C_L)_A}=\frac{\sin\beta_2}{(C_T)_A}=\frac{\sin\alpha_3}{(C_L)_B}=\frac{\sin\beta_3}{(C_T)_B} \tag{11.27}$$

while if the obliquely incident wave is an isometric wave (SV-wave), as shown in Fig. 11.7(b), then we have

$$\frac{\sin\beta_1}{(C_T)_A}=\frac{\sin\beta_2}{(C_T)_A}=\frac{\sin\alpha_2}{(C_L)_A}=\frac{\sin\beta_3}{(C_L)_A}=\frac{\sin\alpha_3}{(C_L)_B} \tag{11.28}$$

where the symbols used to denote the related angles are shown in the figure. Subscripts A and B denote the parameters for material A and B respectively. Similar to Eqs. (11.20) or (11.23), the relationship between the reflection coefficient and the material constants ρ, λ, μ, as well as the incident angle can be determined in the same way as the relationship between the deflection coefficient and the material constants ρ, λ, μ, as well as the incident angle is determined (see Miklowitz, 1978).

It should be noted that the interactions between the reflected unloading waves and the rearward portion (unloading portion) of the incident pressure pulse could induce tensile stress, which may result in the dynamic fracture called spalling, similar to that in the case of normal incidence as discussed previously in Section 3.8. However, the direction of the maximum tension stress is now related to the incident angle. The formation of spalling due to the reflected unloading wave when a pressure pulse (P-wave) is obliquely incident onto a free surface of an isotropic media is schematically shown in Fig. 11.8 (Rinehart, 1975).

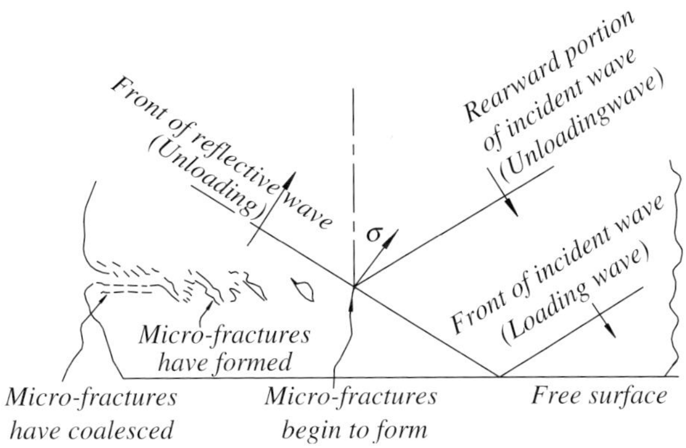

Fig. 11.8. Schematic of the formation of spalling due to the reflection on a free surface of an obliquely incident P-wave.

Obviously, the spalling develops with the reflected process of waves, and the orientation of micro-crack depends on the incident angle.

11.3 Elastic Surface Waves

As discussed previously, when an SV-wave is obliquely incident onto a free surface, if the incident angle β is larger than the critical angle $\beta_{cr} = \sin^{-1}(1/\kappa)$, then the ordinary reflection following Snell's law does not occur. As $\sin\beta > 1/\kappa$, and $\kappa = C_L/C_T > 1$, there is $\sin\alpha = \kappa\sin\beta > 1$, which indicates that α is not a real number, and consequently leads to a so-called **complex reflection**. Such nonordinary reflected P-wave which is a nonuniform wave generated by grazing incidence at a free surface is classified as a kind of surface wave, and will be discussed below.

In the coordinate system shown in Fig. 11.4, if the wave reflection is discussed in terms of the scalar potential φ, the reflected P-wave can be expressed by

$$\varphi_2 = A\exp\{i(-Xk_L\cos\alpha + Yk_L\sin\alpha - \omega t)\} = A\exp\{ik_S(Y - Xctg\alpha - C_s t)\} \tag{11.29}$$

where we have introduced

$$k_S = k_L\sin\alpha, \qquad C_S = \frac{\omega}{k_S} = \frac{C_L}{\sin\alpha} \tag{11.30}$$

As $\sin\alpha > 1$, so

$$C_S < C_L \tag{11.31}$$

On the other side, using the relations of triangle function, from Eqs. (11.30) and (11.31) we obtain

$$ctg^2\alpha = \frac{1}{\sin^2\alpha} - 1 = \left(\frac{C_S}{C_L}\right)^2 - 1 < 0$$

namely, $ctg\alpha$ must be an imaginary number. Introducing a positive real number

$$r_S = \sqrt{1 - \left(\frac{C_S}{C_L}\right)^2}$$

then $ctg\alpha$ can be written as

$$ctg\alpha = \pm i r_S$$

It is noted that if $ctg\alpha = -ir_S$, then it would be physically meaningless and moreover it cannot satisfy the convergence requirement that the wave amplitude should not approach infinity, at an infinite distance from the free surface. Therefore, we only have to take $ctg\alpha = +ir_S$, and then Eq. (11.29) is re-written as

$$\varphi_2 = A\exp(r_S k_S X)\exp(ik_S(Y - C_S t)) = A\exp(r_S k_S X)\exp(i(k_S Y - \omega t)) \quad (11.32)$$

which indicates that the complex reflected P-wave is such a nonuniform wave, which propagates with velocity C_S along the free surface (namely the positive Y-direction in Fig. 11.4), while its amplitude attenuates exponentially with the distance away from the free surface (namely the negative X-direction in Fig. 11.4), and its wave velocity C_S is less than the wave velocity C_L of an ordinary reflected P-wave (refer to Eq. (11.31)).

It can also be proved that when a P-wave is grazing along a free surface (grazing incidence), the reflected P-wave is a nonuniform wave propagating along the free surface, while the reflected SV-wave is still an ordinary reflected wave. But when an SV-wave is grazing along a free surface, both the reflected SV-wave and the reflected P-wave are nonuniform waves propagating along the free surface.

The formation of a nonuniform wave can be interpreted in the following way. Consider a semi-infinite media, the free surface of which is now chosen as the Y–Z plane of coordinate system. A compressive plane wave (P-wave) propagates along the Y direction with wave velocity C_L, as shown in Fig. 11.9 (Rinehart, 1975). The media through which the compressive plane wave passes are under compression. But the media at the free surface dilate transversely due to the Poisson's effect, namely a dilatation wave is correspondingly exited when the compressive plane wave arrives. In other words, each surface-point swept by the compressive plane wave can be regarded as a source of a new dilatation wave, which is a cylindrical dilatation wave propagating with the velocity C_L. For example, when a compressive plane wave successively passes through the points M_0, M_1, M_2, M_3... in the figure, a series of cylindrical dilatation waves are generated at these points. When the compressive plane wave front propagates from $M_0 N_0$ to AB, the series of cylindrical dilatation waves arrive successively at the positions of arc AD_0, AD_1, AD_2, AD_3... in

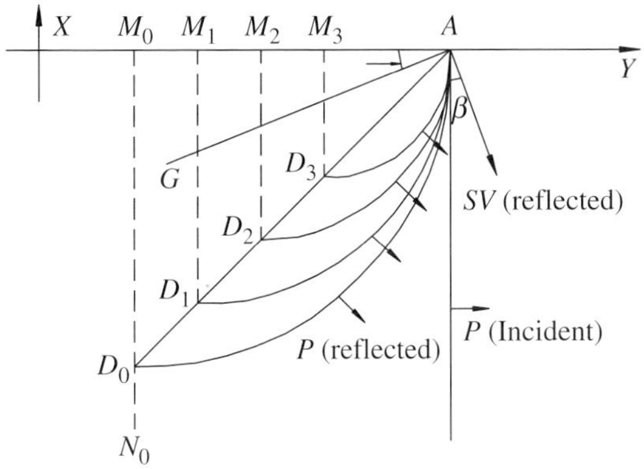

Fig. 11.9. Reflection of nonuniform P-wave due to a grazing incidence of a plane P-wave.

the figure. Because they propagate out from different sources at different times, no envelope surface with respect to those arcs exists, and thus no real wave front can be formed. Moreover, considering that the cylindrical dilatation waves attenuate with the propagating distance due to the cylindrical dispersion effect (see Section 8.2), it can be understood that the particles at the left side of arc AD_0 move not only longitudinally (in Y-direction) but also transversely (in X-direction), and such motion of particles are nonuniformly distributed in coordinate X direction. The plane $\overline{AD_0}$ in Fig. 11.9, namely the plane having 45° with the free surface, is obviously the trajectories of the particles having maximum transverse particle velocity. On the other hand, the particles in the area between the arc AD_0 and the plane wave front AB only move longitudinally. As for the ordinary reflective SV wave, due to the existence of free surface, the position of its wave front is shown as $\overline{AG}$ in the figure, when the incident plane P wave propagates from M_0N_0 to AB. Thus, Fig. 11.9 schematically provides a physical image to illustrate how the ordinary plane SV-wave and the nonuniform P-wave (nonplane wave) are reflected when a plane P-wave is grazing along a free surface (namely in the grazing incidence).

Analogically, it can also be understood that a nonuniform SV wave can be formed along a free surface.

In the same way, under certain conditions, a nonuniform P-wave and a nonuniform SV-wave can be coupled to propagate along a free surface with an identical velocity. This sort of surface waves was first discovered and depicted by Rayleigh (1887), hence called **Rayleigh surface wave**. It plays an important role in the research of seismic waves.

Similar to Eq. (11.29), for a couple of P-wave and SV-wave propagating at a same wave velocity $C_R = \omega/k$, describing in terms of displacement functions φ and ψ, there should be

$$\begin{aligned}\varphi &= A\exp[ik(X + Zctg\alpha - C_R t)] = A\exp(-rZ)\exp[ik(X - C_R t)]\\ \psi &= B\exp[ik(X + Zctg\beta - C_R t)] = B\exp(-sZ)\exp[ik(X - C_R t)]\end{aligned} \tag{11.33}$$

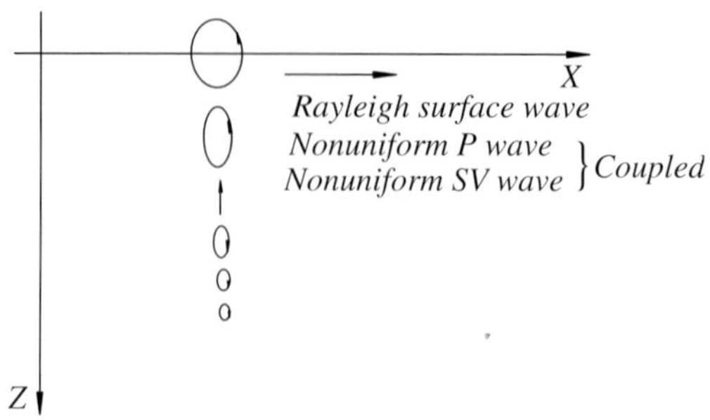

Fig. 11.10. Schematics of Rayleigh surface wave.

where the coordinate system has been changed to that shown in Fig. 11.10 in order to maintain consistency with the customary notations used in the discussion of Rayleigh surface waves, namely the Y and X in Fig. 11.4 and Eq. (11.29) are now replaced by X and $(-Z)$. Note that in the case of nonuniform wave, both $ctg\alpha$ and $ctg\beta$ are imaginary numbers, namely

$$\begin{aligned} ctg^2\alpha &= \left(\frac{C_R}{C_L}\right)^2 - 1 < 1 \\ ctg^2\beta &= \left(\frac{C_R}{C_T}\right)^2 - 1 < 1 \end{aligned} \tag{11.34}$$

then we have

$$C_R < C_T < C_L \tag{11.35}$$

while r and s are all positive real numbers

$$\begin{aligned} r &= k\sqrt{1 - \left(\frac{C_R}{C_L}\right)^2} = \sqrt{k^2 - k_L^2} > 0 \\ s &= k\sqrt{1 - \left(\frac{C_R}{C_L}\right)^2} = \sqrt{k^2 - k_T^2} > 0 \end{aligned} \tag{11.36}$$

Using Hooke's law [Eqs. (7.10) and (11.13a)], the boundary conditions on free surfaces can be expressed in terms of φ and ψ (actually ψ_z, while the subscript z has been ignored for convenience) as

$$\begin{aligned} \sigma_{ZZ}|_{Z=0} &= \left\{(\lambda + 2\mu)\frac{\partial u_Z}{\partial Z} + \lambda\frac{\partial u_Z}{\partial X}\right\}\Bigg|_{Z=0} \\ &= \left\{(\lambda + 2\mu)\left(\frac{\partial^2\varphi}{\partial X^2} + \frac{\partial^2\varphi}{\partial Z^2}\right) + 2\mu\left(\frac{\partial^2\psi}{\partial Z\partial X} - \frac{\partial^2\varphi}{\partial X^2}\right)\right\}\Bigg|_{Z=0} = 0 \end{aligned}$$

$$\sigma_{ZX}|_{Z=0} = \mu\left(\frac{\partial u_X}{\partial Z} + \frac{\partial u_Z}{\partial X}\right)\Bigg|_{Z=0}$$

$$= \mu\left(2\frac{\partial^2\varphi}{\partial X\partial Z} + \frac{\partial^2\psi}{\partial X^2} - \frac{\partial^2\psi}{\partial Z^2}\right)\Bigg|_{Z=0} = 0$$

Substituting Eqs. (11.33) and (11.34) in the above equations, we obtain

$$\begin{aligned} &\left\{(\lambda + 2\mu)r^2 - \lambda k^2\right\} A - i2\mu\, ksB = 0 \\ &i2krA - \left(k^2 + s^2\right) B = 0 \end{aligned} \tag{11.37}$$

The condition that the nontrivial solution of A and B exists is that determinant of coefficients must be zero. Then we obtain the following Rayleigh equation with respect to k^2

$$\left(2k^2 - k_T^2\right)^2 - 4k^2\sqrt{k^2 - k_L^2}\cdot\sqrt{k^2 - k_T^2} = 0 \tag{11.38a}$$

or

$$\left(2 - K_R^2\right)^2 = 4\left(1 - b^2k_R^2\right)^{1/2}\left(1 - K_R^2\right)^{1/2} \tag{11.38b}$$

where $b^2 = C_T^2/C_L^2 = 1/\kappa$ is determined totally by material constants (refer to Eq. 11.20b), while K_R^2 is

$$K_R^2 = \frac{C_R^2}{C_T^2} = \frac{k_T^2}{k^2}$$

Therefore, if k can be solved from Rayleigh equation [Eq. (11.38)], then it is proved that there really exists the surface wave expressed by Eq. (11.33).

Squaring Eq. (11.38), we obtain

$$K_R^2\left\{K_R^6 - 8K_R^4 + 8K_R^2\left(3 - 2b^2\right) - 16\left(1 - b^2\right)\right\} = 0$$

Since the solution of $K_R^2 = 0$ is physically meaningless, the problem is now reduced to solving the following cubic equation with respect to K_R^2

$$f\left(K_R^2\right) = \left(K_R^2\right)^3 - 8\left(K_R^2\right)^2 + 8\left(3 - 2b^2\right)\left(K_R^2\right) - 16\left(1 - b^2\right) = 0 \tag{11.39}$$

Because $f(0) < 0$ and $f(1) = 1$, there must be a real root in the interval $0 < K_R < 1$. This is consistent with Eq. (11.35). For example, if $\nu = 1/4$, $\lambda = \mu$, and $b^2 = 1/3$, then three roots can be solved from Eq. (11.39): $K_{R1} = \left[2 - \left(2/\sqrt{3}\right)\right]^{1/2}$,

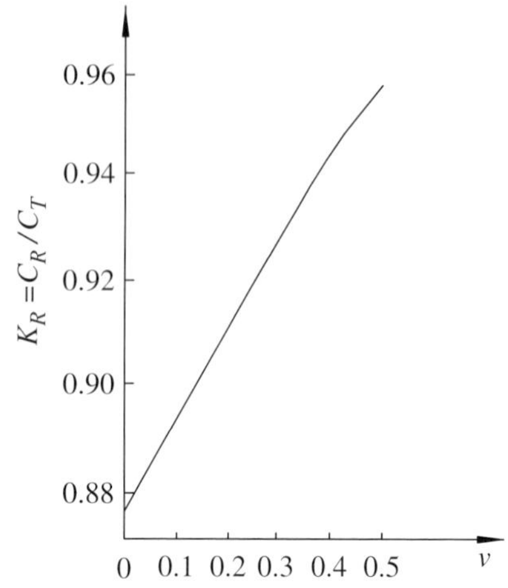

Fig. 11.11. K_R varies as a function of ν.

$K_{R2} = \left[2 + \left(2/\sqrt{3}\right)\right]^{1/2}$, and $K_{R3} = 2$. However, the last two roots conflict with Eq. (11.35). As only the first root is of significance, we obtain

$$C_R = 0.9194 C_T \quad \text{or} \quad k = 1.087 k_T \tag{11.40}$$

Figure 11.11 shows how K_R varies as a function of ν in the range of $0 \le \nu \le 1/2$ (Knopoff, 1952).

The displacement components u_X and u_Z, according to Eq. (11.13a) in the current situation, can be expressed as

$$u_X = \frac{\partial \varphi}{\partial X} - \frac{\partial \psi}{\partial Z}$$
$$u_Z = \frac{\partial \varphi}{\partial Z} + \frac{\partial \psi}{\partial X}$$

Substituting Eq. (11.33) in the above equations and using Eq. (11.37), we have

$$u_X = iAk\left(\exp(-rZ) - \frac{2rs}{k^2 + s^2}\exp(-sZ)\right)\exp[ik(X - C_R t)] \tag{11.41a}$$

$$u_Z = Ak\left(-\frac{r}{k}\exp(-rZ) + \frac{2rk}{k^2 + s^2}\exp(-sZ)\right)\exp[ik(X - C_R t)] \tag{11.41b}$$

Therefore, once K_R (or equivalently k) is determined by Eq. (11.39), the displacement of a Rayleigh surface wave can be calculated by Eq. (11.41). For example, if $\nu = 1/4$, the real parts of Eq. (11.41) are

$$u_X = D(\exp(-0.8475kZ) - 0.5773\exp(-0.3933kZ))\sin(kZ - \omega t)$$
$$u_Z = D\Big(-0.8475\exp(-0.8475kZ)^{-0.8475kZ} + 1.4678\exp(-0.3933kZ)\Big) \times \sin(kX - \omega t) \quad (11.42)$$

where $D = Ak$. At the free surface ($Z = 0$), the above equations are reduced to

$$u_X = 0.433D\sin(kX - \omega t) \quad (11.43a)$$
$$u_Z = 0.62D\cos(kX - \omega t) \quad (11.43b)$$

Therefore, the trajectory of a particle is an anticlockwise ellipse, as shown in Fig. 11.10.

$$\frac{u_X^2}{(0.433D)^2} + \frac{u_Z^2}{(0.62D)^2} = 1$$

At $Z = 0.193L$ ($L = 2\pi/k$ is the wave length), the horizontal displacement is zero, namely $u_X = 0$, only vertical displacement exists. At further depths, $Z/L > 0.193$, the sign of the wave amplitude reverses (an opposite phase). The trajectory of a particle is a clockwise ellipse (shown in Fig. 11.10). If we introduce the dimensionless amplitudes of displacements ($\hat{u}_X/\hat{u}_{Z0}$) and ($\hat{u}_Z/\hat{u}_{Z0}$), where $\hat{u}_{Z0}$ is the amplitude of vertical displacement of particle at free surface, $\hat{u}_X(Z)$ and $\hat{u}_Z(Z)$ are the amplitudes of u_X and u_Z, respectively, then Fig. 11.12 shows how the dimensionless amplitudes of displacements $\hat{u}_X/\hat{u}_{Z0}$ and $\hat{u}_Z/\hat{u}_{Z0}$ exponentially decay with the dimensionless depth Z/L (Viktorov, 1967).

The Rayleigh wave is a kind of surface wave due to the existence of free surface. Besides this kind of wave, other kinds of surface waves may also exist. For instance, nonuniform

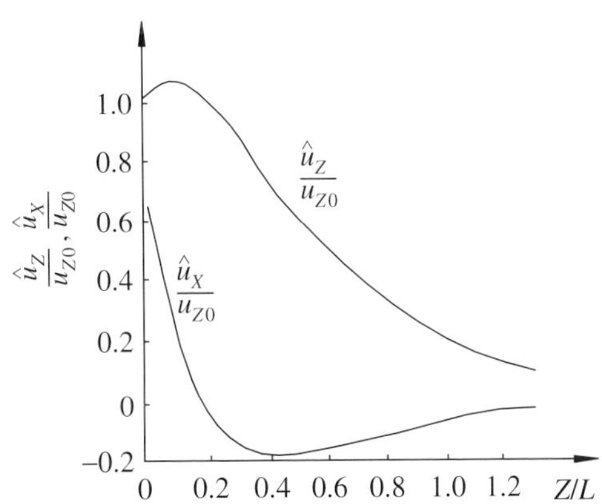

Fig. 11.12. The exponential decay of ($\hat{u}_X/\hat{u}_{Z0}$) and ($\hat{u}_Z/\hat{u}_{Z0}$) with the depth Z/L.

waves propagating along the interface between two media can be similarly formed, such as the so-called Stoneley waves, etc.

Reviewing the previous two sections, comparing with 1D stress wave problems, the complexities of 3D elastic wave problems mainly lie on the conversion of wave mode and the generation of new nonuniform waves during the reflection of waves arriving at the interface. It is easy to understand that the 3D problems involving elastic–plastic waves, visco-elastic waves, or elastic-visco-plastic waves, etc., will be much more complicated. To solve these problems, we have to work with the aid of numerical methods, which will be discussed in the next chapter.

CHAPTER 12

Numerical Methods for Stress Wave Propagation

A main difference in the study of structure response under impact loadings and quasi-static loadings is that the inertia of each infinitesimal element of structure has to be considered, which leads to the research on propagation of stress waves in various forms, whether using accurate methods or simplified methods. Mathematically, it results in solving hyperbolic wave equations. However, only under specially limited conditions, accurate solutions of such equations can be obtained. Such restricted conditions include: geometry of structures, spatial dimensions of motion, material constitutive equations, initial-boundary conditions, etc. For example, even for a relatively simple problem, such as the propagation of waves in a nonlinear elastic bar with finite length, if the nonlinear elastic constitutive equation of the bar is relatively complicated, then it could be very difficult, or sometimes impossible, to obtain the analytical solution of interaction of two simple waves. Then, numerical simulation is an important method to solve such hyperbolic (wave) equations. Especially in the last half a century, accompanying rapid development in computer technology, there are well-developed computational methods, which have been widely used to provide satisfactory approximate solutions even for the most complicated problems, neither subjected to restrictions on the form of material constitutive models, nor subjected to restrictions on initial and boundary conditions. Although mathematicians have only been able to prove the methods being applicable to linear cases, in practice, such methods seem to be equally applicable in complicated nonlinear situations too (e.g. Chou and Hopkins, 1972). At present, there are many numerical methods that can be used to analyze the instantaneous response of structures involving stress wave propagation, regardless of how complicated the material constitutive models and the initial-boundary conditions are.

In this chapter, three numerical methods are briefly introduced, including characteristics method, finite difference method, and finite element method, which have been extensively used to solve the stress wave problems. As a rudiment, only elementary theories of these methods are introduced here. In addition, illustrations and features of these methods are presented by analyzing some simple examples.

Among these three numerical methods, the characteristics method can be regarded as an "optimized" finite difference method. Once the characteristic equations and the corresponding compatibility equations along the characteristics have been determined by manipulating the original wave equation set (hyperbolic partial differential equation set), the problem to solve the wave equation set is transferred to solve these characteristic equations and characteristics compatibility equations by using finite difference numerical methods. Generally, the latter is much simpler than the former. For problems of 1D stress waves, since solving the original hyperbolic **partial** differential equations with two independent variables has been converted into solving simultaneously the corresponding **ordinary** differential equations, the problem is much more simplified in mathematics. Moreover, the characteristics method can provide clear physical pictures of the problem to be solved and is capable to deal with the propagation of strong discontinuous waves directly. So the characteristics method has been widely applied to solve problems of 1D stress waves. However, for general multidimensional problems of stress waves, the corresponding characteristics equations and compatibility equations along characteristics are still quite complicated, so that the applications of characteristics method are limited. Actually, only in some very special situations, characteristics method can be applied to solve 2D or 3D problems. In other words, in the case of 1D stress waves, the advantage of characteristics method is that the numerical integration of the original partial differential equations is converted into numerical calculation of the corresponding ordinary differential equations along characteristics. Obviously the difficulty in using the characteristics method lies in the manipulation of such a conversion.

The basis of finite difference method is to convert the original partial differential equations into the corresponding approximate difference equations, namely, continuous time and space are discretized into small intervals represented by nodes, then using Taylor series to solve the partial differential equations on each node. Therefore, a group of algebraic equations with respect to the unknowns on nodes is obtained, considering the initial and boundary conditions. Following the order of time and space, this group of algebraic equations can be solved step by step. Thus, finite difference method is looking for the approximate solutions of the original accurate partial differential equations. The key attention for this method should be focused on numerical errors, convergence, and stability of this method.

Finite element method is based on the principle of variation. The spatial region (structure), where the continuous functions described by partial differential equations exist, is divided into finite small regions (elements). The variation of unknown functions in each element is depicted by the selected function (shape function). Thus, the unknown functions in the whole region (structure) are discretized and an approximate mathematical–physical model is composed to describe the motion governed by the original partial differential equations. This approximated model is characterized by a group of algebraic equations, and the equations can be solved numerically by means of a computer. Therefore, finite element method is, looking for precise solutions of an approximated model. As the partial differentials with respect to time are included in the wave equation, the finite difference method is usually used to deal with the time variable in finite element method.

However, readers should also realize the disadvantages of numerical methods. First, the real influence of some variables on the results obtained may be concealed by such approximate methods. Second, the numerical errors, convergence, and stability involved in the numerical methods may encumber us to understand and recognize the physical essence of the problem studied. Therefore, it is usually necessary to judge and check physically the numerical results from the view of stress wave theory, so that not only the physical pictures of the obtained solutions can be correctly understood and explained, but also the numerical methods can be more reasonably and effectively used to study the wave propagation. In this sense, the contents of previous chapters on the fundamental theory of stress wave propagation are also important to the readers who mainly work on numerical simulations of structural dynamic response.

12.1 Characteristics Numerical Method

Characteristics method has been accepted extensively and has been applied in the study of elastic and plastic 1D stress waves, because it can be used to simulate accurately, propagation of strong discontinuous waves and it clearly figures out physical processes. Actually, the basic theory and application of characteristics methods used in problems of elastic or elasto-plastic 1D stress waves has been discussed in detail in Sections 2.3 and 2.4, and Sections in Chapter 4. Characteristics methods on visco-elastic and visco-plastic 1D stress waves have been studied in Chapter 6. Characteristics method on elasto-plastic 1D strain waves has been mentioned in Chapter 7. Characteristics method on spherical waves has been analyzed in Chapter 8. In this section, characteristics method is briefly discussed in the view of numerical methods. To obtain more details, readers should refer to the related monographs (e.g. Chou and Hopkis, 1972).

12.1.1 Characteristics numerical method for 1D waves

In governing equations of 1D waves, there are two independent variables, namely spatial variable X and time variable t. From Chapter 2, it is known that governing equations can be represented by two first-order partial differential equations with respect to strain ε and particle velocity v [refer to Eqs. (2.12) and (2.16)]. More generally, assuming $f_1(X, t)$ and $f_2(X, t)$ are unknown functions of two variables, governing equations of wave propagation can be written as

$$\begin{aligned} a_{11}\frac{\partial f_1}{\partial X} + b_{11}\frac{\partial f_1}{\partial t} + a_{12}\frac{\partial f_2}{\partial X} + b_{12}\frac{\partial f_2}{\partial t} &= R_1 \\ a_{21}\frac{\partial f_1}{\partial X} + b_{21}\frac{\partial f_1}{\partial t} + a_{22}\frac{\partial f_2}{\partial X} + b_{22}\frac{\partial f_2}{\partial t} &= R_2 \end{aligned} \tag{12.1}$$

where a_{ij}, b_{ij}, and R_i $(i,j = 1, 2)$ could be functions of f_1, f_2, X, and t.Using characteristics method, the numerical integration on Eq. (12.1) is converted into the integration of four ordinary differential equations, namely two families of characteristics S_1 and S_2 represented by

$$\begin{aligned} \frac{dX}{dt} &= \frac{-b \pm \sqrt{b^2 - 4ac}}{2a} = \tau_1 \\ \frac{dX}{dt} &= \frac{-b - \sqrt{b^2 - 4ac}}{2a} = \tau_2 \end{aligned} \tag{12.2}$$

where

$$a = b_{11}b_{22} - b_{12}b_{21}$$
$$b = a_{21}b_{12} + b_{21}a_{12} - b_{11}a_{22} - b_{22}a_{11}$$
$$c = a_{11}a_{22} - a_{12}a_{21}$$

and two compatibility equations along characteristics represented by

$$A\sqrt{1+\tau_1^2}\frac{df_1}{d\alpha} + \left(B - \frac{C}{\tau_1}\right)\sqrt{1+\tau_1^2}\frac{df_2}{d\alpha} = (\tau_1 b_{21} - a_{21})R_1 - (\tau_1 b_{11} - a_{11})R_2$$
$$A\sqrt{1+\tau_2^2}\frac{df_1}{d\beta} + \left(B - \frac{C}{\tau_2}\right)\sqrt{1+\tau_2^2}\frac{df_2}{d\beta} = (\tau_2 b_{21} - a_{21})R_1 - (\tau_2 b_{11} - a_{11})R_2 \quad (12.3)$$

where $d\alpha$ and $d\beta$ are the increment of arc length along the characteristics S_1 and S_2, respectively, and

$$A = a_{11}b_{21} - b_{11}a_{21}$$
$$B = a_{12}b_{21} - b_{11}a_{22}$$
$$C = a_{12}a_{21} - a_{11}a_{22}$$

Assuming that functions f_1 and f_2 are defined and specified on an arbitrary part QR of a space-like curve as shown in Fig. 12.1, the solutions of partial differential equations in the curving triangular region QPR which is bounded by the curve QR, the rightward characteristic QP and the leftward characteristic RP, can be obtained. Such kind of problem is often named the **initial-value problem** or **Cauchy problem** (referring to Section 2.4). Since the characteristics in the same family cannot intersect, the solutions at $P(x, t)$ depend only on the variables defined in the curving triangle QPR. Such a domain bounded by the space-like curve QR, the rightward characteristic QP, and the leftward characteristics RP

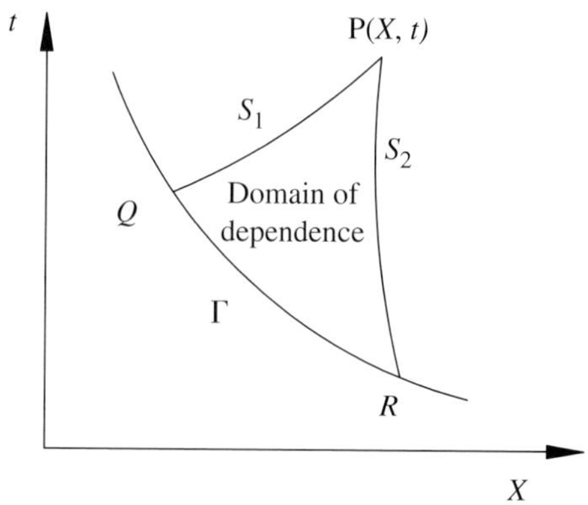

Fig. 12.1. Domain of dependence for $P(x, t)$.

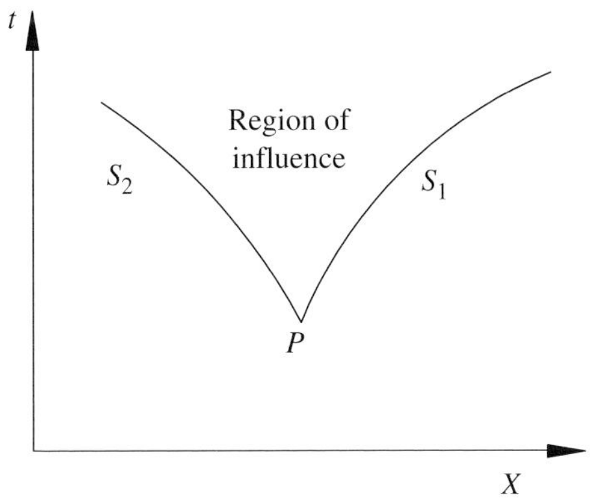

Fig. 12.2. Region of influence of $P(x, t)$.

is called the **domain of dependence** of point $P(x, t)$. The solutions at $P(x, t)$ depend only on the initial values prescribed on the curve Γ which lies between Q and R.

Similarly, it is easy to understand that the **region of influence** of point $P(x, t)$ is an infinite curving triangle which is bounded by the leftward and rightward characteristics passing the point $P(x, t)$, as shown in Fig. 12.2. Obviously, the solutions in the region of influence are influenced by the variables at $P(x, t)$.

As discussed in Section 2.4, if functions f_1 and f_2 are defined and specified on a space-like curve and are also defined and specified on a time-like curve, then the solutions in the region bounded by both the curves can be uniquely determined. Such kind of problems are usually named the **mixed problem** or **Picard problem**.

Taking the 1D wave propagation in a semi-infinite linear elastic bar as a simple example, the steps of numerical integration along characteristics are illustrated as follows. According to Eqs. (2.12) and (2.16), the governing equations with respect to the unknown functions ε and v can be written as

$$\begin{aligned} \frac{\partial v}{\partial X} - \frac{\partial \varepsilon}{\partial t} &= 0 \\ \frac{\partial v}{\partial t} - C^2 \frac{\partial \varepsilon}{\partial X} &= 0 \end{aligned} \tag{12.4}$$

where $C = (E/\rho_0)^{1/2}$ is the elastic wave velocity in bars. Assuming that the bar is initially in an undisturbed state, namely the initial condition is

$$v(X, 0) = \varepsilon(X, 0) = 0 \tag{12.5a}$$

To start a motion of the media, a velocity boundary condition at $X = 0$ is given by

$$v(0, t) = g(t) \tag{12.5b}$$

It is clear that if $a_{11} = 1$, $a_{22} = -C^2$, $b_{12} = -1$, $b_{21} = 1$ in Eq. (12.1) and other coefficients are zero, then Eq. (12.1) is converted to Eq. (12.4). Therefore, the differential equations of characteristics S_1 and S_2 can be obtained from Eq. (12.2)

$$S_1: \frac{dX}{dt} = C \quad \text{and} \quad S_2: \frac{dX}{dt} = -C \tag{12.6}$$

Furthermore, the compatibility equations can be obtained from Eq. (12.3)

$$\frac{dv}{d\alpha} - C\frac{d\varepsilon}{d\alpha} = 0 \tag{12.7}$$

$$\frac{dv}{d\beta} + C\frac{d\varepsilon}{d\beta} = 0 \tag{12.8}$$

where Eq. (12.7) is valid along S_1 and Eq. (12.8) is valid along S_2.

As the velocity C of an elastic wave in bars is a constant, Eq. (12.6) shows that the two families of characteristics S_1 and S_2 are straight lines with slopes $\pm C$ (wave velocity), respectively. Thereby, the grids of characteristics in the X–t plane can be drawn as shown in Fig. 12.3. The size of the grid ΔX is dependent on the precision required for the problem studied, while the time step is $\Delta t = \Delta X/C$.

Our aim is to obtain numerical solutions at every node. Actually three categories of computation are involved, namely, (1) the computation of points on the strong discontinuity (such as the nodes 1, 3, 6, 10... in Fig. 12.3), (2) the computation of points on the boundary (such as the nodes 2, 5, 9... in Fig. 12.3), and (3) the computation of the interior points (such as the nodes 4, 7, 8... in Fig. 12.3).

Computation of points on the strong discontinuous wave front. If there is a velocity-jump on the boundary at $t = 0$, namely if velocity-jump $g(0) \neq 0$ is loaded on the end of a semi-infinite linear elastic bar, then it can be proved that for this linear system, the velocity-jump

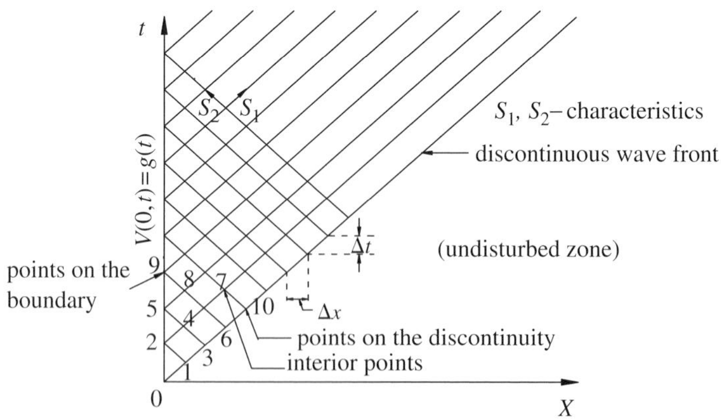

Fig. 12.3. Characteristics of a 1D wave in a semi-infinite linear elastic bar.

will propagate along a characteristic in simple wave zone (refer to Section 2.5), and keeps its amplitude value (a constant intensity of the strong discontinuity) during its propagation along this leading characteristics, namely

$$[v] \equiv v^{+} - v^{-} = g(0) = \text{constant}$$

and according to the compatibility conditions across a strong discontinuous wave front [Eq. (2.55)], strain discontinuity also keeps constant

$$[\varepsilon] = -\frac{1}{C}[v] = -\frac{g(0)}{C}$$

As it is known from Eq. (12.5a) that the particle velocity and strain ahead of the strong discontinuous wave front are all zero, the particle velocity v and strain ε behind the wave front can be solved by using the above two equations directly. Along the rightward leading characteristics S_1, namely behind the strong discontinuous wave front, we have

$$v = g(0), \quad \varepsilon = -\frac{g(0)}{C}$$

When $g(0)$ approaches zero, the strong discontinuous wave front converts to a weak discontinuous wave front, which results in the particle velocity v and the strain ε along the rightward leading characteristic S_1 also approaches zero. Thus, it is concluded that the particle velocity v and the strain ε can be determined uniquely for both a strong discontinuous wave and a weak discontinuous wave.

Computation of points on the boundary. The solutions at an arbitrary point P on the boundary can be obtained by using the boundary condition and the compatibility condition along the leftward characteristics passing this point. In Fig. 12.4, the line QP is the leftward characteristics S_2 connecting the point P and the point Q, at which the solutions have been obtained in previous computation. The compatibility condition along the leftward characteristic S_2 [Eq. (12.8)] should be satisfied, which in a finite difference form can be written as

$$v_P - v_Q + C(\varepsilon_P - \varepsilon_Q) = 0$$

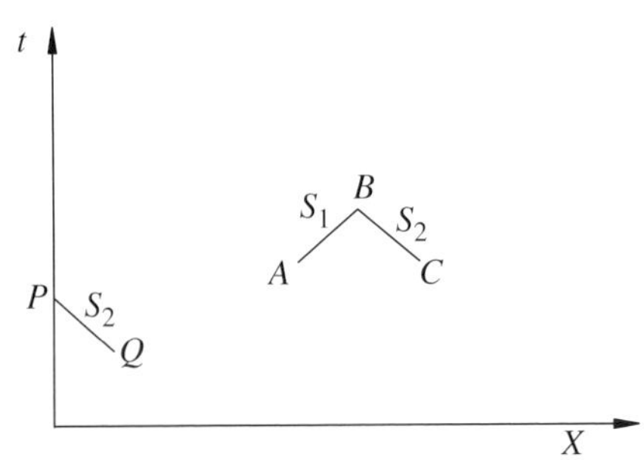

Fig. 12.4. Computation pattern of boundary point P and interior point B.

As the particle velocity at the point P, $v_p = v(0, t_p)$, has been determined by the boundary condition $v(0, t_p) = g(t_p)$, the strain at P on the boundary can be uniquely calculated by the above equation.

Computation of interior points. For an arbitrary interior point B neither on discontinuous lines nor on boundaries, there are two characteristics $S_1(AB)$ and $S_2(CB)$ through the point, as shown in Fig. 12.4. Then, the solutions at the point B can be determined by the compatibility conditions along these two characteristics $S_1(AB)$ and $S_2(CB)$, namely Eqs. (12.7) and (12.8), which in the finite difference form are

$$\text{along } S_1\text{: } v_B - v_A - C(\varepsilon_B - \varepsilon_A) = 0$$

$$\text{along } S_2\text{: } v_B - v_C + C(\varepsilon_B - \varepsilon_C) = 0$$

As solutions at points A and C have been obtained in previous computations, the solutions of v_B and ε_B, at B are determined uniquely.

Therefore, according to the time order of nodes on characteristics grids shown in Fig. 12.3, the solutions of an elastic semi-infinite bar can be obtained from left to right in turn.

For an elastic finite bar, the computation of end nodes is involved due to reflection of stress waves at the bar end. Although it increases the complexity of computation a little bit, the computation on those nodes is, in principle, similar to the computation on the boundary points on the loading boundary as discussed earlier.

However, for elasto-plastic bars, due to the nonlinearity in elasto-plastic constitutive equation, the governing equations are generally quasi-linear. In this situation, the characteristics equations are nonlinear and the slopes of characteristics depend on the dependent variables (f_1 and f_2). Consequently, the characteristics equations and the characteristics compatibility equations have to be solved together for the determination of unknown functions f_1 and f_2. A repeated iteration is sometimes needed to successively approach the solutions which increases difficulties in characteristics numerical methods. In Chapter 4, actually, it has been discussed in detail how to solve such problems of elasto-plastic waves propagating in bars, including the analyses of the interaction of elasto-plastic waves and the propagation of elasto-plastic boundaries, etc. Those are not necessary to be repeated here again.

12.1.2 Characteristic surfaces numerical method for 2D waves

In the previous section, characteristics numerical method is mainly applied to solve the problems of 1D wave propagation, and the problems of 2D waves have not been touched. For 1D problems of wave propagation, the advantages of characteristics method are very clear. The reason is that only one characteristic direction exists, so the problem of numerically integrating the original partial differential equations can be reduced to integrating the ordinary differential equations which involve the directional derivatives along the characteristics. Consequently, the difficulty to obtain the solutions of the governing equations decreases much more. However, for 2D problems that include three independent variables, the problem of numerically integrating the original partial differential equations, under certain conditions, can be simplified to integrating the equations which involve the directional derivatives along two characteristics directions on the characteristic surfaces. Unfortunately, these are still partial differential equations, so that the simplification for solving the equations in 2D situation is not as great as that in 1D situation. Therefore, relatively complex numerical schemes are required for the integration of these equations. In some very special situations and strictly limited conditions, the problem of integrating the original partial differential equations can be converted into integrating the lower-order partial differential equations which involve the characteristic equations along the so-called bicharacteristics on a characteristic cone.

In an example of a 2D linear elastic waves below, the elementary numerical method of characteristic surfaces will be discussed in brief (refer to Chou and Hopkins, 1972).

First, from Eq. (11.13) in Chapter 11, it has been established that the governing equations of 3D linear elastic waves are

$$\frac{\partial^2 \varphi}{\partial t^2} = \frac{\lambda + 2\mu}{\rho_0}\left(\frac{\partial^2}{\partial X^2} + \frac{\partial^2}{\partial Y^2} + \frac{\partial^2}{\partial Z^2}\right)\varphi = C_l^2 \nabla^2 \varphi$$

$$\frac{\partial^2 \boldsymbol{\Psi}}{\partial t^2} = \frac{\mu}{\rho_0}\left(\frac{\partial^2}{\partial X^2} + \frac{\partial^2}{\partial Y^2} + \frac{\partial^2}{\partial Z^2}\right)\boldsymbol{\Psi} = C_T^2 \nabla^2 \boldsymbol{\Psi}$$

where φ denotes the scalar potential of displacement of dilatation waves (irrotational waves), $\boldsymbol{\Psi}$ denotes the vector potential of displacement of distortion waves (isometric waves), C_L and C_T are the corresponding wave velocities of dilatation wave and distortion wave respectively and both of them are constant (refer to Chapter 11). Concisely, the above two equations can be represented by one equation

$$\nabla^2 \phi = \frac{\partial^2 \phi}{\partial X^2} + \frac{\partial^2 \phi}{\partial Y^2} + \frac{\partial^2 \phi}{\partial Z^2} = \frac{1}{C^2}\frac{\partial^2 \phi}{\partial t^2} \tag{12.9}$$

where C is a constant (C could be C_L or C_T, depending on different instances).

If a problem is cylindrically symmetric, it can be converted into a 2D problem. In the cylindrical coordinates (r, θ, Z), Eq. (12.9) is reduced to

$$\frac{\partial^2 \phi}{\partial r^2} + \frac{1}{r}\frac{\partial \phi}{2r} + \frac{\partial^2 \phi}{\partial Z^2} = \frac{\partial^2 \phi}{\partial T^2} \tag{12.10}$$

where $T = Ct$. In practical problems, the first-order partial derivatives of displacement potential ϕ (displacement, etc.) are of much concern. Taking

$$p = \frac{\partial \phi}{\partial r}, \quad q = \frac{\partial \phi}{\partial Z}, \quad S = \frac{\partial \phi}{\partial t} \tag{12.11}$$

then Eq. (12.10) is equivalent to the following first-order partial differential equation set

$$\begin{cases} \dfrac{\partial S}{\partial T} = \dfrac{\partial p}{\partial r} + \dfrac{p}{r} + \dfrac{\partial q}{\partial Z} \\ \dfrac{\partial p}{\partial T} = \dfrac{\partial S}{\partial r} \\ \dfrac{\partial q}{\partial T} = \dfrac{\partial S}{\partial Z} \end{cases} \tag{12.12}$$

Following the "indeterminate line method" used to determine the characteristics in 1D wave propagation (Section 2.3), the characteristic directions for Eq. (12.12) are determined below. Multiplying the above three equations (in Eq. 12.12) with indeterminate coefficients α_1, α_2, and α_3, their summation yields

$$-\alpha_1 \frac{\partial p}{\partial r} + \alpha_2 \frac{\partial p}{\partial T} - \alpha_1 \frac{\partial q}{\partial Z} + \alpha_3 \frac{\partial q}{\partial T} - \alpha_2 \frac{\partial S}{\partial r} - \alpha_3 \frac{\partial S}{\partial Z} + \alpha_1 \frac{\partial S}{\partial T} = \alpha_1 \frac{p}{r} \tag{12.13}$$

We should determine these indeterminate coefficients to make the above equation include only the directional derivatives along the characteristic directions on a common surface. If $\mathbf{e_r}$, $\mathbf{e_z}$, and $\mathbf{e_T}$ are the unit vectors on axis r, Z, and T respectively, Eq. (12.13) can be written in a vector form

$$\mathbf{A_1} \cdot \nabla p + \mathbf{A_2} \cdot \nabla q + \mathbf{A_3} \cdot \nabla S = \alpha_1 \frac{p}{r} \tag{12.14}$$

where the vectors $\mathbf{A_1}$, $\mathbf{A_2}$, $\mathbf{A_3}$ and the gradient operator ∇ are defined below:

$$\begin{aligned} A_1 &= -\alpha_1 \mathbf{e_r} + \alpha_2 \mathbf{e_T} \\ A_2 &= -\alpha_1 \mathbf{e_Z} + \alpha_3 \mathbf{e_T} \\ A_3 &= -\alpha_2 \mathbf{e_r} + \alpha_3 \mathbf{e_Z} + \alpha_1 \mathbf{e_T} \\ \nabla &= \mathbf{e_r} \frac{\partial}{\partial r} + \mathbf{e_z} \frac{\partial}{\partial z} + \mathbf{e_T} \frac{\partial}{\partial T} \end{aligned}$$

The first item on the left side of Eq. (12.14) represents the product of the magnitude of vector $\mathbf{A_1}$ and the directional derivative of p taken in the direction of vector $\mathbf{A_1}$. It can be interpreted similarly for the second and third items.

Similar to the directional derivative method in obtaining the characteristics equations in the 1D problems, it is required that the three derivative directions $\mathbf{A_1}$, $\mathbf{A_2}$, and $\mathbf{A_3}$ must be on a common plane. If the normal unit vector of this plane is denoted as $\boldsymbol{\lambda} \left(= \lambda_\gamma \mathbf{e_r} + \lambda_z \mathbf{e_z} + \lambda_T \mathbf{e_T}\right)$, then

$$\mathbf{A_1} \cdot \boldsymbol{\lambda} = \mathbf{A_2} \cdot \boldsymbol{\lambda} = \mathbf{A_3} \cdot \boldsymbol{\lambda} = 0$$

which in matrix form can be written as

$$\begin{bmatrix} -\lambda_r & \lambda_T & 0 \\ -\lambda_z & 0 & \lambda_T \\ \lambda_T & -\lambda_r & -\lambda_z \end{bmatrix} \begin{bmatrix} \alpha_1 \\ \alpha_2 \\ \alpha_3 \end{bmatrix} = 0 \tag{12.15}$$

A nontrivial solution of indeterminate coefficients α_1, α_2, and α_3 will exist if the determinant of the above coefficient matrix equals zero, namely

$$\lambda_T(\lambda_r^2 + \lambda_z^2 - \lambda_T^2) = 0 \tag{12.16}$$

If $\lambda_T \neq 0$ ($\lambda_T > 0$), assuming that θ is the angle between the r-axis and the projection of $\boldsymbol{\lambda}$ on the plane r–Z, then Eq. (12.16) is reduced to

$$\boldsymbol{\lambda} = \frac{1}{\sqrt{2}}(\cos\theta\, \mathbf{e_r} + \sin\theta\, \mathbf{e_Z} + \mathbf{e_T}) \tag{12.17}$$

This normal vector $\boldsymbol{\lambda}$ of the characteristic surface determines a family of planes with a single parameter θ; their enveloping surface is a cone, which is called the **characteristic cone** (Fig. 12.5). Then the characteristic surface passing through the point O (r_0, Z_0, T_0) is expressed as

$$(r - r_0)^2 + (Z - Z_0)^2 = (T - T_0)^2$$

Assuming that the unit vector $\boldsymbol{\beta}$ is the bicharacteristic direction which is the intersection between the characteristic plane and the characteristic cone, and the unit vector $\boldsymbol{\gamma}$ is tangent to the cone and perpendicular to $\boldsymbol{\lambda}$ and $\boldsymbol{\beta}$, then according to Eq. (12.17), $\boldsymbol{\beta}$ and $\boldsymbol{\gamma}$ can be expressed respectively as

$$\begin{aligned} \boldsymbol{\beta} &= \frac{1}{\sqrt{2}}(-\cos\theta\, \mathbf{e_r} - \sin\theta\, \mathbf{e_Z} + \mathbf{e_T}) \\ \boldsymbol{\gamma} &= \frac{1}{\sqrt{2}}(-\sin\theta\, \mathbf{e_r} + \cos\theta\, \mathbf{e_T}) \end{aligned} \tag{12.18}$$

As shown in Fig. 12.5, at the point i, the intersection between the characteristic cone and the plane parallel to r–Z plane is a circle. It is clear that $\boldsymbol{\gamma}$ is the tangent unit vector of this

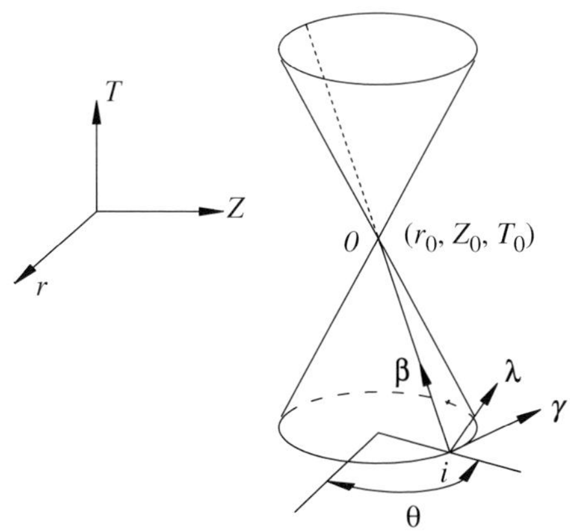

Fig. 12.5. The characteristic cone at the point $O(r_0, z_0, T_0)$.

circle at the point i. Furthermore, from Eq. (12.15), we can also obtain

$$\frac{\alpha_2}{\alpha_1} = \cos\theta \quad \text{and} \quad \frac{\alpha_3}{\alpha_1} = \sin\theta$$

Thus, dividing each side of Eq. (12.13) by α_1, using the above equations, we obtain

$$-\frac{\partial p}{\partial r} + \cos\theta\frac{\partial p}{\partial T} - \frac{\partial q}{\partial Z} + \sin\theta\frac{\partial q}{\partial T} - \cos\theta\frac{\partial S}{\partial r} - \sin\theta\frac{\partial S}{\partial Z} + \frac{\partial S}{\partial T} = \frac{p}{r} \tag{12.19a}$$

Or in a vector form, we have

$$\cos\theta\boldsymbol{\beta}\cdot\nabla p + \sin\theta\boldsymbol{\beta}\cdot\nabla q + \boldsymbol{\beta}\cdot\nabla S = \frac{1}{\sqrt{2}}\left[(\cos\theta\nabla q - \sin\theta\nabla p)\cdot\boldsymbol{\gamma} + \frac{p}{r}\right] \tag{12.19b}$$

Equation (12.19) is the compatibility equation on characteristic surface, it contains only directional derivatives along the bicharacteristics $\boldsymbol{\beta}$ and $\boldsymbol{\gamma}$. If $d\beta$ is the length increment in direction $\boldsymbol{\beta}$ of bicharacteristics defined by θ, and $d\gamma$ is the length increment in direction $\boldsymbol{\gamma}$, then according to the definition of directional derivative, the compatibility equation Eq. (12.19) can be written in the form of ordinary differential equation

$$\cos\theta\frac{dp}{d\beta} + \sin\theta\frac{dq}{d\beta} + \frac{dS}{d\beta} = \frac{1}{\sqrt{2}}\left(\cos\theta\frac{dq}{d\gamma} - \sin\theta\frac{dp}{d\gamma} + \frac{p}{r}\right) \tag{12.20}$$

It should be noted, if the λ_T in Eq. (12.16) equals zero, any plane perpendicular to the plane r–Z is always a characteristic surface. Following the steps presented in previous section, the corresponding compatibility equation can be obtained

$$\sin\theta\left(\frac{\partial S}{\partial r} - \frac{\partial P}{\partial T}\right) + \cos\theta\left(\frac{\partial q}{\partial T} - \frac{\partial S}{\partial Z}\right) = 0$$

Actually, this equation is only another form of the last two equations in Eq. (12.12). If we search a characteristic surface from the original second-order partial differential equation (Eq. 12.9), then λ_T will not appear. Consequently, the characteristic surface perpendicular to the plane r–Z can be neglected. The reason is that three new variables p, q, and S were adopted, which introduced the λ_T.

Steps to obtain numerical solution of the compatibility equation [Eq. (2.20)] will be discussed below. In Eq. (12.20), there are two directional derivatives in directions along $\boldsymbol{\beta}$ and $\boldsymbol{\gamma}$ which increase the difficulty of solving numerically. For further simplification, we will manipulate the compatibility Eq. (12.20). Expanding the right side of Eq. (12.20) and multiplying them with $d\beta$, we obtain

$$\cos\theta dp + \sin\theta dq + dS = \frac{d\beta}{\sqrt{2}}\left[\sin^2\theta\frac{\partial p}{\partial r} - \sin\theta\cos\theta\left(\frac{\partial q}{\partial r} + \frac{\partial p}{\partial Z}\right) + \cos^2\theta\frac{\partial p}{\partial Z} + \frac{p}{r}\right] \tag{12.21}$$

where all differentials denote the increment in the directions of the bicharacteristics defined by $\boldsymbol{\beta}$. Comparing with the partial differential Eq. (12.12), Eq. (12.21) contains neither the partial differential of the dependent variable S, nor the partial differential with respect to the variable T, which results in the compatibility Eq. (12.21) for which it is much easier to obtain numerical solution than Eq. (12.12). Note that the first equation of the bicharacteristics Eq. (12.18) can also be written as

$$\frac{dr}{-\cos\theta} = \frac{dZ}{-\sin\theta} = \frac{dT}{1} \tag{12.22}$$

consequently, the length increment of characteristics can be written as

$$d\beta = \sqrt{dr^2 + dZ^2 + dT^2} = \sqrt{2}dT$$

and then the $d\beta/\sqrt{2}$ in Eq. (12.21) can be replaced by dT. On the other hand, in Eq. (12.21), the parameter θ is an arbitrary value between 0 and 2π. When θ is set to be a specific value θ_I, which means that a bicharacteristics has been selected to connect the points I and J (shown in Fig. 12.6), the compatibility equation in a finite difference form between the two points I and J can be written as

$$\cos\theta_I \Delta p + \sin\theta_I \Delta q + \Delta S = \Delta T\left[\sin^2\theta_I \overline{\frac{\partial p}{\partial r}} - \sin\theta_I\cos\theta_I\left(\overline{\frac{\partial q}{\partial r}} + \overline{\frac{\partial p}{\partial Z}}\right) + \cos^2\theta_I\overline{\frac{\partial p}{\partial Z}} + \frac{1}{2}\left(\frac{p_I}{r_I} + \frac{p_J}{r_J}\right)\right] \tag{12.23}$$

Where $\Delta Y = Y_J - Y_I$, $(Y = p, q, S, T)$, and $\overline{\partial Y/\partial r}$ is defined as the average of partial derivatives along the bicharacteristics between the points I and J. Here, we consider that the state before time J has been determined, in other words, all the variables of every point located in the bottom of the cone which passes through the point J, are known.

The rest of the problem is how to determine the average of partial differential in Eq. (12.23). There are generally two methods (Chou and Hopkins, 1972). One is to assume that

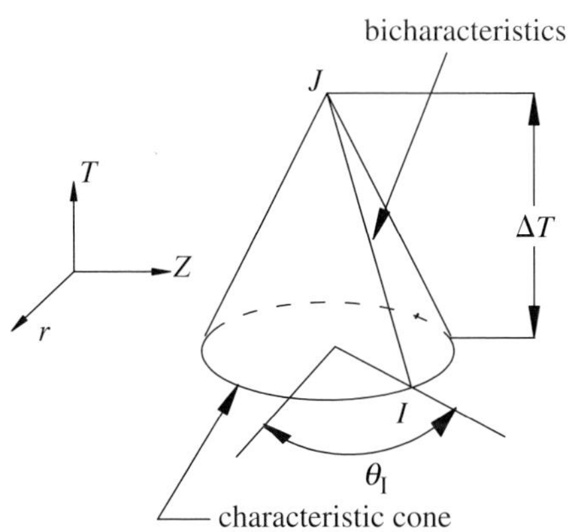

Fig. 12.6. Characteristic cone and the bicharacteristics connecting points I and J.

the dependent variables are continuous in the region between points I and J, then two additional equations can be obtained (Sauerwin, 1967),

$$dp = \frac{\partial p}{\partial r}dr + \frac{\partial p}{\partial Z}dZ + \frac{\partial p}{\partial T}dT \tag{12.24a}$$

$$dq = \frac{\partial q}{\partial r}dr + \frac{\partial q}{\partial Z}dZ + \frac{\partial q}{\partial T}dT \tag{12.24b}$$

Their finite difference forms are

$$\begin{aligned} \Delta p &= \overline{\frac{\partial p}{\partial r}}\Delta r + \overline{\frac{\partial p}{\partial Z}}\Delta Z + \overline{\frac{\partial p}{\partial T}}\Delta T \\ \Delta q &= \overline{\frac{\partial q}{\partial r}}\Delta r + \overline{\frac{\partial q}{\partial Z}}\Delta Z + \overline{\frac{\partial q}{\partial T}}\Delta T \end{aligned} \tag{12.25}$$

Here, the average of partial differential is defined by (Fig. 12.7)

$$\overline{\frac{\partial Y}{\partial y}} = \frac{1}{2}\left(\left.\frac{\partial Y}{\partial y}\right|_I + \left.\frac{\partial Y}{\partial y}\right|_J\right)$$

Thus, we can determine the variables at point J based on the known variables at point I by using Eqs. (12.23) and (12.25). However, only three equations can be obtained based on one characteristics, while there are nine unknowns p_J, q_J, S_J, $\left.\frac{\partial p}{\partial r}\right|_J$, $\left.\frac{\partial p}{\partial Z}\right|_J$, $\left.\frac{\partial p}{\partial T}\right|_J$, $\left.\frac{\partial q}{\partial r}\right|_J$, $\left.\frac{\partial q}{\partial Z}\right|_J$, and $\left.\frac{\partial q}{\partial T}\right|_J$. Hence to solve all the unknowns, six equations are still needed. It is known that every straight line from the bottom of characteristic cone to its top is a bicharacteristics. According to the numerical scheme shown in Fig. 12.7 (Sauerwin, 1967), we can select arbitrarily three bicharacteristics to provide nine equations. Consequently, all nine unknowns can be obtained numerically.

Another numerical method used to determine the average of partial differential in Eq. (12.23) is to adopt the numerical scheme based on four bicharacteristics, which is shown in Fig. 12.8 (Butler, 1962; Clifton, 1967). Corresponding to the four

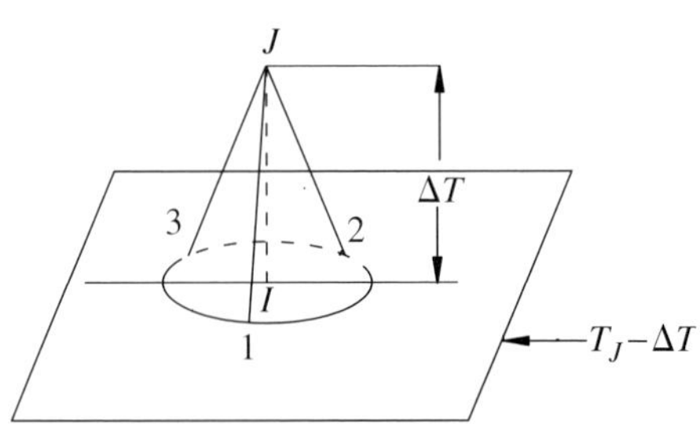

Fig. 12.7. Numerical scheme based on three bicharacteristics.

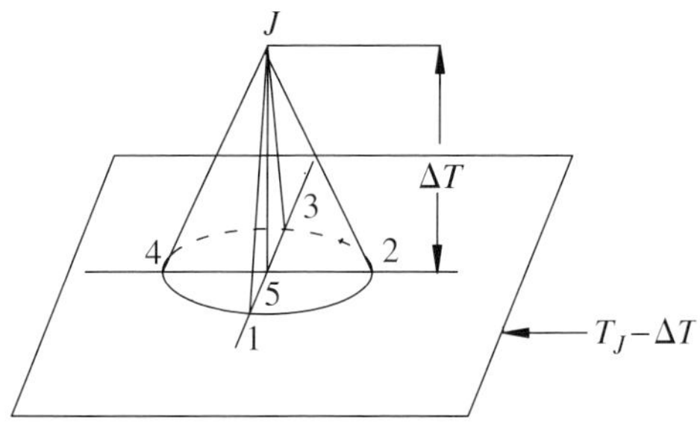

Fig. 12.8. Numerical scheme based on four bicharacteristics.

bicharacteristics, four compatibility equations ($I = 1, 2, 3, 4$) can be obtained according to Eq. (12.23)

$$\cos\theta_I(P_J - P_I) + \sin\theta_I(q_J - q_I) + (S_J - S_I) = \frac{\Delta T}{2}\left[\sin^2\theta_I\left(\left.\frac{\partial p}{\partial r}\right|_J + \left.\frac{\partial p}{\partial r}\right|_I\right) + \cos^2\theta_I\left(\left.\frac{\partial q}{\partial Z}\right|_J + \left.\frac{\partial q}{\partial Z}\right|_I\right) + \frac{P_J}{r_J} + \frac{P_I}{r_I}\right] \tag{12.26}$$

where $\theta_I = 0, \pi/2, \pi, 3\pi/2$. Note that $\sin\theta_I \cdot \cos\theta_I = 0$ here. As all relevant quantities in the cone bottom are known, Eq. (12.26) includes four equations with five unknowns: $p_J, q_J, S_J, \left.\frac{\partial p}{\partial r}\right|_J$, and $\left.\frac{\partial q}{\partial Z}\right|_J$. To solve these equations, an additional equation is needed. Connecting a perpendicular line from point J to the bottom at the intersect point 5 (Fig. 12.8), and integrating the first equations of Eq. (12.12) from points 5 to J, we obtain the following equation

$$S_J - S_5 = \frac{\Delta T}{2}\left[\left.\frac{\partial p}{\partial r}\right|_J + \left.\frac{\partial p}{\partial r}\right|_5 + \frac{P_J}{r_J} + \frac{P_S}{r_S} + \left.\frac{\partial q}{\partial Z}\right|_J + \left.\frac{\partial q}{\partial Z}\right|_5\right] \tag{12.27}$$

Then, using the five equations [Eqs. (12.26) and (12.27)], five unknowns can be solved. The actual steps include to eliminate derivative items $\left.\frac{\partial p}{\partial r}\right|_J$ and $\left.\frac{\partial q}{\partial Z}\right|_J$ first, then using the three equations obtained to solve p_J, q_J, and S_J, the derivatives on the plane $T_J - \Delta T$ of the cone bottom can be determined by using central difference equations.

Therefore, the dependent variables at every point can be computed numerically by using either of the above two methods. But it should be noted that in order to obtain a stable numerical result, the Courant–Friedrichs–Lewy criterion must be satisfied; namely, **for a set of first-order linear hyperbolic equations, the domain of dependence of the difference equations must contain the domain of dependence of the differential equations**. In Butler's method (numerical scheme based on four bicharacteristics), for example, the domain of dependence of the differential equations is a cone, and its intersection with the plane $T_J - \Delta T$ is a circle. But the intersection between the domain of dependence of the difference equations (central difference) and the plane $T_J - \Delta T$ is a right rectangle as shown by the dashed lines in Fig. 12.9. It is stable only when the former is contained

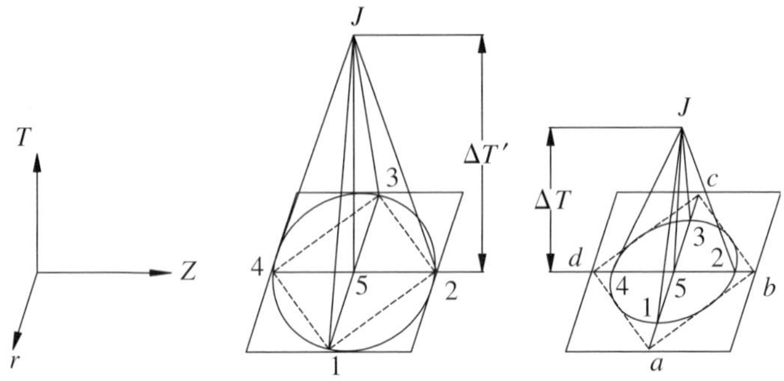

Fig. 12.9. Unstable numerical scheme and critical stable numerical scheme.

in the latter. If time step ΔT is too large, the right rectangle (the domain of dependence of the difference equations) cannot cover the circular area of the differential equations, as shown in Fig. 12.9(a), which results in an unstable numerical result. In this case, the time step ΔT must be reduced to ensure that the rectangular area contains the circular area. The critical state is shown in Fig. 12.9(b). In the numerical scheme of critical stability, although each point (1, 2, 3, and 4) on the cone bottom is not a node, its physical quantities can be obtained by an interpolation from the values at its neighboring nodes (a, b, c, d, 5).

It should be noted that the discussion so far is mainly based on Eq. (12.9). When it is applied to 2D elastic wave problems, in fact, it is known from Eq. (11.13) that elastic waves are actually composed of dilatation waves and distortion waves, propagating at velocities C_L and C_T respectively. When a 2D characteristic surface method is adopted, there are two kinds of characteristic cones generated: a volumetric characteristic cone which corresponds to the propagation of dilatation wave, and a shear characteristic cone which corresponds to the propagation of distortion wave. The whole computation is more complicated than that discussed above. It can be imagined that the computation is much more complicated in multidimensional nonlinear elasto-plastic, and visco-plastic problems. In such cases, other numerical methods can be adopted, including the finite difference method to be discussed in the next section or the finite element method to be discussed later.

12.2 Finite Difference Method

To introduce finite difference method (refer to Chou and Hopkins, 1972; Xu, 1985), we discuss a simplest example of a hyperbolic equation

$$\frac{\partial f}{\partial t} - a\frac{\partial f}{\partial X} = 0 \tag{12.28}$$

with an initial condition

$$f(X,0) = \psi(X), \quad -\infty < X < +\infty \tag{12.29}$$

Actually, Eq. (12.28) is in the same form as the following well-known wave equation.

$$\begin{aligned} \frac{\partial v}{\partial t} - C^2 \frac{\partial \varepsilon}{\partial X} &= 0 \\ \frac{\partial \varepsilon}{\partial t} - \frac{\partial v}{\partial X} &= 0 \end{aligned} \tag{12.30}$$

When such an equation is solved by means of finite difference method, it is first needed to establish the corresponding finite difference format which is compatible with the partial differential equation. Meanwhile, the convergence and stability of these difference equations should be checked. For strong discontinuous waves, artificial viscosity should also be introduced. These aspects will be discussed below.

12.2.1 Establishment of finite difference format

The X–t plane is meshed in rectangular grids by the following two groups of isometric straight lines parallel to the axis

$$\begin{aligned} X = X_k = k\Delta X \qquad & k = 0,\ \pm 1,\ \pm 2, \ldots \\ t = t_j = t_0 + j\Delta t \qquad & j = 0,\ 1,\ 2, \ldots \end{aligned}$$

where ΔX and Δt are step length (spacing) in the X and t directions respectively. From now on, for convenience, the node coordinate (X_k, t_j) is denoted by (k, j) and $f(X_k, t_j)$ is denoted by $f(k,j)$. In finite difference method, partial differential equations are approximated by difference equations. But for the same first-order partial derivative, there are several difference formats. For example, the so-called forward difference quotient is defined by

$$\begin{aligned} \left.\frac{\partial f}{\partial t}\right|_{(k,j)} &= \frac{f(k,j+1) - f(k,j)}{\Delta t} - \frac{\Delta t}{2}\frac{\partial^2 f(k,t_1)}{\partial t^2} \\ \left.\frac{\partial f}{\partial X}\right|_{(k,j)} &= \frac{f(k+1,j) - f(k,j)}{\Delta X} - \frac{\Delta X}{2}\frac{\partial^2 f(X_1,j)}{\partial X^2} \end{aligned} \tag{12.31}$$

and the backward difference quotient is defined by

$$\begin{aligned} \left.\frac{\partial f}{\partial t}\right|_{(k,j)} &= \frac{f(k,j) - f(k,j-1)}{\Delta t} + \frac{\Delta t}{2}\frac{\partial^2 f(k,t_2)}{\partial t^2} \\ \left.\frac{\partial f}{\partial X}\right|_{(k,j)} &= \frac{f(k,j) - f(k-1,j)}{\Delta X} - \frac{\Delta X}{2}\frac{\partial f^2(X_2,j)}{\partial X^2} \end{aligned} \tag{12.32}$$

while the central difference quotient is defined by

$$\left.\frac{\partial f}{\partial t}\right|_{(k,j)} = \frac{f(k,j+1)-f(k,j-1)}{2\Delta t} - \frac{\Delta t^2}{6}\frac{\partial^3 f(k,t_3)}{\partial t^3}$$
$$\left.\frac{\partial f}{\partial X}\right|_{(k,j)} = \frac{f(k+1,j)-f(k-1,j)}{2\Delta X} - \frac{\Delta X^2}{6}\frac{\partial^3 f(X_3,j)}{\partial X^3} \qquad (12.33)$$

where $t_j \leq t_1 \leq t_{j+1}$, $X_k \leq X_1 \leq X_{k+1}$, $t_{j-1} \leq t_2 \leq t_j$, $X_{k-1} \leq X_2 \leq X_k$, $t_{j-1} \leq t_3 \leq t_{j+1}$ and $X_{k-1} \leq X_3 \leq X_{k+1}$. So by taking different difference formats, the different difference equations are obtained. Therefore, it is necessary to study which difference format is more suitable to the partial differential equations to be solved. Here, the "more suitable" means that a shorter computation time is needed under a given accuracy. The accuracy is generally depicted by errors.

For example, substituting the forward difference quotient format Eq. (12.31) in Eq. (12.28), we have

$$\frac{f(k,j+1)-f(k,j)}{\Delta t} + a\frac{f(k+1,j)-f(k,j)}{\Delta X} - R_1(\Delta X,\Delta t) = 0$$

where

$$R_1(\Delta X,\Delta t) = \frac{\Delta t}{2}\frac{\partial^2 f(k,t_1)}{\partial t^2} + \frac{a\Delta X}{2}\frac{\partial^2 f(X_1,j)}{\partial X^2} = 0(\Delta t,\Delta X)$$

is called the error item. When Δt and ΔX are small enough, since $R_1(\Delta X,\ \Delta t)$ is the same order of small quantity as Δt and ΔX, then the difference equations of Eq. (12.28) is

$$f(k,j+1) - f(k,j) + a\xi\left[f(k,j+1) - f(k,j)\right] = 0 \qquad (12.34a)$$

where $\xi = (\Delta t/\Delta X)$. The $R_1(\Delta X,\ \Delta t)$ is named the truncation error of Eq. (12.34). Considering that the initial condition Eq. (12.29) can be expressed as

$$f(k,0) = g(k), \qquad k = 0,\ \pm 1,\ \pm 2,\ldots$$

and combining Eqs. (12.28) and (12.29), we obtain the final difference format

$$f(k,j+1) = f(k,j) - a\xi\left[f(k+1,j) - f(k,j)\right]$$
$$f(k,0) = g(k) \qquad (12.34b)$$

If the partial differential with respect to t is still approximated using forward difference quotient format, but the partial differential with respect to X is approximated using backward difference quotient format, then according to the definition of backward difference quotient format, namely the second equation of Eq. (12.32), the difference formats of Eqs. (12.28) and (12.29) are written as

$$\begin{aligned} f(k,j+1) &= f(k,j) - a\xi\,[f(k,j) - f(k-1,j)] \\ f(k,0) &= g(k) \end{aligned} \tag{12.35}$$

its truncation error $R_2(\Delta X,\ \Delta t)$ is

$$R_2(\Delta X,\ \Delta t) = \frac{\Delta t}{2}\frac{\partial^2 f(k,t_1)}{\partial t^2} - \frac{a\Delta X}{2}\frac{\partial^2 f(X_2,\ j)}{\partial X^2} = 0(\Delta X,\ \Delta t)$$

Similarly, if the partial differential with respect to X is approximated by using central difference quotient format, the corresponding difference formats are

$$\begin{aligned} f(k,j+1) &= f(k,j) - \frac{a\xi}{2}[f(k+1,j) - f(k-1,j)] \\ f(k,0) &= g(k) \end{aligned} \tag{12.36}$$

its truncation error $R_3(\Delta X,\ \Delta t)$ is

$$R_3(\Delta X,\ \Delta t) = \frac{\Delta t}{2}\frac{\partial^2 f(k,t_1)}{\partial t^2} - \frac{a\Delta X^2}{6}\frac{\partial^3 f(X_3,j)}{\partial X^3} = 0(\Delta t,\ \Delta X^2)$$

In the above three difference formats, the partial derivative with respect to t is all approximated by a forward difference quotient. In such a case, when the initial condition $g(k)$ is given, the $f(k,\ 1)$ of grid nodes on the first layer can be calculated. In general, when the $f(k,\ j)$ of the layer j have been known, the $f(k,\ j+1)$ of the layer $j+1$ can be calculated. This difference format is named the **explicit format**.

If we use the backward difference quotient for t, while the central difference quotient for X, then the difference formats for the same problem mentioned above are

$$\begin{aligned} f(k,j+1) &= f(k,j) - \frac{a\xi}{2}[f(k+1,j+1) - f(k-1,j+1)] \\ f(k,0) &= g(k) \end{aligned} \tag{12.37}$$

its truncation error $R_4(\Delta X,\Delta t)$ is

$$R_4(\Delta X,\Delta t) = \frac{\Delta t}{2}\frac{\partial^2 f(k,t_2)}{\partial t^2} - \frac{a\Delta X^2}{6}\frac{\partial^3 f(X_3,j)}{\partial X^3} = 0(\Delta t,\Delta X^2)$$

For the difference formats Eq. (12.37), when the initial value of $f(k,0)$ is known, we cannot directly calculate $f(k,\ 1)$ [and successively up to $f(k,\ j)$], but must solve a group of simultaneous equations to obtain the solution of f at $t = 1$. This difference format is named the **implicit format**. As it is needed to solve simultaneous equations, implicit formats are suitable to the problems with the combining initial-boundary value conditions.

In fact, there are many other difference formats. Readers can refer to the relevant books (e.g. Xu, 1985).

In the above discussions on the difference formats and their truncation errors, it is clear that their truncation error approaches zero when ΔX, $\Delta t \to 0$. This indicates that the limit of a difference format is its corresponding partial differential equation. The difference format which satisfies this condition is called, that it is compatible with the corresponding partial differential equations. From the truncation error R_1 to R_4, it is clear that these difference formats are of first-order of accuracy with respect to Δt. On the other hand, R_1 and R_2 indicate that Eqs. (12.34) and (12.35) are of first-order of accuracy with respect to ΔX, while the R_3 and R_4 indicate that the central quotient formats are of second-order of accuracy with respect to ΔX.

In numerical simulation, the compatibility of difference formats is necessary but not a sufficient condition. The convergence and stability of numerical methods should also be considered and will be analyzed in the following sections.

12.2.2 Convergence of difference formats

Consider a point $P(X, t)$ in the solution region and $X = k\Delta X$, $t = j\Delta t$, if a difference format is used to calculate $f(k, j)$ at the point P, and when Δ, $\Delta X \to 0$, $f(k, j)$ satisfies the following relationship

$$f(k, j) - \hat{f}(k, j) \to 0$$

where $\hat{f}$ represents the accurate solution while f is the numerical solution, then this difference format is regarded to be convergent.

For convenience, in the following discussion on the convergence of difference format, we first investigate the domain of dependence of different difference formats. If we use Eq. (12.34) to calculate f at point P, as shown in Fig. 12.10(a), the value of f at point P depends on the quantities at nodes on the initial line segment BC. Then, the segment BC is called the domain of dependence of the solution of Eq. (12.34) at point P. If we use Eq. (12.35) to calculate f at point P, as shown in Fig. 12.10(b), the segment BC is the domain of dependence of point P. In Fig. 12.10(c) the segment BC is the domain of dependence of the solution of Eq. (12.36) at point P. Here, the domain of dependence means the initial segment where the grid nodes are located. Figure 12.10 also indicates that the domain of dependence is not only dependent on the difference formats but also related to the time being calculated.

Consider the case of $a > 0$ in Eq. (12.28). It is known that the characteristics through point P must be located on the left side of the point, as shown in Fig. 12.10. If this characteristics intersects the line $t = 0$ at point D, then the line PD is the characteristics passing through the point P. If the domain of dependence of a difference equation does not cover the domain of dependence of a differential equation, then the value f_p obtained by difference equation at point P is irrelevant to the quantities at point D. Consequently,

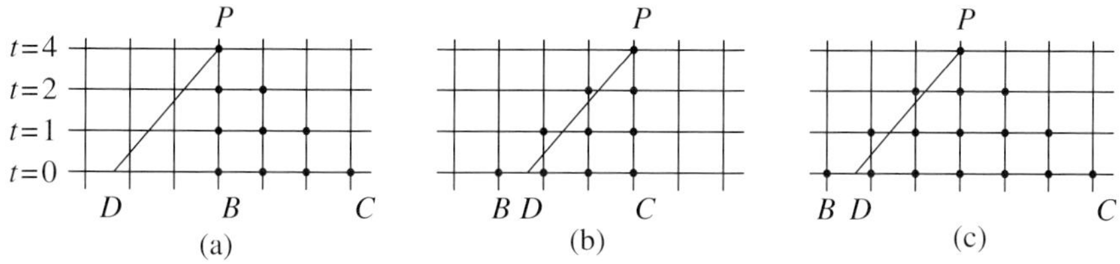

Fig. 12.10. The domain of dependence of three difference formats.

$f_P - \hat{f}_P \to 0$ cannot be satisfied always, namely the convergence of difference format cannot be ensured. Therefore, a necessary condition of convergence of a difference format is to satisfy the Courant condition, namely **the domain of dependence of a difference format must cover that of a differential equation**. When $a > 0$, thereby, from Fig. 12.10, it is clear that (a) the difference format shown in Fig. 12.10(a) (Eq. (12.34)) is not convergent, (b) the Courant condition of convergence for the difference format Eq. (12.35) is

$$0 \le a\Delta t \le \Delta X \qquad (a > 0) \tag{12.38}$$

and (c) the Courant condition of convergence for the difference format Eq. (12.36) is

$$-\Delta X \le a\Delta t \le \Delta X \qquad \text{namely } |a| \le \frac{\Delta X}{\Delta t} \tag{12.39}$$

These conditions should be noticed in difference meshing.

12.2.3 Stability of difference formats

Truncation error is not the only error source in numerically solving initial-value problems using difference formats, because there is also rounding error in every computing step. Computation of initial-value problem is conducted layer by layer. Even though rounding error of each layer is very small, the sum of rounding error accumulates and propagates with increasing layer number, which may be very large. As this accumulated error grows larger and larger, the real solution may be "submerged" by the accumulated error, which results in an instability of numerical results. Thereby, it is very important to analyze the stability of difference formats.

If the coefficients in differential equations are constants, Fourier analysis can generally be adopted to study the stability of difference formats, this is the so-called von Neumann method. The steps of this method will be illustrated by an example of the difference format of Eq. (12.35).

If the error of Eq. (12.35) is written as $\delta(k,j) = f(k,j) - \hat{f}(k,j)$, it is known that $\delta(k, j+1)$ should satisfy

$$\delta(k,j+1) = \delta(k,j) - a\xi\,[\delta(k,j) - \delta(k-1,j)] \tag{12.40}$$

If $\delta(k,j)$ is a harmonic wave with amplitude G^j and frequency w, then $\delta(k, j)$ is written as

$$\delta(k,j) = G^j e^{iwX_k} \text{ (w is an arbitrary real number)} \tag{12.41}$$

Substitution of the above equation into Eq. (12.40) yields

$$G = 1 - a\xi + a\xi e^{-iwX_k} \tag{12.42}$$

Replacing the exponent functions with triangle functions, and using the double-angle relationship of triangle functions, the above equation can be written as

$$G = 1 - 2a\xi \sin^2 \frac{wX_k}{2} - i2a\xi \sin \frac{wX_k}{2} \cos \frac{wX_k}{2} \tag{12.43a}$$

and

$$|G|^2 = 1 - 4a\xi(1 - a\xi) \sin^2 \frac{wX_k}{2} \tag{12.43b}$$

where $|G|$ is named the growing factor. According to Eq. (12.41), when $|G| > 1$, error grows with j exponentially. Therefore, the stability condition is

$$|G| \leq 1 \text{ (for every w)} \tag{12.44}$$

And Eq. (12.43) also indicates that the stability condition of this difference format (Eq. (12.35)) is

$$0 \leq a\xi < 1 \text{ (for $a > 0$)}$$

From this example, it is clear that the main steps in von Neumann's method are: first, assuming the solution of error equation is a form of harmonic wave; second substituting this solution into error equation to obtain the error growing factor; then checking whether the error growing factor is not larger than one. Through these steps, we can judge that the difference format Eq. (12.36) is not stable permanently, while the stability condition of difference format Eq. (12.34) is

$$a < 0 \text{ and } -1 \leq a\xi \leq 0.$$

If we regard the coefficient in the above difference equations a as the disturbance propagation velocity, $a > 0$ represents a wave propagating rightward; $a > 0$ represents a wave propagating leftward; and the $(1/\xi) = (\Delta x/\Delta t)$ in difference equation represents the propagation velocity of difference disturbance, then we know that the difference format Eq. (12.35) is stable for rightward waves and Eq. (12.34) is stable for leftward waves. Whether rightward waves or leftward waves, the stability condition requires that the propagating velocity of a difference disturbance should not be slower than the propagating velocity of a differential disturbance, namely

$$|a| \leq \left|\frac{1}{\xi}\right|$$

This is the so-called Courant condition, which is the necessary condition for stability.

12.2.4 Artificial viscosity

For hyperbolic partial differential equations, smoothness of solutions becomes very prominent once a strong discontinuous wave has been formed. The aim to adopt artificial viscosity is to smoothen the ramp rising edge of a shock wave front, so that a more smooth solution can be obtained. The effect of artificial viscosity in numerical methods to solve problems of shock waves will be discussed as follows by an example of 1D governing equations for nonlinear flow.

In Lagrange coordinates, the conservation equations of mass, momentum, and energy, and state equation of 1D nonviscous fluid are

$$\frac{\partial V}{\partial t} = \frac{1}{\rho_0}\frac{\partial v}{\partial X} \tag{12.45a}$$

$$\frac{\partial v}{\partial t} = -\frac{1}{\rho_0}\frac{\partial p}{\partial X} \tag{12.45b}$$

$$\frac{\partial E}{\partial t} = -\frac{p}{\rho_0}\frac{\partial v}{\partial X} = -p\frac{\partial V}{\partial t} \tag{12.45c}$$

$$p = p(E, V) \tag{12.45d}$$

where $V = 1/\rho$ is the specific volume, ρ_0 is the initial density, p is the pressure, and E is the specific internal energy. The difference equations of Eq. (12.45) can be taken as

$$\begin{aligned}
V\left(k+\tfrac{1}{2}, j+1\right) &= V\left(k+\tfrac{1}{2}, j\right) + \frac{\Delta t}{\rho_0 \Delta X}\left[v\left(k+1, j+\tfrac{1}{2}\right) - v\left(k, j+\tfrac{1}{2}\right)\right] \\
v\left(k, j+\tfrac{1}{2}\right) &= v\left(k, j-\tfrac{1}{2}\right) - \frac{\Delta t}{\rho_0 \Delta X}\left[p\left(k+\tfrac{1}{2}, j\right) - p\left(k-\tfrac{1}{2}, j\right)\right] \\
E\left(k+\tfrac{1}{2}, j+1\right) &= E\left(k+\tfrac{1}{2}, j\right) - \frac{1}{2}\left[p\left(k+\tfrac{1}{2}, j+1\right) + p\left(k+\tfrac{1}{2}, j\right)\right] \\
&\quad \left[V\left(k+\tfrac{1}{2}, j+1\right) - V\left(k+\tfrac{1}{2}, j\right)\right] \\
p\left(k+\tfrac{1}{2}, j+1\right) &= p\left[E\left(k+\tfrac{1}{2}, j+1\right), V\left(k+\tfrac{1}{2}, j+1\right)\right]
\end{aligned} \tag{12.46}$$

The last two equations should be solved simultaneously by iteration for $E(k+\frac{1}{2}, j+1)$, $V(k+\frac{1}{2}, j+1)]$, and $p(k+\frac{1}{2}, j+1)$. This difference format will give a second-order accurate approximation to a smooth solution.

But, if a shock wave appears, the numerical result is as shown in Fig. 12.11 (Richtmyer and Morton, 1967). It can be seen that not only the shock wave velocity is inaccurate, but the numerical solution fluctuates significantly due to the disturbing spurious high-frequency signals in the region near the shock wave front. However, this method is not unstable, since the solution remains bounded when ΔX and Δt decrease with a constant ratio.

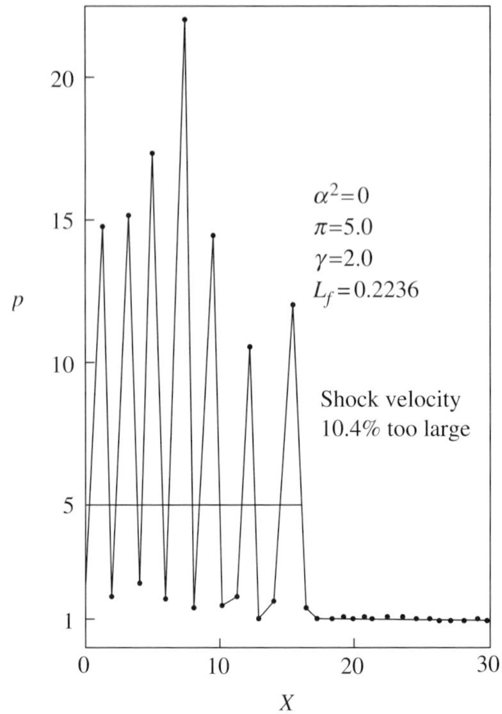

Fig. 12.11. Numerical solution according to Eq. (12.46) without artificial viscosity.

To overcome this difficulty, von Neumann and Richtmyer (1950) introduced the so-called **artificial viscosity**. The basic idea of their method is to replace pressure p in the original differential equations with $p + Q$, where Q is a function of $\partial v/\partial X$ (the spatial derivative of particle velocity with respect to X) and has the form of viscous force. By providing additional dissipation, Q makes all approximate solutions smooth. In other words, the sudden change of pressure p is distributed in several grids so that the solution is smoothed. The Q is suggested in the following form

$$Q = -\frac{(\alpha\Delta X)^2}{V}\frac{\partial v}{\partial X}\left|\frac{\partial v}{\partial X}\right| \tag{12.47}$$

where α is a dimensionless coefficient. It has the effect of smearing the shock wave over an invariable region with a width of about 2–3 ΔX. Because of that,Q is proportional to the square of the spatial derivative of velocity, its amplitude attenuated rapidly to vanish away from the shock wave front, so that the Rankine–Hugoniot conditions can be satisfied across the region of the shock wave (refer to Chapter 7).

After replacing p with $p + Q$ in Eq. (12.45), we can write the corresponding difference equations. For various values of the coefficient α, the numerical solutions of the difference equations considering von Neumann–Richtmyer artificial viscosity are shown in Fig. 12.12

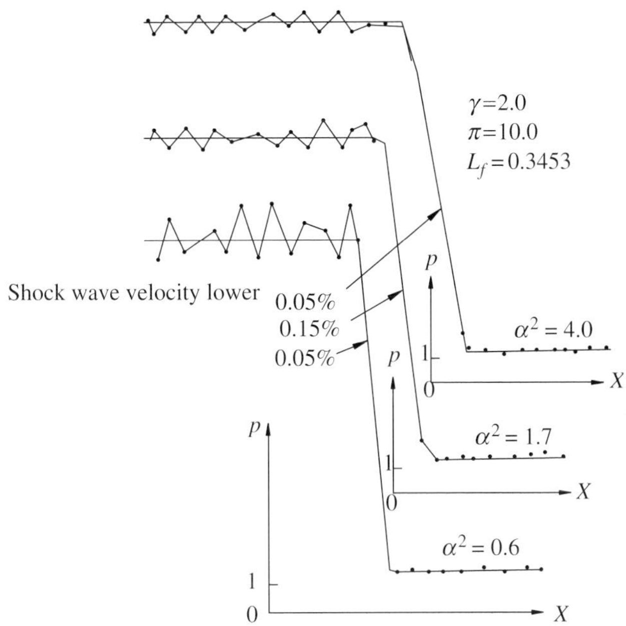

Fig. 12.12. Numerical solution of Eq. (12.45) with artificial viscosity.

(Richtmyer and Morton, 1967). It can be seen that with increasing α, the shock wave is spread over a wider zone and the wave profile is more smooth with less noise.

For shock waves in solids, the influence of selection of artificial viscosity on the numerical solutions is relatively large. The numerical results obtained for shock waves in solids are no longer satisfactory even if von Neumann–Richtmyer's viscosity item (Eq. (12.47)) is adopted. A common method, which is accepted widely, is to add a linear viscous item, namely

$$Q = -\frac{(\alpha\Delta X)^2}{V}\frac{\partial v}{\partial X}\left|\frac{\partial v}{\partial X}\right| + \frac{\beta\Delta X}{V}\frac{\partial v}{\partial X}$$

where $\alpha = 1.5$ and $\beta = 0.06$ usually.

In addition, the history of material deformation should sometimes be considered when using finite difference method. This is because the stress tensor of a solid not only depends on its present state, but also depends on its deformation history. Thereby, the finite difference method must carry and update the quantities that record or valuate the deformation (or loading) history of the material. At present, the most common way is by saving the historic value of the stress (or strain) tensor. Simultaneously, it is often necessary to carry additional numbers to indicate, for instance, the amount of work hardening that has been done, or the extent to which the pores in a porous material have been closed, or whether or not a crack (or spalling) takes place in the media. In these situations, using Lagrange coordinates is much more convenient. If other coordinate systems are used, the equations for these quantities must include convective terms, and assumptions should be introduced such as "continuity of deformation history" and so on.

In the previous discussions on finite difference method, although we only analyzed the examples for 1D wave problems, the elementary principles and methods are still valid for study on multidimensional problems without essential difference. This is because the partial differential governing equations now are established in a multidimensional space, the corresponding difference equations are also multidimensional. Consequently, the dimension-number of grids increases, which results in an enormous increase in the amount of computation.

12.3 Finite Element Method (FEM)

Finite element method has been widely applied in computation of structural mechanics, solid mechanics, fluid mechanics, and thermodynamics. It is an important method to solve partial differential equations. Finite element method is developed on the basis of the variation principle and by adopting the ideas of finite difference formats and combining the block polynomial interpolation. This combination makes finite element method keep the advantages of existing variation methods and the flexibility of finite difference method, and counteract the deficiency of classic variation methods. Therefore, finite element method is an innovation and development of the classic variation methods (refer to Xu, 1985). For example, in ABAQUS/Explicit (a widely used commercial finite element software), the finite difference formats have been adopted to treat the time derivatives in partial differential equations.

In this section, we will introduce the finite element method briefly; for detailed discussion refer to the relevant monographs (e.g. Zienkewicz, 1983; Desai and Kundu, 2001; Shorr, 2004, etc.).

12.3.1 Basic procedures of finite element method

In solid mechanics, we usually need to determine deformation or displacement u of a structure subjected to external loads. As it is difficult to solve displacement of a structure by analytic methods, approximate numerical methods are generally applied. The kernel of finite element method is to segment a continuous object into many small regions, namely **finite elements**, and based on governing equations and variation method using polynomial interpolation, the displacement in each element is solved. The main procedures include the following steps.

Meshing elements (discretization). Meshing an object into many small elements, the points connecting the elements are named by **nodes**. Although the size of elements is not limited, a good meshing must satisfy the requirements of, for instance, the capacity and speed of the computer used on the one hand and the accuracy of computation required on the other hand. In general, the following points must be considered: (1) a vertex of an element is also a vertex of its neighboring element; (2) big obtuse angles and big ratio of element edges should be avoided; (3) regions where the gradient of displacement changes intensively must be meshed with fine grids while regions where the gradient of displacement changes smoothly could be meshed with relatively coarse grids; (4) in order to reduce the bandwidth of a stiffness matrix so as to reduce the amount of computing, the absolute value of the biggest difference between the node-numbers of all adjacent nodes should be possibly small.

With many decades of development of finite element method, many element types have been built. Selection of element types can be based on characteristics of object motion and shapes of object structure. For example, for a 1D object [Fig. 12.13(a)], line element (two-node element) can be selected. For a 2D object [Fig. 12.13(b)], triangle element (three-node element) or quadrangle element (four-node element) can be applied. For a 3D object [Fig. 12.13(c)], hexahedron element (eight-node element) can be adopted.

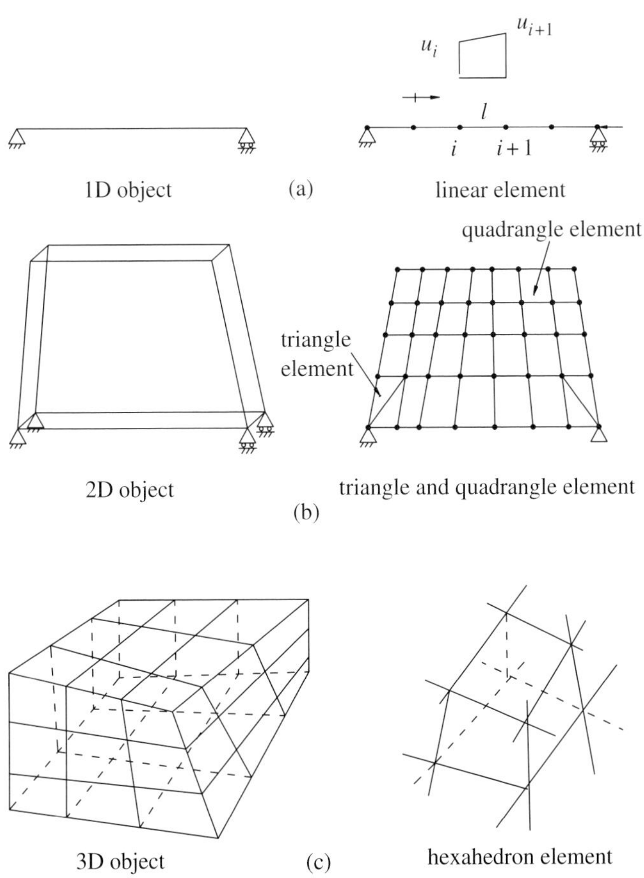

Fig. 12.13. Various types of finite element, (a) 1D element; (b) 2D element; (c) 3D element.

Choosing interpolation functions (shape functions). It is usually assumed that the dependent variables (such as displacement) in each element changes linearly. Mathematically, this means that the actual distribution of dependent variables (unknown functions) is depicted by piecewise linear functions, namely dealt with by a piecewise linear interpolation. This linear function describes the distribution shape of dependent variables in elements, and so is named the shape function. Thus, there are only finite parameters included in shape functions that are needed to figure out. The problem is discretized into an algebraic problem with finite degrees of freedom. In other words, when the function values at nodes of an element are determined, the key problem is converted into how to depict approximately the distribution of this function in the element by using a polynomial interpolation function.

For example, if the unknown function (dependent variable) is displacement u, then a polynomial interpolation function can be written as

$$u = N_1 u_1 + N_2 u_2 + \cdots + N_m u_m \tag{12.48}$$

where u_k ($k = 1, 2, \ldots, m$) are displacements at nodes of this element, N_k ($k = 1, 2, \ldots, m$) are indeterminate interpolation functions (shape functions). Taking the linear element in Fig. 12.13(a) as an example, and assuming the displacements u_i and u_{i+1} in the ith element have been given, then the displacement function $u(x)$ in this element can be expressed by

$$u(x) = \frac{x_{i+1} - x}{x_{i+1} - x_i} u_i + \frac{x - x_i}{x_{i+1} - x_i} u_{i+1}, \qquad x_i \le x \le x_{i+1}$$

Thereby, the problem is simplified to determining displacements at finite nodes only.

Choosing material models. In analyses by finite element method, material models or constitutive equations are absolutely necessary. Constitutive equations of materials are usually determined by mechanical experiments and constitutive theories. For numerical computation relating to large deformation, material models must satisfy the requirement of relevant constitutive theories, such as irrelevant to coordinate systems, nonlinear geometric depiction of large deformation, and nonlinear physical depiction of constitutive relationship, etc. For numerical computation related to dynamic response of structures under explosion/impact loadings, rate-dependence of constitutive equation should also be considered.

Deducing equations of finite elements. Governing equations of finite element are deduced based on the variation principle. Energy method and weighted residual method are the two common methods in the finite element analysis. Equations depicting mechanical behavior of finite elements can be obtained by using either of these methods. The obtained equations are named the equations of finite elements, and they can be written into matrix form as

$$[\mathbf{k}]\{\mathbf{q}\} + [\mathbf{m}]\{\ddot{\mathbf{q}}\} = \{\mathbf{Q}(t)\} \tag{12.49}$$

where $[\mathbf{k}]$ is the stiffness matrix, $\{\mathbf{q}\}$ is the displacement vector, $[\mathbf{m}]$ is the mass matrix, $[\ddot{\mathbf{q}}]$ is the acceleration vector, and $\{\mathbf{Q}\}$ is the force vector at nodes.

Governing equations and boundary conditions. Our target is to determine the mechanical response of the whole object under external loadings. Although Eq. (12.49) is valid for every element, during the deformation process of the structure, the displacement at each node is required to be kept continuous in order to satisfy the continuity requirement of the object. Based on this consideration, Eq. (12.49) is assembled to obtain the following whole governing equations

$$[\mathbf{K}]\{\mathbf{r}\} = \{\mathbf{R}\} \tag{12.50}$$

where $[\mathbf{K}]$ is the whole stiffness matrix, $\{\mathbf{r}\}$ is the vector of assembly displacement at nodes, $\{\mathbf{R}\}$ is the vector of assembly force at nodes.

The mechanical behavior of the object loaded is also influenced by boundary conditions. Considering boundary conditions, the whole governing equations Eq. (12.50) are further modified to

$$[\overline{\mathbf{K}}]\{\bar{\mathbf{r}}\} = \{\overline{\mathbf{R}}\} \tag{12.51}$$

Solving the set of linear equations. From Eq. (12.51), the displacements can be figured out, and consequently, the strain, particle velocity, and stress, etc., can be obtained.

The process of finite element numerical analysis is illustrated as follows by solving the 1D wave equations, as an example.

12.3.2 Examples

As an example, we examine a problem of stress wave propagation in a finite elastic bar under external loading $p(t)$, as shown in Fig. 12.14 (Desai and Kundu, 2001). The motion equation under 1D condition is (refer to Chapter 2)

$$\frac{\partial \sigma}{\partial X} = \rho_0 \frac{\partial^2 u}{\partial t^2} + f(t) \tag{12.52}$$

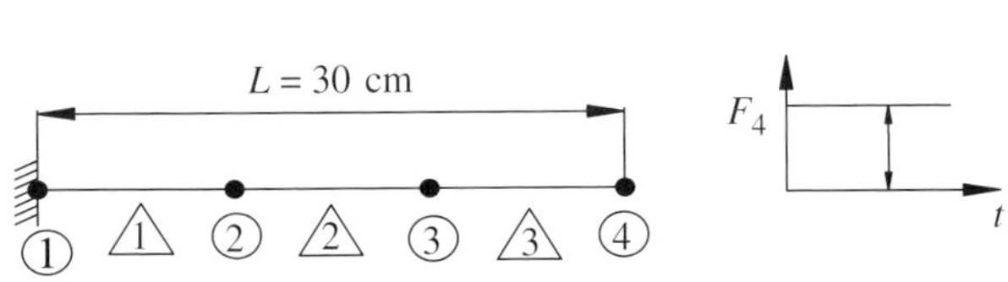

Fig. 12.14. Meshing of a 1D bar for numerical computation (the number in Δ labels elements, the number in O labels nodes).

where $f(t)$ is the body force uniformly distributed in the 1D bar. The 1D elastic constitutive equation is

$$\sigma = E\varepsilon = E\frac{\partial u}{\partial X} \tag{12.53}$$

where E denotes the elastic modulus of the bar material. Combining the above two equations, we obtain the governing equation with respect to displacement $u(X,t)$

$$E\frac{\partial^2 u}{\partial X^2} = \rho_0\frac{\partial^2 u}{\partial t^2} + f(t) \tag{12.54}$$

Meshing the bar uniformly with three line elements (four nodes) is shown in Fig. 12.14. The length of each element is l. The displacements at the two ends of the ith element are denoted by u_i and u_{i+1}. The displacement u of the ith element, by the linear interpolation, can be expressed as

$$u = N_i u_i + N_{i+1}u_{i+1} = [\mathbf{N}]\{\mathbf{q}\} \tag{12.55}$$

where $N_1 = 1 - s$, $N_2 = s$ and $s = (x - x_i)/l$.

Finite element equations can be obtained from virtual work principle. According to this principle, the mechanical state of the ith element should satisfy

$$\int_l \sigma\delta\varepsilon A\,dx = \int_l F\delta u A\,dx + \int_l f\delta u A\,dx \tag{12.56}$$

where $\delta\varepsilon$ and δu denote the virtual strain and virtual displacement, A is the cross-section area, and

$$F = -\rho_0\frac{\partial^2 u}{\partial t^2} \tag{12.57}$$

According to Eq. (12.55), we obtain

$$\varepsilon = \frac{\partial u}{\partial X} = \left(-\frac{1}{l}, \frac{1}{l}\right)\begin{Bmatrix} u_i \\ u_{i+1} \end{Bmatrix} = [\mathbf{B}]\{\mathbf{q}\} \tag{12.58}$$

$$\{\dot{\mathrm{u}}\} = [\mathbf{N}]\{\dot{\mathbf{q}}\} \tag{12.59}$$

$$\{\ddot{\mathrm{u}}\} = [\mathbf{N}]\{\ddot{\mathbf{q}}\} \tag{12.60}$$

Substituting Eqs. (12.57)–(12.60) into Eq. (12.56), and considering that the virtual displacement $\delta\{\mathbf{u}\}$ is arbitrary, we obtain

$$\int_l E[\mathbf{B}]^T[\mathbf{B}]Adx\{\mathbf{q}\} = \int_l \rho_0[\mathbf{N}]^T[\mathbf{N}]Adx\{\ddot{\mathbf{q}}\} + \int_l [\mathbf{N}]^T\{\mathbf{f}\}Adx \tag{12.61}$$

or

$$[\mathbf{k}]_i\{\mathbf{q}\}_i + [\mathbf{m}]_i\{\ddot{\mathbf{q}}\}_i = \{\mathbf{Q}(t)\}_i \tag{12.62}$$

where the stiffness matrix $[\mathbf{k}]$, mass matrix $[\mathbf{m}]$, and stress matrix at nodes $\{\mathbf{Q}\}$ are

$$[\mathbf{k}] = \frac{AE}{l}\begin{bmatrix} 1 & -1 \\ -1 & 1 \end{bmatrix} \tag{12.63a}$$

$$[\mathbf{m}] = \rho_0 Al \int_0^1 \begin{Bmatrix} 1-s \\ s \end{Bmatrix} [1-s, s]\, ds = \frac{\rho_0 Al}{6}\begin{bmatrix} 2 & 1 \\ 1 & 2 \end{bmatrix} \tag{12.63b}$$

$$\{\mathbf{Q}\} = \frac{Alf}{2}\begin{Bmatrix} 1 \\ 1 \end{Bmatrix} \tag{12.63c}$$

Next, we need to assemble the equations of finite element, namely to compose the general governing equations for depicting the response of the whole bar. Considering that there are only four nodes in this example, for convenience to sum up, we extend these second-order matrix to fourth-order matrix, namely

$$[\mathbf{k}]_i = \begin{pmatrix} & \vdots & \vdots & \\ \cdots & k_{i,i} & k_{i,i+1} & \cdots \\ \cdots & k_{i+1,i} & k_{i+1,i+1} & \cdots \\ & \vdots & \vdots & \end{pmatrix}$$

$$[\mathbf{m}]_i = \begin{pmatrix} & \vdots & \vdots & \\ \cdots & m_{i,i} & m_{i,i+1} & \cdots \\ \cdots & m_{i+1,i} & m_{i+1,i+1} & \cdots \\ & \vdots & \vdots & \end{pmatrix}$$

Then the general stiffness matrix and general mass matrix can be written as $[\mathbf{K}] = \sum_{i=1}^{3}[\mathbf{k}]_i$ and $[\mathbf{M}] = \sum_{i=1}^{3}[\mathbf{m}]_i$ respectively, namely

$$[\mathbf{K}] = \frac{AE}{l}\begin{bmatrix} 1 & -1 & 0 & 0 \\ -1 & 1+1 & -1 & 0 \\ 0 & -1 & 1+1 & -1 \\ 0 & 0 & -1 & 1 \end{bmatrix} = \frac{AE}{l}\begin{bmatrix} 1 & -1 & 0 & 0 \\ -1 & 2 & -1 & 0 \\ 0 & -1 & 2 & -1 \\ 0 & 0 & -1 & 1 \end{bmatrix} \tag{12.64a}$$

$$[\mathbf{M}] = \frac{\rho_0 Al}{6}\begin{bmatrix} 2 & 1 & 0 & 0 \\ 1 & 2+2 & 1 & 0 \\ 0 & 1 & 2+2 & 1 \\ 0 & 0 & 1 & 2 \end{bmatrix} = \frac{\rho_0 Al}{6}\begin{bmatrix} 2 & 1 & 0 & 0 \\ 1 & 4 & 1 & 0 \\ 0 & 1 & 4 & 1 \\ 0 & 0 & 1 & 2 \end{bmatrix} \tag{12.64b}$$

Introducing

$$\{\mathbf{r}\} = \begin{pmatrix} u_1 \\ u_2 \\ u_3 \\ u_4 \end{pmatrix}, \qquad \{\mathbf{R}\} = \frac{Alf}{2}\begin{pmatrix} 1 \\ 1 \\ 1 \\ 1 \end{pmatrix}$$

then we obtain the general governing equations below

$$\underset{(4\times4)}{[\mathbf{K}]}\ \underset{(4\times1)}{\{\mathbf{r}\}} + \underset{(4\times4)}{[\mathbf{M}]}\ \underset{(4\times1)}{\{\ddot{\mathbf{r}}\}} = \underset{(4\times1)}{\{\mathbf{R}(t)\}} \tag{12.65}$$

The above equation contains the term of second-order differential of displacement with respect to time (particle acceleration), of which the finite difference formats is

$$\begin{aligned} \{\dot{\mathbf{r}}\}_{t+\Delta t} &= \frac{3}{\Delta t}(\{\mathbf{r}\}_{t+\Delta t} - \{\mathbf{r}\}_t) - 2\{\dot{\mathbf{r}}\}_t - \frac{\Delta t}{2}\{\ddot{\mathbf{r}}\}_t \\ \{\ddot{\mathbf{r}}\}_{t+\Delta t} &= \frac{6}{\Delta t^2}(\{\mathbf{r}\}_{t+\Delta t} - \{\mathbf{r}\}_t) - \frac{6}{\Delta t}\{\dot{\mathbf{r}}\}_t - 2\{\ddot{\mathbf{r}}\}_t \end{aligned} \tag{12.66}$$

Then the general governing equations are further reduced to

$$[\overline{\mathbf{K}}]\{\mathbf{r}\}_{t+\Delta t} = \{\overline{\mathbf{R}}\}_{t+\Delta t} \tag{12.67}$$

where

$$\begin{aligned} [\overline{\mathbf{K}}] &= [\mathbf{K}] + \frac{6}{(\Delta t)^2}[\mathbf{M}] \\ \{\overline{\mathbf{R}}\}_{t+\Delta t} &= \{\mathbf{R}\}_{t+\Delta t} + \frac{6}{(\Delta t)^2}[\mathbf{M}]\left(\{\mathbf{r}\}_t + \Delta t(\dot{\mathbf{r}})_t + \frac{1}{3}(\Delta t)^2\{\ddot{\mathbf{r}}\}_t\right) \end{aligned} \tag{12.68}$$

If the length of the bar is 30 mm (the length of each element $l = 10$ mm), the cross-sectional area $A = 1\ \text{mm}^2$, the Young's modulus is $E = 62.5$ GPa, the density of the bar is $\rho_0 = 2.5 \times 10^3\ \text{kg/m}^3$, then the elastic wave velocity is $C = 5$ km/s, the corresponding time step can be taken as

$$\Delta t = \frac{l}{C} = 2 \times 10^{-6}\ \text{s} \tag{12.69}$$

Assume that the bar is in static state initially, namely the initial condition is

$$u(x,0) = \dot{u}(x,0) = \ddot{u}(x,0) = 0 \tag{12.70a}$$

As to the boundary condition, assume that the bar is subjected to a external load of 1 N at the bar end with the 4th node ($X = 30$ mm),

$$f_4(t) = 1\ \text{N} \tag{12.70b}$$

while the other bar end with the 1st node ($X = 0$) is fixed,

$$u_1(0, t) = 0 \tag{12.70c}$$

According to the above conditions, the relevant items in the governing equations Eq. (12.67) are

$$[\overline{\mathbf{K}}] = 6.25 \times 10^6 \begin{bmatrix} 3 & 0 & 0 & 0 \\ 0 & 6 & 0 & 0 \\ 0 & 0 & 6 & 0 \\ 0 & 0 & 0 & 3 \end{bmatrix} \tag{12.71a}$$

$$\{\overline{\mathbf{R}}\}_{t+\Delta t} = \{\mathbf{R}\}_{t+\Delta t} + 6.25 \times 10^6 \begin{bmatrix} 2 & 1 & 0 & 0 \\ 1 & 4 & 1 & 0 \\ 0 & 1 & 4 & 1 \\ 0 & 0 & 1 & 2 \end{bmatrix}$$
$$\times \left(\{\mathbf{r}\}_t + 2 \times 10^{-6} \{\dot{\mathbf{r}}\}_t + \frac{4 \times 10^{-12}}{3} \{\ddot{\mathbf{r}}\}_t \right) \tag{12.71b}$$

And according to the stress conditions at nodes (Eq. 12.70b), it is known

$$\{\mathbf{R}\}_{t+\Delta t} = \begin{pmatrix} 0 \\ 0 \\ 0 \\ 1 \end{pmatrix} \tag{12.71c}$$

Substituting Eqs. (12.69)–(12.71) into Eq. (12.68), we obtain the general matrix relationship at $t = \Delta t$

$$6.25 \times 10^6 \begin{bmatrix} 3 & 0 & 0 & 0 \\ 0 & 6 & 0 & 0 \\ 0 & 0 & 6 & 0 \\ 0 & 0 & 0 & 3 \end{bmatrix} \begin{pmatrix} u_1 \\ u_2 \\ u_3 \\ u_4 \end{pmatrix} = \begin{bmatrix} 0 \\ 0 \\ 0 \\ 1 \end{bmatrix} \tag{12.72}$$

As $u_1(0, t) = 0$, namely $u_1 = 0$, Eq. (12.72) can be simplified to

$$6.25 \times 10^6 \begin{bmatrix} 6 & 0 & 0 \\ 0 & 6 & 0 \\ 0 & 0 & 3 \end{bmatrix} \begin{bmatrix} u_2 \\ u_3 \\ u_4 \end{bmatrix} = \begin{bmatrix} 0 \\ 0 \\ 1 \end{bmatrix} \tag{12.73}$$

From the above equation, the displacement (its unit is m) of each node at $t = \Delta t$ can be figured out

$$\begin{pmatrix} u_1 \\ u_2 \\ u_3 \\ u_4 \end{pmatrix} = \begin{pmatrix} 0 \\ 0 \\ 0 \\ 5.33 \times 10^{-8} \end{pmatrix} \tag{12.74}$$

Substituting Eqs. (12.74) and (12.70a) into Eq. (12.66), the velocity and acceleration of each node at $t = \Delta t$ can be obtained as

$$\begin{pmatrix} \dot{u}_1 \\ \dot{u}_2 \\ \dot{u}_3 \\ \dot{u}_4 \end{pmatrix}_{\Delta t} = \begin{pmatrix} 0 \\ 0 \\ 0 \\ 0.08\ \text{m/s} \end{pmatrix} \quad \text{and} \quad \begin{pmatrix} \ddot{u}_1 \\ \ddot{u}_2 \\ \ddot{u}_3 \\ \ddot{u}_4 \end{pmatrix}_{\Delta t} = \begin{pmatrix} 0 \\ 0 \\ 0 \\ 8 \times 10^4\ \text{m/s}^2 \end{pmatrix} \tag{12.75}$$

As to the strain and stress of each node at $t = \Delta t$, can be figured out according to Eqs. (12.58) and (12.53). Repeating these steps, we can obtain solutions of Eq. (12.67) at $t = 2\Delta t, 3\Delta t\ldots$ and so on. Consequently, all the unknowns required can be figured out.

The mass matrix shown by Eq. (12.63) represents that mass is uniformly distributed in the bar. In some finite element software, for convenience, mass is usually assumed to be concentrated at the nodes. Thereby, the mass $Al\rho_0$ of a line element is equally shared by two nodes, namely

$$[\mathbf{m}] = \frac{Al\rho_0}{2}\begin{bmatrix} 1 & 0 \\ 0 & 1 \end{bmatrix} \tag{12.76}$$

This means that the mass matrix is a diagonal matrix when it is concentrated at the nodes, which facilitates computation. However, better accuracy can be obtained by taking a uniform mass distribution described by Eq. (12.63). This is shown in the following example.

In Fig. 12.14, let a bar of length 500 mm be equally meshed into 50 elements (the length of an element $l = 10$ mm, totally 51 nodes), the section area $A = 1\ \text{mm}^2$, the Young's modulus $E = 200$ GPa, the density of the bar be $\rho_0 = 8\times10^3\ \text{kg/m}^3$, then the elastic wave velocity is $C = 5\times10^3$ mm/ms. Initial and boundary conditions are

$$u(x,0) = \dot{u}(x,0) = \ddot{u}(x,0) = 0$$
$$u(0,t) = 0, \quad \dot{u}t(L,t) = 1\ \text{mm/ms}$$

In principle, we can obtain the solution to this problem following the same way as the above example. But, due to the node number being 51, the general governing equations is of 51-order. So a computer has to be used to perform this finite element computation.

The numerical results of particle velocity at $t = 0.08$ ms are shown in Fig. 12.15 (Desai and Kundu, 2001). The time steps used in both Fig. 12.15(a and b) are identical, namely $\Delta t = l/C = 0.002$ ms. On comparison, it is clear that the result of the model with uniformly distributed mass is better than that of the model with concentrated mass. The comparison between Fig. 12.15(c and d) also shows this trend, where the identical time step is $\Delta t = 0.5l/C = 0.001$ ms. In addition, Fig. 12.15 shows that the computation results for $\Delta t = l/C$ has higher accuracy than that for $\Delta t = 0.5l/C$.

Because engineering material usually shows more or less viscosity, there are certain damping effects in structure response. In finite element method, it is usually assumed that viscous resistance is proportional to particle velocity (which is equivalent to Newton's

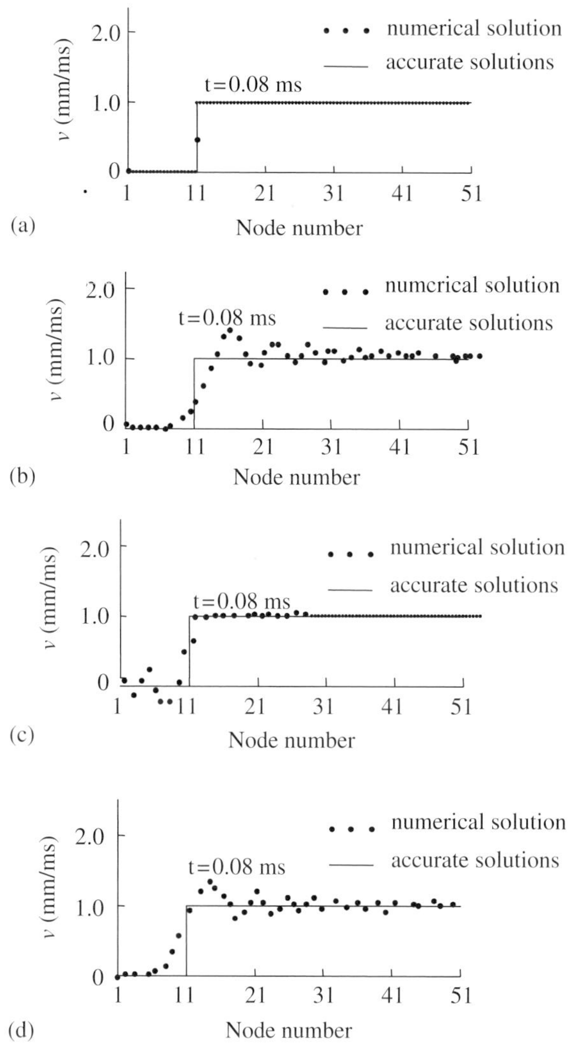

Fig. 12.15. Numerical results obtained by finite element method for 1D elastic waves: (a) $\Delta t = l/C$, mass uniformly distributed; (b) $\Delta t = l/C$, mass concentrated; (c) $\Delta t = 0.5l/C$, mass uniformly distributed; (d) $\Delta t = 0.5l/C$, mass concentrated.

viscosity law), namely it is proportional to $[\mathbf{c}]\{\dot{\mathbf{q}}\}$, where $[\mathbf{c}]$ is a damp matrix. Considering this damping, governing equations become

$$[\mathbf{k}]\{\mathbf{q}\} + [\mathbf{c}]\{\dot{\mathbf{q}}\} + [\mathbf{m}]\{\ddot{\mathbf{q}}\} = \{\mathbf{Q}(t)\} \tag{12.77}$$

The determination of damping property of materials is related to the research on rate-dependent dynamic behavior of materials, which will not be discussed here. Readers can refer relevant books (e.g. Desai and Christian, 1972; Shorr, 2004).

Appendices

Appendix I Conversion Factors for Stress and Pressure

	Pa	dyne/cm^2	bar	atm	kgf/mm^2	ksi
1 Pa =	1	10	10^{-5}	9.8692×10^{-6}	1.0197×10^{-7}	1.4504×10^{-7}
1 dyne/cm^2 =	10^{-1}	1	10^{-6}	9.8692×10^{-7}	1.0197×10^{-8}	1.4504×10^{-8}
1 bar =	10^5	10^6	1	0.98692	1.0197×10^{-2}	1.4504×10^{-2}
1 atm =	1.0133×10^5	1.0133×10^6	1.0133	1	1.0133×10^{-2}	1.4696×10^{-2}
1 kgf/mm^2 =	9.8067×10^6	9.8067×10^7	98.067	96.784	1	1.4223
1 ksi =	6.8947×10^6	6.8947×10^7	68.947	68.046	0.70307	1

Appendix II Characteristic Method used to Solve Quasi-Linear Hyperbolic Partial Differential Equations

We will discuss the following second-order quasi-linear hyperbolic partial differential equation with two independent variables, such as Eq. (2.18) in Chapter 2

$$\frac{\partial^2 u}{\partial t^2} = c^2 \frac{\partial^2 u}{\partial x^2} \tag{A.1}$$

where u is the unknown function, t and x are independent variables, the coefficient c^2 depends on $(\partial u/\partial t)$, $(\partial u/\partial x)$, u, x, and t, namely

$$c^2 = c^2(u_t, u_x, u, x, t)$$

where and hereafter, u_t and u_x are used to denote $(\partial u/\partial t)$ and $(\partial u/\partial x)$, respectively.

A2.1 If u satisfies Eq. (A.1) and the values of u_t and u_x on the curve $c(x, t)$ are given, it is required to determine under what conditions there are infinite second-order derivatives of u.

Along curve c, we have

$$\left.\begin{aligned} du_x &= \frac{\partial^2 u}{\partial x^2}dx + \frac{\partial^2 u}{\partial x \partial t}dt \\ du_t &= \frac{\partial^2 u}{\partial x \partial t}dx + \frac{\partial^2 u}{\partial t^2}dt \end{aligned}\right\} \tag{A.2}$$

where dx and dt are inter-related by the expression of curve c (dx/dt is the tangent slope of curve c), u_x and u_t are known. Therefore, Eqs. (A.1) and (A.2) are algebraic equations set in respect to $(\partial^2 u/\partial t^2)$, $(\partial^2 u/\partial x^2)$, and $(\partial^2 u/\partial t \partial x)$. Only if these equation set are interdependent, there would exist infinite solutions. In other words, a linear combination of two arbitrary equations, e.g. the two equations in (A.2), should be identically equal to the third equation. The linear combination of Eq. (A.2) is

$$du_t + \lambda du_x = \lambda \frac{\partial^2 u}{\partial x^2}dx + (\lambda dt + dx)\frac{\partial^2 u}{\partial x \partial t} + \frac{\partial^2 u}{\partial t^2}dt$$

where λ is an indeterminate coefficient. The condition under which the above equation is identically equal to Eq. (A.1), is

$$\frac{1}{dt} = -\frac{c^2}{\lambda dx} = \frac{0}{\lambda dt + dx} = \frac{0}{du_t + \lambda du_x}$$

From these, we have

$$\left(\frac{dx}{dt}\right)^2 = c^2, \quad du_t = \frac{c^2}{\frac{dx}{dt}}du_x$$

For hyperbolic equations, c^2 is positive, so there are two different real roots. Thus, when the following characteristic relations are satisfied

$$dx = cdt \tag{A.3a}$$

$$du_t = cdu_x \tag{A.3b}$$

or

$$dx = -cdt \tag{A.4a}$$

$$du_t = -cdu_x \tag{A.4b}$$

there exist infinite second-order derivatives of $u(x, t)$. In the above equations, equation a is called the **characteristics differential equation**, while the equation b is called the **compatibility condition along the characteristics**, or the **characteristics differential equation on the $u_t - u_x$ plane**.

A2.2 It is easy to prove that in a certain variables domain, if Eqs. (A.3) or (A.4) are identical equations, then the secondary differentiable function $u(x,t)$ satisfies Eq. (A.1). In fact, if there exist the following identical equations

$$du_t \equiv \pm c du_x, \qquad dx \equiv \pm c dt$$

since $u(x,t)$ is secondary differentiable, they can be immediately rewritten as

$$\frac{\partial^2 u}{\partial t^2}dt + \frac{\partial^2 u}{\partial t \partial x}dx \equiv \pm c\left(\frac{\partial^2 u}{\partial x \partial t}dt + \frac{\partial^2 u}{\partial x^2}dx\right)$$

$$\frac{\partial^2 u}{\partial t^2} \pm c\frac{\partial^2 u}{\partial t \partial x} \equiv c^2\frac{\partial^2 u}{\partial x^2} \pm c\frac{\partial^2 u}{\partial x \partial t}$$

namely, Eq. (A.1) is satisfied. Similarly, it is easy to prove the inverse inference: if the function u, in addition to satisfying Eq. (A.1), simultaneously complies with Eqs. (A.3a) and (A.4a), then Eqs. (A.3b) and (A.4b) along the characteristics are identities.

A2.3 In general, Eqs. (A.3) and (A.4) are not ordinary differential equations. In special cases, if c^2 only depends on u_x and u_t, then Eqs. (A.3b) and (A.4b) are reduced to ordinary differential equations. If there exists secondary differentiable single-valued solution for Eq. (A.1) as well, then simultaneously Eqs. (A.3a) and (A.4a) are reduced to ordinary differential equations. Geographically, Eqs. a of (A.3) and (A.4) and Eqs. b of (A.3) and (A.4) are represented by curves on the x–t plane and the u_x–u_t plane, which are called characteristics on the corresponding planes. Thus, the problem of solving partial differential equation (A.1) is reduced to the problem of solving ordinary differential equations (A.3) and (A.4), namely the problem to obtain the characteristics. These are just the problems discussed in the present book, since c^2 only depends on u_x, $c^2 = c^2(u_x)$.

A2.4 The geographical representation of the above inference is to map the domain G on the the x–t plane to the domain G' on the u_x–u_t plane, and to establish the correspondence between the characteristics C on the $x - t$ plane and the characteristics C' on the u_x–u_t plane, as shown in Fig. A.1. Therefore, the function u, which establishes the correspondence between the intersections of two different series of characteristics on the x–t plane and on the u_x–u_t plane, is the solution of Eq. (A.1).

A2.5 By utilizing such correspondence between the characteristics, usually the following three classes of boundary-value problems can be solved. (1) The Cauchy problem. If u_x

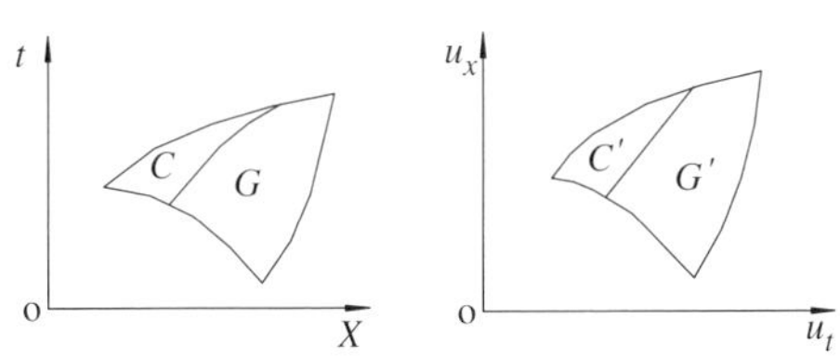

Fig. A.1. The domain G' on u_x–u_t plane is the mapping of the domain G on x–t plane.

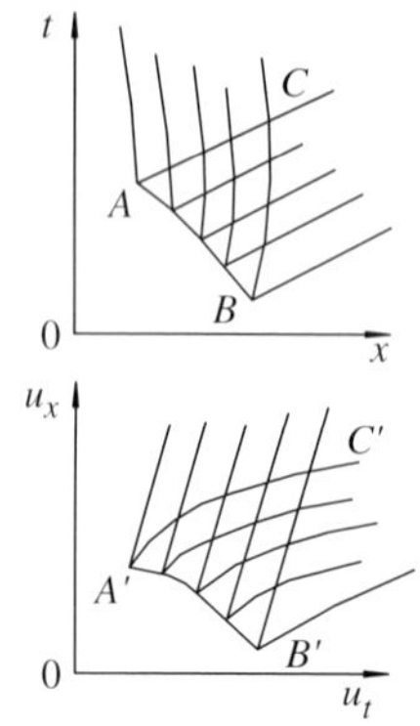

Fig. A.2. Cauchy problem.

and u_t are given on the non-characteristics curve, *AB*, then the single-value solutions in the curved triangle domain *ABC*, as shown in Fig. A.2 can be obtained, where *AC* and *BC* are characteristics. (2) The Darboux problem or the Cauchy characteristics boundary-value problem. If u_x and u_t are given on the characteristics curves *AB* and *AC*, which are not of the same set of characteristics, then the single-value solutions in the curved tetragonum domain *ABCD* as shown in Fig. A.3 can be obtained. (3) The Picard problem or the mixtured problem. If u_x and u_t are given on the characteristics curves *AB* and u_x (or u_t) is given on the non-characteristics curve *AC*, then by utilizing the correspondence between the characteristics the single-value solutions in the curved triangle domain *ABC* as shown in Fig. A.4 can be obtained. Some examples of those three classes of boundary-value problems have been given in Chapters 2 and 4.

A2.6 The second-order partial differential equation (A.1) is equivalent to the following first-order quasi-linear partial differential equation set in respect to u_t and u_x as unknown

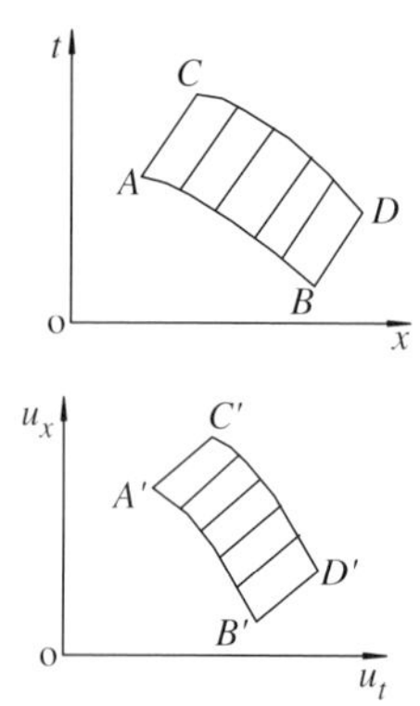

Fig. A.3. Darboux problem.

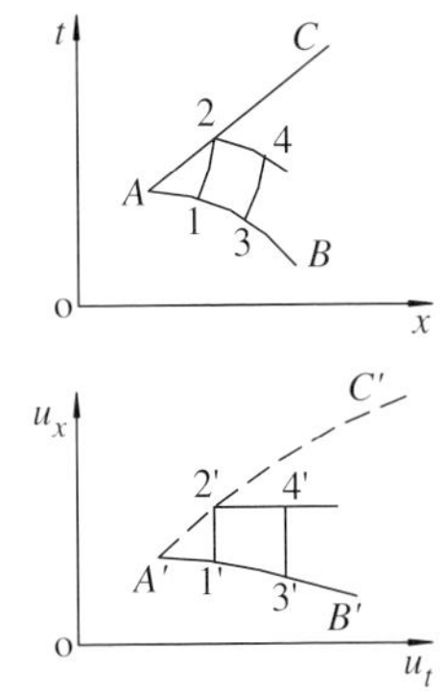

Fig. A.4. Picard problem.

functions [see Eqs. (2.12) and (2.16) in Chapter 2]

$$\begin{cases} \dfrac{\partial u_t}{\partial t} - c^2 \dfrac{\partial u_x}{\partial x} = 0 \\ \dfrac{\partial u_t}{\partial x} - \dfrac{\partial u_x}{\partial t} = 0 \end{cases} \tag{A.5}$$

Starting from Eq. (A.5) the same Eqs. (A.3) and (A.4) can be obtained. Equations (A.5) and (A.2) can be regarded as four algebraic equations with respect to $(\partial u_x/\partial t)$, $(\partial u_t/\partial t)$, $(\partial u_x/\partial x)$, and $(\partial u_t/\partial x)$, of which the solutions obtained by the well-known determinant method are

$$\frac{\partial u_x}{\partial t} = \frac{\Delta_1}{\Delta}, \cdots, \frac{\partial u_t}{\partial x} = \frac{\Delta_2}{\Delta}$$

where

$$\Delta = \begin{vmatrix} 0 & 1 & -c^2 & 0 \\ 1 & 0 & 0 & -1 \\ dx & dt & 0 & 0 \\ 0 & 0 & dx & dt \end{vmatrix}, \quad \Delta_1 = \begin{vmatrix} 0 & 1 & -c^2 & 0 \\ 1 & 0 & 0 & 0 \\ dx & dt & 0 & du_t \\ 0 & 0 & dx & du_x \end{vmatrix}, \cdots$$

If the curve c is a characteristics, then the conditions under which the derivatives of u_x and u_t along this curve are indeterminate (or the solutions are infinite), should be

$$\Delta = \Delta_1 = \Delta_2 = \Delta_3 = \Delta_4 = 0$$

From the determinant operation, we obtain

$$\begin{cases} dx \pm cdt = 0, \\ du_t \pm cdu_x = 0, \end{cases}$$

which are just Eqs. (A.3) and (A.4). Notice that if the u here means the displacement, then the u_x and u_t mean strain ε and velocity v, respectively, and thus Eqs. (A.3) and (A.4) here are just the Eqs. (2.23) and (2.24) in Chapter 2.

Appendix III An Introduction to Self-Similar Movement

A3.1 When we study the movement of continuum media, sometimes we may encounter such a kind of movement, if it is described by dimensionless variables then the time variable t is unseparatively composed with the coordinate variables x, y, z. No matter how t and x (or y, z) vary independently, so long as such composed dimensionless variables are identical, the movement is similar. This is the so-called **similar variable problem**, or **self-similar problem**. Such kind of movement is defined by Sedov, L. E. (Седов, Л. И., 1977) as "if all the dimensionless characteristic variables in the movement of continuum media are the functions of the only dimensionless variable composed of (x/bt^{δ}), (y/bt^{δ}) and (z/bt^{δ}), then such kind of movement is called a self-similar movement, where δ is a constant and b is a constant with a dimension of $LT^{-\delta}$."

Because of one independent variable reduction, such problems are much simplified. In the one-dimensional spatial problems, a partial differential equation is then reduced to an ordinary differential equation. So long as the spatial distribution of the characteristic variable of movement concerned at an instant is known, its spatial distribution at any time can be inferred. So long as how the characteristic variable of movement concerned at a spatial location varies with time is known, how it varies with time at all spatial locations can be inferred.

A3.2 The self-similar movement can be judged by the "Π-theorem" of dimensional analysis.

The "Π-theorem" is stated as following: if a physical relation consisted of n quantities with different dimensions described as

$$f(a_1, a_2, \ldots, a_k, b_{k+1}, b_{k+2}, \ldots, b_n) = 0$$

where the first k quantities are basic quantities which are independent of each other, and the latter n–k quantities are derived quantities, then this physical relation certainly can be completely expressed by n–k dimensionless quantities, $\pi_1, \pi_2, \ldots, \pi_{n-k}$, namely

$$f(\pi_1, \pi_2, \ldots, \pi_{n-k}) = 0$$

Thereupon, we can prove that if all the characteristic quantities of movement are only dependent on $a, b, x, y, z, t, \alpha_1, \alpha_2, \ldots, \alpha_n$, where the dimensions of a and b are independent of each other and different from the length dimension L or the time dimension T, x, y, and z are the spatial coordinates, t the time, α_i the i dimensionless constants (no limit for the number i), then the movement is self-similar. Usually we set the dimensions of a and b as $[a] = ML^K T^S$ and $[b] = LT^{-\delta}$, respectively, where $\delta \neq 0$, K and S are arbitrary numbers, both a and b could be even zero. This proposition can be very easily proved, namely, by choosing $[a]$, $[b]$, and T as basic dimensions, and then using the "Π-theorem."

This proposition can also be simply stated as following: if no characteristic length and characteristic time are included in all the variables involved to determine the movement, then the movement is self-similar. Some examples of such self-similar problems have been given in Chapters 2 and 9.

A3.3 It should be noted that sometimes the self-similar movement is not easy to judge immediately from a simple dimensional analysis, but a further physical analysis is needed. For example, as we discussed in Chapter 10, the deflection of a beam under transverse impact is usually considered as a function of the following variables

$$w = f(x, t, a, v_0)$$

where x denotes the coordinate along the beam axis, t the time, $a = (EI/\rho_0 A_0)^{1/4}$ [see Eq. (10.23) in Chapter 10], v_0 the invariable impact velocity. In such case, by using the "Π-theorem," we have

$$\frac{w}{v_0 t} = f\left(\frac{x}{v_0 t}, \frac{a^2}{v_0^2 t}\right)$$

Obviously, no conclusion of self-similar movement can be found immediately from the above equation. However, if we notice the physical equation of the problem, Eq. (10.14), and further introduce $a^4 = (EI/\rho_0 A_0) = c_0^2 R^2$ as shown in Eq. (10.23), then the basic physical equation can be rewritten as [see Eq. (10.14a) following Eq. (10.23) in Chapter 10]

$$a^4 \frac{\partial^4 w}{\partial x^4} + \frac{\partial^2 w}{\partial t^2} = 0$$

By introducing a new variable $\xi = x/a$, we can finally prove that this problem is self-similar, as discussed in detail in Section 10.3 of Chapter 10.

References

Abramson, H.N., Plass, H.J. and Ripperger, E.A., Stress Wave Propagation in Rods and Beams, *Advance in Applied Mechanics*, Vol. 5, eds. H.L. Dryden and T. von Karman, Academic Press, New York, 1958, 111.

Al-Hassani, S.T.S. and Johnson, W., The Dynamics of the Fragmentation Process of Spherical Bombs, Int. J. Mech. Sci., **11** (1969) 811.

Alter, B.E.K. and Cautis, C.W., Effect of Strain-Rate on the Propagation of a Plastic Strain Pulse along a Lead Bar, J. Appl. Phys., **27** (1956) 1079.

Arenberg, D.L., Ultrasonic Solid Delay Lines, J. Acoust. Soc. Am., **20** (1948) 1.

Bathe, K.J., *Finite Element Procedures in Engineering Analysis*, Prentice-Hall Inc., 1982.

Bejda, J. and Perzyna, P., The Propagation of Stress Waves in a Rate Sensitive and Work-Hardening Plastic Medium, Archwm Mech. Stosow., **16** (1964) 5.

Bell, J.F., Propagation of Plastic Waves in Solids, J. Appl. Phys., **30(2)** (1959) 196.

Bell, J.F., Propagation of Plastic Waves in Pre-Stressed Bar, Nary Contract N6-ONR-243, VIII, Johns Hopkins Univ. Tech. Rept. No. 5 (1951).

Bianchi, G., Some Experimental and Theoretical Studies on the Propagation of Longitudinal Plastic Waves in a Strain-Rate-Dependent Material, *Stress Waves in Anelastic Solids*, eds. H. Kolsky and W. Prager, Springer-Verlag, Berlin, 1964, 101.

Birch, F., The Effect of Pressure upon Elastic Parameters of Isotropic Solids, according to Murnagham's Theory of Finite Strain, J. Appl. Phys., **9** (1938).

Bohnenblust, H.F., A Note of von Karman's Theory of Propagation of Plastic Deformation in Solids, NDRC Memo. A-47M (1942), PB. 32180.

Bohnenblust, H.F., Comments on White and Griffis' Theory of the Permanent Strain in a Uniform Bar due to Longitudinal Impact, NDRC Memo. A-47M (1942), PB. 32180.

Bohnenblust, H.F., Propagation of Plastic Waves. A Comparison of Report NDRC A-29 and R. C. 329, NDRC Memo. A-53M (1942), PB. 20275.

Bohnenblust, H.F., Charyk, J.V. and Hyers, D.H., Graphical Solutions for Problems of Strain Propagation in Tension, NDRC Report A-131 (OSRD No. 1204) (1943).

Bridgman, P.W., *The Physics of High Pressure*, G. Bell and Sons, London, 1949.

Broberg, K.B., *Shock Waves in Elastic and Elastic-Plastic Media*, Avhandling Kungl. Tekn Högsk., Stockholm, 1956.

Brown, A.F.C. and Edmonds, R., The Dynamic Yield Strength of Steel and Intermediate Rate of Loading, Proc. Inst. Mech. Eng., **159** (1948) 11.

Butler, D.S., The Numerical Simulation of Hyperbolic Systems of Partial Differential Equations in Three Independent Variables, Proc. R. Soc. London, **A255** (1962) 232–252.

Campbell, J.D., Dynamic Plasticity: Macroscopic and Microscopic Aspects, Mater. Sci. Eng., **12** (1973) 3.

Chou, P.C. and Hopkins, A.K., eds., *Dynamic Response of Materials to Intense Impulsive Loading*, U.S. Air Materials Lab., Wright Patterson AFB, Ohio, 1972.

Chu Chao-Hsiang, Li Yong-chi and Wang Xiao-jun, Numerical Analyses of Spallation for Steel Plates under Explosive Loading, Appl. Math. and Mech., **2(4)** (1981) 353 (in Chinese).

Chu Chaoshiang, Wang Lili and Xu Daben, A Nonlinear Thermo-Viscoelastic Constitutive Equation for Thermoset Plastics at High Strain-Rates, *Proceedings of the International Conference on Nonlinear Mechanics*, (Oct. 28–31, 1985, Shanghai, China), ed. Chien Wei-Zang, Science Press, Beijing, 92–97.

Clifton, R.J., A Difference Method for Plane Problems in Dynamic Elasticity, Quarterly of Applied Mechanics, **25** (1967) 97–116.

Clifton, R.J. and Bodner, S.R., An Analysis of Longitudinal Elastic-Plastic Pulse Propagation, J. Appl. Mech., **31** (1966) 248–255.

Clifton, R.J. and Ting, T.C.T., The Elastic-Plastic Boundary in One Dimensional Wave Propagation, J. Appl. Mech., **35** (1968) 812–814.

Cole, J.D., Dougherty, C.B. and Huth, J.H., Constant-Strain Waves in Strings, J. Appl. Mech., **20** (1953) 519–522.

Coleman, B.D. and Noll, W., An Approximation Theorem for Functionals with Applications in Continuum Mechanics, Arch. Ratl. Mech. Anal., **6** (1960) 355.

Coleman, B.D. and Noll, W., Foundation of Linear Viscoelasticity, Rev. Mod. Phys., **33** (1961) 239.

Courant, R. and Friedrichs, *Supersonic Flow and Shock Waves*, Interscience Publishers Inc., New York, 1948.

Craggs, J.W., Wave Motions in Elastic-Plastic Strings, J. Mech. Phys. Solids, **2(4)** (1954) 286–295.

Craggs, J.W., The Propagation of Infinitesimal Plane Waves in Elastic-Plastic Materials, J. Mech. Phys. Solids, **5** (1957) 115.

Craggs, J.W., Plastic Waves, *Progress in Solid Mechanics*, Vol. II, eds. I.N. Sneddon and R. Hill, North-Holland Pub. Co., Amsterdam, 1961, Chap. IV.

Cristescu, N., *Dynamic Plasticity*, North-Holland Pub. Co., Amsterdam, 1967.

Davies, R.M., Stress Waves in Solids, Appl. Mech. Rev., **6** (1953) 1.

Davies, R.M., Stress Waves in Solids, Br. J. Appl. Phys., **7** (1956) 203.

De Juhasz, K.J., Graphical Analysis of Impact of Bars Stressed above the Elastic Range, J. Franklin Inst., **248** (1949) 113–142.

Desai, C.S. and Christian, J.T., eds. *Introduction to the Finite Element Method*, Van Nostrand Reinhold, New York, 1972.

Desai, C.S. and Kundu, T., *Introduction of Finite Element Method*, CRC press, New York, 2001.

Donnell, L.H., Longitudinal Wave Transmission and Impact, Trans. ASME, **52** (1930) 153–161.

Duvall, G.E., Shock Waves in Condensed Media, *Physics of High Energy Density*, eds. P. Caldirola and H. Knoepfel, Academic Press, New York, 1971, 7.

Duvall, G.E., Appications, *Dynamic Response of Materials to Intense Impulsive Loading*, eds. P.C. Chou and A.K. Hopkins, U. S. Air Materials Lab., Wright Patterson AFB, Ohio, 1972, 481–516.

Duvall, G.E., Shock Waves and Equations of State, *Dynamic Response of Materials to Intense Impulsive Loading*, eds. P.C. Chou and A.K. Hopkins, U. S. Air Materials Lab., Wright Patterson AFB, Ohio, 1972, 89–122.

Duwez, P.E., Preliminary Experiments on the Propagation of Plastic Deformation, NDRC Report A-33 (1942), PB. 40536.

Duwez, P.E., Physics of Solid-Plastic Flow, J. Aero. Sci., **23** (1956) 435.

Duwez, P.E. and Clark, D.S., An Experimental Study of the Propagation of Plastic Deformation under Conditions of Longitudinal Impact, Proc. ASTM, **47** (1947) 502.

Duwez, P.E., Wood, D.S., Clark, D.S. and Charyk, J.N., The Effect of Stopped Impact and Reflection on the Propagation of Plastic Strain in Tension, NDRC Report A-108 (1942), PB. 202725.

Duwez, P.E., Clark, D.S., Wood, D.S. and Bohnenblust, H.F., The Behavior of Long Beams under Impact Loading, NDRC Report A-216 (OSRD No. 1828) (1943), PB. 18472.

Duwez, P.E., Clark, D.S. and Bonenblust, H.F., The Behavior of Long Beams under Impact Loading, J. Appl. Mech., **17(1)** (1950) 27–34.

Friedman, M.B., Bleich, H.H. and Parnes, R., Spherical Elastic-Plastic Shock Propagation, Proc. ASCE, **91** (1965) 189.

Goldsmith, W., *Impact, The Theory and Physical Behavior of Colliding Solids*, Edward Arnold, London, 1960.

Graff, K.F., *Wave Motion in Elastic Solids*, Clarendon Press, Oxford, 1975.

Green, A.E. and Rivlin, R.S., The Mechanics of Nonlinear Materials with Memory, Part 1, Arch. Rat. Mech. Anal., **1** (1957) 1–21.

Griffis, L., The Behavior of Longitudinal Stress-Waves Near Discontinuities in Bars of Plastic Materials, NDRC Report A-212 (1943), PB. 18476.

Hillier, K.W., A Method of Measuring Some Dynamic Elastic Constants and its Application to the Study of High Polymers, Proc. Phys. Soc. B, **62** (1949) 701–713.

Hopkins, H.G., Dynamic Expansion of Spherical Cavities in Metals, *Progress in Solid Mechanics*, Vol. I, eds. I.N. Sneddon and R. Hill, North-Holland Publ. Co., Amsterdam, 1960, Chap. III.

Hopkinson, J., On the Rupture of Iron Wire by a Blow, Proc. Man. Lit. Phil. Soc., **11** (1872) 40.

Hopkinson, B., The Effect of Momentary Stresses in Metals, Proc. R. Soc., **A74** (1905) 498.

Hopkinson, B., A Method of Measuring the Pressure Produced in the Detonation of High Explosive or by the Impact of Bullets, Philos. Trans. R. Soc., **A213** (1914) 437.

Johnson, W., *Impact Strength of Materials*, Edward Arnold, London, 1972.

Jones, N., *Structural Impact*, Cambridge University Press, Cambridge, UK, 1989.

Jones, O.E., Metal Response under Explosive Loading, *Behavior and Utilization of Explosives in Engineering Design*, New Mexico Sec., ASME, 1972, 125.

Kaliski, S., Nowaski, W.K. and Wlodarczyk, E., On Certain Closed Solution for the Shock Wave with Rigid Unloading, Bull. Acad. Pol. Sci., Serie Sci. Technol., **15** (1967) 5.

Karman, T. von, On the Propagation of Plastic Deformation in Solids, NDRC Report A-29 (OSRD No. 365) (1942).

Karman, T. von and Duwez, P., The Propagation of Plastic Deformation in Solids, J. Appl. Phys., **21** (1950) 987–994.

Karman, T. von, Bohnenblust, H.F. and Hyers, D.H., The Propagation of Plastic Waves in Tension Specimens of Finite Length. Theory and Methods of Integration, NDRC Report A-103 (1942), PB. 18477.

Knopoff, L., On Rayleigh Wave Velocities, Bulletin of the Seismological Society of America, **42** (1952) 307.

Kobayashi, A. and Wang Lili, *Quest for Dynamic Deformation and Fracture of Viscoelastic Solids*, Ryoin Publishers, Japan, 2001.

Kolsky, H., An Investigation of the Mechanical Properties of Materials at Very High Rates of Loading, Proc. Phys. Soc., **B62** (1949) 676.

Kolsky, H., *Stress Wave in Solids*, Clarendous Press, Oxford, 1953.

Lee, E.H., A Boundary Value Problem in the Theory of Plastic Wave Propagation, Quart. Appl. Math., **10** (1952) 335–346.

Lee, E.H., Wave Propagation in Anelastic Materials, *Deformation and Flow of Solids*, ed. R. Grammel, Springer-Verlag, Berlin, 1956, 129.

Lee, E.H., Plastic-Wave Propagation Analysis and Elastic-Plastic Theory at Finite Deformation, *Shock Wave and the Mechanical Properties of Solids*, eds. J.J. Burke and V. Weiss, Syracuse Univ. Press, Syracuse, 1971, 3.

Lee, E.H. and Liu, D.T., Influence of Yield on High Pressure Wave Propagation, *Stress Waves in Anelastic Solids*, eds. H. Kolsky and W. Prager, Springer-Verlag, 1964, 239.

Lee, E.H. and Symonds, P.S., Large Plastic Deformations of Beams under Transverse Impact, J. Appl. Mech., **19** (1952) 308–314.

Lee, E.H. and Tupper, S.J., Analysis of Plastic Deformation in a Steel Cylinder Striking a Rigid Target, J. Appl. Mech., **21** (1954) 63.

Lee, E.H. and Wierzbicki, T., Analysis of the Propagation of Plane Elastic-Plastic Waves at Finite Strain, J. Appl. Mech., **34** (1967) 931.

Liu Xiao-min and Hu Shi-sheng, Two-Dimensional Numerical Analyses of the Dispersion Effects in SHPB with Large Diameter Bars, Experimental Mechanics, **15(4)** (2000) 371–376 (in Chinese).

Liu Xiao-min and Hu Shi-sheng, Wave Propagation Characteristics in Cone Bars Used for Variable Cross-Section SHPB, Explosion and Shock Waves, **20(2)** (2000) 110–114 (in Chinese).

Lockett, F.J., *Nonlinear Viscoelastic Solids*, Academic Press, London, 1972.

Lüdwik, P., *Elemente der technologischen Mechanik*, Springer-Verlag, Berlin, 1909.

Malvern, L.E., The Propagation of Longitudinal Waves of Plastic Deformation in a Bar of Material Exhibiting a Strain-Rate Effect, J. Appl. Mech., **18** (1951) 203.

McQueen, R.G., Marsh, S.P., Taylor, J.W., Fritz, J.N. and Carter, W.J., The Equation of State of Solids form Shock Wave Studies, *High-Velocity Impact Phenomena*, ed. R. Kinslow, Academic Press, New York, 1970, 293.

Menkes, S.B. and Opat, H.J., Broken beams, Exp. Mech., **13** (1973) 480–486.

Miklowitz, T., *The Theory of Elastic Waves and Wave-guides*, North-Holland Publ. Co., Amsterdam, 1978.

Morland, L.W., The propagation of Plane Irrotational Waves through an Elastoplastic Medium, Philos. Trans. R. Soc. London, **A251** (1959) 341.

Morland, L.W., Spherical Wave Propagation in Elastic-Plastic Work-hardening Materials, J. Mech. Phys. Solids, **17** (1969) 371.

Murnagham, F., The Compressibility of Media under Extreme Pressures, Proc. Nat. Acad. Sci., **30** (1994).

Neumann, J. von and Richtmyer, R.D., A Method for the Numerical Calculation of Hydrodynamic Shocks, J. Appl. Phys., **21** (1950) 232–257.

Nonaka, T., Some Interaction Effects in a Problem of Plastic Beam Dynamics, part 1: Interaction Analysis of a Rigid-perfectly Plastic Beams, J. Mech., **34(3)** (1967) 623–630.

Nowaski, W.K., *Stress Waves in Non-elastic Solids*, Pergamon Press Ltd., Oxford, 1978.

Pack, D.C., Evans, W.M. and James, H.J., The Propagation of Shock Waves in Steel and Lead, Proc. Phys. Soc., **60** (1948) 1.

Perzyna, P., The Constitutive Equations for Rate Sensitive Plastic Materials, Quart. Appl. Math., **20** (1963) 321.

Plass, H.J. Jr., A Theory of Longitudinal Plastic Waves in Rods of Strain-Rate Dependent Material, Including Effects of Lateral Inertia and Shear, *Plasticity*, eds. E.H. Lee and P.S. Symonds, Pergamon Press, Oxford, 1960, 453.

Ravichandran, G. and Subhash, G., Critical Appraisal of Limiting Strain Rates for Compression Testing of Ceramics in a Split Hopkinson Pressure Bar. J. Amer. Cer. Soc., **77** (1994) 263–267.

Rayleigh, L., On Waves Propagated along the Plane Surface of an Elastic Solid, Proc. London Math. Soc., **17** (1887) 4.

Rice, M.H., McQueen, R.G. and Walsh, J.M. Compression of Solids by Strong Shock Waves, *Solid State Physics*, Vol. 6, eds. F. Seitz and D. Turnbull, Academic Press, New York, 1958, 1.

Richtmyer, R.D. and Morton, K.W., *Difference Methods for Initial-Value Problems*, Interscience Publishers, New York, 1967.

Rinehart, J.S., *Stress Transients in Solids, Hyper Dynamics*, New Mexico, 1975.

Riparbelli, C., On the Time Lag of Plastic Deformation, *Proc. 1st Midwestern Conf. on Solid Mech.*, Univ. of Illinois, Urbana, 1953, 148.

Sauerwin, H., Numerical Calculations of Multidimensional and Unsteady Flows by the Method of Characteristics, J. Comput. Phys., **1** (1967) 406–432.

Selberg, H.L., Transient Compression Waves from Spherical and Cylindrical Cavities, Arkiv. for Fysik, **5** (1952) 97.

Shorr, B.F., *The Wave Finite Element Method*, Springer-Verlag, Berlin, 2004.

Simmons, J.A., Hauser, F. and Dorn, J.E., *Mathematical Theories of Plastic Deformation under Impulsive Loading*, Univ. of California Press, Berkeley, California, 1962.

Smith, J.C., McCrackin, F.L. and Schiefer, H.F., Stress-Strain Relationships in Yarns Subjected to Rapid Impact Loading, Part V, Texitile Res. J., **28** (1958) 288–302.

Sparrow, C.M., *Table of Integrals*, Edward Brothers, Ann Arbor, Michigan, 1934.

Sternglass, E.J. and Stuart, D.A., An Experimental Study of the Propagation of Transient Longitudinal Deformation in Elastoplastic Media, J. Appl. Mech., **20** (1953) 427.

Tang Zhi-Ping, Chu Chao-Hsiang, Wang Lili and Tian Lan-Qiao, Mechanical Behavior of Epoxy Resin under High Strain Rates, *Second National Conference of Explosive Mechanics,* Chinese Society of Theoretical and Applied Mechanics, Yangzhou, 1981, 4-1-2 (in Chinese).

Taylor, G.I., The Use of Flat-ended Projectiles for Determining Dynamic Yield Stress, I. Theoretical Considerations, Proc. R. Soc., **A194** (1948) 289.

Taylor, G.I., Propagation of Earth Waves from an Explosion, British Official Report RC. 70(1940); *The Scientific Papers of G.I. Taylor, v.1, Mechanics of Solids*, ed. G.K. Batchelor, Univ. Press, Cambridge, 1958, 456.

Taylor, G.I., The Plastic Wave in a Wire Extended by an Impact Load, British Official Report RC. 329(1942); *The Scientific Papers of G. I. Taylor, v. 1, Mechanics of Solids,* ed. G.K. Batchelor, University Press, Cambridge, 1958, 467–479.

Taylor, G.I., Fragmentation of Tubular Bombs, *The Scientific Papers of G. I. Taylor, v. III*, Univ. Press, Cambridge, 1963, 559.

Timoshenko, S. and Goodier, J.N., *Theory of Elasticity*, McGraw-Hill, New York, 1951.

Ting, T.C.T., On the Initial Speed of Elastic-Plastic Boundaries in Longitudinal Wave Propagation in a Rod, J. Appl. Mech., **38** (1971) 441–447.

Ting, T.C.T., Nonexistence of Higher Order Discontinuities Across Elastic/Plastic Boundary in Elastic-Plastic Wave Propagation, in Plasticity and Failure Behavior of Solids, *Memorial Volume Dedicated to the Late Professor Yuriy Nickolaevich Rabotnov*, eds. G.C. Sih, A.J. Ishlinsky and S.T. Mileiko, Kluwer Academic Pub., Dordrecht, 1990, 115–136.

Tuler, F.R., Butcher, B.M., A Criterion for the Time Dependence of Dynamic Fracture, Int. J. Fract. Mech., **4** (1968) 431.

Viktorov, I.A., *Rayleigh and Lamb Waves*, Plenum Press, New York, 1967.

Wang Li-Lih, On Propagation of Elastic-Plastic Waves in Strings, Acta Mech. Sin., **7(3)** (1964) 228–240.

Wang Lili, Tensile Stress Regions Caused by Reverse Plastic Deformation in One Dimensional Strain Elastic-Plastic Compressional Wave Propagation, Explosion and Shock Waves, **2** (1982) 39–44 (in Chinese).

Wang Lili, Dynamic Fracture of Spherical Vessel under Explosive Internal Pressure Loading, Chem. Eng. Machinery, **10(2)** (1983) 11–17 (in Chinese).

Wang Li-li, A Criterion of Thermo-Viscoplastic Instability for Adiabatic Shearing, *Proc. International Symposium on Intense Dynamic Loading its Effects*, eds. Zheng Zhemin and Jing-Ting, Science Press, Beijing, 1986, 359–792.

Wang Lili, Adiabatic Shearing—the Constitutive Instability of Materials Under Impact Loading, in *Progress in Impact Dynamics*, eds. Wang Lili, Yu Tongxi and Li Yongchi, the Press of University of Science and Technology of China, Hefei, 1992, 3–33 (in Chinese).

Wang Lili (Li-Lih WANG), Stress Wave Propagation for Nonlinear Viscoelastic Polymeric Materials at High Strain Rates, Chinese J. Mech., Series A, **19(1)** (2003) 177–183.

Wang Lili, Influences of Stress Wave Propagation upon Studying Dynamic Response of Materials at High Strain Rates, J. Beijing Inst. Technol., **13(3)** (2004) 225–235.

Wang Lili, Chu Chao-Hsiang and Yu Ji-Lin, On Discontinuous Properties of Elastic-Plastic Boundaries in Elastic-Plastic Plane Waves Propagation, Explosion and Shock Waves, **3(1)** (1983) 1–8 (in Chinese).

Wang Li-li, Bao He-sheng and Lu Wei-Xian, The Dependence of Adiabatic Shear Bending on Strain and Temperature, J. Physique, Colloque C3, **49** (1988) 207–214.

Wang Li-li and Bao He-sheng, A Strain-Localization Analysis for Adiabatic Shear Band at Different Environmental Temperatures, *Proc. 6th International Conference*, Mechanical Behavior of Materials, eds. M. Jono and T. Inoue, Pergamon Press, Kyoto, Japan (1991) 1, 479–486.

Wang, Li-Lih (Wang Lili), Field, J.E. and Sun, Q., Dynamic Behaviour of Pre-Stressed High Strength Polymeric Yarns Transversely Impacted by a Blade, *Proceedings of Second International Symposium on Intense Dynamic Loading and Its Effects*, June 9–12, 1992, eds. Zhang Guanren and Huang Shihui, Sichuan University Press, Chengdu, China, 1992, 354–359.

Wang Lili and Hu Shi-Sheng, Wang Xiao-Jun, The Attenuation of Plane Shock Waves Propagation in an Elasto-Plastic Medium, J. China Univ. Sci. Technol., **13(1)** (1983) 90–100 (in Chinese).

Wang Lili and Hu Shisheng, The Amplification of Stress Waves Propagated in a Conical Bar, Journal of Ningbo University (Natural Science & Engineering Edition), **1(1)** (1988) 69–78 (in Chinese).

Wang, L.-L., Huang, D.-J. and Gan, S., Nonlinear Viscoelastic Constitutive Relations and Nonlinear Viscoelastic Wave Propagation for Polymers at High Strain Rates, in Constitutive Relation in High/Very High Strain Rates, *IUTAM Symposium*, Oct. 16–19, 1995, Noda, Japan, eds. K. Kawata, J. Shioiri, Springer-Verlag, Tokyo, 1996, 137–146.

Wang Lili (Li-Lih Wang) and Jones, N., An Analysis of the Shear Failure of Rigid-Linear Hardening Beam Under Impulsive Loading, Acta Mech. Sin. (English Series), **12(4)** (1996) 338–348.

Wang, L.L., Labibes, K., Azari, Z. and Pluvinage, G., Generalization of Split Hopkinson Bar Technique to Use Viscoelastic Bars, Int. J. Impact Eng., **15(5)** (1994) 669–686.

Wang Li-li, Lu Weixian, Hu Shisheng and Tang Zhiping, Study of the Initiation and Development of Adiabatic Shear Bands for a Titanium Alloy Under High Strain Rates, *IUTAM Symposium, Macro and Micro-mechanics of High Velocity Deformation and Fracture*, 1985, Tokyo, Japan, eds. K. Kawata and J. Shioiri, Springer Verlag, Tokyo, 1987, 359–406.

Wang Lili, Pluvinage, G. and Labibes, K., The Influence of Dynamic Constitutive Relation of Polymer at Impact Loading on the Viscoelastic Wave Propagation Character, Journal of Ningbo University (Natural Science & Engineering Edition), **8(3)** (1995) 30–57 (in Chinese).

Wang Li-li, Shi Shao-qiu, Chen Jiang-ying and Huang De-jin, Studies on ZWT Non-Linear Thermo-Viscoelastic Constitutive Relation and its Application, Journal of Ningbo University (Natural Science & Engineering Edition), **13(Supp.)** (2000) 141–149 (in Chinese).

Wang Li-li and Wang Yong-gang, The Important Role Stress Waves Played in the Studies on Dynamic Constitutive Behavior of Materials by SHPB, Explosion and Shock Waves, **24(1)** (2005) 17–25 (in Chinese).

Wang Lili and Yang Liming, The Non-Linear Viscoelastic Constitutive Relation of Solid Polymeric Materials, in *Progress in Impact Dynamics,* eds. Wang Lili, Yu Tongxi and Li Yongchi, The Press of University of Science and Technology of China, Hefei, 1992, 88–116 (in Chinese).

Wang Yong-gang and Wang Li-li, Stress Wave Dispersion in Large-Diameter SHPB and its Manifold Manifestations, J. Beijing Inst. Technol., **13(3)** (2004) 247–253.

Whiffin, A.C., The Use of Flat-ended Projectiles for Determining Dynamic Yield Stress, II, Tests on Various Metallic Materials, Proc. R. Soc., **A194** (1948) 300.

White, M.P. and Griffis, L., The Permanent Strain in a Uniform Bar Due to Longitudinal Impact, NDRC Report A-71(1942), PB. 18480; J. Appl. Mech., **14** (1947), 337–343.

White, M.P. and Griffis, L., The Propagation of Plasticity in Uniaxial Compression, J. Appl. Mech., **15** (1948) 256.

Wood, D.S., On Longitudinal Plane Waves of Elastic-Plastic Strain in Solids, J. Appl. Mech., **19** (1952) 521.

Xu Cui-wei, An Introduction of Computational Methods, Higher Education Press, Beijing, 1985 (in Chinese).

Yang, L.M. and Shim, P.V.W., An Analysis of Stress Uniformity in Split Hopkinson Bar Test Specimens, Int. J. Impact Eng., **31** (2005) 129–150.

Yu Ji-Lin and Wang Lili (Wang Li-Lih), Chu Chao-Hsiang, Propagation Velocity of Elastic-Plastic Boundaries with Weak Discontinuities in a Bar, Kexue Tongbao (A Monthly Journal of Science), **27(5)** (1982) 574 (in Chinese).

Yu Ji-Lin, Wang Lili and Chu Chao-Hsiang, Basic Properties of Elastic-Plastic Boundaries in Stress Wave Propagation in a Bar, Acta Mech. Sol. Sin., **(3)** (1982) 313–324 (in Chinese).

Yu Ji-Lin, Wang Lili and Chu Chao-Hsiang, Determination of Propagation Velocity of Elastic-Plastic Boundaries in a Bar, Acta Mech. Sol. Sin., **(1)** (1984) 16–26 (in Chinese).

Zheng Zhe-min and Xie Bo-min, A Suggestion of Numerical Model for Underground Explosion, Sci. Tech. Research Report No. 00756, Institute of Mechanics, Chinese Academy of Sciences, May, 1965; in ed. Zheng Zhe-min, Collected Works of Zheng Zhe-min, Science Press, Beijing, 2004, 166–190.

Zhou Fenghua, Wang Lili and Hu Shisheng. On the Effect of Non-uniformity in Polymer Specimen of SHPB Tests. J. Exp. Mech., **7(1)** (1992) 23–29.

Zhou Fenghua, Wang Lili and Hu Shisheng, A Damage-Modified Nonlinear Viscoelastic Constitutive Relation and Failure Criterion of PMMA at High Stain Rates, Explosion and Shock Waves, **12(4)** (1992) 333–342 (in Chinese).

Zienkewicz, O.C. and Morgan, K, *Finite Elements and Approximation*, Jone Wiley & Son Inc., New York 1983.

Бидерман, В.Л., Расчеты на ударную нагрузку, Основы современных методов расчедов на прочность в машиностроении, Сб. Под ред. Понамарева, С.Д., Машгиз (1952) (in Russian). [Biderman, V.L., Calculations on Impact Loading, in Ponamareva, S.D. ed. *Basis of Modern Methods for Strength Calculation in Mechanical Engineering*, Mashgiz, 1952.]

Буравцев, А.И., Аналитическиий способ построения волны разгрузки, Вест. Ленингр. Унив., No. 1 (1970), 93 (in Russian). [Buravtsev, A.I., Analytical Method of Building Unloading Waves, Vest. Leningr. Univ., **1** (1970) 93.]

Воронов, Ф.Ф. и Верещакин, Л.Ф., Влияние гидростатического давления на упругия свойства металлов, Физ. Метал. Металловед., 11 (1961), 443; 627 (in Russian). [Voronov, F.F. and Vereshchakin, L.F., Influence of Hydrostatic Pressure on Elastic Properties of Metals, Fiz. Metal. Metalloved., **11** (1961) 443; 627.]

Соколовский, В.В., Распространение упруго-вязко-пластических волн в стержнях, Прик. Мат. Мех., 12, 1948, 261 (in Russian). [Sokolovskii, V. V., Propagation of Elasto-Visco-Plastic Waves in Bars, Prik. Mat. Mekh., 12 (1948) 261.]

Дикович, И.Л., Динамика упруго-пластических балов, Судпромгиз, Ленинград, 1962 (in Russian). [Dikovich, I.L., Dynamics of Elastic-Plastic Bars, Sudpromgiz, Leningrad, 1962.]

Кристеску, Н., О волнах нагрузки и разгрузки, возникающих при движении упругой или пластической нити, Прик. Мат. Мех., 18, 3, 1954, 257 (in Russian). [Cristescu, N., On Loading and Unloading Waves Rising from the Motion of Elastic or Plastic Strings, Prik. Mat. Mekh., **18**(3) (1954) 257.]

Ленский, В.С., Метод построения динамической зависимости между напряжениями и дефомациями по распределению остаточных дефомации, Вестник МГУ., No. 5 (1951) 13 (in Russian). [Lenskii, V.S., A Method of Building Dynamic Stress-Strain Relationship by Distribution of Residual Deformation, Vestnik MGU., **5** (1951) 13.]

Ленский, В.С., Об упруго-пластическом ударе стержня о жесткую преграду, Прик. Мат. Мех., 13 (1949) 165 (in Russian). [Lenskii, V.S., On the Impact of Elastic-Plastic Bar onto Rigid Obstacle, Prik. Mat. Mekh., **13** (1949) 165.]

Надеева, Р.И., Об определении динамической зависимости между напряжениями и дефомациями, Вестник МГУ., No. 10 (1953) 93 (in Russian). [Nadeeva, R.I., On Determination of Dynamic Relation between Stress and Strain, Vestnik MGU., **10** (1953) 93.]

Рахматулин, Х.А., распространении волны разгрузки, Прик. Мат. Мех., 9, 1945, 91–100 (in Russian). [Rakhmatulin, Kh.A., The Propagation of Unloading Waves, Prik. Mat. Mekh., **9** (1945) 91–100.]

Рахматулин, Х.А., косом ударе по гибкой нити с большими скоростями при наличии, Прик. Мат. Мех., 9, 1945, 449–462 (in Russian). [Rakhmatulin, Kh.A., High Speed Oblique Impact on a Flexible String under Tension, Prik. Mat. Mekh., **9** (1945) 449–462.]

Рахматулин, Х.А. и Демьянов, Ю.А., Прочность при интенсивных кратковременных нагрузках, Физматгиз, Москва, 1961 (in Russian). [Rakhmatulin, Kh.A. and Demyanov, Yu., Strength under Intensive Short-Duration Loading, Fizmatgiz, Moscow, 1961.]

Седов, Л.И., Методы подобия и размерности в механике, Изд. Наука, Изд. 8-е, переработанное, 1977 (in Russian). [Sedov, L.I., Methods of Similarity and Dimension in Mechanics, 8 ed., Izd. Nauka, 1977.]

Соколовский, В.В., Распространение упруго-вязко-пластических волн в стержнях, Прик. Мат. Мех., 12, 1948, 261 (in Russian). [Sokolovskii, V.V., Propagation of Elasto-Visco-Plastic Waves in Bars, Prik. Mat. Mekh., **12** (1948) 261.]

Шапиро, Г.С., Продольные колебания стержней, Прик. Мат. Мех., 10, 1946, 597–616 (in Russian). [Shapiro, G.S., Longitudinal Vibration of Bars, Prik. Mat. Mekh., **10** (1946) 597–616.]

Index

1D strain plane wave, 265, 266, 269, 338, 343, 345, 351

Absorption, 77, 219, 229, 230, 231, 236
Adiabatic shear failure, 450
Annealed copper, 429, 431
Artificial viscosity, 497–499
Associate theory, 333
Attenuation characters, 239
Attenuation factor, 229–232, 236, 262

Ballistic pendulum, 74
Basic assumption
  fifth, 403
  first, 400
  fourth, 403
  second, 12, 13, 401
  third, 403
Basic singularity, 170, 181, 182, 184, 185, 195
Bauschinger effect, 160
Bending Wave Theory, 399, 401
Body waves, 456
Boundary point, 173, 236, 238, 242, 243, 245, 246, 343, 481, 482
Branching point, 193, 195
Bridgman equation, 285, 286, 289, 318

Cauchy problem, 513
Centeredwave structure, 31
Central spalling, 92, 93
Characteristic direction, 15, 482, 484
Characteristic line equation, 15, 108
Characteristics differential equation, 260
Characteristics numerical method, 73, 477, 482
Coaxial collision, 57–59, 62, 63, 65, 66
Cold energy, 291
Cold pressure, 291
Cold-rolled low-carbon steel, 429
Complex reflection, 467
Complex wave regions, 105
Condition of mass conservation, 37
Condition of momentum conservation, 38
Conjugate points, 125
Conjugate relationship, 125, 128, 328
Constant value regions, 31, 105
Constitutive viscous dispersion, 45, 233
Contact condition, 379
Continuity condition, 170
Continuous waves, 29, 31, 99, 116, 120, 131, 324, 325
Controlled surfaces, 26
Controlled volume, 26, 27
Convergent wave, 25, 32
Conversion factors
  pressure, 511
  stress, 511
Creep, 219, 225, 226, 252
Creep compliance, 225
Critical impact velocity, 26, 422
Cylindrical wave, 2, 5, 337, 339, 340, 347, 348, 357, 359

Darboux problem, 102, 104, 105, 345, 514
Decreasing hardening material, 25, 26, 29
Diagram method, 148
Dilatation wave, 454, 455, 468, 469, 483, 490
Direct problem, 3, 126, 129, 132, 237, 238, 239, 242–245, 254
Directional derivative method, 14, 15, 484
Dispersion, 45, 229, 231, 233, 406, 409
Dispersion phenomenon
  constitutive viscous dispersion, 45, 233
  nonlinear constitutive dispersion, 45
Dissipation, 40, 41, 234, 238, 252, 299, 301, 324, 361, 498
Distortion wave, 455, 483, 490
Divergent wave, 25
Domain of dependence, 478, 479, 489, 490, 494, 495

Double-wave structure, 31, 39
Dual-wave formation, 279
Dynamic
  compatibility condition, 38, 74, 110, 112, 171, 199, 201, 213, 235, 261, 389, 442
  condition, 11, 304, 365, 369, 370, 377, 378, 380, 383, 390, 412, 437, 448
  fracture, 5, 70, 71, 87–89, 91, 92, 284, 466
    criterion, 87, 88, 91
    strength, 88, 89
Effective influence domain (EID), 248, 249, 254
Effective stress, 271
Elastic beam, 412, 415, 419, 422, 429, 431
Elastic bending waves, 399, 405, 410, 419
Elastic strain softening, 248
Elastic surface waves, 467
Elastic unloading, 106, 107, 109, 168, 19
Elastic unloading hypothesis, 106, 109, 110
Elastic-plastic boundary, 109, 169, 170, 180, 183
Elastic-plastic loading waves, 22, 98, 121, 153
Elastic-visco-plastic theory, 256
Elastic-visco-plastic wave theory, 256, 263, 264
Engineering theory, 13, 42
Entropic equations of state, 289
Equilibrium response, 247–249
Euler coordinate(s), 8, 27, 401
Euler derivative, 8
Euler method, 7, 8, 26
Euler volumetric modulus, 285, 287, 288
Excess transverse deflection failure, 450
Excess transverse shear strain failure, 450

Finite element method (FEM), 500
First kind of inverse problem, 237
First-order partial differential equation, 28, 266, 357, 408, 477, 484
Flexible strings, 5, 6, 363, 364
Fracture-delay, 91
Friction condition, 380
Front spalling, 284

Generalized
  maxwell body, 223
  wave impedance, 70, 72
Geometrical dispersion, 233
Grüneisen
  coefficient, 290, 305, 307, 308
  equation, 289, 292, 309, 310

Head-on loading
  interaction of
    two strong-discontinuous stress waves 69–99
    two weak-discontinuous stress waves 99–102
Head-on unloading, 109, 148–154, 157, 159, 161, 279, 282, 325
High loading rate or high strain rate, 2
Hooke's law, 17, 44, 106, 220, 224, 267, 268, 276, 284, 317, 318, 340, 347, 453, 470
Hopkinson pressure bar (HPBs), 45–47, 74–77
Hugoniot
  curves, 297, 298, 300, 303–308, 310, 312, 314, 316, 322
  elastic limit, 272, 275, 322, 325, 332
  lines, 41
Hydrodynamics approximation, 292, 293, 320–324, 332
Hydro-elasto-plastic media, 317, 321, 325
Hydro-elastoplastic model, 321, 323, 332
Hysteresis loop, 219

Impact with slide, 377, 379–383
Impact without slide, 377, 381
Impedance matching, 69
Implosion, 345
Impulse criterion, 92
Increasing hardening material, 25, 29
Indeterminate line method, 14, 15, 484
Initial value problem, 19, 102, 342, 343, 478, 495
Initialboundary-value-problems (IBVP)
  cauchy problem, 18–20, 23, 102, 478, 513, 514
  picard problem, 18, 20, 23, 102, 104, 105, 123, 138, 479, 514, 515
Initial-value problem or cauchy problem, 478
Instantaneous response, 20, 219, 234, 235, 248, 257, 261–264, 360, 475
Interior point, 237, 242, 244–246, 343, 480–482
Internal collision, 59, 60
Internal reflective waves, 59, 112, 119, 150–152, 159, 283, 350
Intrinsic wave velocity, 9
Inverse problem, 3, 79, 129, 237, 239, 242, 244, 245, 255, 394
Irrotational wave, 269, 455–458, 460, 462, 463, 465, 466, 483

Isometric wave, 55, 455–458, 460, 463, 465, 466, 483
Isotropic hardening hypothesis, 159, 160

Kelvin-Voigt body, 220, 226, 227
Kelvin-Voigt model, 221, 224, 226
Kinematic
compatibility condition, 36, 37, 170, 213, 235, 389, 442
condition, 11, 370, 378, 380, 437, 439, 447
hardening hypothesis, 160

Lagrange
coordinates, 8, 389, 401, 497
derivative, 9
method, 7, 8
velocity, 389
volumetric modulus, 285, 286
Lame solution, 345
Lame's coefficients, 268, 270
Laterally constrained elastic modulus, 269, 335
Lattice kinetic energy, 291
Lattice potential energy, 291
Linear elastic waves, 3, 4, 6, 17–22, 45, 253, 289, 317, 483
Linear hardening material, 25, 41, 110, 116, 171, 215, 271, 350, 376, 385
Linear propositional, 238, 239
Linear visco-elastic body, 220
Loading boundary, 167–169, 180, 190–194
Loading state, 119, 156, 167, 168
Local acceleration, 9
Local linearization method, 146, 328, 330
Longitudinal visco-elastic waves propagating in
kelvin-Voigt bars, 227–229
maxwell bars, 229–230
standard linear solid bars, 230–233
Longitudinal wave, 2

Malvern model, 220, 258, 334
Material coordinate system, 7, 9, 10, 28, 29, 292
Material coordinates, 8, 10, 338, 389
Material
constitutive relationship, 169
derivative, 9, 10, 234, 252
wave velocity, 9, 10, 35, 196, 292, 295
wave velocity (Lagrange wave velocity)
Maxwell
body, 220, 223, 226, 227, 233, 235, 257
model, 221–224, 223
theorem, 36

Meshing elements (discretization), 500
Mie-Grüneisen state equation, 291
Migration acceleration, 9
Mixed problem, 20, 343, 479
Mixed problem or picard problem, 479
Mode conversion angle, 463, 465
Mode coupling, 458
Momentum conservation condition, 57, 67, 96, 171, 210, 343, 391
Murnagham equation, 287–289, 307, 308, 318

Neutral axis, 399–402, 406, 442
Neutral layer (surface), 400
Nonaka's solution, 450
Nondistortion character, 239
Noninstantaneous response, 219, 518

Optimized finite difference method, 475
Ordinary differential equations (ODEs), 14, 15, 251, 476, 477, 482, 513
Over stress or extra stress, 257
Overdrive pressure, 303

Physical plane, 15, 104, 105
Plane bending problem, 15
Plane cross section assumption, 55, 399–403, 405, 406, 442
Plane cross-section, 6, 55
Plastic Bending Waves, 409
Plastic hardening modulus, 31, 271
Plastic shear modulus, 274, 275, 318, 349, 350
Plastic unloading, 276
Poisson's ratio, 267, 461, 465
Polarization conversion, 463
Power series approximate method, 132
Primary theory, 6, 13, 42–44
Principle of superposition, 60, 61
Progressive approximation method, 129, 134, 138
Pursuing loading, 95, 96
Pursuing unloading, 109, 110, 126, 127, 131, 135, 136, 138, 141, 151, 153, 157, 163, 165, 279, 282, 324, 325

Quasi-linear hyperbolic partial differential equations, 511

Rankine-Hugoniot curve, 297
Rankine-Hugoniot relationship, 41

Rate-dependent stress-strain relations, 3
elasto-visco-plastic, 3
visco-elastic, 3
visco-elasto-plastic, 3
Rate-independent nonlinearity, 249
Rate-independent theory, 4, 6, 38, 41, 219, 277
Rayleigh line, 39, 297, 298
Rayleigh surface wave, 6, 469, 472
Reflection coefficient, 68, 81, 461, 462, 464, 466
Reflection loading, 68, 70
Reflection unloading, 68, 70
Reflective fracture, 159
Reflective tensile fracture, 93
Region of influence, 479
Relaxation modulus, 223, 225
Relaxation spectrum, 223, 224
Relaxation time, 222, 223, 247–250, 252
Relaxation-time spectrum, 223
Residual particle velocity, 111, 113, 116, 118, 149, 282
Residual strain, 76, 111, 116, 126, 129, 132, 147, 154, 164, 217
Retard yield phenomenon, 255
Reverse plastic loading, 276
Reverse yield, 159, 160, 276, 279–281, 284, 324, 326, 329, 330
R-H relations, 293, 295–297, 303, 306, 312
Riemann wave, 28
Riemann's invariables, 18
Rigid unloading boundary, 202, 207, 213
Rigid unloading in,
finite Bars 207–212
linear elastic-decreasing hardening plastic bars 202–207
linear elastic-linear hardening plastic bars 200–202
linear hardening plastic bars, 197, 200
Rigid unloading of shock wave propagating in
semi-infinite bars 212–215
finite bars 215–217
Rigid-plastic beam, 432

Scab, 87
Scabbing, 87
Second kind of inverse problem, 237, 239, 394
Secondary plastic zone, 166, 167, 174
Secondary stress waves, 112
Second-order partial differential equation, 13, 14, 228, 266, 486, 514
Self-similar problem, 372, 375, 377, 516, 517
Shear wave, 55, 442, 455
Shock adiabat, 161, 297, 306, 307
Shock adiabatic process, 40
Shock jump condition, 41, 321
Shock secant, 39, 40
Short duration, 1
Similar variable problem or self-similar problem, 516
Simple
torsion wave, 54
wave regions, 105
wave relationship, 19, 21, 22, 26, 121, 124, 128, 139, 165
waves, 21, 90, 99, 102, 121, 147, 203, 329, 332, 345, 475
Singular interface
first-order singular interface, 29
second-order singular interface, 29
Small deformation, 6, 220
Sound impedance, 21, 57, 58, 61, 62
Space-like curve, 19, 20, 478
Spalling, 1, 87, 466
Spalling piece, 87, 89–91
Spatial coordinate, 8, 10, 26, 380, 399, 449, 516
Spatial
coordinate system, 7–10, 28
derivative, 8, 498
wave velocity, 10, 28, 293, 295, 296, 307, 308, 396
Split Hopkinson pressure bar (SHPB), 75–77, 237
Split Hopkinson visco-elastic bar, 237, 238
Standard linear solid model, 226, 227, 239
Stationary discontinuous interface, 113, 283
Straight beam, 400, 433
Strain-rate-dependent theory, 219
Stress intensity, 70, 271, 333
Stress relaxation, 219, 222, 223, 226, 248
Stress wave propagation, 475
Surface waves, 456

Temporal curves, 22
Thermal energy, 40, 291, 292
Thermal equations of state, 289
Thermal pressure, 291, 292
Time-like curve, 20, 479
Timer or flying piece, 74
Timoshenko equation, 408
Total mode conversion, 463
Total reflection, 465
Transmission coefficient, 68, 80
Transverse inertial effect, 13
Triangle stress pulses, 73

Uniformity assumption, 79, 82, 84, 85, 87

Varying boundary problem, 377
Velocity plane, 15
Visco-elastic and visco-plastic1d stress waves, 477
Visco-elastic wave theory, 219
Visco-elastic-plastic wave theory, 219
Visual velocity, 460, 462

Waveform curves, 22
Weak discontinuities, 29

Yielding velocity, 23

Zhu-Wang-Tang (ZWT) nonlinear visco-elastic constitutive equation, 247
ZWT material, 249
π-theorem, 516